MW00844222

MOUNTAIN LEADER BOOK

U.S. MARINE CORPS

Mountain Warfare Training Center

Fredonia Books
Amsterdam, The Netherlands

Mountain Leader Book

by
United States Marine Corps
Mountain Warfare Training Center

ISBN: 1-4101-0884-8

Reprinted from the 2002 edition

Fredonia Books
Amsterdam, The Netherlands
http://www.fredoniabooks.com

SUMMER INSTRUCTOR MANUAL 2002

TABLE OF CONTENTS

UNITED STATES MARINE CORPS
Mountain Warfare Training Center
Bridgeport, California 93517-5001

SML
SMO
02/11/02

LESSON PLAN

MOUNTAIN SAFETY

INTRODUCTION

(5 Min)

1. **GAIN ATTENTION**. The key to mountain safety in a mountainous environment is proper prior planning. Adhering to certain basic principles and predetermined actions will allow an individual or unit to efficiently perform their duties with minimum discomfort and maximum safety.

2. **PURPOSE**. The purpose of this period of instruction is to familiarize the student with the twelve mountain safety considerations and the acronym used to remember them. This lesson relates to all training conducted in a mountainous environment.

3. **INTRODUCE LEARNING OBJECTIVES**

   a. TERMINAL LEARNING OBJECTIVE. In a mountainous environment, execute preventive measures for mountain injuries, in accordance with the references.

   b. ENABLING LEARNING OBJECTIVE. Without the aid of references and given the acronym "BE SAFE MARINE", list in writing the principles of mountain safety, in accordance with the references.

4. **METHOD/MEDIA**. The material in this lesson will be presented by lecture and demonstration. You will practice what you learn during upcoming field training exercises. Those of you who have IRF's please fill them out at the conclusion of this period of instruction.

5. **EVALUATION**. You will be tested later in the course by written and performance evaluations on this period of instruction.

TRANSITION: Having discussed our purposes, let's now look at the planning and preparation of a military operation.

BODY (50 Min)

1. (40 Min) **PLANNING AND PREPARATION**. As in any military operation, planning and preparation constitute the keys to success. The following principles will help the leader conduct a safe and efficient operation in any type of mountainous environment. We find this principle in the acronym "BE SAFE MARINE". Remember the key, think about what each letter means and apply this key in any type of environment.

   B - Be aware of the group's ability.

   E - Evaluate terrain and weather constantly.

   S - Stay as a group.

   A - Appreciate time requirements

   F - Find shelter during storms, if required.

   E - Eat plenty and drink lots of liquids.

   M - Maintain proper clothing and equipment.

   A - Ask locals about conditions.

   R - Remember to keep calm and think.

   I - Insist on emergency rations and kits.

   N - Never forget accident procedures.

   E - Energy is saved when warm and dry.

a. Be Aware of the Group's Ability. It is essential that the leader evaluates the individual abilities of his men and uses this as the basis for his planning. His evaluation should include the following:

   (1) Physical Conditioning. Physical Fitness is the foundation for all strenuous activities of mountaineering. Leaders must be aware of their unit's state of fitness and take in account for the changes in altitude, climate, and amount of time for acclimatization.

   (2) Mental Attitude. Units need to be positive, realistic, and honest with themselves. There needs to be a equal balance here. A "can do" attitude may turn into dangerous overconfidence if it isn't tempered with a realistic appraisal of the situation and ourselves.

(3) Technical Skills. The ability to conduct a vertical assault, construct rope installations, maneuver over snow covered terrain, conduct avalanche search and rescue operations, etc. The more a unit applies these skills increases their ability to operate in a mountainous environment effectively.

(4) Individual skills. At this point, you must choose who is most proficient at the individual skills that will be required for your mission, navigation techniques, security, call for fire, track plans, bivouac site selection, skijoring, etc.

b. Evaluate Terrain and Weather Constantly

(1) Terrain. During the planning stages of a mission, the leader must absorb as much information as possible on the surrounding terrain and key terrain features involved in his area of operation. Considerations to any obstacles must be clearly planned for. Will you need such things as ropes, crampons, climbing gear, skins, etc.

(a) Stress careful movement in particularly dangerous areas, such as loose rock and avalanche prone slopes.

(b) Always know your position. Knowing where you are on your planned route is important.

(2) Weather. Mountain weather can be severe and variable. Drastic weather changes can occur in the space of a few hours with the onset of violent storms, reduced visibility, and extreme changes. In addition to obtaining current weather data, the leader must plan for the unexpected "worst case". During an operation he must diagnose weather signs continually to be able to foresee possible weather changes.

(a) Constantly evaluate the conditions. Under certain conditions it may be advisable to reevaluate your capabilities. Pushing ahead with a closed mind could spell disaster for the mission and the unit.

(b) When in a lightning storm, turn off all radios, stage radios and weapons away from personnel. Have personnel separate in a preferably low-lying area, or around tall natural objects, however personnel should not come into direct contact with trees.

(c) To calculate the approximate distance in miles from a flash of lightning, count in seconds the time from when you see the flash to when you hear the thunder, and divide by five.

c. Stay as a Group. Individuals acting on their own are at a great disadvantage in this environment.

(1) Give adequate rest halts based upon the terrain and elevation, physical condition of the unit, amount of combat load and mission requirements.
(2) Remember to use the buddy system in your group.

(3) Maintain a steady pace that will allow accomplishment of the mission as all members of the unit reach the objective area.

d. Appreciate Time Requirements. Efficient use of available time is vital. The leader must make an accurate estimate of the time required for his operation based on terrain, weather, unit size, abilities, and the enemy situation. This estimate should also take into account the possibility of unexpected emergencies such as injuries or unplanned bivouacs due to severe conditions.

(1) Time-Distance Formula (TDF). This formula is designed to be a guideline and should not be considered as the exact amount of time required for your movement. Furthermore, this formula is for use in ideal conditions:

3 kph + 1 hour for every 300 meters ascent; and/or + 1 hour for every 800 meters descent.

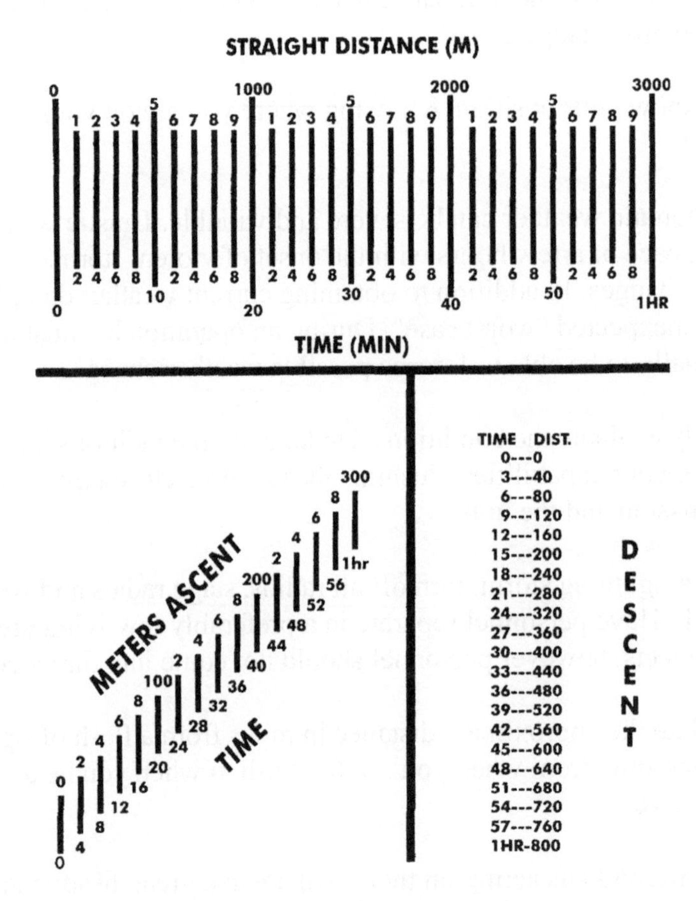

NOTE: The TDF is made for troops on foot in the summertime or troops on snowshoes in the wintertime.

(2) Route Planning. Route cards are not to be used in place of an overlay, but as a tool to be used in route planning. Overlays/Route cards should contain the following information at the minimum:

Unit designation	Unit commander
Number of personnel	Inclusive dates and times of movement
Grid of each checkpoint and bivouac	ETA and ETD
Map references	Azimuth and distances for each leg
Elevation gain/loss per leg	Description of the ground

ROUTE CARD

UNIT I.D.		UNIT COMMANDER		NUMBER OF PERSONNEL		DATE AND TIME	MAP REFERENCE
LEG	AZM	DIST	GRID	ETA	ETD	ELEVATION GAIN/LOSS	DESCRIPTION OF GROUND

Total Elevation Gain _____ = _______ (time)
+ Total Elevation Loss _____ = _______ (time)
+ Total Distance _____ = _______ (time)
= Estimated Total Time: ____________

(3) As in any military operation, route planning and execution are of vital importance. Prior to departure, the unit commander must submit a route card and/or patrol overlay to his higher headquarters and keep a duplicate copy for himself. This preplanned route should be followed as closely as possible, taking into account changes based on the tactical situation.

e. Find shelter during storms, if required. Under certain conditions, inclement weather can provide tactical advantages to the thinking unit commander, but by the same token it can reduce the efficiency of a unit to nil if an incorrect evaluation of the situation is made. Being lost will not directly kill an individual. Starvation takes time, but hypothermia can manifest itself in a matter of hours resulting in death.

(1) If there is a drastic change in the weather, tents should be erected immediately.
(2) If tents are not available, the unit leader should begin locating natural shelter or began building a man-made shelter. Adhering to the following principles will give an individual the best chance to spend a relatively safe bivouac with the prospect of continued effort toward mission accomplishment.

(a) Make shelter. The requirements for expedient shelters and the building procedures will be covered in another chapter. The basic requirement for protection from the elements is essential.

(b) Keep warm. The retention of body heat is of vital importance; any action in which body heat is lost should be avoided. The following points should be considered:

1. Adequate shelter

2. Insulation from the ground using branches, a rucksack, etc.

3. Wear extra clothing.

4. Use extra equipment for insulation.

5. Produce external heat while trying to conserve fuel for future use.

(c) Keep dry. Being wet causes the loss of body heat 32 times faster than when dry. Adequate protection from the elements is of prime importance to prevent the onset of hypothermia.

f. Eat Properly and Drink Plenty of Fluids

(1) Food. The human body can be compared to a furnace, which runs on food to produce energy (warmth). By planning the consumption of food to suit the specific situation, adequate nutrition and extra warmth can be supplied.

(2) Water. The intake of adequate amounts of water will maintain the body in proper working order. Danger from dehydration is as high in mountain regions as in hot dry areas. Loss of liquids is easily seen and felt in hot climates; whereas in the mountains, the loss of body fluids is much less noticeable. High water intake, at least 4 to 6 quarts per day when in bivouac, 6 to 8 quarts per day when active, will help to prevent dehydration.

g. Maintain Proper Clothing and Equipment

(1) Equipment. In the mountains a man should never be separated from his gear. Here are some basic and essential items that should be considered during your planning stage.

(a) Assault load

1. There should be at least one assault load per squad.

(b) Combat load

(c) Existence load.

(d) Map and compass. Every individual in a leadership position and his assistant should carry a map and compass. The maps should be weatherproofed and extra maps should be distributed throughout the unit.

(e) Repair kit. This kit should include those items necessary to do emergency repairs on your equipment.

(f) Survival Kit. Always carried on your person. The contents of a survival kit will be covered in another chapter.

h. Ask Locals About Conditions. An often-overlooked source of information is the indigenous population of an area. Local weather patterns, rock slide/avalanche areas, watering points, and normal routes can all be obtained by careful questioning. The unit leader must obtain current information of the actual conditions along his intended route.

i. Remember to Keep Calm and Think

(1) Having recognized that an emergency situation exists, the following principles should be followed:

(a) Keep calm and do not panic. At this point you must make every effort to conserve body heat and energy.

(b) Think. When an individual is cold, tired, hungry, or frightened he must force himself to organize his thoughts into a logical sequence.

(c) The group must try to help itself by either finding the way back to safety or by preparing shelters and procuring food.

(d) Above all else, the group must act as a tight-knit unit. In emergency situations, individual dissension can cause a total loss of control and unit strength.

(2) If the decision is reached that the group should seek its way back to safety, several possibilities exist. In most situations, the safest approach will be to retrace the route to

the last known point and continue from there. The other course of action is to get a group consensus on the present location and send out a small search party to locate a known point. This party must ensure that they mark their trail adequately to return to the group. If all attempts at finding a way back to known terrain fail, a definite emergency situation exists and actions discussed later in this section must be instituted.

j. Insist on Emergency Rations and Kits. Emergency rations and a survival kit should always be carried.

k. Never Forget Accident/Emergency Procedures

(1) Causes of accidents. The general procedures used to handle accidents differ little in this environment, but several distinct points should be kept in mind. The most frequent causes of accidents are as follows:

(a) Overestimation of physical and technical abilities.

(b) Carelessness.

(c) General lack of observation of one's surroundings.

(d) Lack of knowledge and experience by leaders.

(e) The failure to act as a group.

(f) Underestimation of time requirements to move through mountainous terrain and underestimation of the terrain itself.

(2) Preventive measures. The only truly effective preventive measures for the above lie in the education and experience of leaders at all levels. Too often, leaders sit by watching during training and as a result have no concept of the requirements involved in the mountainous environment. Only by active involvement can a leader gain the knowledge and experience needed to effectively lead in this environment.

(3) General procedures for handling an accident. These require only a good dose of common sense as outlined below.

(a) Perform basic first aid.

(b) Protect the patient from the elements to include insulation on top and bottom.

(c) Evacuate if necessary.

(d) Send for help, if required.

(e) If possible, never send a man for help alone.

(f) Send the following information regarding the accident:

1. Time of accident.

2. Nature and location of accident.
3. Number injured.

4. Best approach route to accident scene.

(4) If one man of a two-man team is injured, the injured man must be given all available aid prior to going for help. If the injured man is unconscious, he should be placed in all available clothing and sleeping gear and anchored if on steep terrain. A note explaining the circumstances, and reassuring him, should be left in a conspicuous spot. This note must also contain the following information:

(a) When you expect to return.

(b) Where you went.

(c) What you did before you left (medication, etc.).

(5) International distress signal.

(a) Six short blasts in 1 minute from person requesting help.

(b) The return signal is three blasts in 1 minute from the respondent.

(6) Other methods of signaling:

(a) Red pyrotechnics.

(b) SOS, (...---...)

(c) "Mayday" by voice communications.

l. Energy is Saved When Warm and Dry. With the previous 11 principles in mind this one should fall right into place. Save your heat and energy by following these steps:

(1) Dress properly.

(2) Eat properly.

(3) Drink properly.

(4) Ensure shelter meets criteria.

(5) Produce external heat (fires, stove, extra clothing, etc.) to save body heat and energy for future use.

(6) Avoid getting wet, this increases body heat loss.

TRANSITION: By applying these 12 principles, you increase your chance for survival.

PRACTICE (CONC)

a. Students will apply mountain safety principles throughout the duration of the course.

PROVIDE HELP (CONC)

a. The instructors will assist the students when necessary.

OPPORTUNITY FOR QUESTIONS (3 Min)

1. QUESTIONS FROM THE CLASS

2. QUESTIONS TO THE CLASS

Q. What is the "B" in Be Safe Marine?

A. Be aware of the group's ability.

Q. What is the "F" in Be Safe Marine?

A. Find shelter before storms, it required.

SUMMARY (2 Min)

a. Safety in the mountains doesn't come naturally, it must be practiced or the results can be devastating. Constant observation and common sense are the keys to success and safety.

b. Those of you with IRF's please fill them out at this time; we will take a short break before the next class.

UNITED STATES MARINE CORPS
Mountain Warfare Training Center
Bridgeport, California 93517-5001

SML
SMO
02/11/02

LESSON PLAN

MOUNTAIN HEALTH AWARENESS

INTRODUCTION

(5 Min)

1. **GAIN ATTENTION**. A study of military history in mountain regions reveals that success and failure rates are measured in terms of the regard held for the environment itself. The man who recognizes and respects these forces of nature can do his job and even use these forces to his advantage. The man who disregards or underestimates these forces is doomed to failure, if not destruction. In the mountains, care of the body requires special emphasis. If men fail to eat properly, or do not get sufficient liquids, efficiency will suffer. Lowered efficiency increases the possibility of casualties, either by high altitude injury or enemy action resulting in failure to accomplish the mission.

2. **OVERVIEW**. The purpose of this period of instruction is to provide the student with the information necessary to help prevent, recognize, and treat various health problems that can arise in a mountainous environment in both winter and summer. This lesson relates to all operations conducted while in a mountainous environment.

INSTRUCTOR NOTE: Have students read learning objectives.

3. **INTRODUCE LEARNING OBJECTIVES**.

    a. TERMINAL LEARNING OBJECTIVE. In a mountainous environment, execute measures for preventing, recognizing and treating illnesses and injuries, in accordance with the references.

    b. ENABLING LEARNING OBJECTIVES(MLC)

        (1) Without the aid of references, list in writing the five ways a body loses heat, in accordance with the references.

(2) Without the aid of references, define in writing dehydration, in accordance with the references.

(3) Without the aid of references, list in writing the ways to prevent dehydration, in accordance with the references.

(4) Without the aid of references, define in writing heat exhaustion, in accordance with the references.

(5) Without the aid of references, define in writing Acute Mountain Sickness (AMS), in accordance with the references.

(6) Without the aid of references, define High Altitude Cerebral Edema (HACE), in accordance with the reference.

(7) Without the aid of references, define High Altitude Pulmonary Edema (HAPE), in accordance with the reference.

(8) Without the aid of references, define in writing hypothermia, in accordance with the references.

(9) Without the aid of references, define in writing frostbite, in accordance with the references.

(10) Without the aid of references, list in writing the major risk factors for frostbite, in accordance with the references.

(11) Without the aid of references, define in writing trench foot, in accordance with the references.

(12) Without the aid of references, define snow blindness, in accordance with the references.

(13) Without the aid of references, state in writing how to prevent C0 poisoning, in accordance with the references.

(14) Without the aid of references, list in writing the ways to dispose of waste, in accordance with the references.

SMO CUE: TC 1

c. <u>ENABLING LEARNING OBJFCTIVES (SMO)</u>

(1) With the aid of references, list orally the ways a body loses heat, in accordance with the references.

(2) With the aid of references, define orally dehydration, in accordance with the references.

(3) With the aid of references, list orally the causes of dehydration, in accordance with the references.

(4) With the aid of references, list orally the ways to prevent dehydration, in accordance with the references.

(5) With the aid of references, define orally heat exhaustion, in accordance with the references.

(6) With the aid of references, define orally Acute Mountain Sickness (AMS), in accordance with the references.

(7) With the aid of references, define orally High Altitude Cerebral Edema (HACE), in accordance with the reference.

(8) With the aid of references, define orally define High Altitude Pulmonary Edema (HAPE), in accordance with the reference.

(9) With the aid of references, define orally hypothermia, in accordance with the references.

(10) With the aid of references, define orally frostbite, in accordance with the references.

(11) With the aid of references, list orally the major risk factors for frostbite, in accordance with the references.

(12) With the aid of references, define orally trench foot, in accordance with the references.

(13) With the aid of references, define orally snow blindness, in accordance with the references.

(14) With the aid of references, state orally how to prevent C0 poisoning, in accordance with the references.

(15) With the aid of references, list orally the ways to dispose of waste, in accordance with the references.

4. **METHOD/MEDIA**. The material taught in this lesson will be presented by lecture and practical application method. You will practice what you have learned during upcoming field training exercises. Those of you with IRF's please fill them out at the end of this period of instruction.

5. **EVALUATION**.

a. Mountain Leaders - You will be tested on up coming written exam.

b. Unit Operations - You will be tested on up coming oral exam.

TRANSITION: Are there any questions over the purpose, learning objectives, how the class will be taught, or how you will be evaluated? Let's first talk about how the body is affected by heat.

BODY (125 Min)

1. (5 Min) **GENERAL**. The body's ability to adjust to a harsh environment is greatly controlled by its physical condition, which is influenced by fitness, nutrition, water intake and other factors. Acclimatization is also vital, but to acclimatize properly, a person must force himself to enter the outdoor environment and work in it, which requires a healthy attitude. The importance of a healthy attitude cannot be stressed enough.

   a. Physical Fitness. This is fundamental to successfully perform in a mountainous environment. The more fit an individual is, the easier it is for him to move, to carry heavy loads, and to adapt to the stresses of the mountains.

   b. Nutrition. An adequate number of calories must be consumed on a daily basis. Consuming 4 MRE or 1 RCW/day provides the required vitamins, and calories.

   c. Water Intake. Obviously, a person cannot exist for long without adequate water intake. In the mountains, water intake becomes even more important than in the lowlands.

2. (5 Min) **HEAT GENERATION**. The human body can be compared to a constantly running furnace. To run efficiently, it must always maintain a temperature within a small range. The food we eat is our fuel. The amount of energy in the food we eat is measured in calories. Typically, about 25% of the food we take in is used by the body to rebuild itself, while the remaining 75% is used to produce heat to maintain that small temperature range (usually from 96'F to 99'F, with the average being 98.6'F). This generated heat is then distributed to various parts of the body. In the mountains at high altitude, especially in the winter, several other points need to be remembered:

   a. The Diet.

      (1) Caloric intake. At sea level, when in garrison and not physically active, a Marine typically requires about 2,000 calories per day. But at altitude, in the mountains and especially in the winter, the need for calories more than doubles to at least 4,500 calories per day. This is because one is much more physically active, and because of cold, calories needed for heat generation greatly increases.

      (2) Carbohydrate intake. Carbohydrates are simple foods like sugars, bread, rice, and pasta. In the mountains an increase in carbohydrate intake is recommended because at high altitude these tend to taste better and are more easily converted into heat energy by the body.

      (3) Hot meals and hot wets. These should be consumed whenever possible.

b. The Body at rest produces Heat at a specific Rate. With physical activity, heat production increases.

(1) Moderate Exercise. This can increase heat production up to 5 - 6 times the normal rate, and can be tolerated for long periods of time.

(2) Moderate Shivering. This can increase heat production by 20 times normal, but only for a few minutes. Shivering is not an efficient means of heat production as it quickly leads to exhaustion.

(3) Intense Shivering. When muscle activity is at its maximum rate, as in intense shivering, heat production can increase by up to 50 times normal. However, exhaustion follows within 30 seconds.

3. (10 Min) **FIVE WAYS THE BODY LOSES HEAT**

a. Radiation. The direct heat loss from the body to its surroundings. If the surroundings are colder than the body, the net result is heat loss. A nude man loses about 60% of his total body heat by radiation. Specifically, heat is lost in the form of infrared radiation. Infrared targeting devices work by detecting radiant heat loss.

b. Conduction. The direct transfer of heat from one object in contact with a colder object.

(1) Most commonly conduction occurs when an individual sits or rests directly upon a cold object, such as snow, the ground, or a rock. Without an insulating layer between the Marine and the object (such as an isopor mat), one quickly begins to lose heat. This is why it's important to not sit or sleep directly on cold ground or snow without a mat or a pack acting as insulation.

(2) Metals, like rocks, conduct heat very rapidly.

(3) Water conducts heat away from the body 25 times faster than air.

c. Convection. The loss of heat to the atmosphere or liquid such as water in the following manner:

(1) Air and water can both be thought of as "liquids" running over the surface of the body. Water or air, which is in contact with the body, attempts to absorb heat from the body until the body and air or water is both the same temperatures. However, if the air or water is continuously moving over the body, the temperatures can never equalize and the body keeps losing heat.

(2) Most commonly one encounters convection through the wind-chill effect. Whether walking, skiing, or moving in open vehicles, wind must be taken into account to determine the effective temperature experienced by the unprotected body. For example: At a 32'F air temperature, the effective temperature, or wind-chill, with a 25

mph wind is actually 3'F. At 5'F, a 25-mph wind will give a wind-chill of 37'F below zero (-37'F)!

(a) However, if dressed properly (with the appropriate protective layers) wind-chill effects are minimized, except for areas of exposed skin. If Marines are cold because of wind-chill, it means that they are not dressed properly and that is a leadership failure.

d. Evaporation. Heat loss from evaporation occurs when water (sweat) on the surface of the skin is turned into water vapor. This process requires energy in the form of heat and this heat comes from the body.

(1) This is the major method the body uses to cool itself down. This is why you sweat when you work hard or PT. One quart of sweat, which you can easily produce in an hour of hard PT, will take about 600 calories of heat away from the body when it evaporates.

e. Respiration. When you inhale, the air you breathe in is warmed by the body and saturated with water vapor. Then when you exhale, that heat is lost. That is why breath can be seen in cold air. Respiration is really a combination of convection (heat being transferred to moving air by the lungs) and evaporation, with both processes occurring inside the body.

4. (5 Min) **PHYSICAL RESPONSES TO HEAT.** When the body begins to create excess heat, it responds in several ways to rid itself of that heat.

a. Initially, the blood vessels in the skin expand, or dilate. This dilation allows more blood to the surface where the heat can more easily be transferred to the surroundings.

b. Soon afterwards, sweating begins. This contributes to heat loss through convection and evaporation.

TRANSITION: Now that we have discussed how heat can be lost, are there any questions? Now let's talk about how the body reacts to cold.

5. (5 Min) **PHYSICAL RESPONSES TO COLD.** Almost the opposite occurs as with heat.

a. First, blood vessels at the skin surface close down, or constrict. This does two things:

(1) Less blood goes near the surface of the body so that less heat is lost to the outside.

(2) More blood goes to the "core" or the center of the body, to keep the brain, heart, lungs, liver, and kidneys warm. This means fingers and toes tend to get cold.

b. If that is not enough to keep the body warm, the next step is shivering. Shivering is reflexive regular muscular contractions, this muscular activity causes heat production. As mentioned before, shivering can only last for a short time before exhaustion occurs. With shivering you will either warm up, as usually occurs, or continue to get colder and start to become hypothermic. Hypothermia will be discussed later.

c. Another effect of more blood flow going to the body's inner core is that the kidneys are "fooled" into thinking that the body has more blood than it really does. The kidneys respond by producing more urine, and this can contribute to dehydration, which we will talk about next.

TRANSITION: Thus far, we have talked about the body when it is responding normally, are there any questions? Now let's look at things when they go wrong.

6. (5 Min) **DEHYDRATION**

a. Definition: Dehydration is defined as a deficit of total body water.

b. Causes of dehydration. Dehydration is the most common illness seen, both in the winter and in the summer. Ultimately, the reason someone becomes dehydrated is:

(1) Excessive loss.

(a) Urination. Increased as a response to the cold and high altitude.

(b) Cold, dry air. In most mountainous areas, like the high Sierras, the air is often cold and always dry. Thus, inhaled air must be humidified and warmed by the body and this takes water.

(c) Strenuous activity. Marines in the mountains are always involved in strenuous activity, and this leads to large amounts of sweating, even in the winter.

(d) Coffee and tea. These are mild diuretics; that is, they stimulate the kidneys to produce excess urine.

(2) Inadequate intake.

(a) Thirst. Thirst is not a good indicator of your state of hydration, especially in a high altitude, mountainous environment. *If you are thirsty, it is too late, you are already dehydrated.*

(b) Water inaccessibility. It is possible, though not likely, that adequate amounts of water may not be available. This should only be a factor if you are in a survival situation.

c. Symptoms of Dehydration

(1) Headache.

(2) Nausea.

(3) Dizziness.

(4) Fainting.

(5) Constipation.

(6) Dry mouth.

(7) Weakness.

(8) Lethargy.

(9) Stomach cramps.

(10) Leg and arm cramps.

d. Signs of Dehydration

(1) Swollen tongue.

(2) Dark urine.

(3) Low blood pressure.

(4) Rapid heart rate, greater then 100 beats per minute.

e. Prevention of Dehydration

(1) Since it is impossible to limit the water that you lose (except by limiting your coffee and tea consumption), you must then ensure adequate intake.

(a) Drink a minimum of 6 - 8 quarts per day.

(b) Watch the color of your urine. Try to keep it crystal clear. The more yellow it gets the more you need to drink.

(c) Do not rely on thirst as an indicator.

f. Field Treatment of Dehydration.

(1) Oral Fluids. Give at least 6 quarts per day.

(2) Intravenous fluids. Severe cases may require fluid by IV.

g. Leadership. A healthy Marine should *NEVER* become dehydrated. It is an entirely preventable condition. In the Israeli Army, when a soldier becomes dehydrated, his platoon commander and platoon sergeant are court-martialed. It is the responsibility of small unit leaders to ensure their men have adequate access to water and that they drink it. Failure to prevent dehydration in a healthy Marine is a failure in leadership.

7. (5 Min) **HEAT CRAMPS**

a. Definition: Heat cramps are painful spasms of skeletal muscle as a result of excessive loss of body salt.

b. Cause. Sweat is composed of water and salt. When a Marine is involved in strenuous activity that leads to excessive sweating, and replaces the lost water but *not* the lost salt, a salt imbalance within the body may result. This salt imbalance may then lead to muscle cramps.

c. Symptoms of Heat Cramps.

(1) Muscle cramps in the arms, legs, or abdomen.

d. Prevention of Heat Cramps. Prevention is always better than treatment.

(1) Avoid overheating by proper ventilation.

(2) Eat correctly. There is no need to take salt tablets as long as proper diet is maintained. MRE's and RCW's contain more than enough salt.

e. Field Management for Heat Cramps

(1) Have the victim stop moving. (Rest)

(2) Gentle massage of the affected muscles may help relieve the spasm.

(3) Stretch-out the muscle.

(4) Ensure the victim is adequately hydrated. Replace the victim's salt by adding either 1 salt tablet or 1 tablespoonful of table salt (from an MRE accessory packet) to a quart of water. Have the victim sip the salted water over a period of a few hours.

8. (5 Min) **HEAT EXHAUSTION**

a. Definition: Heat exhaustion occurs when body salt losses and dehydration from sweating are so severe that a person can no longer maintain adequate blood pressure.

b. Causes of Heat Exhaustion. Heat exhaustion is really a severe form of dehydration combined with or as a result of strenuous physical activity. Blood is made up of mostly of water. When a large amount of water is lost in the form of sweat, the amount of blood volume in the body drops. When the blood volume drops low enough, in combination with tough physical exercise, heat exhaustion results.

c. Symptoms of Heat Exhaustion .

(1) Headache.

(2) Nausea.

(3) Dizziness.

(4) Fatigue.

(5) Fainting.

d. Prevention of Heat Exhaustion

(1) It is the same as for heat cramps. Dress properly with adequate ventilation to avoid overheating. Dress comfortably cool.

e. Field Management for Heat Exhaustion

(1) Lay the victim down, with his feet higher than his head. Insulate the victim from the cold ground with an isopor mat.

(2) Ensure that he is well ventilated. Unzip his parka or take it off, until he feels cool. Make sure he doesn't get too cold.

(3) Fluids. If he is awake and not vomiting, he may be given fluids by mouth. Usually 3 quarts at a minimum are required.

9. (5 Min) **HEAT STROKE**

a. Definition: Also known as Sunstroke, it is defined as a failure of the body's cooling mechanisms that rid the body of excessive heat build up.

NOTE: Heat stroke is LETHAL in up to 40% of cases, while a majority of those who live may suffer permanent brain damage.

b. Risk Factors for Heat Stroke. Heat stroke occurs when the body is unable to rid itself of excess heat, such as when exercising in a hot, humid environment. Typically the air in a mountainous environment is cool and dry; nonetheless, heat stroke can and does occur in the mountains, even in the winter. The elevation of body temperature levels is usually greater then 103'F.

c. Symptoms of Heat Stroke. In the majority of cases, the onset of heat stroke is sudden, and the victim becomes delirious or comatose, before he begins to complain of symptoms. However, approximately 20% of victims will complain of:

(1) Headache.

(2) Nausea.

(3) Dizziness.

(4) Fatigue.

d. Signs of Heat Stroke.

(1) Usually, victims are delirious or comatose.

(2) Pupils maybe pinpoint.

(3) Flushed skin may or may not be present.

(4) Rectal temperatures of 103'F or greater.

(5) Sweating. It is often taught that in Heat Stroke, sweating is absent. **THIS IS UNTRUE!!!** Sweating often is present in sunstroke, so do not assume a victim does not have heat stroke simply because he is sweating. As with hypothermia, the only way to absolutely diagnose a victim as having heat stroke is with a rectal thermometer. Anybody with abnormal behavior (hallucinations, bizarre behavior, confusion, etc.) and a rectal temperature of 103'F or greater has heat stroke, until proven otherwise.

e. Preventive Measures for Heat Stroke. For the most part, the same principles apply as with Heat Exhaustion. That is, drink 6 - 8 quarts of water per day and keep as well ventilated as possible. When the temperature and humidity are high, however, physical activity must be reduced.

f. Field Management for Heat Stroke. Remember that heat stroke is a true life and death emergency. The longer the victim remains overheated, the more likely it's irreversible.

(1) Reduce heat immediately by dousing the body with large amounts of cool water or by applying wet, cool towels to the neck, the groin, chest, and armpits. If cold packs are available then use them.

(2) Maintain an open airway.

(3) Remove as much of the victim's clothing as possible.

(4) Give him nothing by mouth.

(5) When his rectal temperature has dropped below 102F, you may discontinue cooling. Be sure to recheck the temperature every 5 minutes, if his temperature rises to 103F or greater, begin re-cooling.

(6) Casevac immediately!

TRANSITION: Now that we have discussed problems with the body, are there any questions? Now let us talk about problems of altitude.

10. (5 Min) **ACUTE MOUNTAIN SICKNESS (AMS)**

a. Definition: AMS is a self-limited illness resulting from the rapid exposure of an un-acclimatized individual to high altitude.

b. Risk Factors for AMS. Anyone ascending rapidly from sea level to over 7,000 feet may develop AMS. Approximately 25% of individuals who ascend rapidly to 8,000 - 9,000 feet will develop AMS. Virtually all un-acclimatized persons who rapidly ascend to 11,000 - 12,000 feet will develop AMS. Factors which will increase your chance of developing AMS or make it worse are overexertion at altitude and dehydration. The cause of AMS, or altitude illness in general, is not well understood. However, it is known that the lower levels of oxygen's barometric pressure found at high altitude leads to a state of hypoxia, which means low levels of oxygen in the blood. The way in which the body responds to this hypoxia can lead to AMS or other altitude illness. Symptoms of AMS will usually occur 6 - 48 hours after reaching altitude.

c. Symptoms of AMS.

(1) Headache, the most common symptom, may be severe.

(2) Nausea.

(3) Decreased appetite.

(4) Difficulty sleeping; due to irregular breathing.

(5) Weakness, loss of coordination.

(6) Easily fatigued.
(7) Dizziness.

(8) Apathy.

d. Signs of AMS.

(1) Rapid breathing or an irregular breathing pattern.

(2) Rapid pulse.

(3) Vomiting.

e. Prevention of AMS.

(1) The best prevention of AMS is a staged ascent. For Marines going to altitude from sea level, the following ascent rates should be adhered to:

ALTITUDE	RATE OF ASCENT
<8,000 ft	Unlimited rest for 48 -hrs @ 8,000 ft, then proceed
8,000 - 10,000 ft	Rest for 24 hrs @ 10,000 ft, and proceed
10,000 - 14,000 ft	1,000 ft per day.
>14,000 ft	500-1,000 ft per day

(2) After 48 hours with no symptoms of altitude illness, proceed no higher than 10,000 ft. Remain at 10,000 for 24 hrs and then if no symptoms of altitude illness are present, proceed no faster than 1,000 feet per day.

(3) Above 14,000 ft, ascend no faster than 500 -1000 feet per day. If no symptoms occur after 48 hrs at a given altitude, it is safe to assume you can ascend, but remember, there are no steadfast rules or guarantees.

(4) Certain drugs, can be used to treat or even prevent AMS. These drugs are used ONLY under the direction of a medical department personnel. These drugs include diamox.

f. Treatment of AMS

(1) Light duty.

(2) Fluids - ensure adequate fluid intake. (AMS is a fluid retention condition, be careful not to over hydrate)

(3) Drugs for AMS:

(a) Tylenol or aspirin can be given for the headache.

(b) Diamox as prescribed.

(4) **DESCEND**. Most cases of AMS should resolve with 2 - 3 days of the above measures. However, if this does not occur or if the symptoms are severe or worsening, then a descent of 1,000 - 3,000 feet should greatly improve the condition of the victim. He can

re-ascend after several days.

11. (5 Min) **HIGH ALTITUDE CEREBRAL EDEMA (HACE)**

a. Definition: HACE is a high altitude illness, which is characterized by swelling of the brain.

b. Symptoms/Signs of HACE

(1) Usually the symptoms of AMS are also present. In fact, the two may look exactly alike. However, because HACE is lethal and AMS is not, you must be able to distinguish the two. Testing the victim's balance easily does this. Simply have him walk heel to toe (just like a field sobriety test). A HACE victim will have difficulty executing this maneuver, where an AMS victim will not.

(2) Other symptoms may include:

(a) Bizarre behavior.

(b) Hallucinations.

(c) Confusion.

(d) Excessive fatigue.

(e) In severe cases--coma.

c. Prevention of HACE. The preventive measures for HACE are the same as for AMS.

d. Treatment of HACE. Immediate descent is mandatory!

(1) Drugs prescribed by medical personnel include decadron (a steroid), diamox, and oxygen if available.

(2) A device called a Gamow Bag. This is a man-portable (14 lb.) hyperbaric chamber. A HACE (or HAPE) victim is placed in the bag and zipped up. Using a foot pedal operated pump; the pressure in the bag is increased. This simulates a decrease in altitude. Altitude "decreases" of up to 6,000 feet may be achieved. However, the use of this bag should only be reserved for emergencies when rapid descent is delayed. This is very labor intensive and only a temporary measure.

(3) Remember: ***HACE KILLS*** if not treated.

12. (5 Min) **HIGH ALTITUDE PULMONARY EDEMA (HAPE)**

a. Definition: HAPE is a high altitude illness, which is characterized by filling of the lungs with fluid.

b. Risk Factors for HAPE. They are the same as for AMS and HACE, except HAPE is rarely seen below 10,000 feet.

c. Symptoms of HAPE

(1) The two key symptoms to look for are:

(a) A cough that persists even at rest. Cough will be initially dry and the progress over several hours to days to produce a pink frothy sputum.

(b) Severe shortness of breath, which also persists even at rest.

(2) Other symptoms of AMS are also usually present and include:

(a) Disorientation.

(b) Fainting.

d. Signs of HAPE

(1) Cool, clammy skin.

(2) Rapid breathing.

(3) Rapid, weak pulse.

(4) Blue lips.

(5) Undue fatigue.

e. Prevention of HAPE. The same as for AMS, slow graded ascent.

f. Treatment of HAPE

(1) The best treatment is descent-rapid descent, to as low as possible, preferably to sea level.

(2) The victim should be given 100% oxygen by mask if available.

(3) Medical can prescribe a drug called nifedipine. Diamox may also be given.

(4) A device called a Gamow Bag. This is a man-portable (14 lb.) hyperbaric chamber. A HAPE (or HACE) victim is placed in the bag and zipped up. Using a foot pedal operated pump; the pressure in the bag is increased. This simulates a decrease in altitude. Altitude "decreases" of up to 6,000 feet may be achieved. However, the use of

this bag should only be reserved for emergencies when rapid descent is delayed. This is very labor intensive and only a temporary measure.

(5) Remember: ***HAPE KILLS*** if not treated.

TRANSITION: These are the high altitude injuries that you are going to see. Hopefully you will not fall victim to them during training or otherwise. Another aspect of the high altitude is the much colder temperatures; therefore, we need to be aware of the potential for cold weather injuries. Let's talk about Hypothermia.

13. (10 Min) **HYPOTHERMIA**

a. Definition. Hypothermia is defined as the state when the body's core temperature falls to 95'F or less.

b. A core temperature is the temperature at the center of the body. Taking an oral or armpit (auxiliary) temperature is not an accurate way to determine body core temperature. The best way to measure core temperature in the field is to take the temperature rectally. This must be done with a special low range rectal thermometer, which should be carried by all officers, SNCO's and corpsmen. These thermometers are available through the federal stock system. (FSN 6515-00-139-4593)

c. Commonplace Misconceptions.

(1) Exposure. At times you may hear that an individual has "exposure". While the term is usually used in reference to hypothermia, it is without real meaning and should not be used to describe hypothermia.

(2) Extreme cold. It is a common belief that extreme cold is needed for hypothermia to occur. In fact, most cases occur when the temperature is between 30'F and 50'F. This temperature range is quite common in the Fall, Winter, and Spring months.

d. Causes of Hypothermia. The ways in which the body generates and loses heat has been discussed earlier. Quite simply, hypothermia occurs when heat loss from the body exceeds the body's ability to produce heat. Contributing factors include:

(1) Ambient temperature. Outside air temperature.

(2) Wind chill. This only affects improperly clothed individuals.

(3) Wet clothing.

(4) Cold water immersion.

(5) Improper clothing.

(6) Exhaustion.

(7) Alcohol intoxication, nicotine and drugs such as barbiturates and tranquilizers.

(8) Injuries. Those causing immobility or major bleeding, major burn and head trauma.

e. Signs and symptoms of Hypothermia

(1) The number one sign to look for is altered mental status ; that is, the brain is literally getting cold. These signs might include confusion, slurred speech, strange behavior, irritability, impaired judgment, hallucinations, or fatigue.

(2) As hypothermia worsens, victims will lose consciousness and eventually slip into a coma.

(3) Shivering. Remember that shivering is a major way the body tries to warm itself early on, as it first begins to get cold. Shivering stops for 2 reasons:

(a) The body has warmed back up to a normal temperature range.

(b) The body has continued to cool. Below 95'F shivering begins to decrease and by 90F it ceases completely.

(c) Obviously, continued cooling is bad. So if a Marine with whom you are working, who was shivering, stops shivering, you must determine if that is because he has warmed up or continued to cool.

(4) A victim with severe hypothermia may actually appear to be quite dead, without breathing or a pulse. However, people who have been found this way have been successfully "brought back to life" with no permanent damage. So remember, ***you are not dead until you are warm and dead***.

f. Prevention of Hypothermia

(1) Obviously, prevention is always better (and much easier) than treatment.

(2) Cold weather clothing must be properly warm and cared for.

(a) Keep your clothing as dry as possible.

(b) If your feet are cold, wear a hat. Up to 80% of the body's heat can escape from the head.

(3) Avoid dehydration. Drink 6 - 8 quarts per day.

(4) Eat adequately. At least 4,500 calories per day.

(5) Avoid fatigue and exhaustion. A Marine in a state of physical exhaustion is at increased risk for hypothermia.

(6) Increase levels of activity as the temperature drops. Do not remain stationary when the temperature is very low. If the tactical situation does not permit moving about, perform isometric exercises of successive muscles.

(7) Use the buddy system to check each other for signs/symptoms of hypothermia.

g. Treatment of Hypothermia.

(1) Make the diagnosis.

(2) Prevent further heat loss.

(a) Remove the victim from the environment where he became hypothermia, that is, bring him into the BAS, a tent, a snow cave, etc.

(b) As soon as possible, remove the victim's cold, wet clothes.

(3) Insulate the victim.

(a) First, wrap the victim in a vapor barrier liner (VBL). A VBL will prevent heat loss as a result of evaporation and slow down heat loss from convection. The easiest way to do this in the field is by wrapping the victim in plastic trash bags. (Be sure not to cover the face.)

(b) Next, place the victim in a sleeping bag.

(4) Re-warm the victim

(a) The easiest way to do this in the field is to zip two sleeping bags together. Place the victim in the zipped up bags with 2 stripped volunteers. While this may not agree with Marines, it could save the victims life.

(b) In addition to the two stripped volunteer's place warmed materials on either side of the victim's neck, armpits, and his groin. Items such as warmed rocks, bags of warm water, or heat packs can be used. Be advised the warmed materials should not be hot, and the stripped volunteers should be in contact with the items as well. A hypothermia victim may not be able to tell if his skin is burning, but the volunteers will.

(5) Evacuate the victim. A casualty evacuation may not be possible due to the tactical situation, weather, or other factors. However, the sooner a victim can be evacuated, the better. ***Severe hypothermia is a medical emergency.***

(6) Other Points to Remember.

(a) Fluids. If the victim is mildly hypothermic, he may be given hot wets. Otherwise give him nothing by mouth.

(b) Avoid, if possible, excessive movement of the victim, as his heart may stop beating if it is jarred.

(c) Major Wounds. Apply first aid to major wounds first, before attempting to re-warm the victim. Re-warming a victim who has bled to death does little good.

(d) Never give alcohol to hypothermia victims.

(e) Even after you have started re-warming a victim, he must be constantly monitored. Don't forget about him.

TRANSITION: Not only can cold temperatures cause casualties through hypothermia, but cold can also lead to frostbite.

14. (10 Min) **FROSTBITE**.

a. Definition: Frostbite is the actual freezing of tissues.

b. Risk Factors of Frostbite. The high-risk areas are fingers, toes, nose, cheeks, and ears.

(1) Three Major Risk Factors. There are many factors that cause frostbite, of which three stand out as contributing to the majority of injuries:

(a) Improper clothing or improper care of clothing. This is a major factor in frostbite.

1. Wearing gloves when mittens should be worn.

2. Failure to dry gloves or liners after they have become wet.

3. Wet clothing of any kind.

4. Improper footwear, such as wearing summer combat boots when VB or ski/march boots should be worn.

5. Improper care of footwear. Failing to remove boots at night, sleeping with boots on, or failing to dry boots when they become wet.

6. Wearing boots which are too tight.

7. The proper use of cold weather clothing, as well as its proper maintenance in the field, is dependent on small unit leadership. Small unit leaders must ensure

that their men are adequately clothed, as well as the clothing being adequately maintained.

(b) Dehydration. This is another major contributing factor in frostbite. Marines who are well hydrated are much better equipped to fight off frostbite.

(c) Poor diet or starvation. This is another major contributing factor in frostbite. Remember that the body can be thought of as a furnace, and that the fuel is food. When food intake is low, there is less fuel to feed the furnace and the risk for frostbite goes up.

(2) Other factors that contribute to frostbite.

(a) Outside temperature. Obviously, the colder it is, the greater the risks.

(b) Snow or Ground temperature. The snow temperature can be 30'-40' colder than the air temperature.

(c) Wind chill. As mentioned previously, wind-chill should not be a factor with properly dressed Marines.

(d) Cold metals. Never touch very cold metals with bare flesh. Use contact gloves.

(e) Petroleum products. Fuels and oils freeze at a much lower temperature then water. Spilling cold fuel (such as white gas, gasoline, etc.) on bare skin can cause immediate, severe frostbite.

(f) Exhaustion. The body's natural defense mechanisms in general are lowered when you are exhausted.

(g) Hypothermia.

(h) Race/Place of Birth. African Americans and those from the south are at increased risk for frostbite.

(i) Other factors include prolonged immobility (as when sitting in an ambush position), wounds with blood loss, previous cold injury, and tobacco use.

c. <u>Signs and Symptoms of Frostbite.</u>

(1) Signs of Frostbite.

(a) The skin may appear red, white, yellow, gray, blue, frosty, or even normal.

(b) The skin may feel woody or firm.

(c) The joints may be stiff or immobile.

(d) The affected part may feel like a block of wood or even ice.

(e) Pulses may or may not be present.

(2) Symptoms of Frostbite. A victim may complain of:

(a) Tingling.

(b) Burning.

(c) Aching cold.

(d) Sharp pain.

(e) Increased warmth.

(f) Decreased sensation.

(g) No sensation at all. The victim may describe the affected part as clumsy, lifeless, bulky or club like.

d. Classification of Frostbite. Like burns, frostbite has been divided into 1st, 2nd, 3rd, and 4th degrees. But it is much easier to divide it up into Frosting, Superficial Frostbite, and Deep Frostbite.

(1) Frosting is something we all have experienced at one time or another. It is when some part of the body (toes, fingers, or nose usually) becomes painfully cold, but does not freeze. It is a harmless condition and the affected part returns to normal with re-warming.

(2) Superficial Frostbite. This is when the skin freezes, but not the tissue beneath (such as muscle, nerves, and bone).

(a) Skin appears red, gray, or even blue, and has a waxy feel to it.

(b) Pulses will be present, but decreased.

(c) The sensation of pain and light touch may be absent, but deeper sensations such as pressure will be intact.

(d) The joints will be mobile, but stiff.

(e) Movement of the part by the victim will be possible, although it may be difficult.

(3) Deep Frostbite.

(a) Initially, the skin may appear the same as above.

(b) Pulses will not be present.

(c) The skin will feel woody, firm or even rock hard.

(d) Tissues below will feel doughy or hard.

(e) All sensation will be absent.

(f) Skin will not move easily or not at all.

(g) Joints will be stiff or immobile.

(h) Movement of the affected part will be minimal or absent.

(4) It is often difficult to say exactly how severe a case of frostbite is until several weeks have passed. Therefore, it is wise to assume the worst.

NOTE: Frostbite may be present in different degrees in the same affected part, for example: a frostbitten hand may have deep at the fingers, superficial at the palm and frosting at the wrist.

e. Prevention of Frostbite.

(1) Frostbite is an entirely preventable injury. Obviously, there is little one can do about the weather, but Marines can ensure that the other risk factors that can lead to frostbite are minimized. The best way to prevent frostbite is to prevent the three major risk factors: Improper clothing or improper care of clothing, dehydration, and starvation.

(a) Dress in layers. Keep comfortably cool. If you begin to become uncomfortable, add layers. If your hands or feet become uncomfortable, do not ignore them - you may have to add more layers, or you may want to change socks or gloves.

(b) Keep clothes dry. This is vitally important. If your boots, socks, or gloves get wet, then dry them. This may mean you have to change socks up to 4 - 5 times a day (especially with Vapor Barrier boots). If your gloves or liners are wet, warm and dry them. Do not continue to wear wet clothing.

(c) Dress properly. This may seem obvious, but Marines have gotten frostbite because they did not. If the wind is blowing, wear the correct protective layer. Always have a balacalava or watch cap available, and if it's cold - wear it. If your fingers are getting cold in gloves, wear mittens.

(d) Avoid Dehydration. When you become dehydrated, the amount of blood available to warm your fingers and toes goes down, greatly increasing your risk of frostbite.

(e) Avoid Starvation. Remember - Food is Fuel - and the body uses that fuel to make heat. When you are low on fuel, you will be low on heat.

(2) U.S. Marines are the toughest people in the world. However, Mother Nature is tougher. If you notice your fingers or toes are getting cold even after you have tried to warm them, do not ignore it. Let your leaders know. Ignoring the problem will not make it go away, it will only get worse.

(3) Small Unit Leaders must ensure that preventive measures are taken. Like dehydration, frostbite results from ***failure of leadership.***

f. Field Management for Frostbite. Only frosting should be treated in the field, all others should be evacuated immediately. If you don't know, assume the worst and evacuate.

(1) Treatment of Frosting. This is easily done in the field using the 15 minute rule. Frosting will revert to normal after using this technique of body heat re-warming. Hold the affected area, as described below, skin to skin for 15 minutes. If the affected area does not return to normal, assume a frostbite injury has occurred and report it to your seniors.

(a) Re-warm face, nose, ears with hands.

(b) Re-warm hands in armpits, groin or belly.

(c) Re-warm feet with mountain buddies armpits or belly.

(d) ***DO NOT RUB ANY COLD INJURY WITH SNOW-EVER!***

(e) Do not massage the affected part.

(f) Do not re-warm with stove or fire, a burn injury may result.

(g) Loosen constricting clothing.

(h) Avoid tobacco products.

(2) Treatment of Superficial or Deep frostbite. Any frostbite injury, regardless of severity, is treated the same – evacuate the casualty and re-warming in the rear. Unless the tactical situation prohibits evacuation or you are in a survival situation, no *consideration should be given to re-warming frostbite in the field.* The reason is something-called freeze – thaw – re-freeze injury.

(a) Freeze – Thaw – Re-freeze injury occurs when a frostbitten extremity is thawed out, then before it can heal (which takes weeks and maybe months) it freezes again.

This has devastating effects and greatly worsens the initial injury.

(b) In an extreme emergency it is better to walk out on a frostbitten foot than to warm it up and then have it freeze again.

(c) Treat frozen extremities as fractures - carefully pad and splint.

(d) Treat frozen feet as litter cases.

(e) Prevent further freezing injury.

(f) Do not forget about hypothermia. Keep the victim warm and dry.

(g) Once in the rear, a frostbitten extremity is re-warmed in a water bath, with the temperature strictly maintained at 101'-108'F.

15. (5 Min) **TRENCHFOOT/IMMERSION FOOT**

a. Definition. This is a cold - wet injury to the feet or hands from prolonged (generally 7 - 10 hours) exposure to water at temperatures above freezing.

b. Causes of Trenchfoot/Immersion Foot. The major risk factors are wet, cold and immobility.

c. Signs and Symptoms of Trenchfoot/Immersion Foot

(1) The major symptom will be pain. Trench foot is an extremely painful injury.

(2) Trench foot and frostbite are often very difficult to tell apart just from looking at it. Often they may both be present at the same time. Signs include:

(a) Red and purple mottled skin.

(b) Patches of white skin.

(c) Very wrinkled skin.

(d) Severe cases may leave gangrene and blisters.

(e) Swelling.

(f) Lowered or even absent pulse.

(3) Trench foot is classified from mild to severe.

d. Prevention of Trench foot/Immersion Foot is aimed simply at preventing cold, wet and immobile feet (or hands).

(1) Keep feet warm and dry.

(2) Change socks at least once a day. Let your feet dry briefly during the change, and wipe out the inside of the boot. Sock changes may be required more often.

(3) Exercise. Constant exercising of the feet whenever the body is otherwise immobile will help the blood flow.

e. Treatment of Trench foot/Immersion Foot

(1) All cases of trench foot must be evacuated. It cannot be treated effectively in the field.

(2) While awaiting evacuation:

(a) The feet should be dried, warmed, and elevated.

(b) The pain is often severe, even though the injury may appear mild; it may require medication such as morphine.

(3) In the rear, the healing of trench foot usually takes at least two months, and may take almost a year. Severe cases may require amputation. *Trench foot is not to be taken lightly.*

TRANSITION: We have just discussed the definition, signs and symptoms, prevention, and treatment of trench foot. Are there any questions? The next injury, which is easy to acquire at high altitude, is snow blindness.

16. (5 Min) **SNOW BLINDNESS**

a. Definition. Sunburn of the cornea.

b. Causes of Snow Blindness. There are two reasons Marines in a winter mountainous environment are at increased risk for snow blindness.

(1) High altitude. Less ultraviolet (UTV) rays are filtered out, UV rays are what cause snow blindness (as well as sunburn). So at altitude, more UV rays are available to cause damage.

(2) Snow. The white color of snow reflects much more LTV rays off of the ground and back into your face.

c. Signs and Symptoms of Snow Blindness

(1) Painful eyes.

(2) Hot, sticky, or gritty sensation in the eyes, like sand in the eyes.

(3) Blurred vision.

(4) Headache, may be severe.

(5) Excessive tearing.

(6) Eye muscle spasm.

(7) Bloodshot eyes.

d. Prevention of Snow Blindness. Prevention is very simple. Always wear sunglasses, with UV protection. If sunglasses are not available, then field expedient sunglasses can be made from a strip of cardboard with horizontal slits, and charcoal can be applied under the eyes to cut down on reflection of the sun off the snow.

e. Treatment of Snow Blindness

(1) Evacuation, when possible.

(2) Patch the eyes to prevent any more light reaching them.

(3) Wet compresses, if it is not too cold, may help relieve some of the discomfort.

(4) Healing normally takes two days for mild cases or up to a week for more severe cases.

17. (5 Min) **CARBON MONOXIDE POISONING**

a. Definition. Carbon Monoxide (CO) is a heavy, orderless, colorless, tasteless gas resulting from incomplete combustion of fossil fuels. CO kills through asphyxia even in the presence of adequate oxygen, because oxygen-transporting hemoglobin has a 210 times greater affinity for CO than for oxygen. What this means is that CO replaces and takes the place of the oxygen in the body causing Carbon Monoxide poisoning.

b. Signs/Symptoms. The signs and symptoms depend on the amount of CO the victim has inhaled. In mild cases, the victim may have only dizziness, headache, and confusion; severe cases can cause a deep coma. Sudden respiratory arrest may occur. The classic sign of CO poisoning is cherry-red lip color, but this is usually a very late and severe sign, actually the skin is normally found to be pale or blue.

(1) CO poisoning should be suspected whenever a person in a poorly ventilated area suddenly collapses. Recognizing this condition may be difficult when all members of the party are affected.

c. Treatment. The first step is to immediately remove the victim from the contaminated area.

(1) Victims with mild CO poisoning who have not lost consciousness need fresh air and light duty for a minimum of four hours. If oxygen is available administer it. More severely affected victims may require rescue breathing.

(2) Fortunately, the lungs excrete CO within a few hours.

d. Prevention. Prevention is the key. Ensure that there is adequate ventilation when running vehicle engines, operating stoves in closed spaces (tents), or when cooking over open flames.

18. (5 Min) **PERSONAL HYGIENE.** The five most important areas of personal hygiene are:

a. Body

(1) The body should be washed frequently in order to minimize the chances of small cuts and scratches developing into full blown infections and as a defense against parasitic infections.

(2) A daily bath or shower consisting of soap and hot water is ideal. However, when this is not possible you should:

(a) Give yourself a sponge bath using soap and water, making sure particular attention is given to body creases i.e., armpits, groin area, face, ears, and hands.

(b) If water is in extremely short supply, you should take an air bath. To do this:

1. Remove all clothing and hang it up to air.

2. Expose the body for two hours to sunlight, which is ideal, but the effects will basically be the same if done indoors or during an overcast day. **BE CAREFUL NOT TO SUNBURN.**

b. Hair

(1) Should be cleaned frequently.

(2) Should be inspected at least once a week for parasites.

c. Fingernails

(1) Should be trimmed to prevent accidentally scratching yourself.

(2) Should be kept clean to prevent harborage areas for bacteria.

d. Feet Are your primary source of transportation.

(1) The feet should be inspected frequently for:

(a) Blisters.

(b) Infections. Bacterial and fungal.

(2) They should be kept dry by:

(a) Frequent sock changes (one to three times daily) in conjunction with:

1. Foot powders.

2. Antiperspirants.

e. Oral Hygiene. The mouth and teeth should be cleaned at least daily to prevent tooth decay and gum disease.

(1) Ideally, cleaning should be done with:

(a) Toothbrush.

(b) Toothpaste.

(c) Dental floss.

(2) If these items are not available, the following methods can be used:

(a) Make a chew stick from a clean twig about 8" long and about finger width. Chew one end until it becomes frayed and brush-like, and then brush the interior and exterior surfaces thoroughly.

(b) The gums should be stimulated at least once a day by rubbing them vigorously with a clean finger.

(c) Field expedient dental floss can be made from the inner strands of a 10 inch section of paracord.

TRANSITION: Now that we have discussed personal hygiene, are there any questions? Since we have in mind our own personal hygiene, let's consider another item which can effect our health.

19. (5 Min) **WATER PURIFICATION.** Water purification simply consists of removing or destroying enough impurities to make water safe to drink. Giardia cysts are an ever-present danger in clear mountain water, even though this water appears clean and safe to drink.

a. Boiling is the oldest way of disinfecting water. Recent studies have shown that the old recommendation of boiling water for 10 minutes and adding 1 minute of boiling for each 1000 feet in elevation is not necessary and wasteful of limited fuel supplies. The studies found that the thermal death point of microorganisms is reached in shorter time at higher altitudes, while lower temperatures are effective with a longer contact period. Therefore the minimum critical temperature needed to render water safe to drink is well below the boiling point at elevation. With these findings in mind it can now be safely said that once water is brought to a boil, and allowed to cool it is disinfected and safe to drink. For an extra margin of safety the Wilderness Medical Society recommends boiling water for 11minute no-matter what altitude you're at to render water safe to drink.

TRANSITION: Are there any questions over water purification? Now we will discuss what the human animal produces in the area of waste products and how to dispose of them.

20. (5 min) **WASTE DISPOSAL.** Waste should be disposed of by burning, burying, or hauling it away.

a. The importance of waste disposal cannot be overemphasized. It serves to:

(1) Eliminate harborage areas for rodents and vermin.

(2) Preclude an attractant for rodents and vermin.

(3) Prevent a source of pathogenic contamination.

b. Two basic types:

(1) Organic wastes

(a) Human waste - burn or haul away to a designated waste pit area.

(b) Urine - Use only assigned, marked areas away from food and water sources.

(c) Edible garbage - Burn. Do not leave exposed for animals, vermin, or the enemy.

(2) Non-organic wastes

(a) Papers -burn.

(b) Metals - haul away or bury.

(c) Liquids – burn or bury.

TRANSITION: Improper disposal of waste is hazardous in more ways than one. Are there any questions at this time?

PRACTICE

a. Students will practice what was taught in upcoming field evolutions.

PROVIDE HELP

a. The instructors will assist the students when necessary.

OPPORTUNITY FOR QUESTIONS (3 Min)

1. QUESTIONS FROM THE CLASS

2. QUESTIONS TO THE CLASS

Q. What are the five ways a body loses heat?

A. (1) Radiation
(2) Conduction
(3) Convection
(4) Evaporation
(5) Respiration

Q. What are the two causes of dehydration?

A. (1) Inadequate intake
(2) Excessive loss

Q. What is the best treatment for HAPE and HACE?

A. Descent.

SUMMARY (2 Min)

a. During this period of instruction we have covered the way a body loses heat, dehydration, heat injuries, high altitude sickness, hypothermia, frostbite, trench foot, snow blindness, carbon monoxide poisoning, and personal hygiene.

b. Those with IRF's please fill them out at this time. We will now take a short break.

UNITED STATES MARINE CORPS
Mountain Warfare Training Center
Bridgeport, California 93517-5001

SML
SMO
02/11/02

LESSON PLAN

COLD WEATHER AND MOUNTAIN LEADERSHIP PROBLEMS FOR THE SMALL UNIT LEADER

INTRODUCTION

(3 Min)

1. GAIN ATTENTION. Leadership is a vital aspect of military operations in all environments, particularly in the extreme conditions that we normally encounter in mountainous operations.

2. PURPOSE. The purpose of this period of instruction is to emphasize the vital role of leadership in the conduct of successful operations and to promote among leaders at all levels an understanding of the problems common to units operating in a mountainous environment. This lesson relates to all of the training that you will receive here at MWTC.

3. METHOD / MEDIA. The material in this lesson will be presented by lecture.

4. EVALUATION. This period of instruction has no learning objectives but students will be evaluated throughout their stay at MWTC.

TRANSITION: Your Marines will find themselves up against many new and challenging problems-but none that a properly trained Marine cannot overcome.

BODY

(25 Min)

1. **SUPPORTINING PUBLICATIONS**. In addition to your manuals, MCRP 3-35.1a Small Unit Leaders Guide to Cold Weather Operations and MCRP 3-35.2a Small Unit Leaders Guide to Mountain Operations, both in draft form, are great sources of knowledge for the small unit leader. These manuals may be downloaded from the doctrine division's page on the USMC home page.

a. Further more, the Mountain Warfare Training Center has authored two Marine Corps Warfighting Publications, one on MAGTF Cold Weather Operations (MCWP 3-35.1), and one on Mountain Operations (MCWP 3-35.2).

b. Also authored by MWTC were three X-Files (Cliff Assault, Mule Packing, and Water Procurement), the Military Nordic Ski Manual, and the Assault Climbers Guide.

c. The aforementioned manuals and publications discuss specific techniques and procedures for operating in a cold weather / mountainous environment. Your small unit leaders should become very familiar with these publications.

TRANSITION: All leadership traits and principles are of importance to leaders assigned to units operating in high altitude areas. The extreme conditions so often found are the cause of some peculiar problems for leaders.

2. UNDERSTANDING THE COLD WEATHER ENVIRONMENT

d. The Russo-Finnish War: Conducted during the winter of 1939-1940 for 105 days. The Soviets outnumbered the Finnish 40-1 in personnel, 30-1 in aircraft and 100-1 in tanks and artillery. The soviets suffered somewhere between 200,000, according to Field Marshal Mannerheim, and 2 million, according to Secretary Khrushchev, casualties. The Finnish's ability to operate freely in the AO against the road bound Soviet Army was the main reason why the Soviet 7 day war lasted 105 days. The use of motti tactics, the destruction of Soviet Combat Service Support, and lack of Soviet winter training are cited keys to Finnish success. Inevitably the Finns capitulated when their economy was no longer able to support the war effort.

e. U.S. Army WWII: The U.S. Army suffered 84,000 cold weather injuries during World War II. 90% of these injuries were in the infantry. On average, each case required 87 days in the hospital.

f. German Army: During Operation Barbarosa (the invasion of the Soviet Union) the German Army suffered 100,000 cold weather injuries. 15,000 required a limb to be amputated.

g. Korean War: MWTC was established as a result of environmental injuries sustained by USMC forces during the winter drive to the Yalu River (Chosin Campaign). Cold weather injuries accounted for 10 percent of all injuries during the Korean War, most occurring during the winter of 1950-51.

h. Falklands War: After 25 days of combat, the Argentine Army surrendered. Without getting into the feats of the British Royal Marines and Paras, 75% British forces suffered from immersion foot or other cold weather injuries. Considering all factors, if the campaign were extended, this could have posed a serious problem to the British.

TRANSITION: Good leadership at any time can make the difference between life and death for your subordinates, but this is especially true in climates and locales that can be so hostile, such as the mountains.

3. **UNDERSTANDING THE MOUNTAIN ENVIRONMENT**.

   i. Italian Campaign WWII: Considered to be the soft under belly of Europe, Italy proved to be a meat grinder for General Mark Clark's 5th Army, a drain on material, and frustration for Allied command. After the Anzio landing, the allied forces quickly became bogged down with advances measured in yards. It was not until the 10th Mountain Division arrived on scene did the allies begin to make significant advances and break into the Po valley. Mules were used extensively by allied forces to transport gear up the mountainous terrain.

   j. Russian Afghanistan War: Soviet Doctrine was a major factor in the Soviet defeat in Afghanistan. With their infantry tied to their vehicles, Assault and Offensive Air Support not coordinated with maneuver elements and focus on controlling the cities, Soviet forces were easy targets for the Mujahedin who controlled rural Afghanistan. The Soviets were unable to effectively hunt this guerrilla in his terrain.

   k. Chechnya: Although in the last two years, the Russians appear to be winning the war against the Islamic Extremists in Chechnya, the initial campaign resembled the debacle in Afghanistan. Notably, the Russian inability to sustain their forces logistically, resulted in numerous environmental injuries as well as the rampant spread of disease. Typhus, dysentery and hepatitis posed serious problems to Russian operations during the initial campaign. Cold weather clothing and equipment was also a contributing factor to cold weather injuries as most of their clothing was made of cotton.

TRANSITION: Understanding the mountain and its environment will aid the leader in making good decisions. There are keys to good leadership that you should know.

4. **KEYS TO GOOD LEADERSHIP**. There are four keys to good leadership that will be discussed in this chapter.

   a. **Pre-environmental training.** Most casualties sustained in mountainous or cold weather environments are due to Marines and Sailors not being physically and mentally prepared for the environment. Based off your units TEEP, you should begin your work-up as soon as possible. As you develop your unit's training package, you need to set goals for what you expect the individual and unit to be able to accomplish prior to deployment. Additionally, you will need to keep in mind what assets you will have to support you prior to and once deployment has occurred. Pre-environmental training can be accomplished by your unit Mountain Leaders or by an MTT from the Mountain Warfare Training Center. At a minimum, pre-environmental training should encompass the following:

1. Physical Fitness: This should entail moving heavy loads over medium distances at a quick pace. Loads should consist of a Combat Load, individual and crew-served weapons initially. Continue to increase weight and speed as the physical fitness program develops. Also, circuit courses combining cardio-vascular training as well as iso-tonic exercises serves to prepare individuals.

2. Individual Training. Individual training should initially focus on the use of clothing and equipment for this environment, then on cross training of weapon systems. Most cold weather injuries result from the improper wearing of cold weather clothing.

3. Unit Training. Establish your SOPs while you are still in garrison and refine them once you are in the environment; it will save you time and frustration in the environment. TDG's are great tools, as well as historical studies, for preparing subordinates prior to them getting in the environment. Finally, get use to working without communications and key leaders. This will better prepare your subordinates to assume higher's mission.

b. **Prepare for increases casualties**. A 500 man infantry Battalion normally suffers 15 – 30 injuries during summer operations and 30 – 45 injuries during the winter while training at Bridgeport. These injuries are normally exaggerated twisted ankles and Marines trying to avoid training.

1. Understanding that you will take increased casualties, as illustrated in the historical examples, you need to make sure that you are capable of still accomplishing the mission despite personnel shortfalls. As an example, it takes 3 Marines to transport and carry a 60 mm mortar system. If you lose one or two Marines from that section, you are only able to employ two systems, cutting your ID fires by 1/3 at the company level. You need to ensure your 0311s are just as capable at employing that weapon system.
2. Another example, is the RTO the only Marine who knows how to operate the AN/PRC-119, if so, you are out of luck if you lose him. Can you do it? After all, in addition to your 3 Squad Leaders, it is your primary weapon system as a Platoon Commander.

3. Also, have you trained your squad leaders to assume the platoon commanders job, if the answer is yes, have you tested them? Can your FO run your FIST Team? Finally, are you Marines capable of being the Marine on site to provide essential First Aid? After all, the buddy aid does come before corpsman aid.

4. Casualties will significantly hinder the ability to accomplish the mission. Depending on the CASEVAC plan, to move one casualty from the point

of injury to the casualty collection point may take the Marine's entire squad. Leaders must ensure the CASEVAC plan is supportable at all levels.

c. **Understanding your capabilities and limitations**. In the estimate of the situation (METT-TSL) we talk about the enemy's capabilities and limitations, which is derived from his composition, disposition and strength, which leads us to assume what we think his most probable and most dangerous course of action will be. Understanding our capabilities and limitations is what most leaders fail to do in this environment. Time and time again, leaders at all levels fail to plan operations in the environment that are within their personnel and equipment's capabilities and limitations. The classic example of this is the platoon commander who is assaulting and consolidating on the objective alone, except for the RTO whom he managed to drag along by the handset, and wondering why his platoon is still a terrain feature away.

1. The fact is you cannot achieve success in this environment if you do not realistically plan your operation from crossing the line of departure to consolidation, re-supply and CASEVAC. You must account for everything.

2. Past performance documentation will indicate future capabilities: How many times do we have to re-invent the wheel? For example, every time we conduct a movement, we create a route card. What do we normally do with that route card? Throw it away when we are done. How many leaders have taken the time estimate that he uses during planning, compared that to actual performance, and created a time estimate formula that better reflects his unit's ability?Realistic training results in accurate assessment: In training, the more things you fairy dust, the more you detract from an accurate assessment of your unit. How many Battalions have ever required their Marines to conduct CASEVACS to the BAS? How can you ever determine if the CASEVAC plan would work if you do not test it? The sad thing is, most units will learn that it will fail when they are in a real world situation.

d. **Controlling the situation**. When Marines are tired or cold, the first thing they lose is individual discipline. This problem is exacerbated in this environment because Marines will always be tired or cold as it is the nature of the environment. With that, Marines get lazy and the first thing that goes is continuing actions and a tactical mindset.

1. Individual and unit discipline: Marines are on patrol, as they start getting tired, fewer and fewer Marines, including the leaders, execute continuing actions (looking at the ground. kneel or prone), the laziness expands, re-entry of friendly lines is administrative, they get back to their tents and throw their packs on the ground and go to sleep gear

maintenance is not conducted… Marine wakes up with immersion foot…

2. Leadership presence: Leadership presence through inspections and supervision is required at all levels. Discipline at levels must be exercised. This includes not only self-discipline, but the enforcement of discipline on Marines for violations of orders and SOPs. This must be the norm from the beginning. Once you let one transgression slide, it will open the floodgates. Leaders who do not make the mark need to be dealt with swiftly and sternly. DO NOT BE AFRAID TO FIRE ANY LEADER WHO IS NOT MAKING THE MARK. HE WILL GET MARINES INJURED OR KILLED IN THIS ENVIROMENT!

3. Keeping the Marines informed: This goes a long way to keeping Marines involved in the mission. The more a Marine is involved in the mission, the better he performs.

TRANSITION: As in any environment there are certain trends in the mountains that will affect all leaders and their subordinates.

5. **COMMON LEADERSHIP PROBLEMS**. There are five specific problems that you as a leader will face in this environment. Those problems are: cocooning, group hibernation, loss of personal contact and communication, time and space planning, and sustainment.

   a. Cocooning. Turning inward on yourself and being more concerned about your own comfort than the environment around you. It results in diminished situational awareness and having no tactical mindset. Basically, you are cold or tired and all you care about is getting some sleep or getting warm. We have all seen this at one time or another, except in this environment it is endemic. Marines are not use to operating in a cold weather environment. How do we deal with this?

      1. Environmental training. Teach the Marines how to properly wear the clothing and equipment that is issued to them. Are they wearing every piece of clothing they have and are still freezing? Well the problem is probably that their clothing is not loose and layered. Or, are they wearing cotton utilities that are soaking wet instead of gore-tex?

      2. Physical activity. When Marines cocoon, they focus in on themselves. Give them a mission, which causes them to focus on something else. There is always something that needs to be accomplished; like improving the defensive positions. Or, rotate Marines out of security more frequently Increase the frequency of security patrols. Basically, Marines cocoon when their leadership does not supervise. This is a continual problem that needs to be dealt with.Time. Ultimately, most Marines will learn to deal with the environment the longer they are in it. However, poor time

management and making Marines stand around only contributes to the problem. Also, taking your Marines out of the field after four days accomplishes nothing but building a four-day mindset.

b. Group hibernation. Cocooning ultimately leads to group hibernation or collective cocooning. The training schedule will say Night Company Movement. During the day, the temperature was cold and toward evening it started snowing and it was rather windy. As the Marines are new to the environment, the environment obviously mentally affects them. The Company Commander, who is suffering the same as the Marines, wants desperately to go to his tent. He confers with the XO, 1st SGT and Company Gunny, they want to go to the tent as well. They all agree to cancel training for the "safety" of the Marines. That is collective cocooning.

1. Remedy. As the Leader, you have to be the sounding board. If you know that training is not going to endanger the safety of the Marines, explain it to the Commander. If that does not work, bring it to higher's attention.

c. Loss of personal contact and communication. A by-product of cocooning and group hibernation is the loss of personal contact and communication. Both cocooning and group hibernation degrade the ability of the chain of command to operate. Because Marines find sanctuary in the tent or where ever they throw themselves down, they do not want to leave it, and that becomes their world. Squad leaders no longer show up to the Platoon Sergeant's meeting, instead he sends one of his fire team leaders with some excuse why he could not be there. Additionally, the leaders tour the areas less frequently and as a result don't talk to their Marines. They have no idea what their state of mind is or how they are doing physically. Word no longer gets passed and people have no idea what is going on. This results in an informational and emotional isolation. Marines become apathetic to the world around them. He no longer knows what the mission of his adjacent unit is, nor cares. This significantly degrades from the moral of the unit as well as their ability to function as a cohesion unit, in the maneuver sense.

1. Remedy. Insist on the proper use of the chain of command, talk to the Marines at the lowest level, and supervise the Marines as much as possible. Check on the Marines manning the M-240G at 0230 to see how he is doing.

d. Time and space planning. This is another area where leaders and units consistently fail in this environment. What people need to understand is that everything takes longer, from breaking down tents to moving from A to B. In Camp Pendleton it may take a platoon 10 minutes to break down and pack up their tents, but here it will take at least double that time.

1. Mobility. Mobility is affected by the terrain and altitude (flat or mountains, at sea level or 2000 meters) the weather (34 degrees and

raining, sunny or 12 inches of snow) the physical condition of the Marines and their moral. All these factors are going to determine how well your Marines will move a distance over time.

2. Planning. Very rarely do units cross the LD on time for a morning movement. This is because the leaders do not adequately back-plan from LD to pull pole to reveille. As talked about during knowing your capabilities and limitations, realistic assessment of mobility capabilities is essential to mission accomplishment. Furthermore, you need to issue timely warning orders to ensure your subordinate leaders have enough time to accomplish their tasks as well.

e. Sustaining your Marines. When it comes down to it, you accomplished your mission, good job, but are you ready to continue the fight? We all know that we can push our Marines to the breaking point to accomplish the mission, but if we do that we are actually failing, especially in this environment?

1. Realistic Mission Planning. Terrain selection, which minimizes the impact on the Marines, the gear selection essential to mission accomplishment, and the selection of personnel that will not hinder the mission. With all that Marines locate close with and destroy the enemy. However, inadequate mission planning results in your Marines having to suffer the wraths of poor terrain selection, extra / needless weight on their backs and making up for the Marine who cannot carry himself. Your Marines are spent long before they reach the objective.

2. Logistical Support. You need to ensure that your plan is logistically supportable. You can come up with some great big blue arrow plan of attack, but if you cannot get chow and water to your Marines, it is worthless. Furthermore, if your Marines are spent on the approach march, what good are they in the attack?

3. Conservation of energy is key. What it comes down to is, once your Marines are spent, it takes time for them to recover. Re-hydration takes at least 6 hours minimum and it takes almost 24 hours for your body to recover energy from food. The time it takes you to move a distance has to be at a pace sustainable by the Marines. Getting to the objective still ready to fight is key in this environment. Especially true in this environment is that the more starved your body is, the harder it will be for the immune system to fight off disease.

TRANSITION: The mountains are a very dangerous place to survive and fight. They present many dangers to overcome. Are there any questions?

PRACTICE **(CONC)**

a. Students will apply mountain leadership principles throughout duration of the course.

PROVIDE HELP **(CONC)**

a. The instructors will assist the students when necessary.

OPPORTUNITY FOR QUESTIONS **(3 Min)**

1. QUESTIONS FROM THE CLASS

2. QUESTIONS TO THE CLASS

Q. What are 2 of the key points of good leadership?

A. (1) Pre environmental training.
(2) Prepare for increased casualties.

Q. What are the five leadership problems peculiar to mountain operations?

A. (1) Cocooning.
(2) Group hibernation.
(3) Loss of personal contact and communications.
(4) Time and space planning.
(5) Sustaining your marines.

SUMMARY (2 Min)

a. This period of instruction has discussed keys to good leadership, including: pre-environmental training, prepare for increased casualties, understanding your capabilities and limitations, controlling the situation. Also discussed were common leadership problems: cocooning, group hibernation, loss of personal contact and communication, time and space planning, and sustaining your Marines.

b. Those of you with IRF's please fill them out at this time and turn them in. We will now take a short break.

UNITED STATES MARINE CORPS
Mountain Warfare Training Center
Bridgeport, California 93517-5001

WML
WMO
02/07/02

LESSON PLAN

MOUNTAIN WEATHER

INTRODUCTION (5 Min)

1. **GAIN ATTENTION**. Normally, as Marines in a temperate climate we think of bad weather as possibly a tactical advantage, giving us concealment in order to move undetected. But in a cold weather mountainous environment, bad weather could be devastating if not properly prepared. For you as a Mountain Leader, it is crucial that you be able to understand the fundamentals of meteorology and determine what to expect from incoming weather patterns.

2. **OVERVIEW.** The purpose of this period of instruction is to introduce you to weather patterns, the invisible aspects of weather, and some visible clues to help you forecast the weather. This lesson relates to all operations in the mountains.

INSTRUCTORS NOTE: Have students read learning objectives.

3. **INTRODUCE LEARNING OBJECTIVES**

   a. TERMINAL LEARNING OBJECTIVE. In a mountainous environment and 24 hours prior to weather conditions occurring, forecast weather, in accordance with the references.

   b. ENABLING LEARNING OBJECTIVES. (MLC)

      (1) Without the aid of references and given a list of air masses, describe in writing the types of air masses, in accordance with the references.

      (2) Without the aid of references, list in writing the ways that air could be lifted and cooled beyond its saturation point, in accordance with the references.

(3) Without the aid of references, list in writing the types of clouds, in accordance with the references.

(4) Without the aid of references, describe in writing the types of clouds, in accordance with the references.

(5) Without the aid of references, list in writing the cloud progression of fronts, in accordance with the references.

c. ENABLING LEARNING OBJECTIVES. (SMO)

(1) With the aid of references and given a list of air masses, list orally the types of air masses, in accordance with the references.

(2) With the aid of references, list orally the ways that air could be lifted and cooled beyond its saturation point, in accordance with the references.

(3) With the aid of references, list orally the types of clouds, in accordance with the references.

(4) With the aid of references, describe orally the types of clouds, in accordance with the references.

(5) With the aid of references, list orally the cloud progression of fronts, in accordance with the references.

4. **METHOD/MEDIA**. The material in this lesson will be presented by lecture with the use of computer visual aid and/or turn chart. You will practice what you learned during upcoming field training exercises. Those of you with IRF's please fill them out at the end of this period of instruction.

5. **EVALUATION**.

a. MLC - You will be tested by a written exam.

b. SMO - You will be tested by a performance tests later in the course.

TRANSITION: Are there any questions over the purpose, learning objectives, how you will be taught, or how you will be evaluated? Let's start by looking at some of the tactical or backcountry considerations and general info you need to be aware of.

BODY (100 Min)

1. (5 Min) **THE WEATHER IN GENERAL**

   a. How well you are able to see the enemy, terrain and Marines around you is greatly affected by the type of weather you might have to deal with. Visibility affects your ability to navigate over the terrain as well as being able to identify the enemy upon contact.

   b. Route selection over snow covered mountainous terrain, is affected by the avalanche potential the weather brings with it.

   c. While you are traveling or sitting in your Patrol base ask these questions concerning the weather. What is the weather currently? What was the weather recently and when did it change last? When is the forecasted next change coming? Having access to this information can help you better prepare before going to the field.

   d. The earth is surrounded by the atmosphere, which is divided, into several layers. The world's weather systems are in the troposphere, the lower of these layers. This layer reaches as high as 40,000 feet.

   e. Dust and clouds in the atmosphere absorb or bounce back much of the energy that the sun beams down upon the earth. Less than one half of the sun's energy actually warms the earth's surface and lower atmosphere.

   f. Warmed air, combined with the spinning (rotation) of the earth, produces winds that spread heat and moisture more evenly around the world. This is very important because the sun heats the Equator much more than the poles and without winds to help restore the balance, much of the earth would be impossible to live on. Where the air cools, you can get clouds, rain, snow, hail, fog, frost, etc.

   g. The weather that you find in any place depends on many things. How hot the air is, how moist the air is, how it is being moved by the wind, and especially, is it being lifted or not?

TRANSITION: Are there any questions over the general information? Now let's take a look at how wind effects our weather.

2. (10 Min) **WINDS**. The uneven heating of the air by the sun and rotation of the earth causes winds. Much of the world's weather depends on a system of winds that blow in a set direction. This pattern depends on the different amounts of sun (heat) that the different regions get and also on the rotation of the earth.

   a. Above hot surfaces, air expands (air molecules spread out), and move to colder areas where it cools and becomes denser, and sinks to the earth's surface. This forms a circulation of air from the poles along the surface or the earth to the Equator, where it rises and moves towards the poles again.

b. Once the rotation of the earth is added to this, the pattern of the circulation becomes confusing.

c. Because of the heating and cooling, along with the rotation of the earth, we have these surfaces winds. All winds are named from the direction they originated from:

(1) Polar Easterlies. These are winds from the polar region moving from the east. This is air that has cooled and settled at the poles.

(2) Prevailing Westerlies. These winds originate from approximately 30 degrees North Latitude from the west. This is an area where prematurely cooled air, due to the earth's rotation, has settled back to the surface.

(3) Northeast Trade winds. These are winds that originate from approximately 30 degrees North from the Northeast. Also prematurely cooled air.

d. Here are some other types of winds that are peculiar to mountain environments but don't necessarily affect the weather:

(1) Anabatic wind. These are winds that blow up mountain valleys to replace warm rising air and are usually light winds.

(2) Katabatic wind. These are winds that blow down mountain valley slopes caused by the cooling of air and are occasionally strong winds.

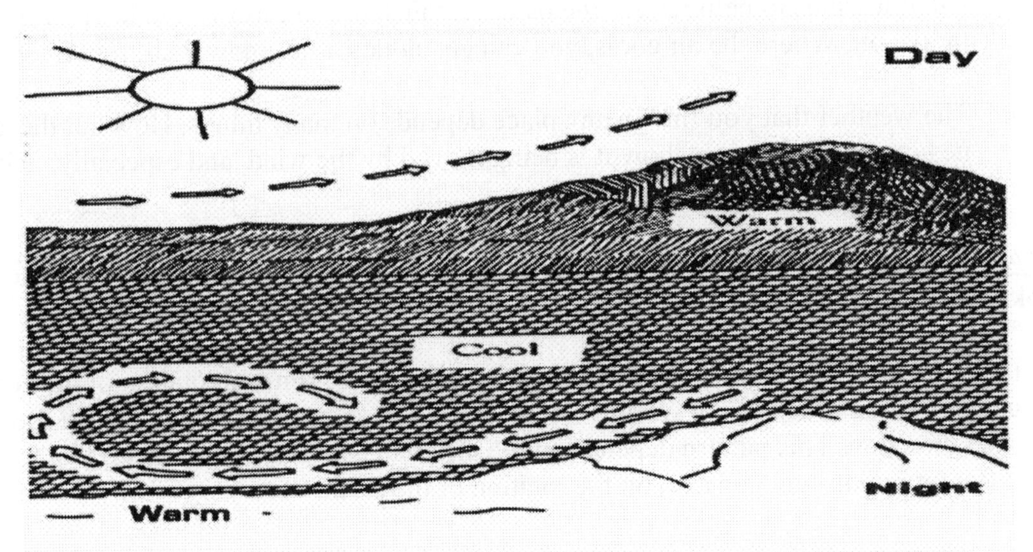

Air flows up the mountain during the day and down the mountain at night.

e. Jet Stream. A jet stream can be defined as a long, meandering current of high speed winds near the tropopause (transition zone between the troposphere and the stratosphere) blowing from generally a westerly direction and often exceeding 250 miles per hour. The jet stream results from:

(1) Circulation of air around the poles and Equator.

(2) The direction of airflow above the mid latitudes.

(3) The actual path of the jet stream comes from the west, dipping down and picking up air masses from the tropical regions and going north and bringing down air masses from the polar regions.

NOTE: The average number of long waves in the jet stream is between three and five depending on the season. Temperature differences between polar and tropical regions influence this. The long waves influence day to week changes in the weather; there are also short waves that influence hourly changes in the weather.

f. Wind Speed. Determined by Sir Francis Beaufort in 1805. Combined with the air temperature will determine the wind chill index. The Wind chill is the actual temperature of the air on exposed skin.

THE BEAUFORT SCALE	WIND SPEED (MPH)
0-1	Smoke Rises Straight up; Calm
1-3	Smoke Drifts
4-7	Wind Felt on Face; Leaves Rustle
8-12	Leaves and Trigs Constantly Rustle; Wind Extends Small Flags
13-18	Dust and Small Paper raised; Small Branches Moved
19-24	Crested Wavelets form on Inland Waters; Small Trees Sway
25-31	Large Branches Move in Trees
32-38	Large Trees Sway; Must Lean to Walk
39-46	Twigs Broken from Trees; Difficult to Walk
47-54	Limbs Break from Trees; Extremely Difficult to Walk
55-63	Tree Limbs and Branches Break
64 on up	Widespread Damage with Trees Uprooted

WIND CHILL INDEX									
TEMP		40	30	20	10	0	-10	-20	-30
WIND MPH	5	37	27	16	6	-5	-15	-26	-36
	10	28	16	2	-9	-22	-31	-45	-58
	15	22	11	-6	-18	-33	-45	-60	-70
	20	18	3	-9	-24	-40	-52	-68	-81
	25	16	0	-15	-29	-45	-58	-75	-89
	30	13	-2	-18	-33	-49	-63	-78	-94
	35	11	4	-20	-35	-52	-67	-83	-98
	40	10	-4	-22	-36	-54	-69	-87	-101

TRANSITION: Now that we have a better understanding of the winds, are there any questions? Let's discuss the air masses and their effects on weather.

3. (5 Min) **AIR MASSES**. As we know, all of these patterns move air. This air comes in parcels known as "air masses". These air masses can vary in size from as small as a town to as large as a country. These air masses are named from where they originate.

   a. Maritime. Over water.

   b. Continental. Overland.

   c. Polar. Above 60 degrees North.

   d. Tropical. Below 60 degrees North.

   e. Combining these give us the names and description of the four types of air masses:

      (1) Continental Polar (CP). Cold, dry air mass.

      (2) Maritime Polar (MP). Cold, wet air mass.

      (3) Continental Tropical CT). Dry, warm air mass.

      (4) Maritime Tropical (MT). Wet, warm air mass.

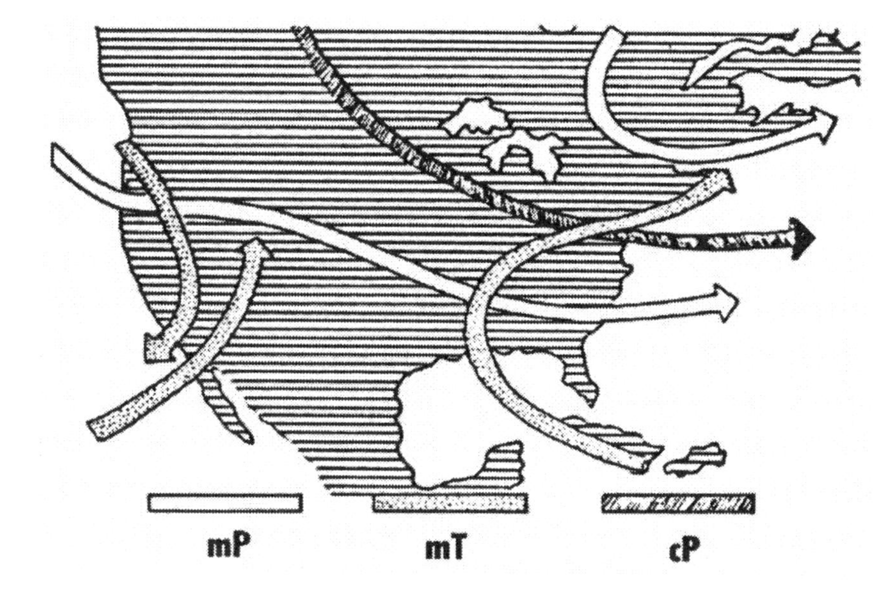

Movement of air masses: Maritime Polar (mP), Maritime Tropical (mT), and Continental Polar (cP).

f. The thing to understand about air masses, they will not mix with another air mass of a different temperature and moisture content. When two different air masses collide, we have a front, which will be covered in more detail later in this period of instruction.

TRANSITION: Now that we have an understanding of meteorology, we can now look at the reason why we receive precipitation.

4. (10 Min) **HUMIDITY**. Humidity is the amount of moisture in the air. All air holds water vapor, although it is quite invisible.

   a. Air can hold only so much water vapors, but the warmer the air, the more moisture it can hold. When the air has all the water vapor that it can hold, the air is said to be saturated (100% relative humidity).

   b. If the air is then cooled, any excess water vapor condenses; that is, it's molecules join to build the water droplets we can see.

   c. The temperature at which this happens is called the "condensation point". The condensation point varies depending on the amount of water vapor and the temperature of the air.

   d. If the air contains a great deal of water vapor, condensation will form at a temperature of 68°F. But if the air is rather dry and does not hold much moisture, condensation may not form until the temperature drops to 32°F or even below freezing.

e. Adiabatic Lapse Rate. The adiabatic lapse rate is the rate that air will cool (-) on ascent and warm (+) on descent. The rate also varies depending on the moisture content of the air.

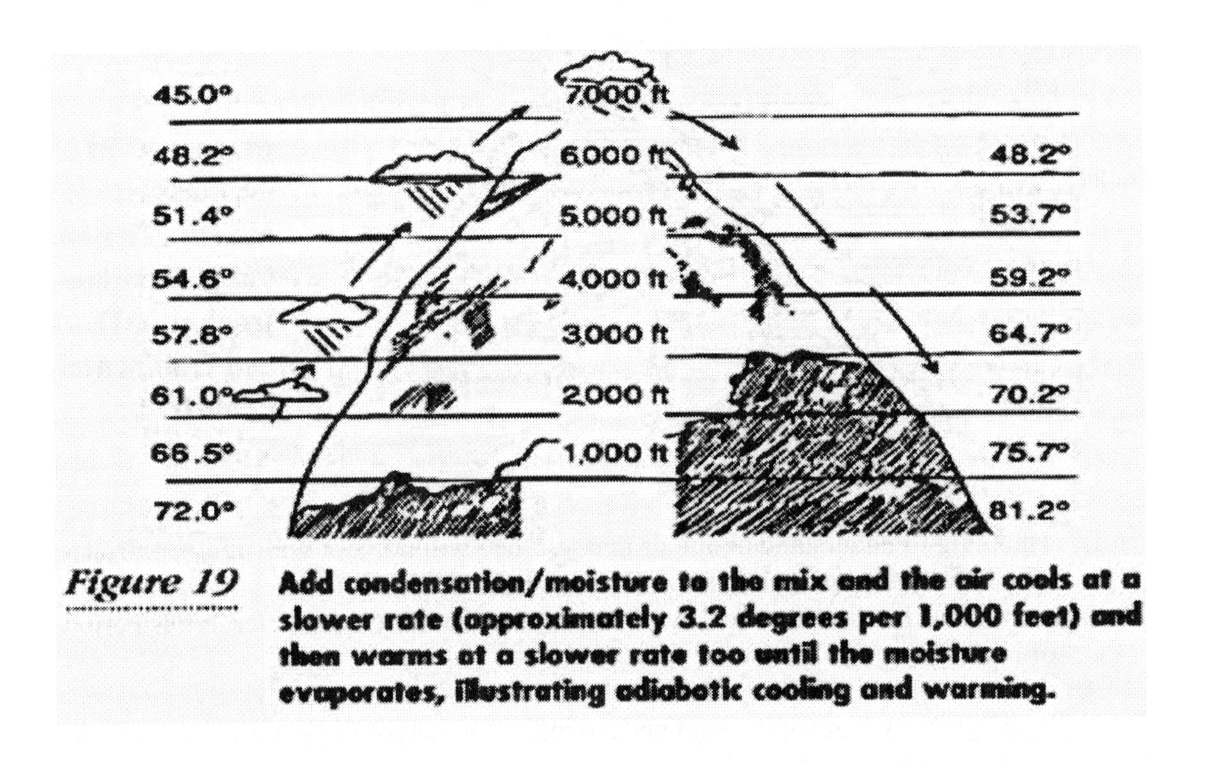

Figure 19 **Add condensation/moisture to the mix and the air cools at a slower rate (approximately 3.2 degrees per 1,000 feet) and then warms at a slower rate too until the moisture evaporates, illustrating adiabatic cooling and warming.**

(1) Saturated Air = 3.2 degrees F per 1,000 feet.

(2) Dry Air = 5.5 degrees F per 1,000 feet.

NOTE: For military planning purposes, 4 degrees F should be used.

TRANSITION: Are there any questions over humidity? Next, we'll talk about the effects of pressure on weather.

5. (10 Min) **PRESSURE**

a. All of these factors are related to air pressure, which is the weight of the atmosphere at any given place. The lower the pressure, the more likely it is to rain and have strong winds.

b. In order to understand this we can say that the air in our atmosphere acts very much like a liquid.

c. Areas with a high level of this liquid would exert more pressure on the Earth and be called a "high pressure area".

d. Areas with a lower level would be called a "low pressure area".

e. In order to equalize the areas of high pressure it would have to push out to the areas of low pressure.

f. The characteristics of these two pressure areas are as follows:

(1) High pressure area. Flows out to equalize pressure.

(2) Low pressure area. Flows in to equalize pressure.

g. The air from the high-pressure area is basically just trying to gradually flow out to equalize its pressure with the surrounding air; while the low pressure is beginning to build vertically. Once the low has achieved equal pressure, it can't stop and continues to build vertically; causing turbulence, which results in bad weather.

NOTE: When looking on the weather map, you will notice that these differences in pressure resemble contour lines. These contour lines are called isobars and are translated to mean equal pressure area.

h. Isobars. Pressure is measured in millibars or another more common measurement is inches mercury.

i. Fitting enough, areas of high pressure are called ridges and areas of low pressure are called troughs or depressions.

j. As we go up in altitude, the pressure (or weight) of the atmosphere decreases.

EXAMPLE: At 18,000 feet in elevation it would be 500 millibars vice 1,013 millibars at sea level.

TRANSITION: Now that we have discussed pressure, are there any questions? Let's take a look at the affects of Lifting and Cooling.

6. (10 Min) **LIFTING/COOLING**. As we know, air can only hold so much moisture depending on its temperature. If this air is cooled beyond its saturation point, it must release this moisture in one form or another, i.e. rain, snow, fog, dew, etc. There are three ways that air can be lifted and cooled beyond its saturation point.

a. Orographic uplift. This happens when an air mass is pushed up and over a mass of higher ground such as a mountain. Due to the adiabatic lapse rate, the air is cooled with altitude and if it reaches its saturation point we will receive precipitation.

OROGRAPHIC UPLIFT

b. Convection effects. This is normally a summer effect due to the sun's heat re-radiating off of the surface and causing the air currents to push straight up and lift air to a point of saturation.

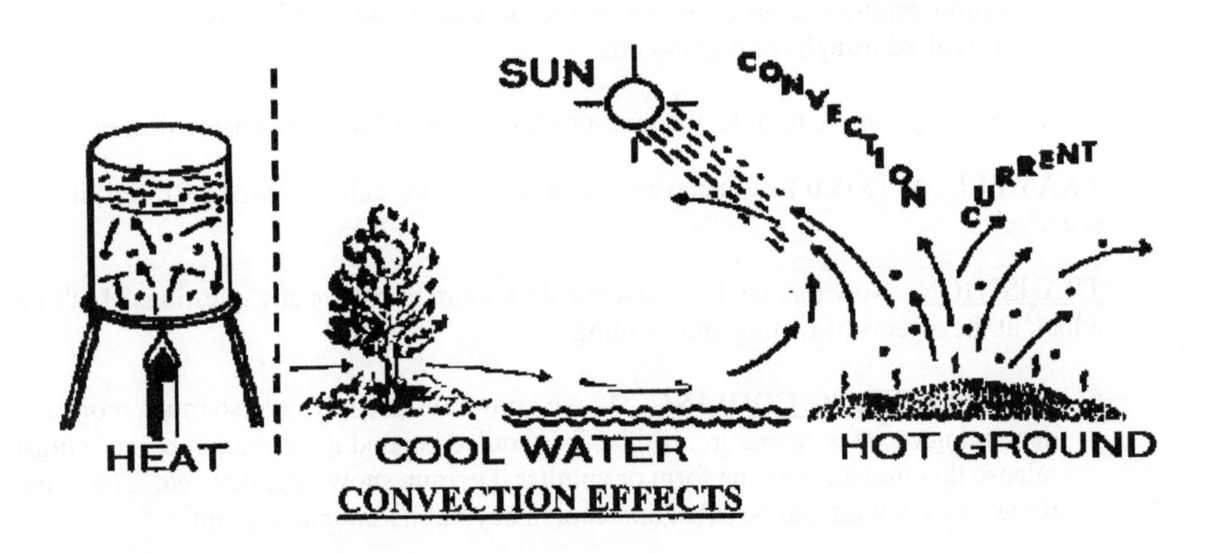

CONVECTION EFFECTS

c. Frontal lifting. As we know when two air masses of different moisture and temperature content collide, we have a front. Since the air masses will not mix, the warmer air is forced aloft, from there it is cooled and then reaches its saturation point. Frontal lifting is where we receive the majority of our precipitation. A combination of the different types of lifting is not uncommon.

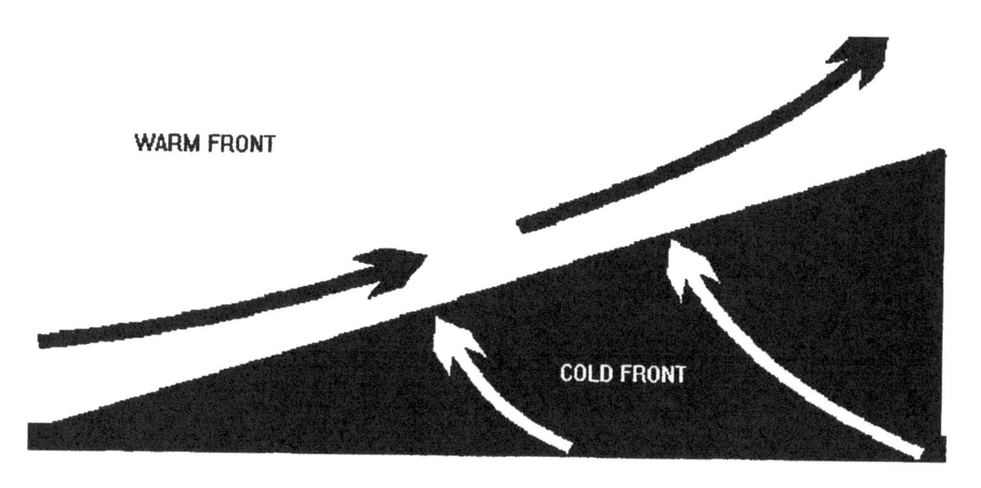

FRONTAL LIFTING

TRANSITION: Are there any questions over lifting and cooling of air masses? So far we've talked about are the invisible aspects of weather, now let's talk about the visible aspects of weather. The first one being clouds.

7. (15 Min) **CLOUDS**. Anytime air is lifted or cooled beyond its saturation point (100% relative humidity), clouds are formed. Clouds are one of our signposts to what is happening. Clouds can be described in many different ways, they can also be classified by height or appearance, or even by the amount of area covered, vertically or horizontally.

   a. Cirrus. These clouds are formed of ice crystals at very high altitudes (usually 20,000 to 35,000 feet) and are thin, feathery type clouds. These clouds can give you up to 24 hours warning of approaching bad weather, hundreds of miles in advance of a warm front. Frail, scattered types, such as "mare-tails" or dense cirrus layers, tufts are a sign of fair weather but predictive may be a prelude to approaching lower clouds, the arrival of precipitation and the front.

   b. Cumulus. These clouds are formed due to rising air currents and are prevalent in unstable air that favors vertical development. These currents of air create cumuliform clouds that give them a piled or bunched up appearance, looking similar to cotton balls. Within the cumulus family there are three different types to help us to forecast the weather:

      (1) Cotton puffs of cumulus are Fair Weather Clouds but should be observed for possible growth into towering cumulus and cumulonimbus.

      (2) Towering cumulus are characterized by vertical development. Their vertical lifting is caused by some type of lifting action, such as convective currents found on hot summer afternoons, or when wind is forced to rise up the slope of a mountain or possibly the lifting action that may be present in a frontal

system. The towering cumulus has a puffy and "cauliflower-shaped" appearance.

(3) Cumulonimbus clouds are characterized in the same manner as the towering cumulus, form the familiar "thunderhead" and produce thunderstorm activity. These clouds are characterized by violent updrafts, which carry the tops of the clouds to extreme elevations. Tornadoes, hail and severe rainstorms are products of this type of cloud. At the top of the cloud, a flat anvil shaped form appears as the thunderstorm begins to dissipate.

c. Stratus. Stratus clouds are formed when a layer of moist air is cooled below its saturation point. Stratiform clouds form mostly in horizontal layers or sheets, resisting vertical development. The word "stratus" is derived from the Latin word "layer". The stratus cloud is quite uniform and resembles fog. It has a fairly uniform base and a dull, gray appearance. Stratus clouds make the sky appear heavy and will occasionally produce fine drizzle or very light snow with fog. However, because there is little or no vertical movement in the stratus clouds, they usually do not produce precipitation in the form of heavy rain or snow.

d. As previously stated, clouds are formed when air is lifted to a point where it cools to its saturation point. We also know that frontal lifting affects our fronts, which produce the largest portion of our precipitation.

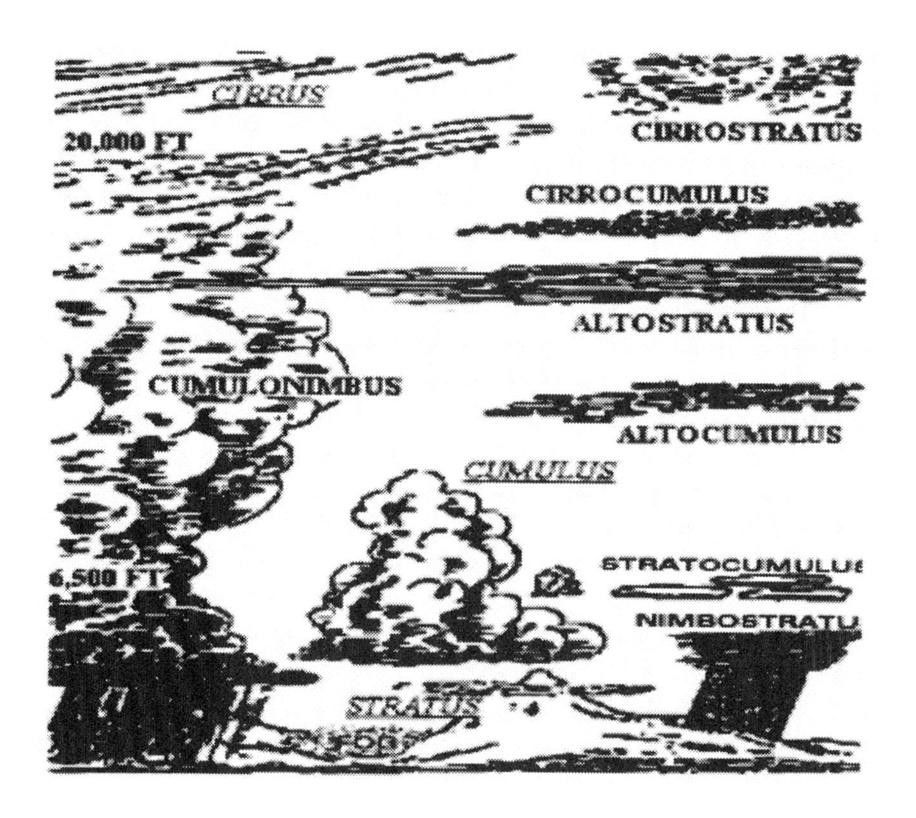

TRANSITION: Are there any questions over clouds? Let's look at the different types of fronts and the progression of clouds that accompany them.

8. (10 Min) **FRONTS**. Fronts often happen when two air masses of different moisture and temperature content interact. One of the ways we can identify this is happening is by the progression of the clouds.

   a. Warm Front. A warm front occurs when warm air moves into and over a slower (or stationary) cold air mass. Since warm air is less dense, it will rise naturally so that it will push the cooler air down and rise above it. The cloud you will see at this stage is cirrus. From the point where it actually starts rising, you will see stratus. As it continues to rise, this warm air is cooled by the cold air, and it receives moisture at the same time. As it builds in moisture, it darkens becoming "nimbus-stratus", which means rain from thunder clouds. At that point, some type of moisture will generally fall.

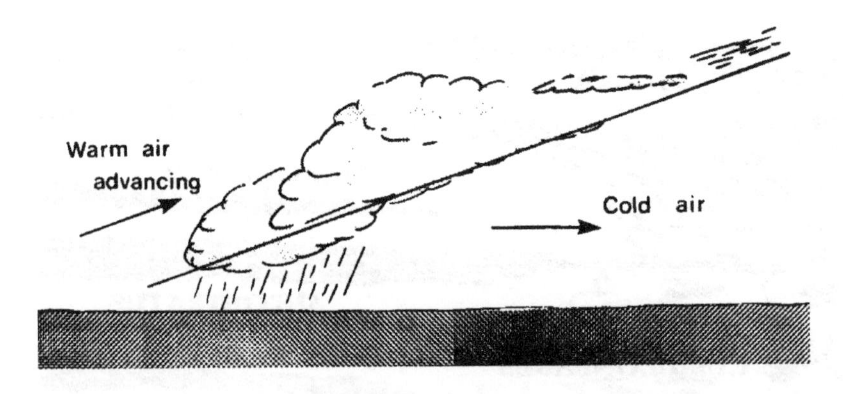

WARM FRONT

b. Cold Front. A cold front occurs when a cold air mass (colder than the ground that it is traveling over) overtakes a warm air mass that is stationary or moving slowly. This cold air, being denser, will go underneath the warm air, pushing it higher. Of course, no one can see this "air", but they can see clouds and the clouds themselves can tell us what is happening. The cloud progression to look for is cirrus to cirrocumulus to cumulus and, finally, to cumulonimbus.

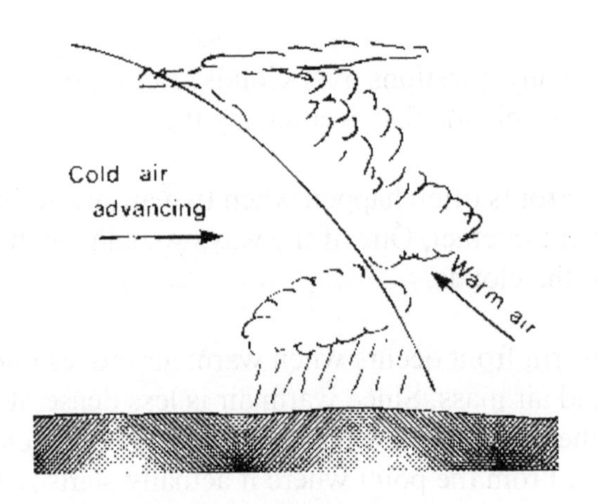

COLD FRONT

c. Occluded Front. Cold fronts move faster than warm ones so that eventually a cold front overtakes a warm one and the warm air becomes progressively lifted from the surface. The zone of division between cold air ahead and cold air behind is called a "cold occlusion". If the air behind the front is warmer than ahead, it is a warm occlusion. Most land areas experience more occlusions than other types of fronts. In the progression of clouds leading to fronts, orographic uplift can play part in deceiving you of the actual type of front, i.e. progression of clouds leading to a warm front with orographic cumulus clouds added to these. The progression of clouds in an occlusion is a combination of both progressions from a warm and cold front.

TRANSITION: Are there any questions over fronts? The clouds and their progression are our basis to our forecast. But there are other indicators to put in before making our decision.

9. (5 Min) **USING PRESSURE AS AN INDICATOR.** A very important factor of telling us what might happen is the pressure. As we know, low pressure or dropping pressure normally indicates deteriorating weather whereas high pressure usually gives us more good weather or clearing of bad weather. There are a couple of ways to monitor our pressure and are as follows:

   a. The Barometer. A barometer could be described as a pan of mercury with a tube leading out of the pan. Pressure from the atmosphere causes the mercury to rise in the tube.

      (1) The tube is marked in millibars and the station that's reading these millibars will know how much it should rise for that location. Once again, if it rises more than normal, it would be considered a high pressure reading.

   b. The Altimeter. Another means that is used to measure pressure is an altimeter, which is commonly used by mountaineers. It works like this:

INSTRUCTOR NOTE: Show an altimeter to the students.

      (1) As you rise in elevation the pressure becomes less, thus allowing the needle in the altimeter to rise. If the needle rises without you rising with it, there is less pressure in the atmosphere than before and thus, a low pressure area.

   c. Contrail Lines. A basic way of identifying a low pressure area is to note the contrail lines from jet aircraft. If they don't dissipate within two hours, that indicates a low pressure area in your area. This usually occurs about 24 hours prior to an oncoming front.

   d. Lenticulars. These are optical, lens-shaped cumulus clouds that have been sculpted by the winds. This indicates moisture in the air and high winds aloft. When preceding a cold front, winds and clouds will begin to lower.

TRANSITION: Are there any questions over using pressure as an indicator? Last of all, let's see how we can use some signs from nature to help us predict the weather.

10. (10 Min) **USING SIGNS FROM NATURE**. These signs will give you a general prediction of the incoming weather conditions. Try to utilize as many signs together as possible, which will improve your prediction. All of these signs have been tested

with relative accuracy, but shouldn't be depended on 100%. But in any case you will be right more times than wrong in predicting the weather. From this we can gather as much information as needed and compile it along with our own experience of the area we are working in to help us form a prediction of incoming weather. The signs are as follows:

a. A spider's habits are very good indicators of what weather conditions will be within the next few hours. When the day is to be fair and relatively windless, they will spin long filaments over which they scout persistently. When precipitation is imminent, they shorten and tighten their snares and drowse dully in their centers.

b. Insects are especially annoying two to four hours before a storm.

c. If bees are swarming, fair weather will continue for at least the next half day.

d. Large game such as deer, elk, etc., will be feeding unusually heavy four to six hours before a storm.

e. When the smoke from a campfire, after lifting a short distance with the heated air, beats downward, a storm is approaching. Steadily rising smoke indicates fair weather.

f. “Red sky at night is a sailor's delight and red sky in the morning is a sailor's warning”. This poem is correct in only some places of the world. When the sun rises in the morning and there is moisture present, the sky will be red. If the wind is moving west to east, that moisture has already past. This does not mean that it will not rain, it just means that the moisture making the sky red is already past. When the sun sets in the west, and there is moisture in the sky, the sunset will be red. If the winds are moving west to east, it means that the moisture in the west making the sky red will move east and possibly form as clouds later.

g. A gray, overcast evening sky indicates that moisture carrying dust particles in the atmosphere have become overloaded with water; this condition favors rain.

h. A gray morning sky indicates dry air above the haze caused by the collecting of moisture on the dust in the lower atmosphere; you can reasonably a fair day.

i. When the setting sun shows a green tint at the top as it sinks behind clear horizon, fair weather is probable for most of the next 24 hours.

j. A rainbow in the late afternoon indicates fair weather ahead. However, a rainbow in the morning is a sign of prolonged bad weather.

k. A corona is the circle that appears around the sun or the moon. When this circle grows larger and larger, it indicates that the drops of water in the atmosphere are evaporating and that the weather will probably be clear. When this circle shrinks by the hour, it indicates that the water drops in the atmosphere are becoming larger, forming into clouds, rain is almost sure to fall.

l. In fair weather, air currents flow down streams and hillsides in the early morning and start drifting back up towards sunset. Any reversal of these directions warns of a nearing storm.

m. When the breeze is such that the leaves show their undersides, a storm is likely on the way.

n. It is so quiet before a storm, that distant noises can be heard more clearly. This is due to the inactivity of wildlife a couple of hours before a storm.

o. When in the mountains, the sight of morning mist rising from ravines is a good sign of clear weather the rest of the day.

p. A heavy dew or frost in the morning is a sign of fair weather for the rest of the day. This is due to the moisture in the atmosphere settling on the ground in the form of precipitation such as rain snow etc.

TRANSITION: It is very important that we pay attention to all of the clues that are given to us so that we are not surprised by bad weather.

PRACTICE (CONC)

a. Students will forecast weather.

PROVIDE HELP (CONC)

a. The instructors will assist the students when necessary.

OPPORTUNITY FOR QUESTIONS (3 Min)

1. QUESTIONS FROM THE CLASS

2. QUESTIONS TO THE CLASS

Q. What are the three types of clouds?

A. (1) Cirrus
(2) Stratus
(3) Cumulus

Q. What the three types of fronts?

A. (1) Warm front
(2) Cold front
(3) Occluded front.

Q. What are the three types of wind patterns?

A. (1) Polar Easterlies
(2) Prevailing Westerlies
(3) Northeast Tradewinds

SUMMARY (2 Min)

a. Now that we have the basic knowledge to forecast the weather, you will find that you are not always right in your forecast. Though with time and experience in forecasting, your degree of accuracy will definitely improve.

b. Those of you with IRF's please fill them out and turn them into the instructor. We will now take a short break.

UNITED STATES MARINE CORPS
Mountain Warfare Training Center
Bridgeport, California 93517-5001

SML
SMO
02/11/02

LESSON PLAN

SUMMER MOUNTAIN WARFIGHTING LOAD REQUIREMENTS

INTRODUCTION (5 Min)

1. **GAIN ATTENTION**. Marines operating in a mountainous environment will be carrying specialized, individual and team gear needed to survive the elements and negotiate the difficult terrain. This equipment, plus the gear they need to wage war, will be carried on their backs. This period of instruction will cover the different types of loads that Marines will be required to carry and also provide some general guidelines on how to properly carry it.

2. **OVERVIEW**. The purpose of this period of instruction is to introduce the student to the different levels of gear requirements in accordance with uniform codes. This lesson relates to mountain walking and tactical evolutions.

INSTRUCTOR NOTE: Have the students read the learning objectives.

3. **INTRODUCE LEARNING OBJECTIVES**.

   a. TERMINAL LEARNING OBJECTIVES. In a summer mountainous environment, pack for tactical movements, in accordance with the reference.

   b. ENABLING LEARNING OBJECTIVES. (MLC)

      (1) Without the aid of references, list in writing the pocket items required to be carried by each individual, in accordance with the references.

      (2) Without the aid of references, list in writing the required items for the Assault Load, in accordance with the references.

      (3) Without the aid of references, list in writing the required items that are added to the Assault Load to make the Combat Load, in accordance with the references.

c. ENABLING LEARNING OBJECTIVES. (SMO)

(1) With the aid of references, state orally the pocket items required to be carried by each individual, in accordance with the references.

(2) With the aid of references, state orally the required items for the Assault Load, in accordance with the references.

(3) With the aid of references, state orally the required items that are added to the Assault Load to make the Combat Load, in accordance with the references.

4. **METHOD/MEDIA**. The material in this lesson will be presented by lecture and demonstration. You will practice what you have learned during upcoming field training exercises. Those of you with IRF's please fill them out at the end of the lesson and turn them in.

5. **EVALUATION**.

a. MLC - You will be tested later in the course by written and performance evaluations on this period of instruction.

b. SMO - You will be tested by a verbal evaluation.

TRANSITION: Are there any questions over the purpose, learning objectives, how the class will be taught, or how you will be evaluated? The first thing we need to talk about is the uniform requirements and the necessary gear worn on your person and carried in your pockets.

BODY (40 Min)

1. (5 Min) **BASIC UNIFORM REQUIREMENTS**.

a. Marines operating in mountainous terrain will generally wear the standard issued utility uniform. However, due to the wide range of temperatures, and sudden changes in weather in the mountains, Marines will find the need to be continually adding and removing layers. The amount of clothing worn may vary depending upon the severity of the weather, the activity level of the Marine, and the individual metabolism of the Marine. Every person in that unit should still maintain the same outer camouflage layer.

b. As part of the basic uniform, each man will be required to have in his possession, at all times, seven required pocket items. These seven items should be carried in the pockets of your utility uniform:

(1) Pocketknife

(2) Whistle

(3) Pressure Bandage
(4) Chapstick and sunscreen

(5) Sunglasses

(6) Survival Kit

(a) Fire starting items

(b) Signaling items

(c) Food gathering items

(d) Water procuring items

(e) First aid items

(f) Shelter items

(7) Notebook with pen/pencil

c. Some additional items that should be carried in your pockets at all times:

(1) Contact gloves

(2) Paracord (10 meters)

(3) Flashlight with tactical lens and spare batteries

(4) Chemlights

TRANSITION: Are there any questions over the pocket items? Next, we will discuss the items that make up the basic equipment required to survive and accomplish our mission.

2. (10 Min) **ASSAULT LOAD**.

a. The Assault Load is equipment in addition to the basic uniform requirements. It is carried in the load-bearing vest (LBV), butt pack and the pack system. This is the equipment carried for short duration missions such as security patrols or during the assault. It is also carried at all times when away from the bivouac site.

(1) An extra insulating layer (Fleece jacket, wooly pully, etc.)

(2) Protective layer (ECWCS parka and trousers)

(3) LBV with 2 quarts of water and first aid kit

(4) Helmet

(5) Rations for the time away from your bivouac site

(6) Extra socks and gloves

(7) Cold weather hat or balaclava

(8) Individual mountaineering gear

a. Sling rope

b. Carabiners

c. Rappelling Gloves

(9) Specialized mountaineering equipment

(10) Mission essential gear

(a) T/O weapon w/accessories (sling, magazines, cleaning gear, bayonet/K-bar, and basic allowance of ammunition)

(b) *Extra ammunition, demolitions, and pyrotechniques

(c) *Optical gear (binoculars, night vision devices, etc.)

(d) *Communications equipment (field phones, spare batteries, etc.)

(e) *Navigational equipment

NOTE: Mission essential gear items indicated with an * are spread-loaded throughout the unit as the mission dictates. Also, it may be required for designated personnel (such as RTO's) to carry the assault load in the large Vector pack vice in the small assault pack.

TRANSITION: Are there any questions over the assault load? The next level of equipment is the Combat Load.

3. (5 Min) **COMBAT LOAD**. The Combat Load is the equipment carried for longer duration missions. The following items are in addition to the items already being carried in the Assault Load:

a. Sleeping bag with bivy bag

b. Isopor mat

c. Stove

d. Fuel bottle

e. Thermos

f. Poncho (for expedient shelters or medevac purposes)

TRANSITION: Are there any questions over the combat load? The final individual load is the Existence Load.

4. (5 Min) **EXISTENCE LOAD**. The Existence Load is to be packed for longer duration combat missions, where it is necessary to replace worn out items. These items could be uniforms, boots, personal hygiene gear, etc. The Existence Load should be brought to the forward elements once the situation allows.

TRANSITION: Now that we discussed the different loads we are required to carry, are there any questions? Let's now cover some general guidelines for packing our gear.

5. (5 Min) **PACKING CONSIDERATIONS**. Because most Marines are familiar with how to pack a pack, these are general guidelines only.

a. When not wearing your protective layer or insulating layers, keep it handy at the top of the pack. When taking a break during a movement, you will be able to quickly don a layer to prevent getting chilled.

b. Keep your stove and fuel bottle in the outside pockets of your pack. They may leak and soak your equipment with fuel if stored inside your pack.

c. Keep the climbing rope in the rope bag. This will reduce the rope's exposure to petroleum products, ultra-violet light, cuts, and abrasions.

d. Ice ax is placed through the ice ax loops with spike up, adze outboard, and secured in place.

e. Crampons are attached to the outside of the pack with the tips covered and facing outboard.

TRANSITION: Now that we have discussed the general guideline for packing, are there any questions? If you have none for me, then I have some for you.

PRACTICE (CONC)

a. Students will carry the different fighting loads throughout the upcoming field evolutions.

PROVIDE HELP (CONC)

a. The instructors will assist the students when necessary.

OPPORTUNITY FOR QUESTIONS (3 Min)

1. QUESTIONS FROM THE CLASS

2. QUESTIONS TO THE CLASS

Q. What are the 7 required pocket items required to be carried by the individuals?

A. (1) Pocket Knife
(2) Whistle
(3) Pressure Bandage
(4) Notebook with pen /pencil
(5) Chapstick and Sunscreen
(6) Sunglasses
(7) Survival kit and Rations

Q. What are the 6 items that are added to the Assault Load to make the Combat Load?

A. (1) Sleeping Bag
(2) Isopor Mat
(3) Stove
(4) Fuel Bottle
(5) Thermos
(6) Poncho

SUMMARY (2 Min)

a. During this period of instruction we have covered the basic fighting load requirements for mountain operations and some general guidelines for packing your packs.

b. Those of you with IRF's please fill them out at this time. We will now take a short break.

UNITED STATES MARINE CORPS
Mountain Warfare Training Center
Bridgeport, California 93517-5001

SML
SMO
ACC
03/31/01

LESSON PLAN

MARINE ASSAULT CLIMBER'S KIT

INTRODUCTION

(5 Min.)

1. **GAIN ATTENTION.** As in any military operation requiring special skills, such as mountaineering, some special equipment requirements particular to the mission must be covered. The Marine Corps has adopted a Marine Assault Climbers Kit (MACK) for this very reason.

2. **OVERVIEW.** The purpose of this period of instruction is to familiarize the students with the MACK, it's components, how to inspect for serviceability, preparing the MACK for use, and maintenance procedures.

3. **METHOD/MEDIA.** The material in this lesson will be presented by lecture and demonstration.

4. **EVALUATION.** This period of instruction has no learning objectives and students will be evaluated throughout their stay at MWTC.

TRANSITION: Are there any questions over the purpose, learning objectives, how the class will be taught, or how you will be evaluated? Mountain operations will present tremendous obstacles for your Marines to overcome, by preparing them with the proper gear and equipment; you will reduce the risk involved and increase your chances of success.

BODY (60 Min.)

1. (5 Min) **DESCRIPTION.** The Marine Assault Climber's Kit (MACK) is a comprehensive collection of climbing equipment that enables a Marine rifle company (reinforced), approximately 200 men with organic equipment, to negotiate an average 300-foot vertical danger area. The kit contains sufficient climbing equipment to outfit four 2-man climbing teams plus the additional items necessary to supply the remainder of the Rifle Company. The climbing teams use their equipment to conduct 2-party climbs over vertical obstacles and establish various rope installations to facilitate the movement of the remainder of the company. Marines that engaged in training and combat operations in mountainous areas having rugged compartment terrain and steep slopes would use the MACK. Certain items contained in the MACK will also be used during training and combat operations in urban environments for scaling vertical obstacles such as buildings.

TRANSITION: Now that we have discussed what the MACK is, are there any questions? Let's discuss the components of the MACK.

2. (5 Min) **SL-3 COMPONENTS.** Four containers hold all the items contained in the MACK, and have features that facilitate the organization and accountability of MACK items. Each container protects the contents from degradation due to sunlight and moisture during storage periods of up to 5 years. The lid's interior has a permanently affixed list of the components and quantities stored within that container. Container #1 contains the climbing team equipment. Containers #2, #3 and #4 contain the company climbing equipment. A manual for care/maintenance of SL-3 components is included with each MACK. Refer to the SL-3 components list for current quantities and items (see attached appendix).

INSTRUCTOR NOTE: Have students turn to MACK SL-3 components in appendix 1 and go through the list.

TRANSITION: Now that we have covered what a MACK contains, are there any questions? Let's discuss how to keep the MACK ready to go.

3. (5 Min) **SERVICEABILITY.** Any item that becomes unserviceable or shows excessive signs of wear must be replaced immediately. With the exception of the rope bag and climbing rack bag, no attempt should be made to repair the components of the MACK. Any damaged or broken components should be disposed of using standard supply procedures. Replenishment procurement to the Source of Supply should be accomplished using standard MILSTRIP process. Kit components not available through the MILSTRIP process may be procured through local purchase. Minimum strengths and description are covered in the SL-3 list, any brand that meets the function, strength and description can be purchased.

TRANSITION: Now that we have discussed serviceability, let us now talk about safety.

4. (5 Min) **SAFETY.** A Marine NCO, SNCO or Officer who has received formal military mountaineering training must supervise any Marine using components of the MACK. This formal instruction must be provided by the Marine Corps Mountain Warfare Training

Center's Summer Mountain Leader's Course. Marines with the school code M7A on their BIR/BTR are qualified. The current Summer Mountain Leader, designated by the unit commander, is responsible for inventory, periodic serviceability checks, ordering replacement items through supply officer and supervising issue and recovery of items to ensure accountability and proper storage SOPs.

TRANSITION: Now that we have covered safety, are there any questions? Let's discuss how to prepare the MACK for use.

5. (15 Min) **PREPARING THE MACK FOR USE**.

a. Company Equipment:

(1) Use the electric rope cutter to cut/whip one 15-18 foot sling rope for each Marine in the company, using the dynamic rope. Each Marine will also receive one non-locking carabiner and one locking carabiner.

NOTE: Finish one complete spool before cutting another spool.

(2) Cut the static rope for the mission at hand. Some spools are already 300 feet and 165 feet (50m); others may be 600-foot spools. Cut one-inch tubular tape for static anchor cord (15-25 foot lengths). When static rope becomes unserviceable, cut out good sections of 15-25 foot lengths for static anchor cords. Cut 7mm nylon cord for use in tightening systems (3-6 foot lengths). Do not cut 7mm cord for company personnel to use as anchors because of the relatively low tensile strength (static rope or tape should be used only).

b. Team Equipment:

(1) Rope: The dynamic climbing rope is olive drab and already in 165 foot (50m) lengths.

(2) Cordage: Cut/whip 7mm nylon cord for use in tightening systems and anchors as part of each teams rack. Cut/whip one inch tubular tape of varying lengths for use as web runners. Tie loops using the water/tape knot, and range the size of the web runners from four to forty- eight inches. Tubular tape can also be used for anchors.

(3) Racks: Use the 5.5 mm kevlar cord to wire the hexcentrics, secure the ends by tying a triple fisherman's knot. A knife will be needed to cut this cord because it will not burn. Use the electric rope cutter to whip the nylon sheath around the kevlar core. The knife blade will dull quickly when used to cut through the kevlar core. Use the kevlar cord for hexcentrics only (it is not pliable enough for use as utility cord, nor is there a large quantity provided). Tie a small loop of 7mm cord on the stitch plate (belay device) in the appropriate hole (to keep it from running down the rope during use and for racking it to the harness).

(4) Silencing the rack can be done by wrapping vinyl tape around the non-locking carabiner bodies and the large (size 7-11) hexcentrics (ensure the tape is only one layer thick so that it will not interfere with the safe function of the item). Nut picks can be taped or dummy-corded so that they do not rattle on a carabiner.

c. Tailor the preparation of the MACK to the mission, terrain, and size of using unit. Keep as much unused rope and cordage as possible for use as backup. Maintain a log of rope/cordage usage, and replace after two seasons of use or when unserviceable. Inspect frequently for serviceability (before, during and after use).

TRANSITION: Are there any questions concerning the preparations of a MAC Kit? If not, let's begin talking about after actions.

6. (5 Min) **AFTER ACTIONS**.

a. Clean and dry all MACK components according to the respective instructions in the MACK's care and maintenance manual. Most importantly, ensure that all items are thoroughly DRY before returning items to the container for storage. One wet or damp item will spread its moisture to all other items in that container, causing mildew, rot, rust, etc. If carabiners are being oiled for long term storage, do not place them in the same container as any of the ropes/cordage.

TRANSITION: The MACK will only be as good as the effort spent on properly caring and maintaining this equipment. Time spent doing this is time well spent. Are there any questions?

PRACTICE (CONC)

a. Students will apply this knowledge throughout the duration of the course.

PROVIDE HELP (CONC)

a. The instructors will assist the students when necessary.

OPPORTUNITY FOR QUESTIONS (3 Min.)

1. QUESTIONS FROM THE CLASS

2. QUESTIONS TO THE CLASS

Q. What must be done to any item that becomes unserviceable or shows excessive signs of wear?

A. Replace it immediately.

Q. What is the kevlar cord provided in the MACK used for?

A. It is used for wiring the hexcentrics.

SUMMARY (2 Min)

a. What we have just discussed will ensure that we as Mountain Leaders will be able to accomplish those missions requiring the use of the MACK, and minimize placing our Marines in jeopardy.

b. Those of you with IRF's please fill them out at this time and turn them in. We will now take a short break.

UNITED STATES MARINE CORPS
Mountain Warfare Training Center
Bridgeport, California 93517-5001

SML
SMO
ACC
02/11/02

LESSON PLAN

NOMENCLATURE AND CARE OF MOUNTAINEERING EQUIPMENT

INTRODUCTION

(5 Min)

1. **GAIN ATTENTION**. Ropes, carabiners, and chocks are used to aid you and your unit when operating in a mountainous area. Their proper use can make your movement over cliffs, deep chasms and mountain rivers much easier. Use of this equipment incorrectly can inhibit your ability to move efficiently and in some cases cause serious injury or death to members of your unit.

2. **OVERVIEW**. The purpose of this period of instruction is to introduce you to the types of equipment used here and how to care for it so as to prevent its untimely failure. This lesson relates to all climbing and installation work that you perform here.

INSTRUCTOR NOTE: Have students read learning objectives.

3. **INTRODUCE LEARNING OBJECTIVES**

   a. TERMINAL LEARNING OBJECTIVE. In a summer mountainous environment, maintain mountaineering equipment, in accordance with the references.

   b. ENABLING LEARNING OBJECTIVES (SML) and (ACC)

      (1) Given a diagram of a steel-locking carabiner and without the aid of references, label the parts of a carabiner, in accordance with the references.

      (2) Without the aid of references, list in writing the advantages of nylon rope, in accordance with the references.

      (3) Without the aid of references, list in writing the disadvantages of nylon rope, in accordance with the references.

(4) Without the aid of references, state in writing how much strength is lost when a rope is wet, in accordance with the references.

(5) Without the aid of references, list in writing the types of protection, in accordance with the references.

c. ENABLING LEARNING OBJECTIVES (SMO)

(1) Given a diagram of a steel-locking carabiner and without the aid of references, label the parts of a carabiner, in accordance with the references.

(2) Without the aid of references, list in writing the advantages of nylon rope, in accordance with the references.

(3) Without the aid of references, list in writing the disadvantages of nylon rope, in accordance with the references.

(4) Without the aid of references, state in writing how much strength is lost when a rope is wet, in accordance with the references.

(5) Without the aid of references, list in writing the types of protection, in accordance with the references.

4. **METHOD/MEDIA**. The material in this lesson will be presented by lecture and demonstration. You will practice what you have learned during upcoming field training exercises. Those of you with IRF's please fill them out at the end of the lecture.

5. **EVALUATION**.

a. SML – You will be tested by a written exam.

b. ACC – You will be tested by a written exam.

c. SMO – You will be tested by an oral exam.

TRANSITION: Does anyone have any questions on the purpose, learning objectives, how the class will be taught, and how you will be evaluated? We will begin by discussing the ropes used in the MAC kit.

BODY (65 Min)

1. (10 Min) **ROPES.** All ropes used in the military must meet UIAA standards or U. S. Federal Test Standard 191A. Most ropes have a 5-year shelf life and maximum 2-year service life.

a. Static. Black in color.

(1) Construction. Kernmantle

(2) Minimum tensile strength. 7500 lbs.

(3) Maximum elongation. 1.5%

(4) Diameter. 11mm

(5) Sizes.

(a) 165 ft ± 5 ft

(b) 300 ft ± 10 ft

(6) Usage. Rescue operations and bridging where a low amount of elongation is desirable under a working load.

b. Dynamic. Olive Drab in color.

(1) Construction. Water-resistant treated Kernmantle to reduce friction.

(2) Minimum tensile strength. 6500 lbs.

(3) Maximum elongation. 6%

(4) Diameter. 10.5mm and 11mm

(5) Size. 165 ft ± 5 ft.

(6) Usage. For lead climbing/party climbing.

c. Maxim Dry Rope. Olive Drab or Multi-Colored.

(1) Construction. Water-repellent treated Kernmantle.

(2) Minimum tensile strength. 3472 lbs.

(3) Maximum elongation. 6%

(4) Diameter. 9mm

(5) Size. 150 ft ± 5 ft

(6) Usage. For glacier travel/ice climbing

d. Gold Line II.

(1) Construction. Eight strand braided nylon plymor.

(2) Minimum tensile strength. 4500 lbs.

(3) Maximum elongation. 20%

(4) Diameter. 11 mm

(5) Size. 300 ft or 600 ft spools

(6) Usage. Sling Ropes and litters only.

Three strand twisted or laid rope.

Static, kernmantle rope showing core (kern) and sheath (mantle).

Dynamic, kernmantle rope used for rock climbing.

Double braid rope showing the braided sheath and braided core.

INSIDES AND OUTSIDE SHEATHS OF VARIOUS TYPES OF ROPES

NOTE: Sling ropes are made from 15 foot lengths of plymor or dynamic rope ONLY. Twenty five foot practice coils should be constructed with static rope, but dynamic rope can be used.

2. (5 Min) **<u>ADVANTAGES/DISADVANTAGES</u>**

a. Advantages of Nylon Rope

(1) High strength to weight ratio.

(2) Good energy absorption in dynamic ropes.

(3) Flexible.

(4) Rot resistant, not affected by frost.

b. Disadvantages of Nylon Rope

(1) Low melting point. Nylon fuses at 400°F and melts at 480°F.

(2) Susceptible to abrasions and cuts.

(3) Affected by chemicals and light.

c. Advantages of Manila Rope

(1) Easily gripped.

(2) Hard wearing.

(3) Does not deteriorate in heat.

d. Disadvantages of Manila Rope

(1) Heavy, kinks, especially when wet. Absorbs water and swells.

(2) Burns at +300°F.

(3) Edible by rodents.

3. (3 Min) **GENERAL INFORMATION**

a. Nylon rope stretches under tension and will rupture at between 30% and 70% elongation depending on construction.

b. Nylon rope loses as much as 30% strength when wet.

c. Temperatures as low as 250°F will damage a nylon rope.

4. (5 Min) **NYLON WEBBING**

a. The type of nylon webbing available is tubular. Tubular nylon webbing is very strong and flexible. All rules that apply to nylon rope apply to tubular nylon webbing. The size of nylon webbing used is:

(1) 1 inch tubular nylon. Tensile strength approximately 4,000 - 4,500 lbs., depending on the manufacturer.

b. Pre-sewn Spectra Runners. Tensile strength approximately 5,500 lbs.

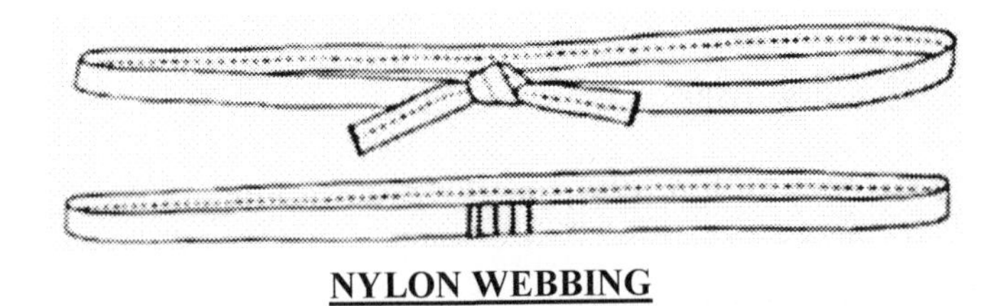

NYLON WEBBING

NOTE: These are minimum strengths. Some manufactures make even stronger webbing.

TRANSITION: We have just discussed general information in nylon webbing, are there any questions? Not only do we use ropes, but we also use carabiners in our installations, we will discuss the types of carabiners used:

5. (5 Min) **CARABINERS**. Also commonly known as snaplinks. Both locking and non-locking are used.

a. Purpose. Carabiners are used for the following purposes:

(1) To attach ropes or runners to pieces of protection.

(2) To attach the rappel rope to the rappel seat for seat-hip rappels or for crossing rope bridges.

(3) To attach the individuals safety rope to a safety line on a rope installation.

(4) To form field expedient pulley systems.

b. Nomenclature of a non-locking carabiner

(1) Gate

(2) Gate pivot pin.

(3) Locking pin.

(4) Body.

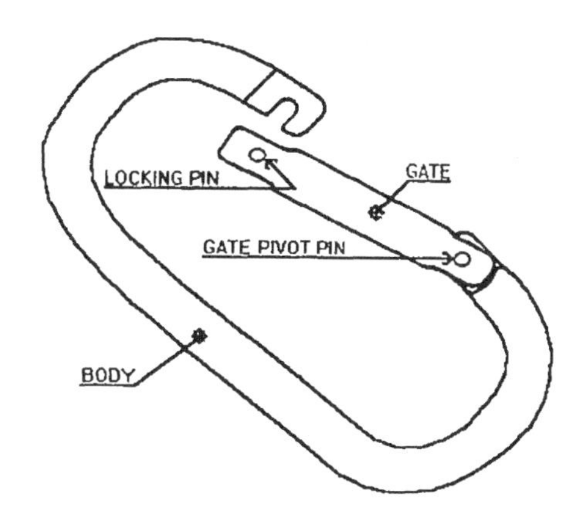

NON-LOCKING ALUMINUM CARABINER

c. Nomenclature of a locking carabiner

(1) Gate.

(2) Gate pivot pin.

(3) Locking notch.

(4) Locking nut.

(5) Body.

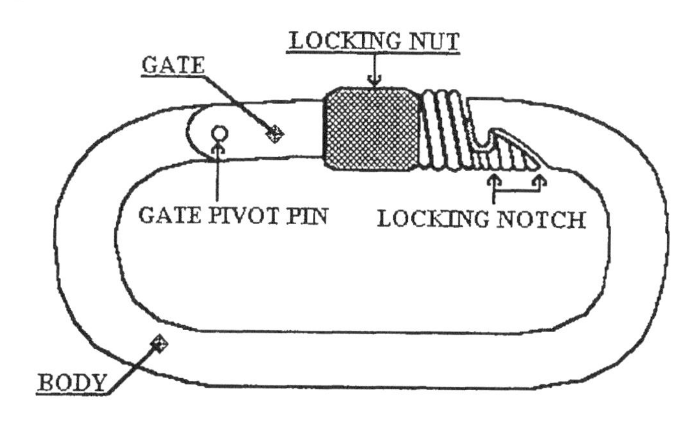

STEEL LOCKING CARABINER

d. There are two types of carabiners used. The two types and their characteristics are:

(1) Steel locking carabiners

(a) Large steel locking "D" (various manufacturers): Minimum tensile strength of 5,500 lbs.

(b) Steel-locking oval Stubai 82 is not in the MAC Kit and obsolete. However, it is being used at MWTC to save money, even though they are beyond the service life. Tensile strength of only 3,300 lbs.

(2) Aluminum non-locking carabiners

(a) Aluminum non-locking oval (various manufacturers): minimum tensile strength of 4,200 lbs.

e. Serviceability Check for a Carabiner. The following steps are used for you to check a carabiner for serviceability:

(1) The gate snaps shut with no friction and with no gap between the locking pin and locking notch.

(2) There is no excessive side to side movement of the gate.

(3) The pivot pin is tight.

(4) The locking pin is tight.

(5) The locking nut travels freely and locks securely.

(6) There are no cracks or flaws in the metal.

NOTE: The weakest part of a carabiner is the gate. If an engraver is used to mark a carabiner, it should be applied to the gate and not the load bearing side.

f. Preventive Maintenance for a Carabiner.

(1) Remove all dirt, moisture and grime.

(2) Lubricate with tri-flow graphite and clean off thoroughly.

NOTE: Whenever you use a locking carabiner ensure that the locking nut is always locked down (tightened).

6. (2 Min) **CARE OF THE CARABINER**. Do not drop the carabiner as this may result in either actual damage to the carabiner or in dirt getting into the workings of the carabiner and damaging it.

TRANSITION: We have covered the parts, the strength, and care of carabiners; are there any questions? Now lets move on to chocks.

7. (10 Min) **PROTECTION**

a. Purpose

(1) Protection or pro is used to protect climbers as they ascend a cliff face. This is accomplished by wedging them into cracks and openings in the rock and securing the

rope to them. Since they can be slid into the rock without banging, they are not noisy to install and are very suitable for a tactical situation that may require silence.

(2) A disadvantage of pro is that it is directional. That is, when it is installed it is wedged into a crack and is meant to take a strain in a specific direction. If you climb above your pro and inadvertently pull the rope, you may pull your pro all or part way out of the crack that you installed it in.

b. Types

(1) Stoppers. These have a wedge-shaped structure and are designed to be used in small cracks. They come in twelve sizes, ranging from widths of 0.16 inches (# 1) to 0.90 inches (# 12). The sides of the wedged portion are slightly beveled, enabling the climber to insert the same stopper into a crack two different ways.

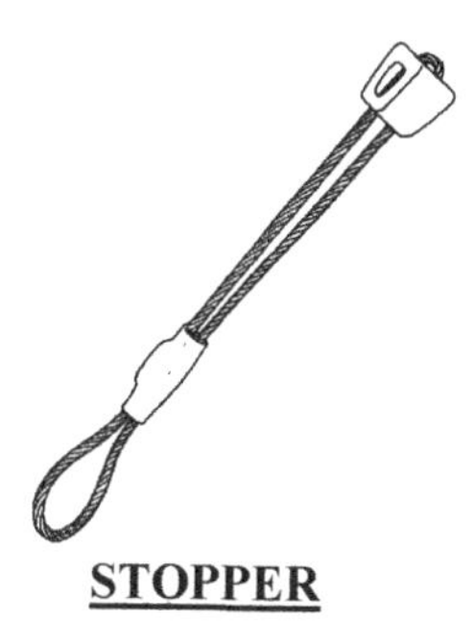

STOPPER

(2) Hexcentrics. These chocks have a six-sided structure shaped like a hexagon; the sides being of unequal width, which allows the same chock to be inserted in different size cracks depending on which way it is inserted. These chocks come in various sizes and are used in larger cracks that stoppers are too small for.

HEXCENTRIC

(3) Spring Loaded Camming Devices (SLCD). Spring loaded camming devices are a unique solution for shallow, horizontal or vertical cracks, thin "tips" cracks and narrow pockets where other types of protection can't be placed. This is an advantage over rigid caroming devices, which can only be placed in a vertical crack.

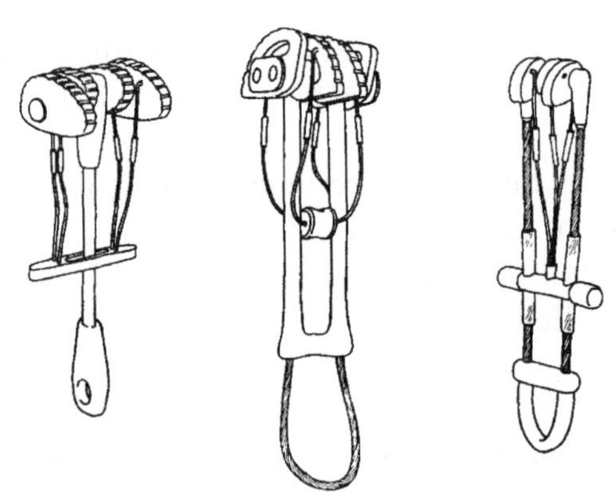

SPRING LOADED CAMMING DEVICES

c. Serviceability Check

(1) Stoppers and Hexcentrics

(a) Check holes used for stringing chocks for burrs that could damage the cord the chocks are strung with.

(b) Check accessory cord for wear, fraying, rupture of the outer sheath, and knot.

(c) If wired, check wires for frays that could damage climbing rope.

(d) If wired, check soldered (or otherwise joined) area for cracks or looseness.

(e) Check nut for splits or cracks.

(2) SLCD's

(a) Ensure wires leading from trigger to cam are not bent or frayed.

(b) Check to ensure that cam movement is free and easy and for contraction and expansion by pulling and releasing trigger.

(c) Check that runner is not frayed and no stitches have popped.

(d) If it is stiff or corroded by sea water, spray with Tri-Flow Graphite and clean off thoroughly.

d. Strengths

(1) The strength of a chock depends on the manufacturers specifications and on the type and size of material used for the sling (rope, webbing, or wire). Below are the strengths of some of the types of chocks used at MWTC.

TYPE	STRENGTH
#1 / #2 Stopper - wired	350 kg (approx. 770 lbs.)
#3 - #5 Stopper - wired	650 kg (approx. 1430 lbs.)
#6 - #12 Stopper - wired	1,100 kg (approx. 2420 lbs.)
#1 - #3 Hexcentrics - wired	1,100 kg (approx. 2420 lbs.)

(2) Hexcentrics #4 thru #10 are only strung with Type I cord. Type II cord is used for Prusik cordage. The following are the cord specifics:

TYPE	DIAMETER	CONSTRUCTION	STRENGTH	LENGTH
I	5.5mm	Spectra Kevlar	4,400	150 ft ± 10 ft
II	7mm	kernmantle	2,200	300 ft ± 10 ft

(3) SLCD's vary in strength depending on the manufacturer, as well as the size.

TYPE	MIN STRENGTH
Camming Device Large	2,400 lbs.
Camming Device Small	2,600 lbs.

e. Care of Pro

(1) When placing pro, insure that the cord or wire does not rub against the rock.

(2) Do not drop pro, which may deform it, or the accessory cord that the pro is strung with, which may lead to deterioration of the cord.

8. (10 Min) **OTHER EQUIPMENT USED IN MOUNTAINEERING**

a. CAVING LADDER. Constructed of stainless steel cables and aluminum crossbars. Several ladders can be connected together by the use of two large steel rings on each end of the ladder.

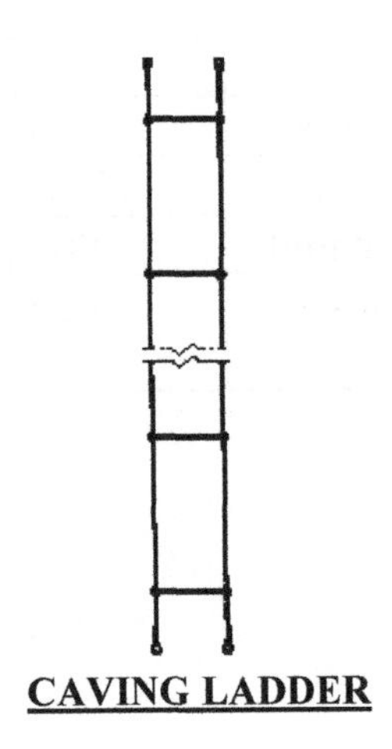

CAVING LADDER

b. ASCENDER. Easily placed and removed from a rope with one hand and allows the rope to run through it in one direction while it grips in the other. A safety device is incorporated to ensure that the cam only releases the rope when the trigger is pressed and out of position.

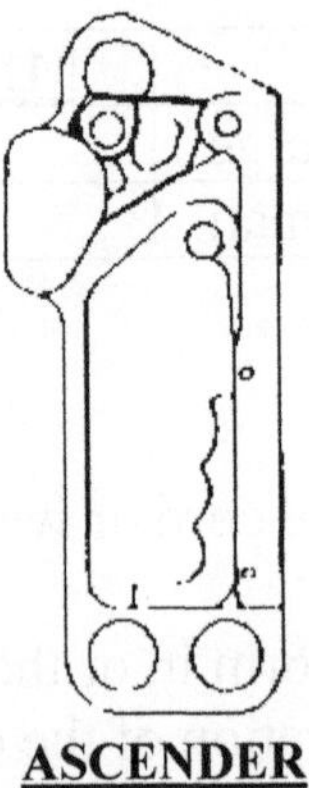

ASCENDER

c. RESCUE PULLEY. Has two independent side plates that enable a user to insert the rope onto the wheel without having to thread the rope through. The pulley is large enough to accommodate a 1/2 inch rope and has an eyehole large enough to accommodate two steel locking carabiners.

RESCUE PULLEY

d. HELMETS. Worn to protect the head from falling rocks and from hitting rocks should you fall. Check helmet for cracks or chips and a good chinstrap. The climbing helmet used at MCMWTC is the Joe Brown Light Weight Helmet.

e. BELAY DEVICES.

(1) Uses. Various devices used as a rappelling or belay device. They should be used with a locking carabiner only.

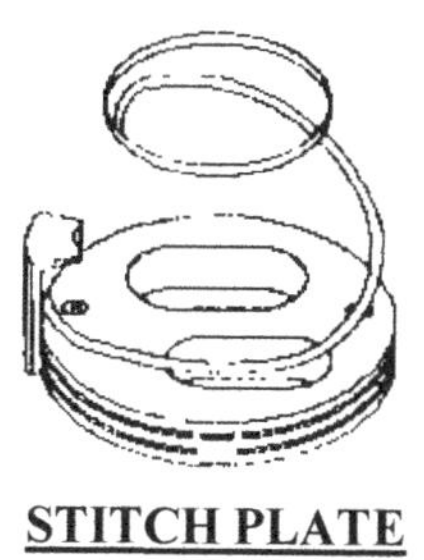

STITCH PLATE

(2) Construction. Made from a heat-dissipating aluminum alloy so device is cool to touch after a fast running rope has passed through it. (User and rope friendly). Care must be taken not to bang or throw these devices onto any hard surface or damage may occur.

9. (5 Min) **RACKING EQUIPMENT**

a. Protection, quick draws and other equipment should be organized on the climbing harness. Gear should be silenced and should not interfere with the climber's movement, i.e. a web runner gets caught on the climbers foot, or on rocks and vegetation as he approaches the cliff face. The gear should be easily accessible for either hand, and the climber should know where it is on the harness. This will be covered in more detail in PLACING PROTECTION.

CLIMBING RACK

PRACTICE (CONC)

a. Students will properly care and maintain mountaineering equipment.

PROVIDE HELP (CONC)

a. The instructors will assist the students when necessary.

OPPORTUNITY FOR QUESTIONS (3 Min.)

1. QUESTIONS FROM THE CLASS

2. QUESTIONS TO THE CLASS

Q. What are the five parts of a locking carabiner?

A. (1) Gate
(2) Gate pivot pin
(3) Locking notch
(4) Body
(5) Locking nut (locking carabiner only)

Q. Name the three types of pro used.

A. (1) Stoppers
(2) Hexcentrics
(3) Spring Loaded Camming Devices (SLCD)

SUMMARY (2 Min)

a. During this period of instruction we have covered, ropes, carabiners, chocks, and helmets. We have discussed their nomenclature, care/maintenance and purposes.

b. Those of you with IRF's please fill them out and turn them in. We will now take a short break until our next class.

UNITED STATES MARINE CORPS
Mountain Warfare Training Center
Bridgeport, California 93517-5001

SML
SMO
ACC
02/11/02

LESSON PLAN

ROPE MANAGEMENT

INTRODUCTION (5 Min)

SMO INSTRUCTOR NOTE: This class is taught outside the gym. Have a rope corral setup that is large enough for a company. You will need at least two assistants to demonstrate and help check knots.

1. **GAIN ATTENTION**. Movement over mountainous terrain in a summer mountainous environment cannot always be accomplished by an individual, a party, or a unit without the use of special equipment. The trained military mountaineer soon learns the value of his rope, and what it means if that rope is either not available or has become unserviceable due to abuse. If the rope is treated properly and with care, the rope could save your life.

2. **OVERVIEW.** The purpose of this period of instruction is to familiarize the student with the basics of rope management, especially those aspects of terminology, considerations in the care of rope, methods of coiling ropes, when to use each coil, the types of knots used, and the classification of knots. This lesson relates to all installations and climbing performed here at MWTC.

INSTRUCTOR NOTE: Have students read learning objectives.

3. **INTRODUCE LEARNING OBJECTIVES**

   a. TERMINAL LEARNING OBJECTIVE. In a summer mountainous environment, utilize rope management, in accordance with the references.

   b. ENABLING LEARNING OBJECTIVES (MLC) and (ACC)

      (1) Without the aid of references and given a list of rope terminology, define in writing, in accordance with the references.

      (2) Without the aid of references, list in writing the rope care considerations, in accordance with the references.

(3) In a summer mountainous environment and given a climbing rope, execute the methods of coiling a rope, in accordance with the references.

(4) In a summer mountainous environment and given a sling rope and a designated amount of time while blindfolded, tie the mountaineering knots, in accordance with the references.

c. ENABLING LEARNING OBJECTIVES (SMO)

(1) With the aid of references and given a list of terms used in rope work, state orally the terminology, in accordance with the references.

(2) With the aid of references, state orally the considerations in the care of rope, in accordance with the references.

(3) In a summer mountainous environment and given a climbing rope, execute the methods of coiling a rope, in accordance with the references.

(4) In a summer mountainous environment and given a sling rope and a designated amount of time, tie the mountaineering knots, in accordance with the references.

4. **METHOD/MEDIA**. The material in this lesson will be presented by lecture and demonstration. You will practice what you have learned during upcoming field training exercises. Those of you with IRF's please fill them out at the conclusion of this lesson.

5. **EVALUATION**.

a. MLC - You will be tested later in the course by written and performance evaluations on this period of instruction.

b. ACC - You will be tested later in the course by written and performance evaluations on this period of instruction.

c. SMO - You will be tested later on by a verbal and performance examination. A timed knot test will also be administered also.

TRANSITION: Are there any questions over the purpose, learning objectives, how the class will be taught, or how you will be evaluated? Now that we know what is expected, it is important for all to know and use the same terminology so that everyone is sure of what is being said to prevent confusion and mistakes. Let us start with some common rope terminology.

BODY (80 Min)

1. (5 Min) **TERMS USED IN ROPE WORK**

a. Bight. A simple bend in the rope in which the rope does not cross itself.

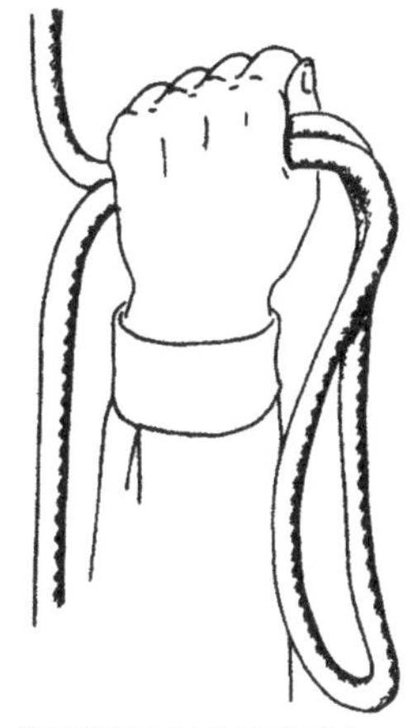

BIGHT OF ROPE

b. Loop. A simple bend in the rope in which the rope does cross itself.

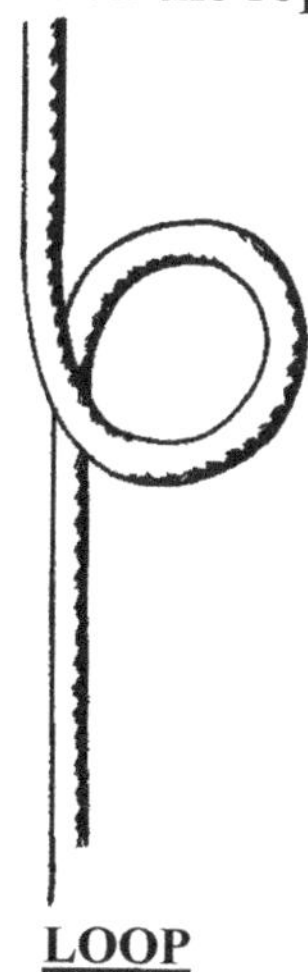

LOOP

c. Half Hitch. A loop which runs around an object in such a manner as to bind on itself.

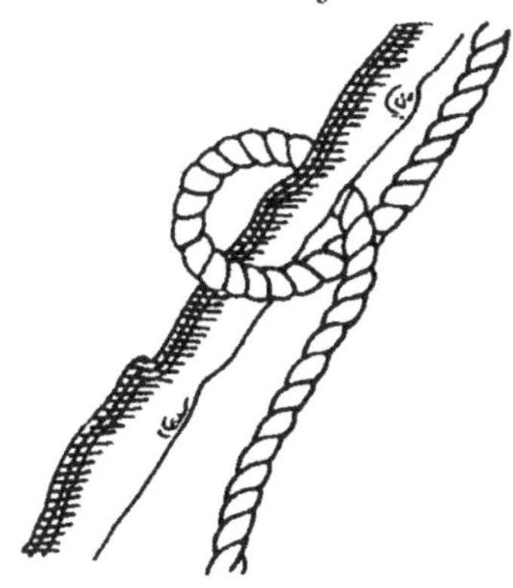

HALF HITCH

d. Standing End. The part of the rope which is anchored and cannot be used, also called the static end.

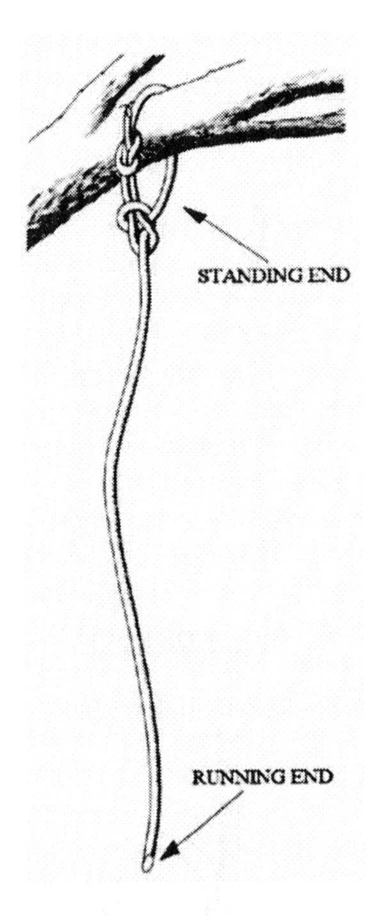

e. Running End. The free end of the rope which can be used.

f. Lay. The same as the twist of the rope. (Applies only to hawser laid ropes, such as manila.)

g. Pigtail. The short length left at the end of a rope after tying a knot or coiling a rope. It may or my not be tied off with a secondary knot, depending on the circumstance.

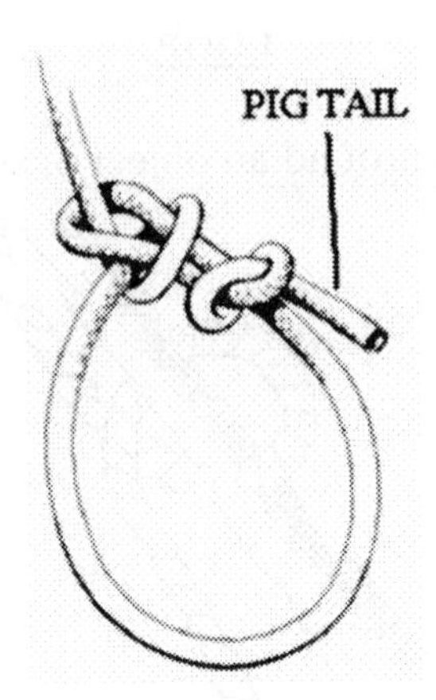

PIG TAIL

h. Stacking (or Flaking). Taking off one wrap at a time from a coil, and letting it fall naturally to the ground.

FLAKED OUT ROPEON THE GROUND

i. Dressing the knot. This involves the orientation of all of the knot parts so that they are properly aligned, straightened, or bundled, and so the parts of the knot look like the accompanying pictures. Neglecting this can result in an additional 50% reduction in knot strength.

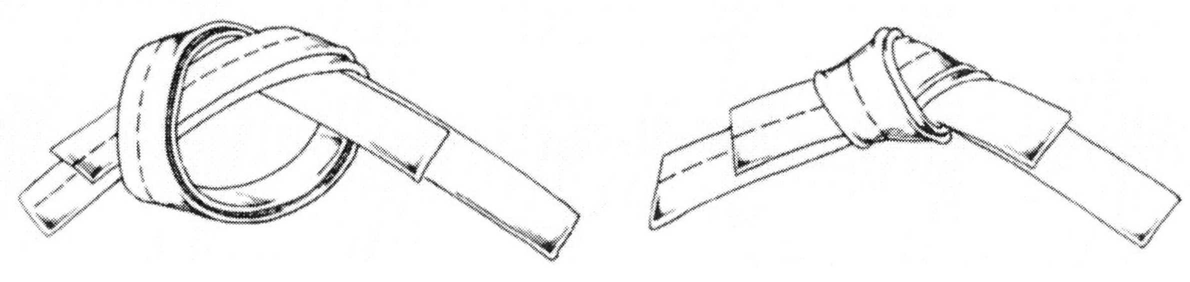

NON-DRESSED DOWN WATER TAPE KNOT **DRESSED DOWN WATER TAPE KNOT**

j. Setting the knot. This involves tightening all parts of the knot so that all of the rope parts bind upon other parts of the knot so as to render it operational. A loosely tied knot can easily deform under strain and change character.

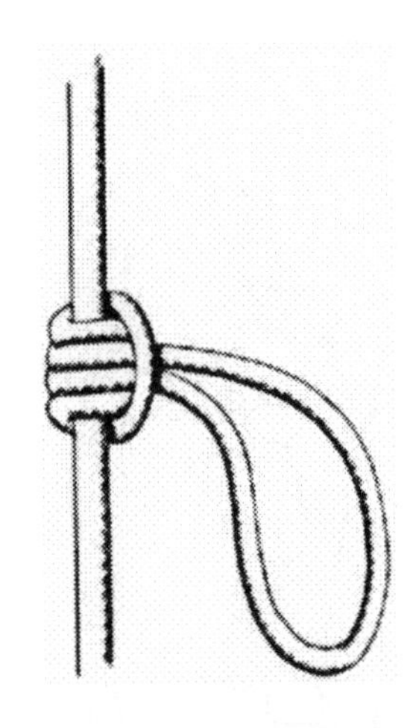

A FULLY SET PRUSIK KNOT

TRANSITION: Are there any questions over the terminology? Since the rope is the climbers lifeline it deserves a great deal of care and respect. Next we will talk about the considerations for the care of the rope and rope log.

2. (10 Min) **CONSIDERATIONS FOR THE CARE OF ROPE**

   a. The rope should not be stepped on or dragged on the ground unnecessarily. Small particles of dirt will get into and through the sheath causing unnecessary wear to the rope within.

   b. The rope should never come in contact with sharp edges of any type. Nylon rope is easily cut, particularly when under tension. If a rope must be used around an edge which could cut it, then that edge must be padded or buffed using fire hose if available, or several small sticks.

   c. Keep the rope as dry as possible. If it should become wet, hang it in large loops, above the ground, and allow it to dry. A rope should never be dried out by an open flame, or be hung to dry on metal pegs, as this will cause rust to get in the rope thus rendering it unserviceable.

   d. Never leave a rope knotted or tightly stretched longer than necessary.

   e. When using rope installations, never allow one rope to rub continually against another.

NOTE: With manila ropes this will cause the rope to fray, whereas nylon ropes can melt under the friction that this causes.

   f. The rope should be inspected prior to each use for frayed or cut spots, mildew, rot or defects in construction.

   g. Mark all climbing ropes at their midpoints to facilitate establishing the midpoint for a procedure requiring you to use the middle of the rope. The rope should be marked with a bright colored adhesive tape.

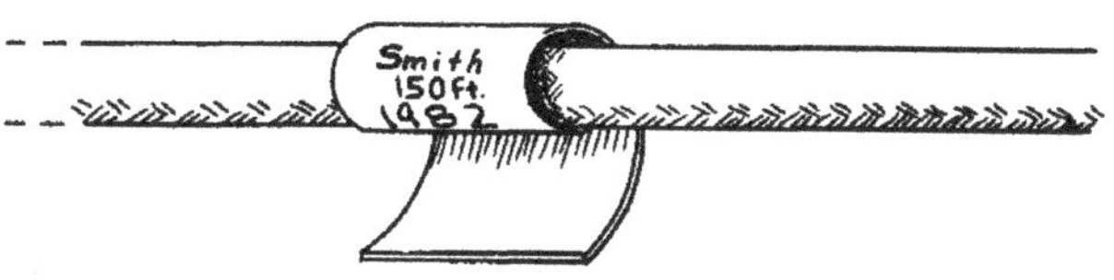

MARKING THE MIDDLE OF THE ROPE

h. The rope should not be marked with paints or allowed to come in contact with oils or petroleum products for these products will weaken it.

i. A climbing rope should NEVER be used for any other purpose except for mountaineering, i.e., towing vehicles.

j. The ends of a new rope or ends caused by a cut should be cut with the rope cutter contained in the MACK and marked with a serial number.

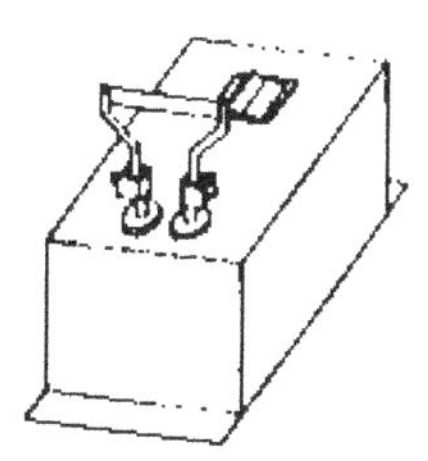

ROPE CUTTER

k. The rope should never be subjected to high heat or flame as this can significantly weaken it.

l. To clean rope use mild soap and rinse thoroughly with water. A rope washer can be used to clean or rinse the rope.

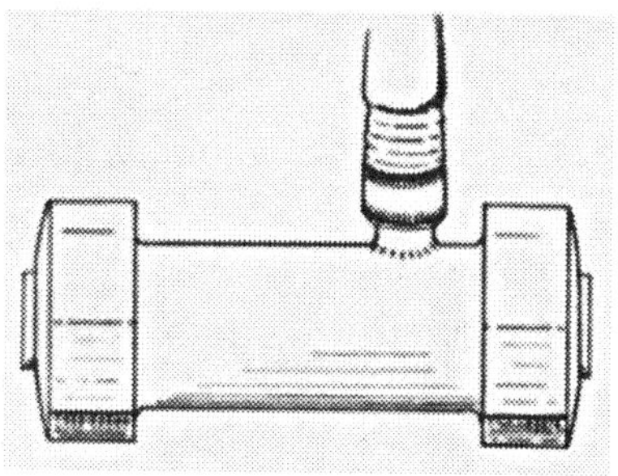

ROPE WASHER

m. When not in use, ropes should be coiled and hung on wooden pegs rather than on nails or any other metal object. They should be stored in a cool place out of the direct rays of the sun.

n. When in areas of loose rock, the rope must be inspected frequently for cuts and abrasions.

o. Always maintain an accurate Rope Log whenever using a rope.

p. Ropes 300-600 foot in length should be Mountain Coiled.

3. (10 Min) **INSPECTION OF ROPE**

a. All ropes have to be inspected before, during, and after all operations. Kernmantle rope is harder to inspect than a laid rope. I.e. green line. The Assault Climber must know what to look and feel for when inspecting a rope. Any of the below listed deficiencies can warrant the retirement of a rope.

(1) Excessive Fraying. Indicates broken sheath bundles or PIC breakage.

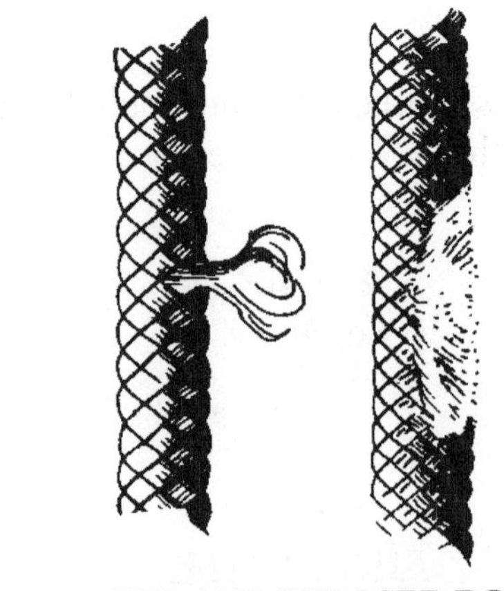

DAMAGED AND EXPOSED ROPE

(2) Exposed Core Fibers. Indicates severe sheath damage. (When you can see the inner core fibers)

(3) Glossy Marks. Signify heat fusion damage, also called a booger.

(4) Uniformity of Diameter / Size. May indicate core damage, noted by an obvious depression (hour glass) or exposure of white core fibers protruding from the sheath (puff).

(5) Discoloration. A drastic change from the ropes original color may indicate chemical change or damage.

(6) Stiffness or Soft Spots. Could signify core damage.

NOTE: Dynamic ropes measuring between 10mm and 12mm are marked at each end of its pigtails with a number "1" indicating that the rope is UIAA approved for single rope lead climbing. Dynamic ropes measuring between 8mm and 9mm are marked at each end of its pigtails with a "1/2" number indicating that two of these ropes are required to conduct a lead climb.

4. (5 Min) **ROPE LOG.** The purpose of the Rope Log is to maintain an accurate record for the use of each rope contained within the Marine Assault Climbers Kit (MACK). Due to the turnover of personnel, and the fact that no one person may have the same rope twice, the rope log is used to ensure the safe use, serviceability, and account of each rope.

a. Serial number. By assigning each rope a serial number, responsible units can determine information about a rope. As soon as the ropes are cut to the desired length for their intended purpose, each rope will be assigned a serial number by the responsible unit. That rope should then be labeled with that serial number in some permanent manner. The best method for this is as soon as the ends of the new rope have been whipped and fused, mark both ends of the rope with the serial number and then dip the ends of the rope, in effect, laminating the serial number to the rope ends.

(1) A rope serial number has five parts. The proper format for a rope serial number is as follows:

1. Type of rope: (S) - Static or (D) - Dynamic.

2. Last two digits of the year the rope was manufactured. Each rope has a shelf life of two years, after that it must not be used for any mountaineering purpose.

3. Four-digit number for that individual rope which is assigned by the responsible unit, and should be assigned sequentially as new ropes are issued out.

4. The length and diameter of the rope. The length may be recorded in feet, and the diameter in millimeters.

5. Responsible unit code. Example: S-96-0001-150/11mm-F2/8

   Example meaning: This rope is a 150 foot, 11mm, static rope, manufactured in 1996. It is the first rope issued by Fox Co 2/8.

b. Recording Information in a Rope Log. Once a new rope has been serialized, the rope log for that rope should be started up. At a minimum, it should contain the following information:

(1) The rope serial number.

(2) Manufacturer. Depending on the manufacturer, ropes of the same type and diameter may vary in tensile strength, stretch factor, and durability.

(3) Date of manufacture. Five years from this shelf life date, ropes are considered to have reached their normal expiration date, and should be destroyed.

(4) Date in service. This is to be recorded for tracking purposes to establish how long a rope is in service. Two years is considered maximum rope life in service.

(5) Each time a rope is used, the using unit is responsible to record how that rope was used and how much use the rope received. Additionally, before checking out a rope and

prior to turning it back in, the rope must be inspected by qualified personnel, and initial in the "Inspected By" block of the rope log.

NOTE: ANY TIME A DYNAMIC ROPE IS SUBJECTED TO A FALL FACTOR 2, THAT ROPE SHOULD NOT BE USED AGAIN FOR MOUNTAINEERING.

DYNAMIC					
SERIAL #					
MANUFACTURER					
DATE OF MANUFACTURE					
DATE IN SERVICE					
INDICATE THE DATE AND NUMBER OF EACH SPECIFIC USE					
CLIMBS	**FALLS (FACTOR)**	**TOP ROPES**	**RAPPELS**	**REMARKS**	**INSPECTED BY**

ROPE LOG

TRANSITION: Are there any questions over the care of a rope or rope log? Now let's discuss the various ways of coiling a rope.

5. (10 Min) **COILING A ROPE**. There are two types of rope coils frequently used at MWTC. The Mountain Coil and the Butterfly Coil.

   a. Mountain Coil. This coil is useful for carrying the rope over a pack or over a climber's shoulder and neck. It can be used for short time storage. The mountain coil can be tied in the following manner:

      (1) Sit down with your leg bent at a 90-degree angle, heel on the deck. Starting at one end, the rope is looped around the leg in a clockwise fashion, going over the knee and under the boot sole until the entire rope is coiled.

      (2) If coiling a 150-foot rope, use only one leg and offset the other, when coiling a 300-foot rope, use two legs and keep them together.

      (3) With the starting end of the rope, form a 12-inch bight on the top of the coils.

      (4) Uncoil the last loop and along the top of the coils, wrap 4-6 times towards the closed end of the bight.

      (5) The end of the rope being wrapped is then placed through the closed end of the bight.

      (6) The running end of the bight is then pulled snuggly to secure the coil.

(7) To prevent the coil from unraveling, the two pigtails are tied together with a square knot.

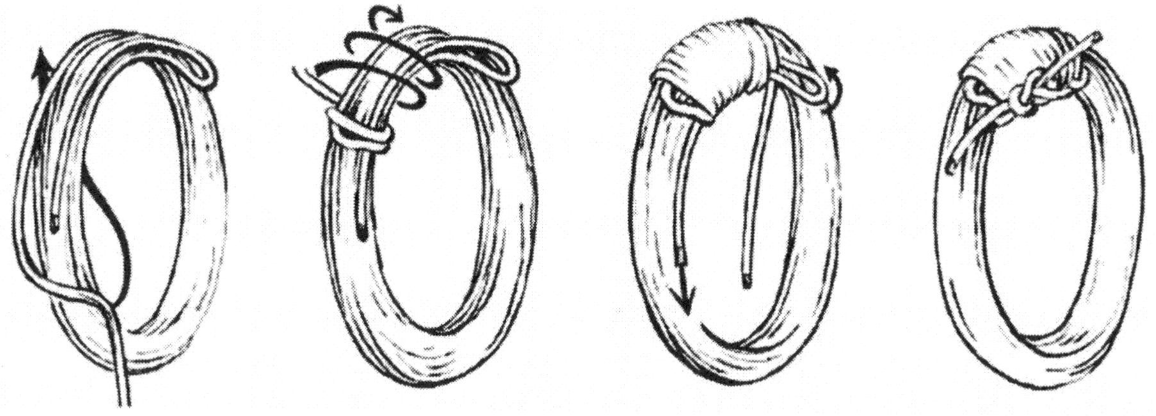

MOUNTAIN COIL

b. Butterfly Coil. This method is used for carrying a rope when the individual needs to have maximum use of his upper body, (i.e. while climbing), without the encumbrance of a large rope coil hanging across his chest.

(1) Coiling the Butterfly Coil

(a) Step 1: Find the middle of the rope, then form a three foot bight laying both ropes in the upraised palm at the two foot point.

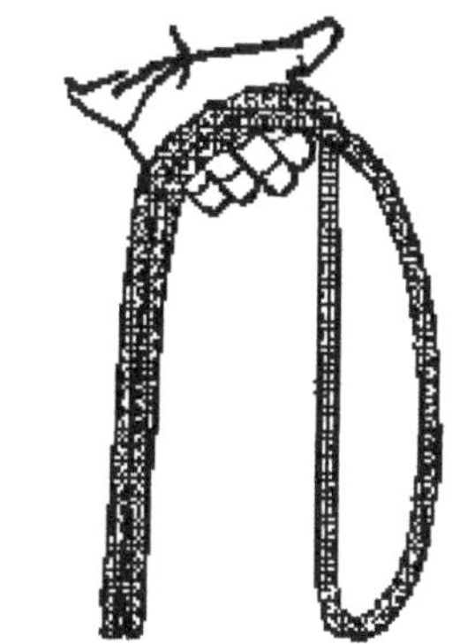

STARTING THE BUTTERLFY COIL

(b) Step 2: Form another two-foot bight with the running end. Place the rope at the two-foot bight along side on top of the original bight ensuring the running end is on the same side as the original bight.

(c) Step 3: Continue making two foot bights, laying them alternately into your palm until there is only six to eight feet remaining. At that point, begin wrapping the two pigtails horizontally four to six times at the mid way point of the ropes in a bight from bottom to top.

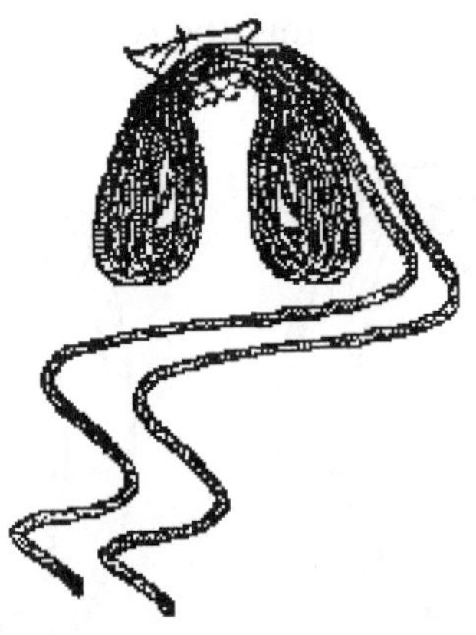

TWO FOOT BIGHTS ON BOTHS SIDES WITH 6 TO 8 FEET REMAINING

(d) Step 4: After completing your wraps, form a bight with the remaining pigtail and then thread it underneath your palm and upwards to one-foot above the coiled rope.

(e) Step 5: With the remaining pigtail, thread it through the one-foot bight in step four.

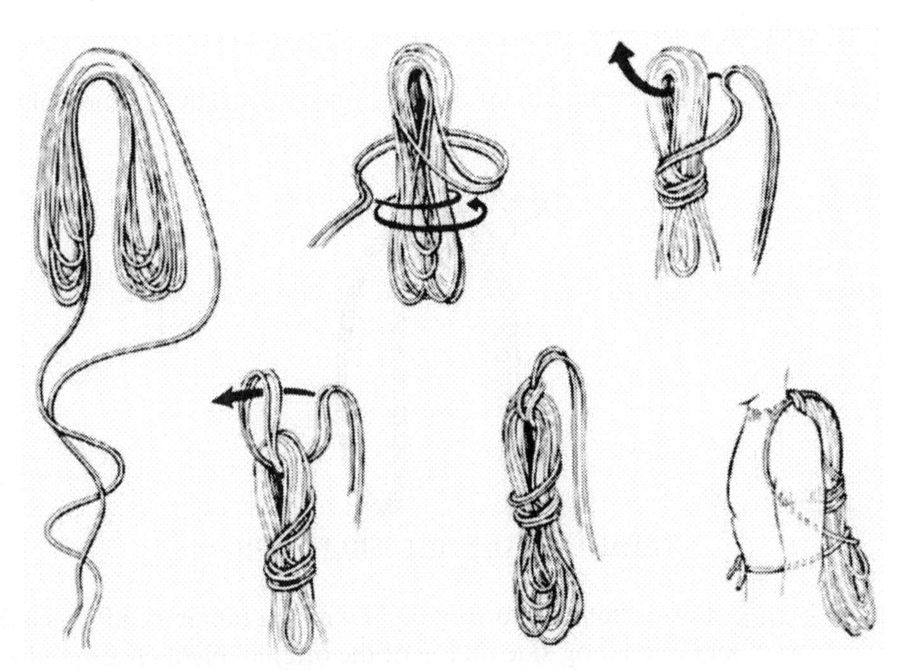

TYING AND CARRYING THE BUTTERFLY COIL

(2) Carrying the Butterfly coil. Separate the running ends, placing the coil in the center of the back of the carrier, then fun the two ends over his shoulders so as to form shoulder straps. The running ends are then brought under the arms, crossed in the back over the coil, brought around the body of the carrier and tied off with a square knot at his stomach.

TRANSITION: Are there any questions over coiling and securing a rope? Now let's discuss how to throw the rope down the cliff face.

6. (5 Min) **ROPE THROWING**. To insure that the rope will not get tangled when deployed, certain steps must be taken.

   a. With a stacked rope, anchor off the standing end.

   b. Take the opposite end of the rope and make 6-8 coils and place them in your strong arm. These wraps will serve as a throwing weight that you can aim.

   c. 10-15 feet from the strong-arm coils create a second set of 6-8 wraps and place them in your weak arm.

   d. From the edge of the cliff, sound off with the command "STAND-BY FOR ROPE". Just before you release the coils, sound off with the command "ROPE". At that time drop the weak arm coils from the cliff.

   e. While taking aim, throw your strong-arm coils overhand or sidearm hard enough to hit your intended target.

   f. If the throw was misdirected due to wind, tree, etc., reorganize and attempt to re-deploy the rope.

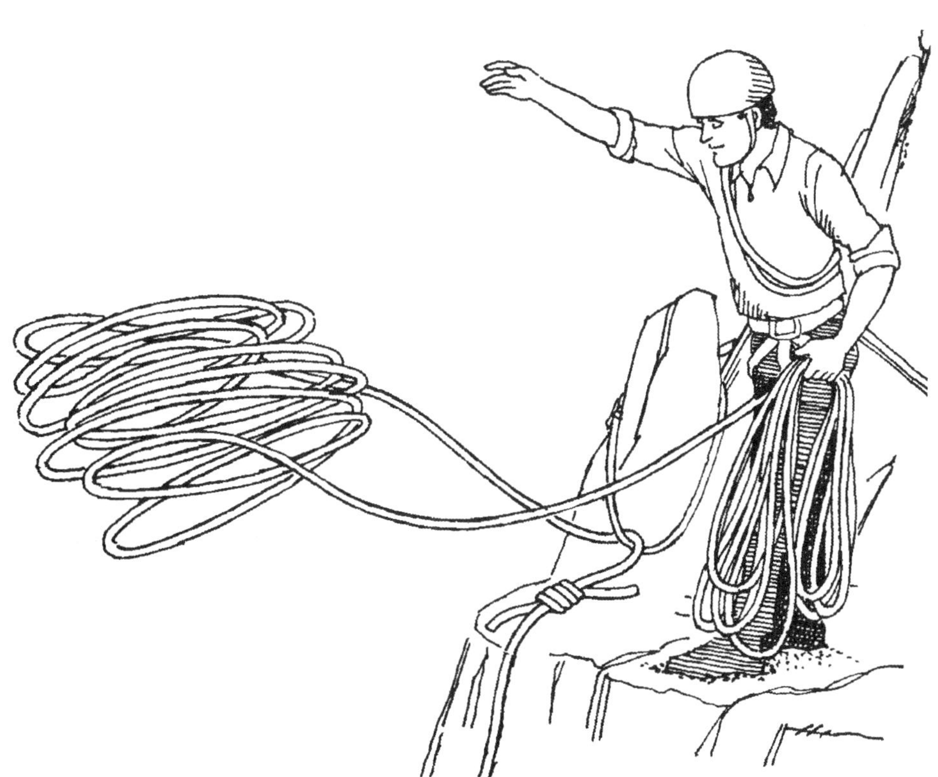

THROWING THE ROPE

TRANSITION: Are there any questions over throwing the rope? Take a 5 minute break and then get around the rope corral with your sling rope. Now we will discuss mountaineering knots, their uses, and how to tie them.

SMO INSTRUCTOR NOTE: Tell the company that you will cover as many knots as time allows. All other knots will be taught by that company's instructor team throughout the core package.

6. (25 Min) **MOUNTAINEERING KNOTS**.

a. **Class I** – *End of the Rope Knots*

(1) Square Knot. Used to tie ends of two ropes of equal diameter together. It should be secured by overhand knots on both sides of the square knot.

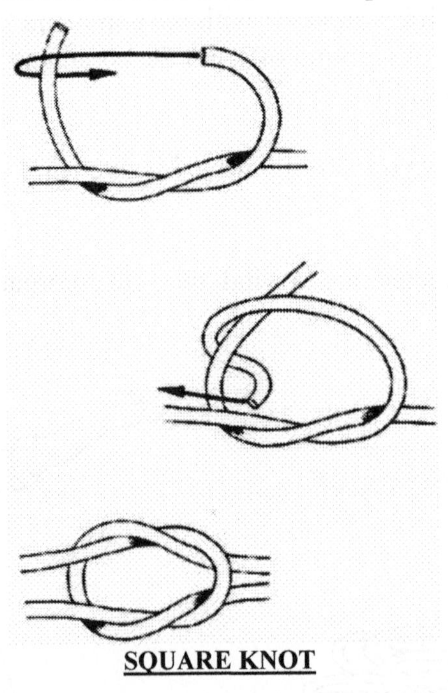

SQUARE KNOT

(2) Double Fisherman's Knot. It is a self-locking knot used for tying two ropes of equal diameter together. It can be tightened beyond untying.

DOUBLE FISHERMAN'S KNOT

(3) Water/Tape Knot. Used to secure webbing or tape runners. It is constructed by tying an overhand knot (without twists) in one end of the tape, and threading the other end of the tape through the knot from the opposite direction. After the knot is dressed down, each pigtail should be a minimum of two inches long.

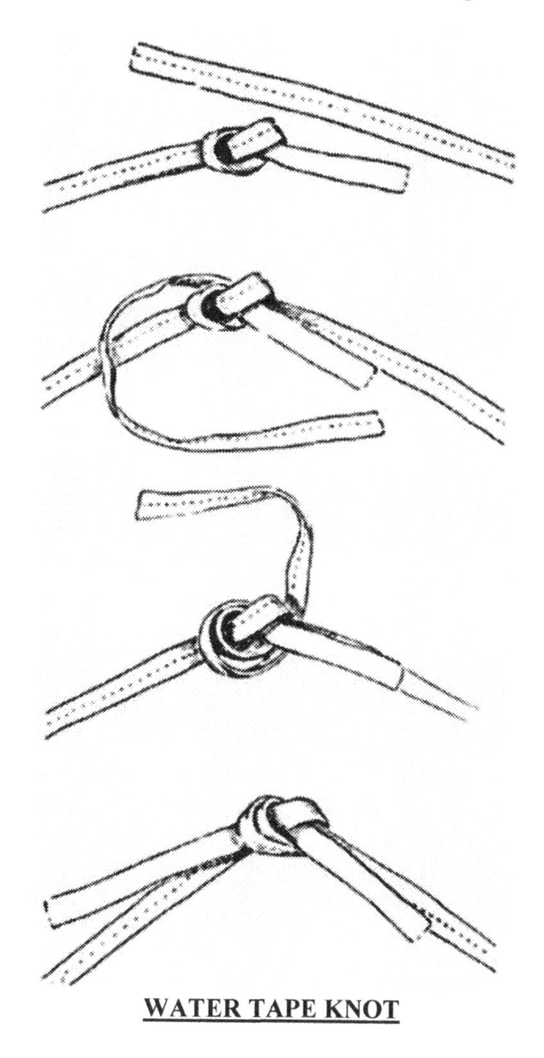

WATER TAPE KNOT

(4) Double Sheet Bend. Used to tie the ends of two or more ropes of equal or unequal diameter together.

SHEET BEND

b. **Class II** - *Anchor Knots*

(1) Bowline. Used to tie a fixed loop in the end of a rope. This knot is always tied with the pigtail on the inside and secured with an overhand knot.

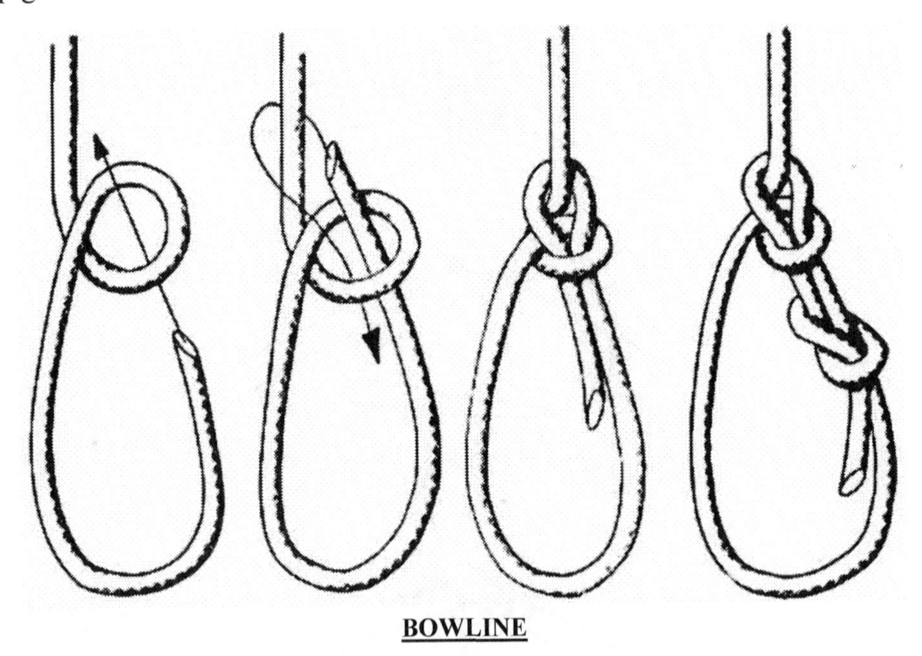

BOWLINE

(2) Round Turn with Two Half Hitches or a Bowline. A loop which runs around an object in such a manner as to provide 360 degree contact and may be used to distribute the load over a small diameter anchor. It will be secured with two half hitches or a bowline.

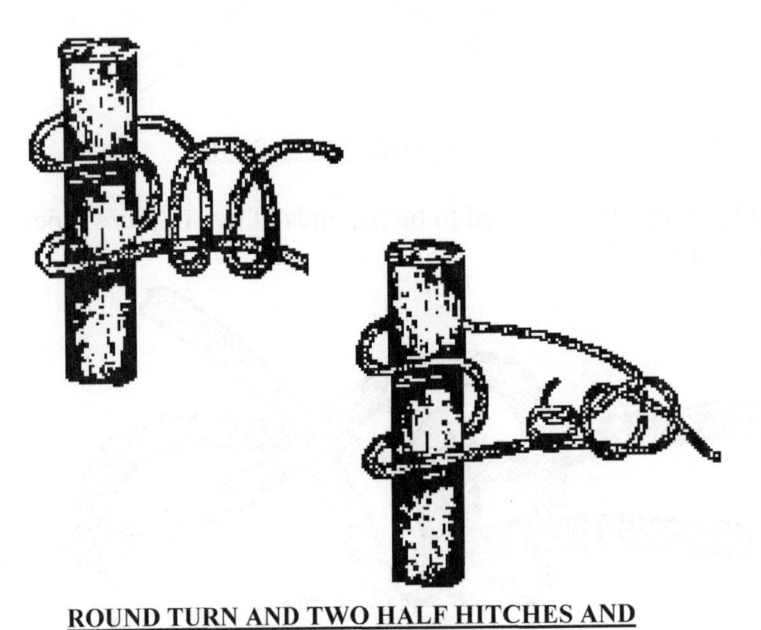

ROUND TURN AND TWO HALF HITCHES AND

ROUND TURN AND A BOWLINE

(3) Clove Hitch. This knot is an adjustable hitch. It could be considered a middle-of-the-rope anchor knot at the end-of-the-rope when used in conjunction with a bowline or round turn and two half hitches.

(1) Around-the-object Clove Hitch.

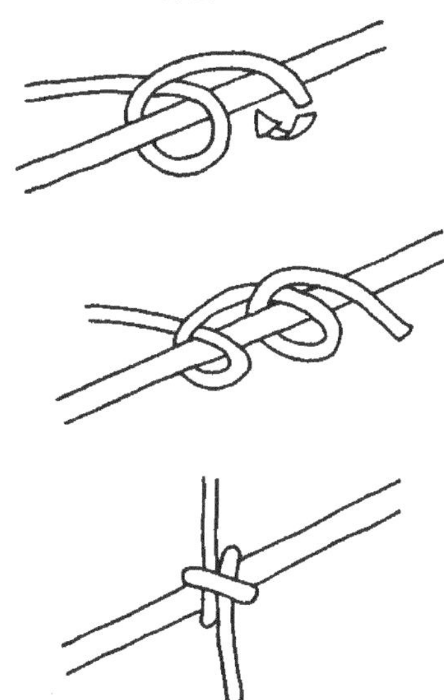

AROUND THE OBJECT CLOVE HITCH

(2) Over-the-object Clove Hitch.

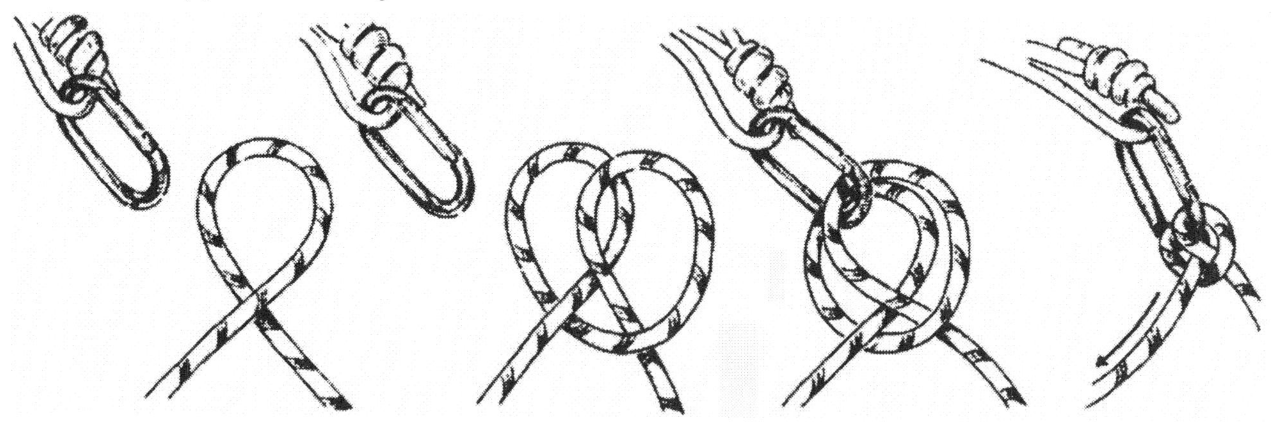

OVER THE OBJECT CLOVE HITCH

c. **Class III -** *Middle-of-the-rope*

(1) Figure-of-Eight-Loop. This is a strong knot that can be readily untied after being under load.

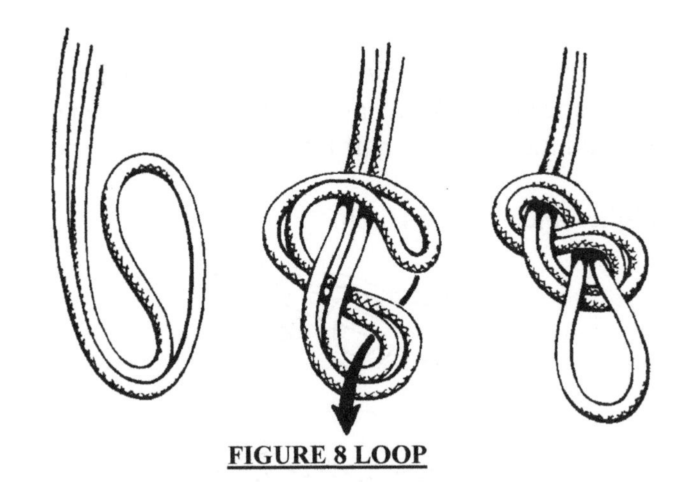

FIGURE 8 LOOP

(2) Double Figure of Eight Loop. The double Figure of Eight Loop is a strong knot and the double loop reduces the wear and strength loss from the rope bending around the carabiner by splitting the load between the two loops.

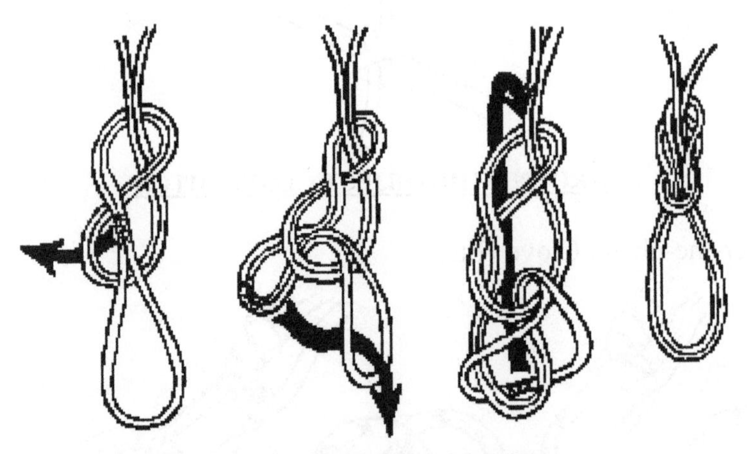

DOUBLE FIGURE 8 LOOP

(3) Two Loop Bowline. This knot can be used to construct a self-equalizing belay or to tie the middle person in on three people on a rope.

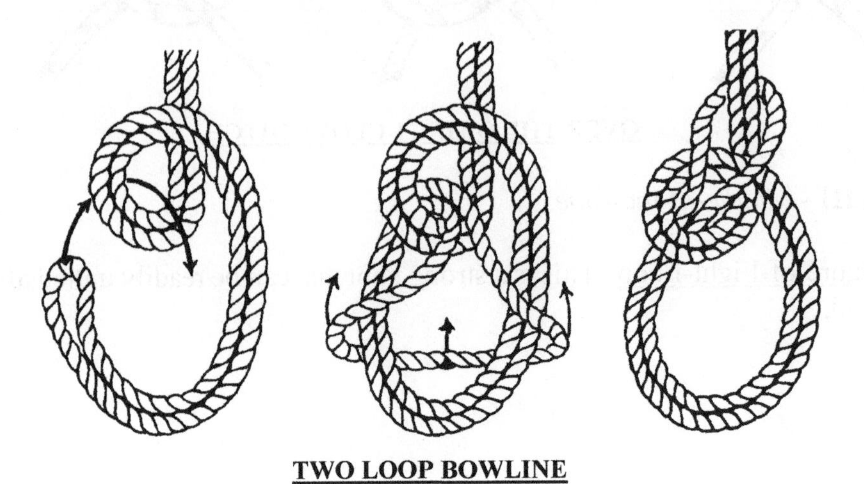

TWO LOOP BOWLINE

d. **Class IV** - *Special Knots*

(1) Prusik Knot. This knot functions by introducing friction that can be alternately set and released. For best results, tie the knot with a smaller diameter cord on a larger diameter cord. If slippage occurs, more wraps maybe used.

(a) Middle-of-the-Rope Prusik. This knot is created with an endless loop also known as a Prusik Cord. Do not tie this knot with tape due to less friction.

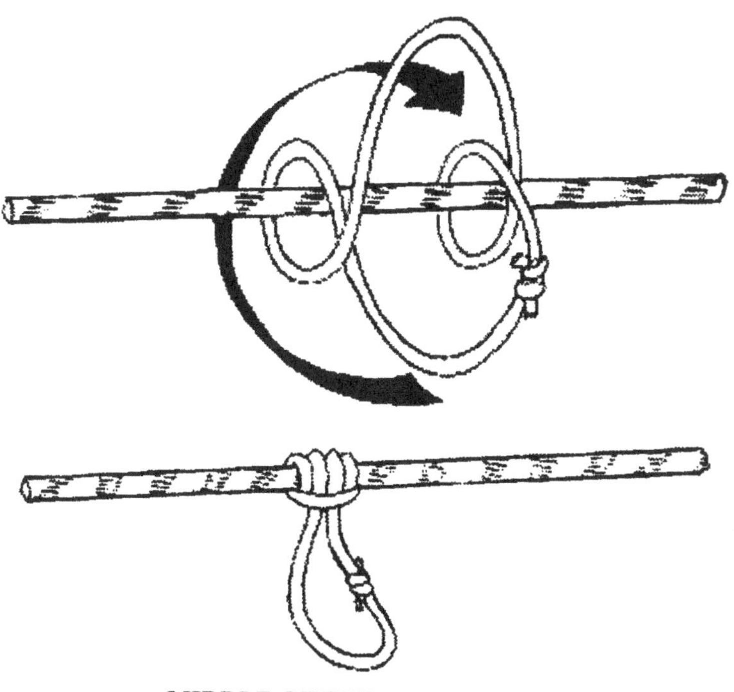

MIDDLE OF THE LINE PRUSIK

(b) End-of-the-Rope Prusik. This knot is always secured with a bowline. Do not tie this knot with tape due to less friction.

END OF THE LINE PRUSIK

(c) French Prusik. This can be constructed with a Prusik Cord or a single strand of cord with figure-of-eight loops on each end.

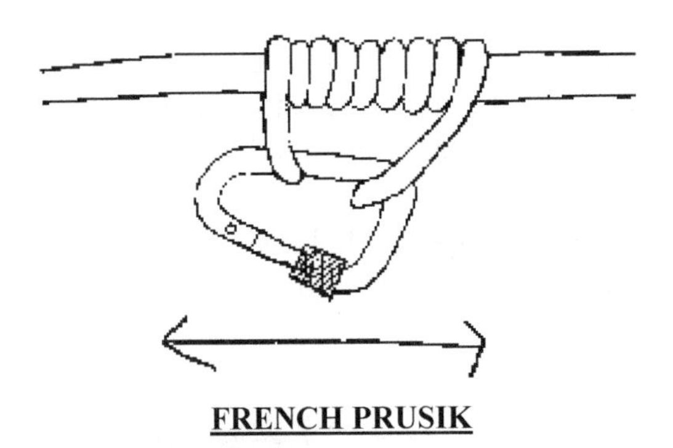

FRENCH PRUSIK

(2) Retraced Figure of Eight. Used to tie the end of the climbing rope into a harness or swammi wrap. The pigtail may or may not be secured with an overhand.

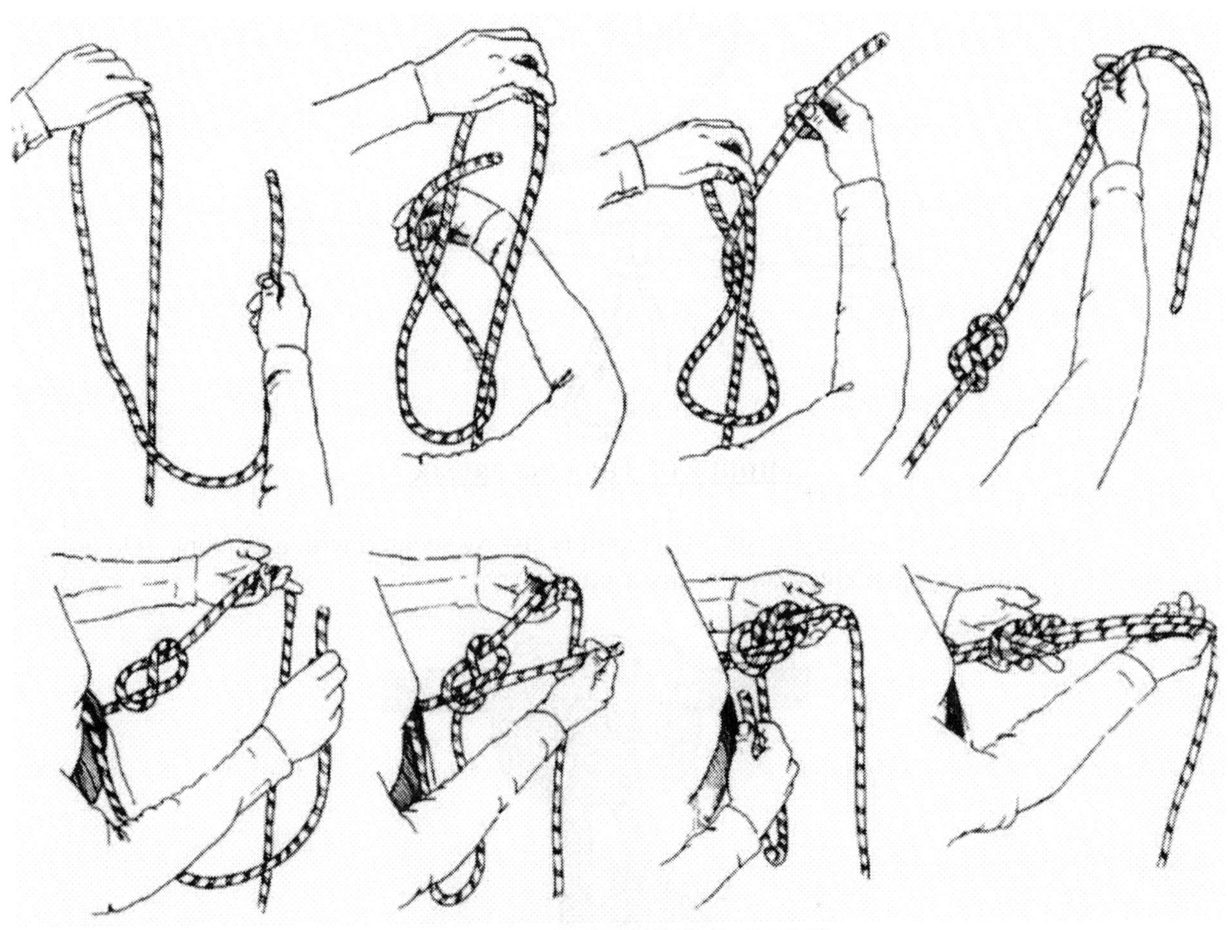

RETRACED FIGURE 8 IN HARNESS

(3) Directional Figure-Of-Eight. When tied and tensioned is applied to the standing and running ends of the rope, the knot will not pull apart. The loop will point toward the direction of pull.

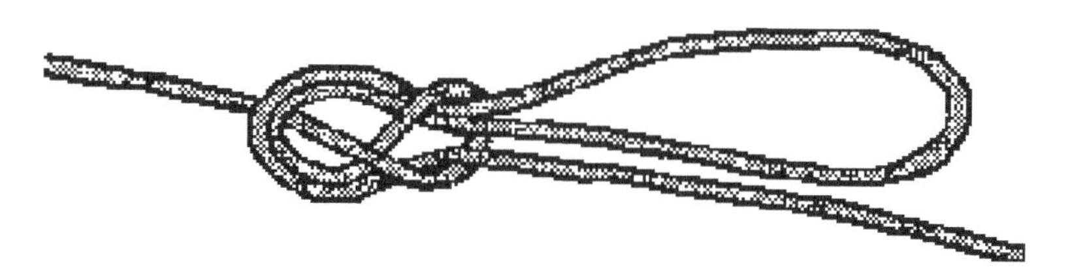

DIRECTIONAL FIGURE 8

(4) Slip Figure 8. This knot is used for retrievable anchors and fixed rope installations for the ease of untying the knot.

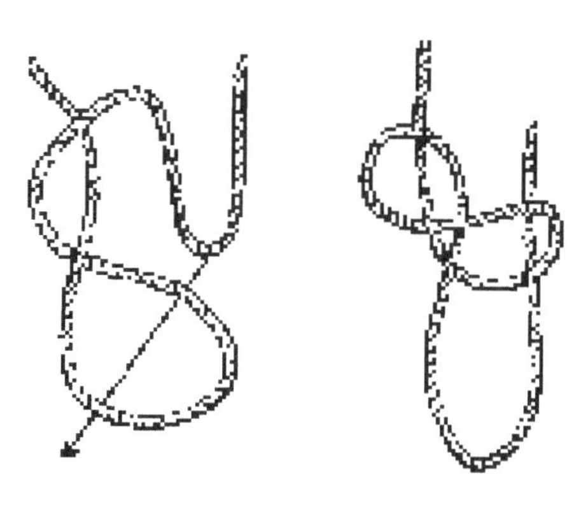

SLIP FIGURE 8

(5) Kliemheist. This is a friction knot.

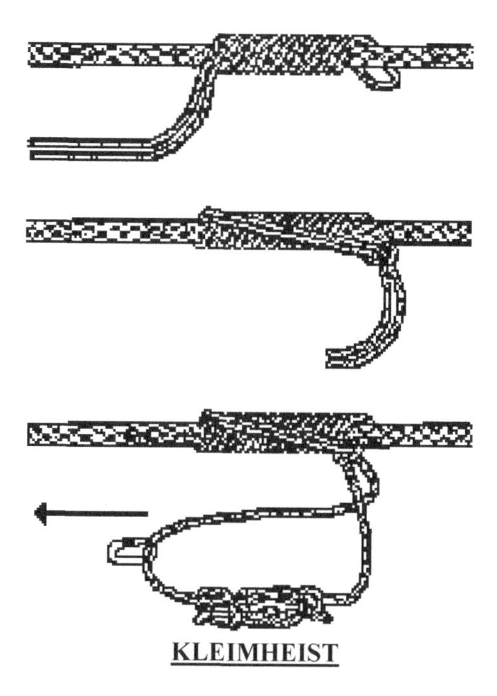

KLEIMHEIST

(6) Overhand Knot. Can be used to secure primary knots on itself.

OVERHAND

(7) Munter Hitch. This is a simple hitch in the rope that is clipped into a carabiner to put friction on the line.

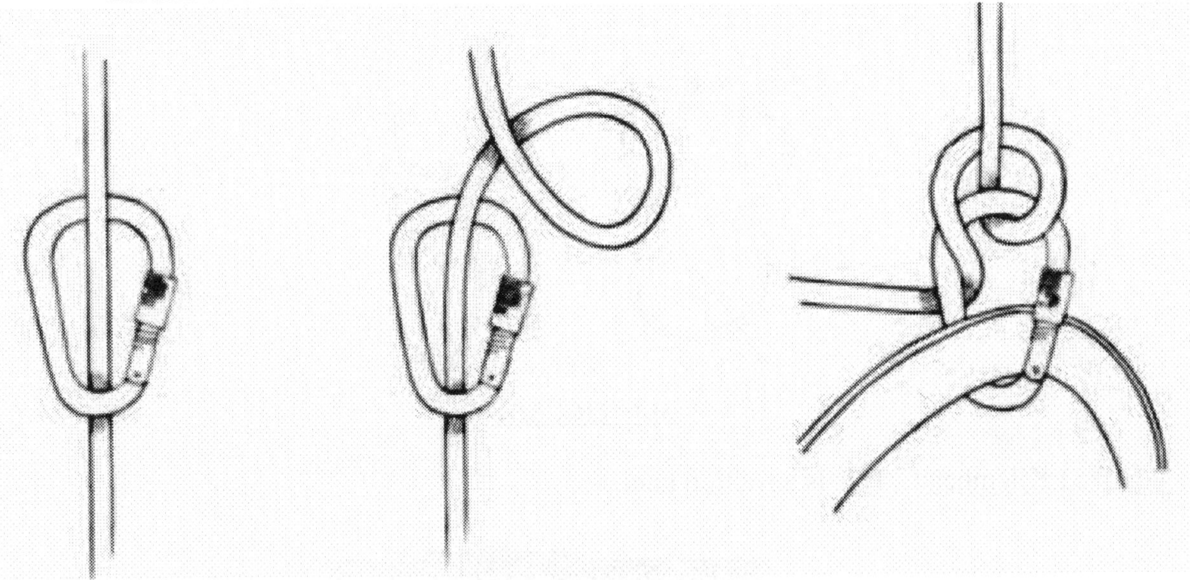

MÜNTER HITCH

(8) Timber Hitch. A Timber Hitch is used to fix a rope to a pole or equivalent for hoisting or towing purposes. It has the capability of casting off easily.

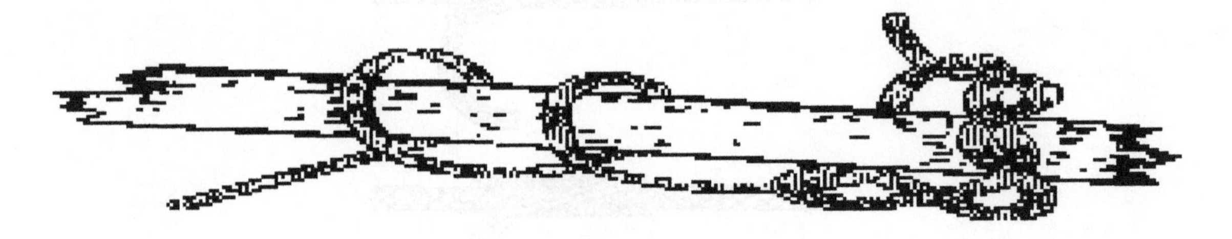

TIMBER HITCH

(9) Mariner's Hitch. This knot's advantage over the prusik is that it can be released under load.

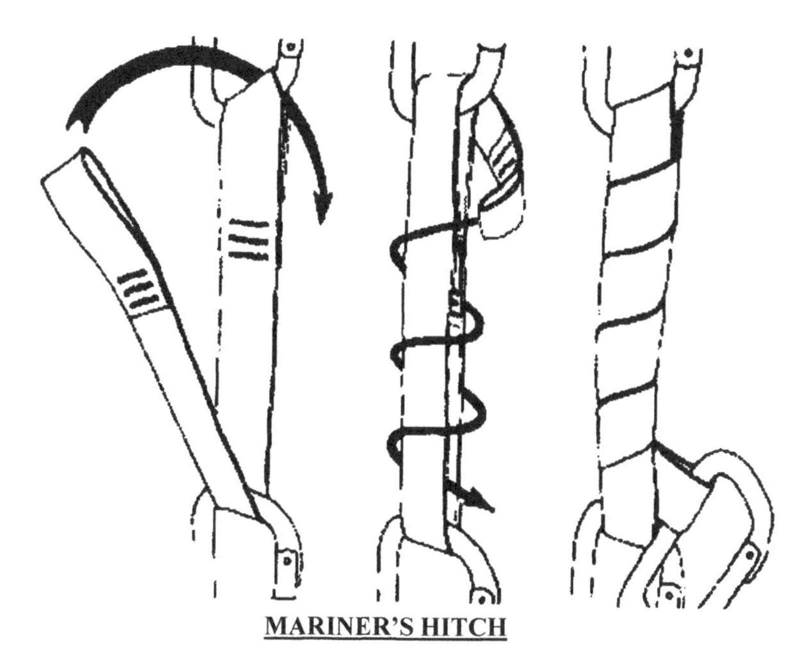

MARINER'S HITCH

(10) Krägur Knot. This friction knot is used to prevent unnecessary slippage when tied onto a rope of the same diameter.

KRÄGUR KNOT

(11) Rappel Seat. This is used as an expedient support harness for rappelling, crossing rope bridges, etc. It is constructed as follows:

(a) Center the sling rope on the left hip.

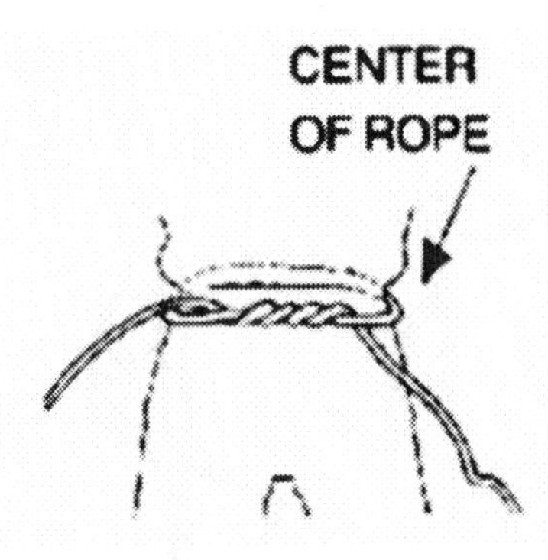

(b) Wrap the sling rope around the waist and tie at least one half hitch around itself (preferably two) in the front.

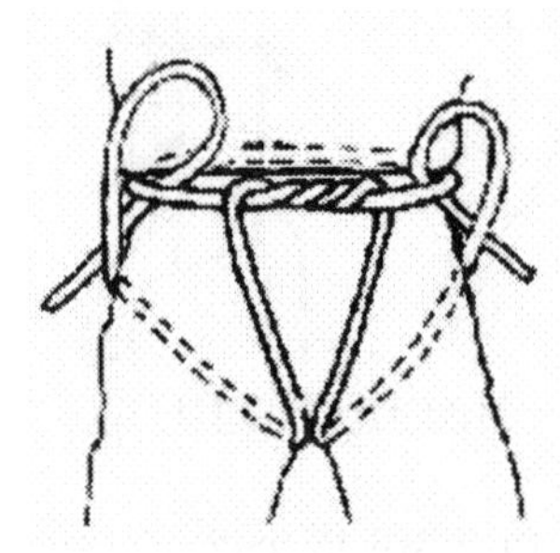

(c) Bring the running ends down through the legs, up over the buttocks, and over the original waist wrap, down between the waist wrap and the waist, and over itself, forming a bight. Cinch this up tightly.

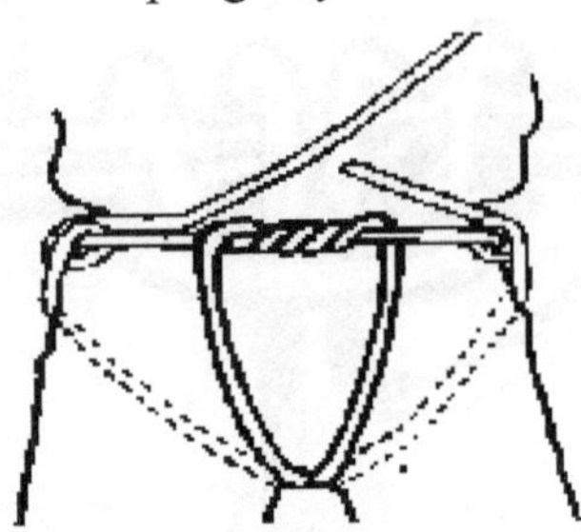

(d) Now tie a square knot with two overhand knots on the left hip ensuring that the overhand knots encompass all the wraps.

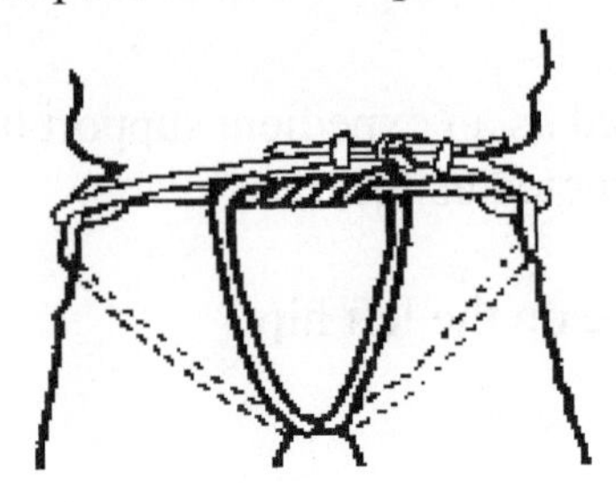

(e) Tuck any excess rope into a pocket.

(12) Bowline on a coil. Used by the first and last men on a climbing rope to tie into the rope. An overhand knot is used behind the knot. It distributes the force of a fall over a larger area of a climber's waist and is preferable to a single bowline around the waist. The bowline on a coil can also be used to take up excess rope. The bowline on a coil should have 4-6 wraps around the waist.

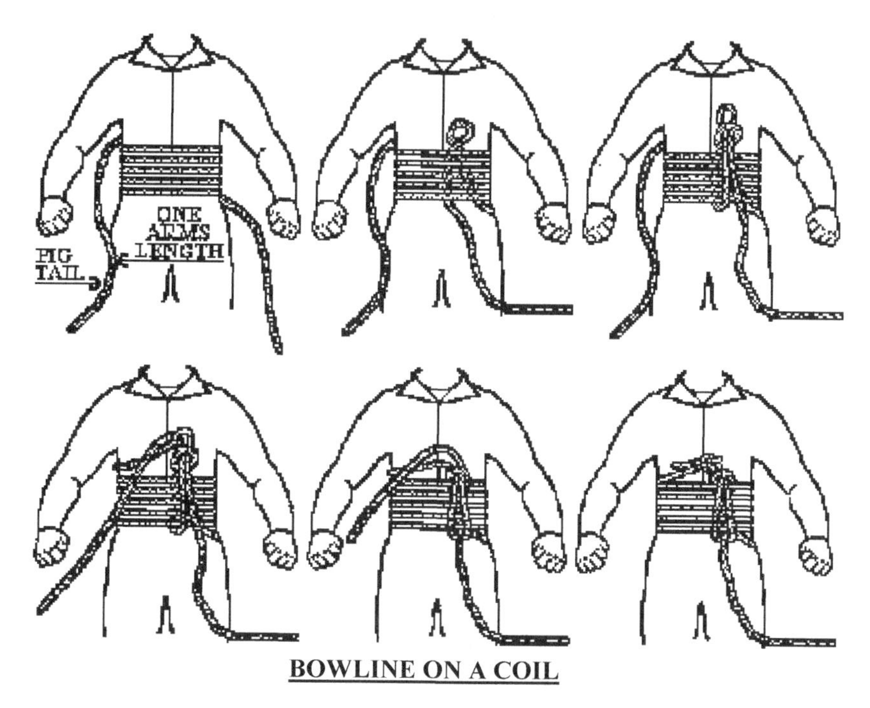

BOWLINE ON A COIL

(13) Swami wrap (Swami belt). The swami wrap (four to six wraps of rope or nylon webbing tied around the waist) popular with rock climbers who wanted to increase the working length of their climbing ropes. It also offers some improvement over the bowline on the coil in distributing the forces sustained in a fall over a larger area of the midsection of the body, but does not eliminate the problem of suffocation when hanging. The Swami wrap is constructed as follows:

(a) Take a sling rope and find the middle.

(b) Place the middle of the sling rope in the small of your back.

(c) Bring both pigtails to the front and continue to wrap around your body until both pigtails are about 15-20 inches long.

(d) Tie the pigtails on the left side with a square knot and two overhand knots.

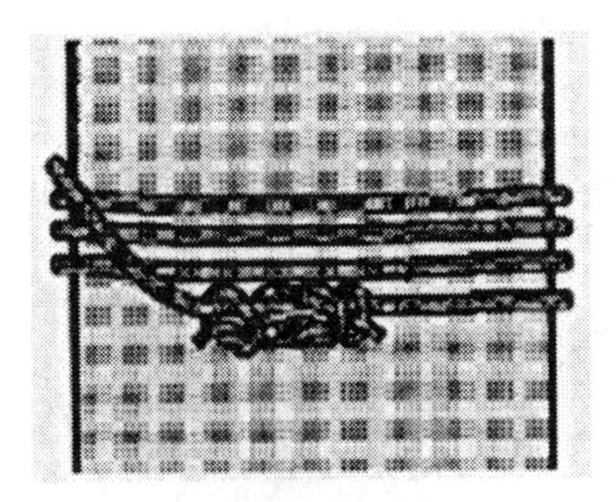

SWAMMI WRAP

(14) Three Loop Bowline. This knot will provide three bights, each of which can be adjusted against the others.

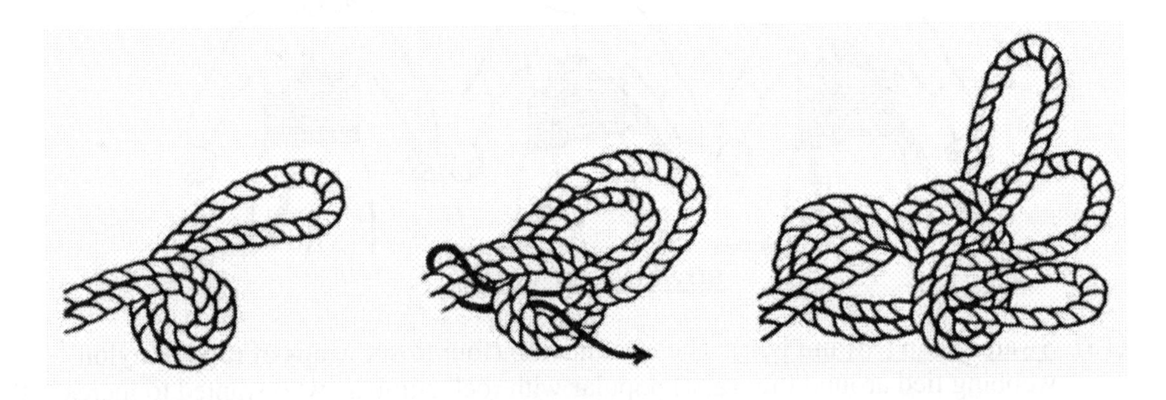

THREE LOOP BOWLINE

TRANSITION: Now that we have discussed the different types of knots and their classifications let us now talk about the relative strength of the knots.

7. (5 Min) **KNOT STRENGTH.** Rope, cordage and webbing is strongest when loaded in a straight line. When bending a rope or web to create a knot, the strength of the rope is reduced All knots should be dressed properly for maximum effective use.

KNOT	RELATIVE STRENGTH OF KNOT
No Knot	100%
Figure 8	75-80%
Bowline	70-75%
Double Bowline	70-75%
Double Fisherman	65-70%
Water Tape Knot	60-70%
Clove Hitch	60-65%
Overhand Knot	60-65%
Square Knot	45%

TRANSITION: Now that we have discussed the relative strength of knots, let's discuss the criteria of the knot test.

8. (5 Min) **KNOT TESTING TIME LIMIT.** The following times must be met to pass the knot tying portion of this course.

KNOT	UNIT OPERATIONS TIME LIMIT	SML, ACC TIME LIMIT (BLINDFOLDED)
Square knot	60 Seconds	30 Seconds
Double Fisherman's Knot	60 Seconds	30 Seconds
Water/Tape Knot	60 Seconds	30 Seconds
Round Turn and a Bowline	60 Seconds	30 Seconds
Round Turn and Two Half Hitches	60 Seconds	30 Seconds
Clove Hitch (around the object)	60 Seconds	30 Seconds
Munter Hitch	60 Seconds	30 Seconds
Slip Figure 8	60 Seconds	30 Seconds
Figure 8 Loop	60 Seconds	30 Seconds
Directional Figure 8	60 Seconds	30 Seconds
End of the Line Prusik	90 Seconds	45 Seconds
Retrace Figure 8	90 Seconds	45 Seconds
Rappel (Swiss) Seat	2 Minutes	60 Seconds

INSTRUCTOR NOTE: Explain the knot, demonstrate, and then have the students tie the knot. Instruct the students not to clear the knot until they have been checked by an instructor.

TRANSITION: Now that we have discussed rope terminology, considerations for care and maintenance, coiling of the ropes, mountaineering knots and their strengths are there any questions? If you have none for me, then I have some for you.

PRACTICE (CONC)

a. Students will practice proper rope management.

PROVIDE HELP (CONC)

a. The instructors will assist the students when necessary.

OPPORTUNITY FOR QUESTIONS (3 Min)

1. QUESTIONS FROM THE CLASS

2. QUESTIONS TO THE CLASS

Q. What are the two types of rope coils and in what situation is each used?

A. (1) Mountain Coil - used in movement from point to point when frequent use of the rope is anticipated.

(2) Butterfly Coil - used when you have to carry the rope and require maximum freedom of movement of your arms.

SUMMARY (2 Min)

a. During this period of instruction we have discussed the various considerations of rope management, including the mountaineering knots that you will be using.

b. Those of you with IRF's please fill them out at this time and turn them in to the instructor. We will now take a short break and then you will be guided to your next class by one of your company instructors.

UNITED STATES MARINE CORPS
Mountain Warfare Training Center
Bridgeport, California 93517-5001

SML
SMO
02/11/02

LESSON PLAN

MOVEMENT IN MOUNTAINOUS TERRAIN

INTRODUCTION (5 Min)

1. **GAIN ATTENTION**. Mountain operations could occur in almost any country in the world, thus creating a need for basic knowledge of military mountaineering and mountain operations. To aid in mobility, help lower the possibility of injury, and make you knowledgeable enough about the mountains to operate effectively, you must learn the fundamentals of terrain walking, mountain terminology and distance estimation.

2. **OVERVIEW**. The purpose of this period of instruction is to familiarize the student with the principles of movement in mountainous terrain by discussing fundamentals, conduct, types of slopes, and distance estimation. This period of instruction will relate to all movement that you conduct in the mountains.

INSTRUCTOR NOTE: Have students read learning objectives.

3. **INTRODUCE LEARNING OBJECTIVES**

   a. TERMINAL LEARNING OBJECTIVES. In a summer mountainous environment, execute movement in mountainous terrain techniques, in accordance with the references.

   b. ENABLING LEARNING OBJECTIVES. (MLC) and (ACC)

      (1) Without the aid of references, state in writing the fundamental principles of movement in mountainous terrain, in accordance with the references.

      (2) Without the aid of references, describe in writing each type of slope, in accordance with the references.

   c. ENABLING LEARNING OBJECTIVES. (SMO)

      (1) With the aid of references, state orally the fundamental principles of movement in mountainous terrain, in accordance with the references.

(2) With the aid of references, describe orally each type of slope, in accordance with the references.

4. **METHOD/MEDIA**. The material in this lesson will be presented by lecture and demonstration. You will practice what you have learned in upcoming field training exercises. Those of you with IRF's please fill them out at the end of this period of instruction.

5. **EVALUATION**

a. MLC - You will be tested later in the course by written and performance evaluations.

b. SMO - You will be tested on T- 11 by a verbal test.

TRANSITION: Are there any questions over the purpose, learning objectives, how the class will taught, or how you will be evaluated? First, let's cover the fundamental principles of movement in mountainous terrain.

BODY (45 Min)

1. (5 Min) **FUNDAMENTALS OF MOVEMENT IN A MOUNTAINOUS TERRAIN**. There are four fundamental principles to conserve energy while moving in a mountainous environment. They are as follow:

a. Maintain your body weight over your feet at all times.

b. Utilize micro-terrain when possible.

c. Stepping over rather than on top of obstacles such as large rocks and fallen trees will help avoid fatigue.

d. Make use of the rest step. The aim of the rest step is to rest the body weight and load onto the skeletal frame, hence relaxing the muscles between each step. The use of the rest step over a given distance can reduce the energy expended by up to 50%.

REST STEP

TRANSITION: Are there any questions over the fundamentals? Let's look at the conduct for movement in a mountainous terrain.

2. (5 Min) **CONDUCT OF MOVEMENT IN MOUNTAINOUS TERRAIN**.

   a. Mountain Pace. Mountain pace is defined as the pace at which all members within the group can move together for the duration of the walk.

   b. Ventilation. The members of the movement should begin by being dressed fairly cool, then be given a chance to ventilate after 10- 15 minutes of movement with a ventilation break.

   c. Information. Information about the overall route and individual legs should be given at each stop.

   d. Factors. Factors that may affect the frequency and length of rest stops are the physical condition of the Marines, load weights and terrain conditions. Generally, a rest stop of five minutes should be taken every 30 minutes of movement.

TRANSITION: Are there any questions over the conduct? Let us now look at different types of slopes that are found in a mountainous environment.

3. (5 Min) **TYPES OF SLOPES**. There are four basic types of slopes that you will encounter in a mountainous environment. They are:

   a. Hard Ground. Hard ground is a slope of firmly packed dirt, with vegetation that will not give way under a man's step.

   b. Grassy. A grassy slope is a covered with scattered clumps of grass.

   c. Scree. A scree slope consists of small rocks and gravel, which have collected below a rock ridge or cliff. The size of the rocks varies from sand sized to pieces about the size of a man's fist.

   d. Talus. A talus slope is similar to a scree slope except that the pieces of rock are fist sized or larger.

TRANSITION: Now that we have identified the four types of slopes, are any questions? Let's discuss the techniques to be used when ascending and descending each type.

4. (5 Min) **HARD GROUND**

   a. Ascending.

      (1) The knees should be locked on every step in order to rest the leg muscles. This is particularly important on long steep slopes when carrying a heavy pack.

(2) If the slope is gentle you may be able to walk straight up it.

(3) Steep slopes should be traversed rather than climbed straight up. When we use the term "traverse" we, mean that you zigzag up the slope, moving in a combination vertically and horizontally.

(a) When turning at the end of a traverse, always step off in the new direction with the uphill foot. This prevents crossing the feet and a possible loss of your balance.

(b) While traversing, roll your ankles away from the hill to maintain full sole contact.

(4) If the terrain is narrow enough to make traversing impractical, the French Technique can be utilized.

b. Descending.

(1) It is usually easiest to come straight down a hard ground slope without traversing.

(2) Keep your back straight and your knees bent so that they act as shock absorbers for each step.

5. (5 Min) **GRASSY SLOPES**

a. Ascending. Step on the uphill side of each clump or mound of grass, as the ground will tend to be more level in these spots.

b. Descending. It is best to traverse when descending a grassy slope due to the uneven nature of the ground. It is easy to build up too much speed and trip if one attempts to go straight down.

(1) The Hop-Skip Step can be used on this type of slope. In this technique, the lower leg takes all of the weight and the upper leg is used for balance only.

6. (5 Min) **SCREE SLOPES**

a. Ascending.

(1) Avoid ascending scree slopes whenever possible, as they are very tiring and difficult to climb.

(2) When ascending is necessary, hard ground principles apply with the addition that each step must be picked carefully and placed slowly so that the foot will not slide down when weight is placed on it.

(a) Kick in the toe of the upper foot so that a step is formed in the scree.

(b) After determining that the step is stable, carefully transfer weight from the lower foot to the upper foot, and repeat the process.

b. Descending.

(1) Never run, as this may cause you to lose control, resulting in possible injury to yourself or others.

(2) Utilizing the plunge step is the best method to come straight down a scree slope.

7. (5 Min) **TALUS SLOPE**

a. Ascending. Always step on the top of and uphill side of the rocks whether ascending, descending, or traversing. This prevents the rocks from tilting and rolling downhill.

b. Descending. When descending, you must step on the top and uphill side of the rocks. (Same technique as ascending a talus slope).

(1) When a group is descending together, they should be as close to each other as possible (approximately one arm's length apart) and one behind the another, to prevent possible injury caused by the momentum of a dislodged rock.

TRANSITION: Are there any questions? If there none for me, the have some for you.

OPPORTUNITY FOR QUESTIONS (3 Min)

1. QUESTIONS FROM THE CLASS

2. QUESTIONS TO THE CLASS

Q. What are the four fundamental principles for movement in mountainous terrain?

A. (1) Body weight centered.
(2) Utilize micro terrain.
(3) Step over obstacles.
(4) Make use of the rest step.

Q. What are the four types of slopes?

A. (1) Hard ground
(2) Grassy Slope
(3) Talus
(4) Scree

SUMMARY (2 Min)

a. During this period of instruction we have discussed the principles for movement in mountainous terrain.

b. Those of you with IRF's please fill them out at this time and turn them in to the instructor We will now take a short break.

UNITED STATES MARINE CORPS
Mountain Warfare Training Center
Bridgeport, California 93517-5001

SML
SMO
ACC
02/11/02

LESSON PLAN

MOUNTAIN CASUALTY EVACUATIONS (SUMMER)

INTRODUCTION

(5 Min)

1. **GAIN ATTENTION.** Harsh terrain and adverse weather conditions are inherent to a mountainous environment. Under these conditions personnel with relatively minor injuries can become casualties which require evacuation. By using the techniques and equipment outlined here, these casualties can be evacuated safely and efficiently.

2. **OVERVIEW.** The purpose of this period of instruction is to introduce the student to the techniques used in evacuating casualties in a mountainous environment. This will be accomplished by discussing the general considerations, one and two man carries, expedient l litters, regular litters, pre-rigging litters, and cliff evacuations. This lesson relates to conducting a cliff assault. (SMO & ACC)

INSTRUCTOR NOTE: Have students read learning objectives.

3. **INTRODUCE LEARNING OBJECTIVES**

   a. TERMINAL LEARNING OBJECTIVES. In a summer mountainous environment and given a simulated/actual casualty, evacuate casualties, in accordance with the references.

   b. ENABLING LEARNING OBJECTIVES. (SML) and (ACC)

      (1) Without the aid of references, list in writing the general considerations for Casevac procedures, in accordance with the references.

      (2) Without the aid of references, construct an expedient litter, in accordance with the references.

      (3) SQUAD: In a summer mountainous environment, and without the aid of references, secure a simulated/actual casualty in a SKED litter, in accordance with the references.

      (4) SQUAD: In a summer mountainous environment, and without the aid of references,

attach a belay line to a SKED litter for movement up/down moderate slopes, in accordance with the references.

(5) Without the aid of references, list in writing the site selection considerations for lowering a victim, in accordance with the references.

(6) SQUAD: In a summer mountainous environment, without the aid of references, and given a simulated/actual casualty, conduct a CASEVAC using the barrow-boy method, in accordance with the references.

c. ENABLING LEARNING OBJECTIVES (SMO)

(1) With the aid of references, state orally the general considerations for Casevac procedures, in accordance with the references.

(2) In a summer mountainous environment, and with the aid of references, construct an expedient litter, in accordance with the references.

(3) SQUAD: In a summer mountainous environment, and with the aid of references, secure a simulated/actual casualty in a SKED litter, in accordance with the references

(4) SQUAD: In a summer mountainous environment, and with the aid of references, attach a belay line to a SKED litter for movement up/down moderate slopes, in accordance with the references.

(5) With the aid of references, state orally the site selection considerations for lowering a victim in accordance with the references.

(6) SQUAD: In a summer mountainous environment, with the aid of references, and given a simulated/actual casualty, conduct a CASEVAC using the barrow-boy method, in accordance with the references.

4. **METHOD/MEDIA.** The material in this lesson will be presented by lecture and demonstration method. You will practice what you have learned in upcoming field training exercises. Those of you with IRF's please fill them out at the end of this period of instruction.

5. **EVALUATION**

a. SML - You will be tested later in the course by written and performance evaluations on this period of instruction.

b. SMO - You will be evaluated by an oral exam and a performance exam on T- 11.

TRANSITION: Are there any questions over the purpose, learning objectives, how the class will be taught, or how you will be evaluated? Any disabled Marine requiring Casevac must be treated with basic first aid, along with considerations in conducting the actual evacuation properly.

BODY (60 Min)

1. (5 Min) **GENERAL CONSIDERATIONS** The general considerations are a set of guidelines that can be used no matter how serious the casualty is. They are remembered by a simple acronym, **A PASS NGG.**

    a. Apply Essential First Aid. (i.e. splints, pressure bandage, etc.)

    b. Protect the Patient from the Elements. Provide the casualty with proper insulation from the ground. Ensure that he is warm and dry. If there are any natural hazards (i.e. rock fall, lighting, etc.) either move the casualty as quickly as possible or ensure that he is well protected.

    c. Avoid Unnecessary Handling of Patient.

    d. Select Easiest Route. Send scouts ahead if possible, to break trails.

    e. Set Up Relay Points and Warming Station. If the route is long and arduous, set up relay points and warming stations with minimum amount of medical personnel at warming stations to:

        (1) Permit emergency treatment. Treat for shock, hemorrhage, or other conditions that may arise.

        (2) Reevaluate the patient constantly. If patient develops increased signs of shock or other symptoms during the evacuation, he may be retained at an emergency station until stable.

    f. Normal Litter Teams Must Be Augmented in Arduous Terrain.

    g. Give Litter Teams Specific Goals. This litter teams job is extremely tiring, both physically and mentally. The litter teams must be given realistic goals to work towards.

    h. Gear. Ensure all of the patient's gear is kept with him throughout the evacuation.

TRANSITION: Now that we have covered the considerations, are there any questions? The basic carries that are taught in the EST manual (fireman's carry, two-hand, four-hand, and the poncho litters) are still viable of transporting an injured man; however, we need to know a few more that can aid us during a medevac.

2. (10 Min) **EXPEDIENT LITTERS**. There are five types of expedient litters that we will talk about. They are the sling rope carry, rope coil carry, pole carry, the alpine basket and the

poncho litter.

a. Sling Rope Carry. The sling rope carry requires two men and a 15-foot sling rope. One as the bearer, and an assistant to help in securing the casualty to the bearer. Conscious or unconscious casualties my be transported in this manner:

(1) Bearer kneels on all fours and the assistant places casualty face down on bearer's back ensuring the casualty's armpits are even with the bearer's shoulders.

(2) He then finds the middle of the sling rope and places it between the casualty's shoulders and the ends of the sling rope are run under the casualties armpits, crossed and over the bearer's shoulders and under his arms.

(3) Then the ropes are run between the casualty's legs, around his thighs, and tied with a square knot with two overhands just above the bearer's belt buckle.

(4) Ensure the rope is tight. Padding, when available, should be placed where the rope passes over the bearer's shoulders and under the casualties thighs.

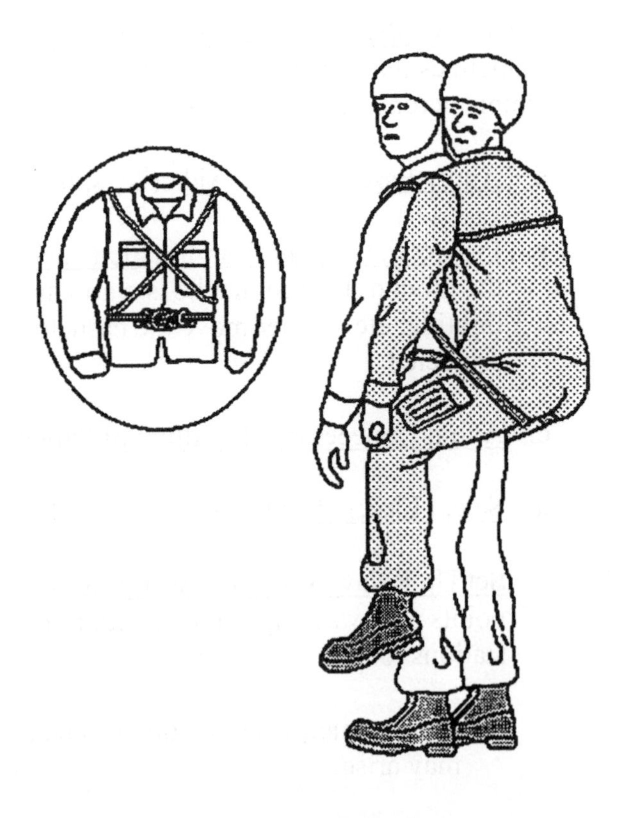

SLING ROPE CARRY

b. The Rope Coil Carry. This requires a bearer and a rope coil. It can be used to carry a conscious or unconscious casualty.

(1) Position the casualty on his back.

(2) Separate the loops of the mountain coil into two approximately equal groups.

(3) Slip ½ of the coil over the casualty's left leg and ½ over his right leg so that the wraps holding the coil are in the casualty's crotch, the loops extending upward the armpits.

(4) The bearer lies on his back between the casualty's leg and slips his arms the loops. He then moves forward until the coil is extended. When using the rope coil the bearer ties the coil to himself vice slipping his arms through the loops.

ROPE COIL CARRY

(5) Grasping the casualty's right or left arm, the bearer rolls over, rolling to the casualty's uninjured side, pulling casualty onto the bearer's back.

(6) Holding the casualty's wrists, the bearer carefully stands, using his legs to lift up and keeping his back as straight as possible.

NOTE: The length of the coils on the rope coil and the height of the bearer are to be considered. If the coils are too long and the bearer happens to be a shorter person, it will require the coils to be uncoiled and shortened. If this is not done, then the casualty will hang too low on the bearer's back and make it a very cumbersome evacuation. A sling rope harness can be used around the victim's back and the bearer's chest, which will free the bearer's hands.

c. Pole Carry. The pole carry method is a field expedient method and should be considered as a last resort only, when narrow ledges must be traversed; vegetation limits the bearers to a narrow trail. This method is difficult for the bearers and uncomfortable for the casualty. Two bearers, four sling ropes and a 12 foot pole - 3 inches in diameter, are required for this carry.

(1) The casualty is placed on his back in a sleeping bag or wrapped in a poncho or blanket, then placed on an insulated pad.

(2) One sling rope is placed under the casualty below the armpits and tied with a square knot across the casualty's chest.

(3) The second sling rope is tied in the same manner at the casualty's waist.

(4) The third sling rope is placed at the casualty's legs below the knee.

(5) The fourth sling rope is tied around the ankles.

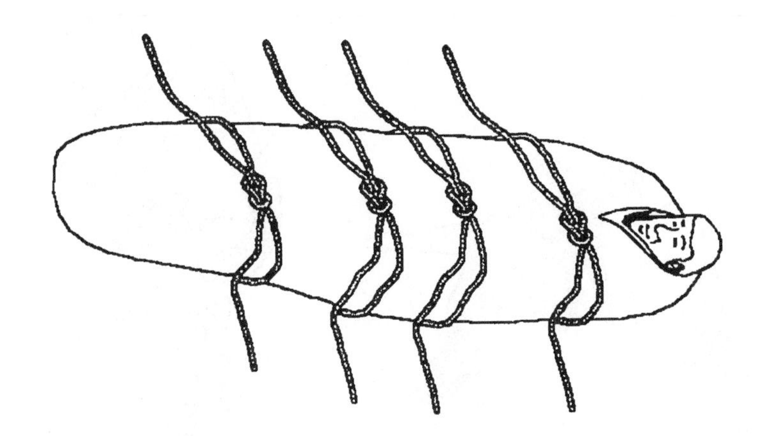

(6) The pole is placed along the casualty's length and secured using square knots with two overhands with the ends of the sling ropes. The square knots should be so tight that the overhands are tied onto themselves

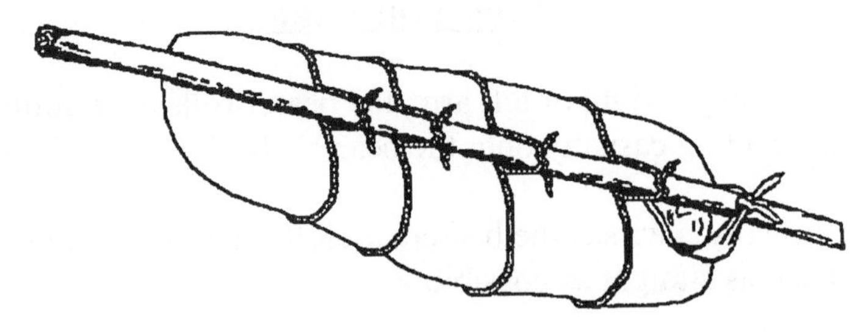

(7) The casualty should hang below the pole, as close to the pole as possible, to prevent swinging during movement.

(8) The casualty's head may be supported using a triangular bandage or a cartridge belt passed around the pole.

(9) For additional support and of movement, two additional bearers may be required, as well as a mountain coil.

(a) Mountain coil is split into two equal coils.

(b) Place knot of mountain coil under casualty's lower back.

(c) Additional bearers slip into each half of the hasty coil one on each side of casualty, aiding in support and movement of the casualty.

d. Alpine Basket. To belay the alpine basket, the pre-rigs are attached to the bights formed coming through the loops.

(1) If barrow boy is to be used, the procedure previously discussed will be adhered to.

(2) If barrow boy is not used, then a tag line from the bottom must be implemented to keep the casualty away from the cliff face on the decent.

(3) Construction of the Alpine Basket:

(a) Start by making the same amount of bights as in the rope litter, but start from one end and tie a figure-of-eight loop to run the first bight through.

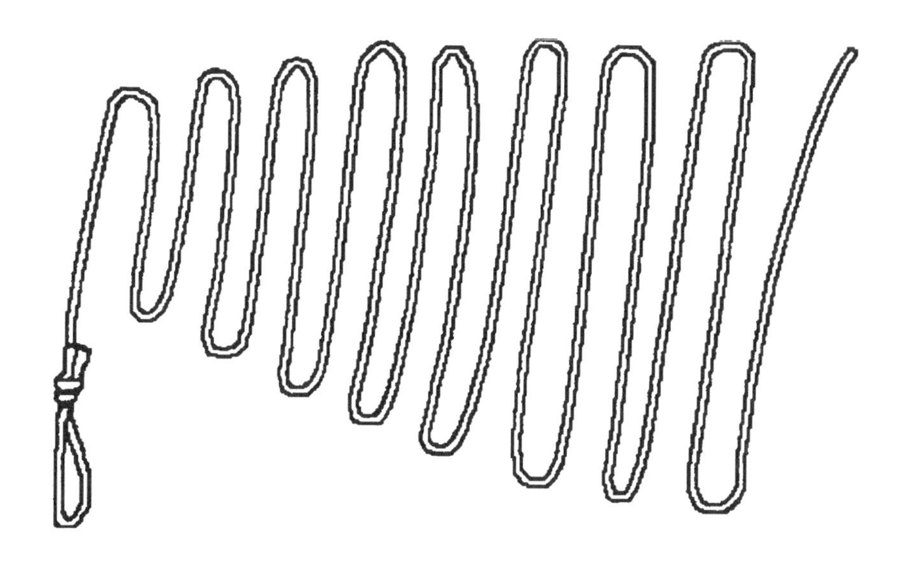

(b) Place padding i.e., isopor mat, on top of the bights and then lay the casualty on the padding and bights.

(c) Start at the casualty's feet and pull the first bight up around the casualty's ankles and through the figure-of-eight loop tied into the starting end of the rope.

(d) Go to the opposite side of the casualty and pull up the second bight and pull it through the loop formed by the bight that was pulled through the figure-of-eight.

(e) Continue until you get to the casualty's armpits, bring the second to the last bight up over the casualty's shoulder and into the bight and then bring the last bight up over the casualty's other shoulder and into the last bight formed.

(f) Secure the last bight with a round turn and two half hitches leaving a big enough bight to tie a figure eight at the end.

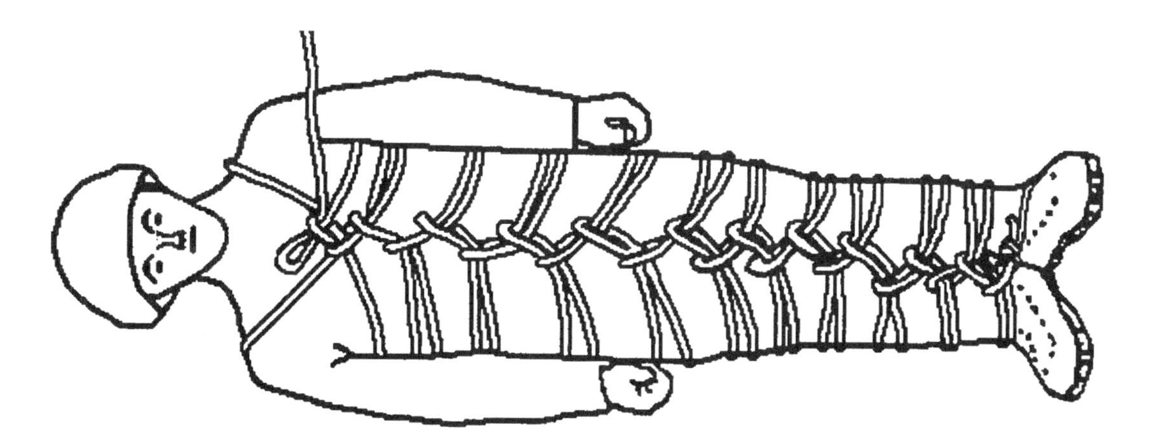

THE ALPINE BASKET

e. Poncho Litter. A poncho, poncho liner, bivy bag or similar piece of material may be used. In addition you will need six individuals with sling ropes.

(1) Lay poncho litter flat on the ground.

(2) Select six rocks about the size of a golf ball. Place one rock in each of the corners and one in the middle on each side in the middle of the litter. The rocks are placed on the under side of the poncho or like material. If a bivy bag is used the casualty should be zipped inside the bivy bag. The rocks should then be arranged in the same manner only on the inside below the zipper.

(3) Tie the sling rope together with an overhand knot. Take the middle of the rope and secure it around the rock with a clove hitch.

(4) An isomat may be laid on the poncho to help make the litter firmer.

(5) The casualty is then placed in the litter. The sling ropes are adjusted by feeding the pigtails of the over hand knot through itself to adjust for length. The loop is then put over the inboard shoulder of the carriers. Insure that the casualty is carried level.

TRANSITION: Are there any questions over the expedient litters? For casualties with more serious injuries, or causalities that may occur in moderate to vertical terrain a more ridged litter may need to be employed.

3. (10 Min) **LITTERS**. There are two kinds of litters that are used for casualty evacuations in moderate to vertical terrain. The SKED litter and the Stokes litter. Each has a variation of procedures for securing a casualty and rigging the litters for either raising or lowering. Let's first look at the SKED litter.

a. SKED Litter. The SKED litter is constructed of thin foam padding, straps and grommets.

(1) Securing a Casualty to the SKED

(a) First unroll the litter. The litter must be re-rolled the opposite way to allow the litter to lay flat. Then lay the litter next to the casualty.

(b) If the casualty has any possibility of a spinal injury the Oregon Spine Splint must be used. Secure the splint to the casualty by use of the color-coded buckles. Experienced medical personal are recommended if spinal immobilization is necessary.

(c) Once the casualty is on the SKED use the four body straps to secure the casualty to the litter. Unless injuries prevent, the casualty's arms should be at his sides to prevent further injuries to himself or the rescuers.

(d) Once the casualty is secured with the body straps the feet straps must be secured. The feet straps are secured last to ensure the casualty is in the proper position on the SKED. The feet can be positioned in three ways. The first position is feet together with the straps running on the outside of the feet. The second position is feet apart with the straps running on the inside of the feet. The last position is the feet stacked. This is the most uncomfortable position and not recommended for casualties with possible spinal cord injuries. This position is formed by placing the heel of one foot on top of the toes of the other. This position will only be used for a casualty in confined spaces. Once the feet are positioned the feet straps must be secured. Start by bending the feet end of the SKED to form a plat form for the feet. Then loop the feet straps through the second grommets on each side.

(e) The last thing to do is to form the head end to protect the casualty's head. If possible the casualty should wear a helmet. Form the head end tying the pull strap up and secure it to the first body straps.

(2) Casevacing a Casualty. There are many ways to move a casualty once in the SKED. However the medevac team must keep the general considerations in mind. The two methods that we will talk about next are the simplest and require the least amount of additional rigging

(a) The first way is to drag the casualty by the drag strap located at the head end of the SKED. You can also use the SKED's carrying bag as a harness in conjunction with the pull strap and towing harness. If additional people are required, cordage can be added to the pull strap or the front carrying handles.

(b) The second way is to carry the casualty using the carrying handles. By using the set of four removable webbing the litter team can be augmented. To do this each pieces of webbing is tied to make an endless loop. Then pass a bight of the loop through one of the grommets to create additional handles.

(3) Rigging the SKED for Vertical Terrain and Helicopter Lift. In this type of terrain special requirements must be taken to raise or lower a casualty in the SKED.

(a) Rigging the SKED for a Vertical Employment. A vertical raise or lower is used when moving a casualty on steep earth to avoid any further injury to the casualty. On vertical terrain the vertical raise or lower can be used if the terrain is not uniformed or there is a chance of rock fall. Ensure that the casualty's head is always above his feet.

1. Identify the 30-foot piece of cordage that comes with the SKED. Then tie a figure eight in the middle of the rope. If the rope is worn or missing, the same process can be done with two sling ropes.

2. Next pass each end of the rope through the grommets at the head of the SKED. Leaving approximately 1-2 feet of rope between the stretcher and the knot.

3. Continue to feed each end through the grommets and the carrying handles

towards the foot end of the SKED. Pass the ends of the rope through the last grommets at the foot end and secure the two ends with a square knot without over hands.

4. Bring the pigtails up and over the casualty's feet and pass the ends through the carrying handles towards the middle of the casualty. Then tie a square knot with two over hands.

(4) Rigging the SKED for a Horizontal Employment. A horizontal raise or lower is preferred on uniformed vertical terrain. The horizontal employment allows the rescuer to assist the casualty on either a raise or lower. It also allows the rescuer the ability to monitor the casualty's condition and can easily treat the casualty if the need arises.

(a) Identify the two 4 inch nylon straps. They should be two lengths; one four inches shorter than the other. The shorter strap should be marked HEAD STRAPS.

(b) Insert one end of the head strap into a slot near the head of the litter. Then wrap the rest of the straps under the SKED and pass the other end through the opposing slot. Do the same at the foot end of the SKED with the other strap. Ensure that the strap runs smoothly under the SKED.

(c) Connect the strap ends with the large locking carabiner that comes with the SKED. If the carabiner is worn or missing, opposing stubai 85 locking carabiners will suffice.

b. Stokes Litter. The stokes litter is a litter that is constructed of metal tubing with a plastic covering. The litter is formed in a rectangular basket shape with mesh attached to the frame. Using the stokes for an evacuation (as with any evacuation) it should be padded for the casualty.

(1) Securing the casualty to the stokes litter. In the event that the "seat belts" are missing from the stokes litter, sling ropes can be used to the lash the casualty. The steps involved are:

(a) Tie two sling ropes together using square knots and two overhands.

(b) Tie a stirrup hitch around ankles and feet, feed the two pigtails through the right angles of the stokes. Do not cross the ropes at the ankles.

(c) Lace the sling rope towards the casualty's head by passing the rope through the right angles (not over the top of the rails of the stokes).

(d) Secure the ends of the sling ropes by tying a clove hitch with two half hitches on the thick vertical bar located by the victim's shoulder.

TRANSITION: Know that we have talked about the two different litters, let us now discuss how to attach a belay line to the litter.

6. (5 Min) **YOSEMITE PRE-RIG**. The title of this method, pre-rig implies the meaning ready to use, but in this case the rig must be constructed. Its purpose is for attaching a belay line to a litter and to make the litter easily adjusted. This is the one method that is used to secure a litter to a belay line.

   a. Construction. This method normally requires four sling ropes. The steps are: (1) Using one sling rope, tie a figure-of-eight loop with one tail.

      (1) Take the remaining tail and run it through the window of the stokes litter or in the stirrups of the collapsible litter.

      (2) Tie a kragur knot onto the same sling rope.

      (3) Repeat steps (1), (2), and (3) with the three remaining sling ropes.

      (4) Suspend the litter to ensure that the comers are balanced.

TRANSITION: Now, let's cover the ways to make descents and the belaying methods.

7. (10 Min) **ASCENTS OR DESCENT OVER STEEP TO MODERATE SLOPES.** When the litter team is ascending or descending a slope they must consider the potential for further injury to the casualty or to themselves. If the risk of injury is high a belay line may be used to prevent injury to the casualty and the rescuers.

   a. Preparing Casualty for Ascents or Decent over Steep to Moderate Terrain. This procedure will be depending on several things. Initially, site selection should contain the following features.

      (1) Suitable anchor points.

      (2) Clearance for casualty along the route

      (3) Loading and unloading points.

   b. Additional considerations.

      (1) The casualty will always be rigged for vertical employment when on steep to moderate terrain.

      (2) The smoothest possible route must be selected.

      (3) Ensure that the casualty's head is above his feet.

c. Rescuers positions. There are two methods that can be used for the rescuers for moving a casualty in steep to moderate terrain.

(1) Two to four men will position themselves on each side of the litter. They can then carry the litter by the carrying handles. In steep terrain a second belay line may be used to assist the rescuers. We will discuses the belay line later in the chapter.

(2) The Caterpillar method will require as many personnel as possible. The personnel will split in half and position themselves on each side of the litter forming a tunnel. As the litter is raised or lowered each member will hand the litter to the next member in the tunnel. As the litter passes each person in the tunnel he will peel off and assume the lead either at the top or bottom of the tunnel. This will continue until the litter reaches its desired destination.

d. Belay Line. For belaying of a casualty, one rope will be used and from the top using one of two methods depending on the application.

(1) Body Belay. This method should only be used over moderate terrain. The belay man will establish a sitting position behind a suitable anchor (i.e., rock, tree, etc.) and pass the standing end of the rope behind his back. The running end of the rope will feed out from the belay man's right side. A figure of eight loop is tied to the end of the running end of the rope. It is then attached to the litter's figure eight loop with a locking carabiner. The belay man will then remove all of the slack between himself and the litter. The standing end of the rope should be stacked on the belay man's left side and run through his left side. As the casualty is lowered, the belay man will feed the rope from behind his back allowing it to run through his right hand. If the belay man needs to stop the casualty, he will clench the rope in his left hand, and bring the rope to the center of his chest.

(2) Direct Belay. This method is the safest for either raising or lowering a casualty in either moderate to steep terrain.

(a) To Lower a Casualty. First a swami wrap will be tied around a suitable anchor point. Two locking carabiners will be clipped into all of the wraps of the swami wraps, gates up. A figure of eight loop is tied into the end of the static rope and attached to the litter with a locking carabiner. After all the slack has been taken up between the litter and the anchor, the rope must be tied through an appropriate belay device. The belay device is attached to the anchor through one of the two locking carabiners on the anchor. A safety (French) prussic will be tied to the running end of the rope and clipped into the second locking carabiner on the anchor. While the casualty is being lowered, one person will control the rope running through the belay device. The safety prussic will be controlled by a second person. Should the primary belay man lose control, the person operating the safety prussic simply lets go and the prussic will bind onto the rope, stopping the casualty.

(b) To Raise a Casualty. The anchors will be established in the same manner as discussed in lowering the casualty with one minor change. The one change is that instead of running the rope through a belay device, the rope will only run through a locking carabiner. The load will be raised by the use of a mule team. The mule must consist of as many people as possible. The mule team will raise the load in as straight a line from the anchor as possible. If the space does not permit a ninety-degree angle away from the anchor is also an option. The mule team will walk backward until the last man reaches his limit of advance. Once he reaches that limit he will peel off the end and return to the front of the mule team. This process is continued until the casualty reaches the top. If the load becomes unmanageable, the safety prussic will be allowed to bind on the rope while the mule team repositions themselves. If the person operating the safety prussic can not see the casualty a Point NCO will be in charge of communicating with the mule team.

(c) To Belay the Rescuers. If the route is too steep or the footing is poor the rescuers may need some assistance either on the raising or lowering of a casualty. If this is the case a separate belay line will be established for the rescuers. The anchor and the belay line are established in the same manner. The same anchor can be used if it is suitable for the load. The rescuers will then tie either around the chest bowlines or swami wraps. A figure eight loop will be tied into the end of the static rope and connected to the bottom rescuer with a locking carabiner. The other rescuers will connect themselves to the same rope with middle of the line prussic. They will be connected in this manner so that they can adjust their position to the casualty.

8. (7 Min) **BARROW BOY**. A barrow boy is no more than an assistant to the litter on vertical to near vertical cliff faces. The barrow boy can be used for either the Stokes or the SKED litters. For this situation the Stokes litter should only be used in the horizontal position. However the SKED can be employed in either the horizontal or the vertical positions.

NOTE: For safety purpose at MWTC, two ropes will be utilized.

a. Rigging the Barrow Boy.

(1) First the rescuer must ensure that a suitable anchor has been established, a proper belay has been constructed, and that a safety prussic has been constructed.

(2) Then the rescuer must ensure that if an A-frame is used that it has been constructed and anchored properly.

(3) Next the rescuer will tie a rappel seat on. (a sit harness can also be used) Then he will ensure that an around the body bowline is tied onto the casualty. A figure eight will be tied on to the end to act as the casualty's safety.

(4) Then after running the running end of the rope though the carabiners or pulley of the A-frame he will tie a figure eight loop at the end of the static rope. He will then

attach the figure eight loop to his hard point with a locking carabiner. Then he will tie a middle of the line prussic above the figure eight and attach it to the same locking carabiner in his hard point. This is called the adjustment prussic. It is used to adjust the position of the Barrow Boy in relation to the litter. Next he will take six to eight feet of slack from the end of the line figure eight and tie a directional figure of eight with direction of pull down. (Note: the prussic should be between the end of the line figure of eight and the directional figure of eight.) The directional figure of eight is the attaching point for the litter and the casualty's safety.

(5) Once the rescuer and the litter are secured, the belay man must take all the slack out of the system. The rescuers will the maneuver the litter through the apex of the A-frame with the help of the point NCO.

(6) Once onto the cliff face the rescuer will then position him self with his adjustment prussic so that he can be of most assistance to the litter on the raise or lower. The rescuer will pull the litter out away from the cliff face so that the casualty rides smoothly up or down the cliff face.

(7) The Point NCO will be in charge of the belay men or the mule team. He will also communicate with the rescuer about the rate of speed, if the rescuer need to be stopped along the route, and when he reaches the top or bottom of the cliff.

9. (5 Min) **TANDEM LOWERING**. The tandem lowering system can be used for the walking wounded, POW's, or more serious casualties when situation would not permit using the barrow boy.

NOTE: For safety purposes at MWTC, two ropes will be utilized.

a. The assistant to the casualty should first tie a rappel seat on himself and then assist the casualty with his.

b. The assistant will take two belay lines and tie a end of the line figure 8 loop, and clip this into his rappel seat.

c. A directional figure 8 will be tied approximately 12 inches up the rope from the figure 8 loop (with the loop pointed down) and this will be clipped to the casualty's rappel seat.

d. If needed, adjustment prussic cords should be tied above the casualty's directional figure 8. These adjustment cords should be attached in the same manner as the first man down a rappel.

e. The casualty and the assistant will lower as one, with the assistant helping on the way down the cliff.

10. (3 Min) **OTHER CONSIDERATIONS**. All of the techniques we have discussed for the evacuation of a casualty from top to bottom can also be used on a suspension traverse or rope

bridge, with a slight variation in the belay line. Two belay lines may be used for rope bridges and the suspension traverse, if they are available. No matter what type of litter is used, the individuals involved in the evacuation must ensure that the head is always uphill or not lower than the feet.

TRANSITION: What we have just covered are other considerations for Casevac, are there any questions.

PRACTICE (CONC)

a. Students will evacuate a simulated/actual casualty.

PROVIDE HELP (CONC)

a. The instructors will assist the students when necessary.

OPPORTUNITY FOR QUESTIONS (3 Min)

1. QUESTIONS FROM THE CLASS

2. QUESTIONS TO THE CLASS

Q. What are the eight considerations when conducting a mountain casualty evacuation?

A. (1) Apply essential first aid.
(2) Protect the patient from the elements.
(3) Avoid unnecessary handling of patient.
(4) Select easiest route.
(5) Set up relay points and warming stations.
(6) Normal litter teams specific goals to work towards.
(7) Give litter teams specific goals to work towards.
(8) Keep gear with casualty.

Q. What are the four site selections for lowering a CASEVAC victim?

A. (1) Suitable anchor points.
(2) Suitable Loading and unloading platforms.
(3) Clearance for casualty along the route.
(4) Anchor points for A-Frame, if used.

SUMMARY (2 Min)

a. During this period of instruction we have covered some of the general considerations to include expedient litters, collapsible litters, stokes litter and the barrow-boy.

b. Those of you with IRF's please fill them out at this time and turn them in to the instructor. We will now take a short break.

UNITED STATES MARINE CORPS
Mountain Warfare Training Center
Bridgeport, California 93517-5001

SML
SMO
ACC
03/31/01

LESSON PLAN

NATURAL AND ARTIFICIAL ANCHORS

INTRODUCTION

(5 Min)

1. **GAIN ATTENTION**. All military mountaineering activities that utilize ropes require that the ropes be anchored (secured) to something. If you wish to utilize a rope installation safely and in the correct manner, you must know what type of anchors are available for your use. At the conclusion of this period of instruction you will be able to identify and construct a suitable anchor.

2. **OVERVIEW**. The purpose of this period of instruction is to introduce the students to the various methods used to anchor a rope to both natural and artificial devices. This lesson relates to all other lessons utilizing anchors.

INSTRUCTOR NOTE: Have the students read learning objectives.

3. **INTRODUCE LEARNING OBJECTIVES**.

   a. TERMINAL LEARNING OBJECTIVE. In a summer mountainous environment, establish an anchor system in accordance with the references.

   b. ENABLING LEARNING OBJECTIVES. (SML) and (ACC)

      (1) Without the aid of references, list in writing the types of anchors, in accordance with the references.

      (2) Without the aid of references, describe in writing how many pieces of protection are used for an artificial anchor when installing a two rope, high-tension system, in accordance with the references.

      (3) Without the aid of references, describe in writing how many pieces of artificial protection are used in constructing a two-rope high-tension installation, in accordance with the references.

(4) In a summer mountainous environment, establish a natural anchor point, in accordance with the references.

(5) In a summer mountainous environment, establish an artificial anchor system, in accordance with references.

c. ENABLING LEARNING OBJECTIVES. (SMO)

(1) With the aid of references, list orally the types of anchors, in accordance with the references.

(2) With the aid of references, list orally how many pieces of protection are used for an artificial anchor when installing a two rope, high tension system, in accordance with the references.

(3) With the aid of references, list orally how many pieces of protection are used for artificial anchors when installing a one rope, non-high tension system, in accordance with the references.

(4) In a summer mountainous environment, establish a natural anchor point, in accordance with the references.

(5) In a summer mountainous environment, establish an artificial anchor system, in accordance with references.

4. **METHOD/MEDIA**. The material in this lesson will be presented by lecture and demonstration. Those of you with IRF's please fill them out at the end of this period of instruction.

5. **EVALUATION**.

a. MLC - You will be tested later in the course by written and performance evaluations on this period of instruction.

b. ACC - You will be tested later in the course by written and performance evaluations on this period of instruction.

c. SMO - You will be evaluated by an oral and performance evaluation later on.

TRANSITION: Are there any questions on the learning objectives, how I will be presenting this period of instruction, or how and when you will be evaluated. Let's now discuss the two types of anchors.

BODY (45 Min)

1. (5 Min) **TYPES OF ANCHORS**. There are two types of anchors that we use. These two types are:

   a. Natural

   b. Artificial

TRANSITION: We will now take a look at considerations for anchors.

2. (5 Min) **GENERAL CONSIDERATIONS FOR ANCHORS**. Anytime you employ natural or artificial anchors; there are special considerations that you must apply. These considerations apply both to the anchor itself and to the material being used to build the anchor. Some examples of these considerations are:

   a. Whether using natural or artificial anchors, the installing unit must insure that the anchor is suitable for the load.

   b. The anchor position must be relative to the direction of pull on the anchor.

   c. The angle between the anchor points should not exceed 90 degrees. This is to ensure that no added stress is put upon the anchors, as well as the equipment being used to construct the anchor.

      (1) To decrease the angle between anchor points, materials (i.e. sling ropes, web runners, prusik cord, etc.) could be used to extend the anchor which will decrease the angle between anchor points.

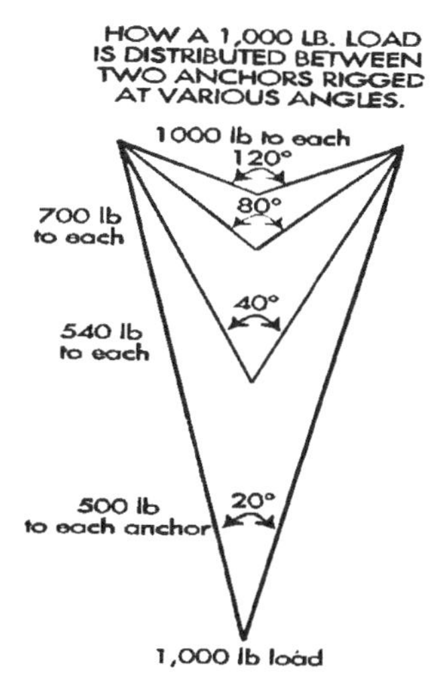

TRANSITION: We have just talk about considerations for anchors. Are there any questions? Now we will talk about Natural Anchors.

3. (15 Min) **NATURAL ANCHORS**. A natural feature is the preferred type of anchor point. Some examples and considerations are as followed:

a. Types of natural anchors.

(1) Trees.

(a) Select a tree that has not been chopped, burned or is rotten.

(b) The tree should be at least 6" in diameter and strong enough to support the intended load.

(c) Trees growing on rocky terrain should be treated with suspicion, since the roots normally are shallow and spread out along a relatively flat surface.

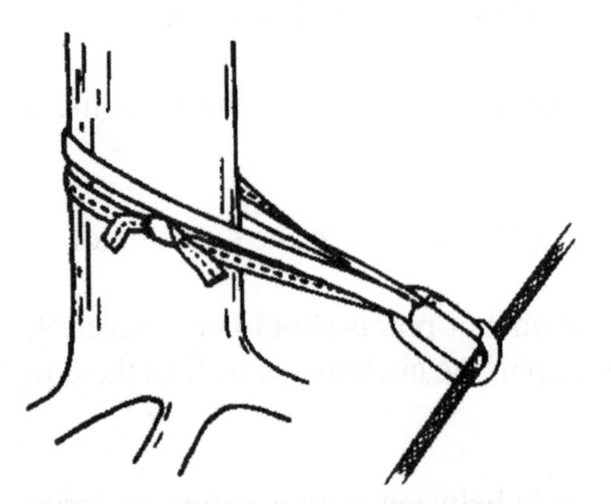

RUNNER USED AROUND A TREE

(2) Shrubs and Bushes.

(a) Select a shrub or bush that is alive and is not brittle, charred or loose.

(b) To avoid leverage, locate the central root and construct the anchor as near to the base as possible.

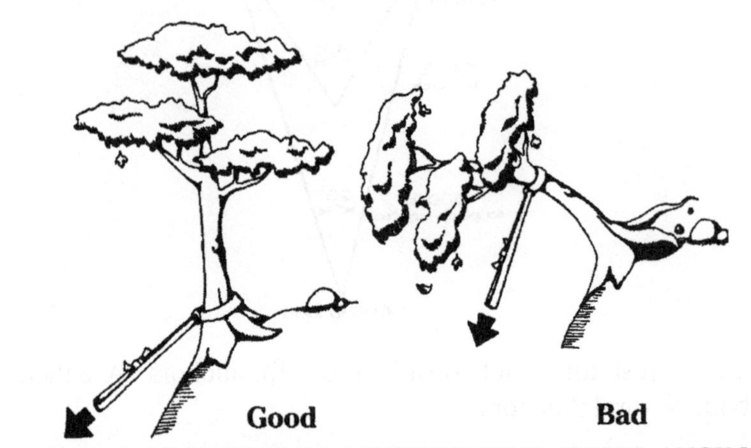

PROPER AND IMPROPER PLACEMENT ON NATURAL ANCHORS

(3) Rocks and Boulders.

(a) Stability is of prime importance when considering a rock or boulder for an anchor. It must be strong enough and secure enough for the intended load.

(b) All surfaces of the rock or boulder should be inspected for any rough or sharp points. These areas must be padded to protect the rope from being abraded or cut.

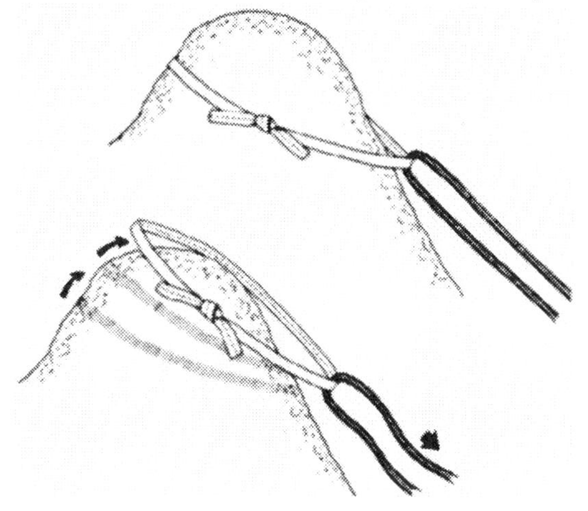

PROPER AND IMPROPER PLACEMENT OF A RUNNER OVER A BOULDER

(4) Spikes and Flakes.

(a) To check stability of a spike or flake, thump it with the heel of your hand. Anything that sounds hallow is suspicious.

(b) They should be checked for cracks or other signs of weathering that may impair their firmness.

(c) Sharp edges must be padded to protect the rope against cuts and abrasion.

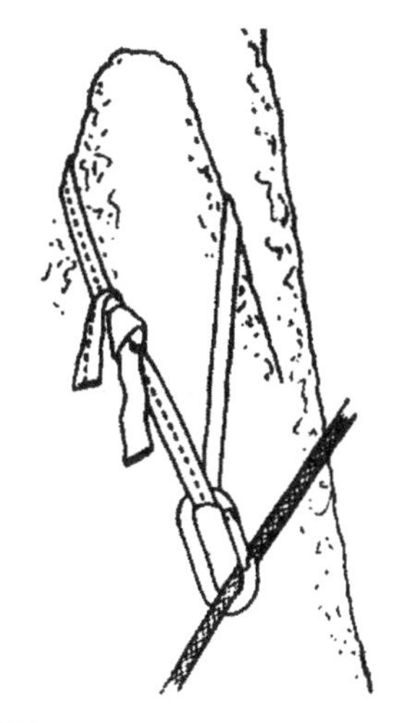

RUNNER OVER A FLAKE

(5) Threads and Chockstones.

(a) A thread is when the rock weathers or cracks to form a hole in the main wall. A chockstone is a rock wedged in a crack.

(b) Check a thread by thumping it with the heel of your hand. Make a visual inspection for cracks and weather. Common sense will prevail for choosing this type of anchor.

(c) When choosing a chockstone, ensure that it has substantial contact with the crack and that the stone's symmetry corresponds with the intended direction of pull.

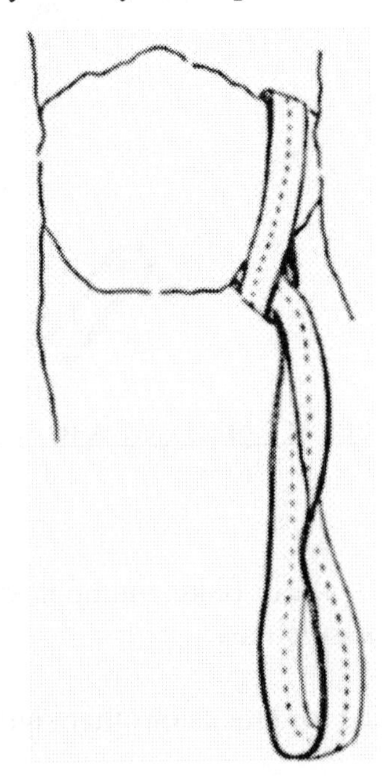

RUNNER IN A GIRTH HITCH OVER A CHOCKSTONE

b. Types of Natural High Tension Anchors. The following are types of Natural High Tension Anchor Systems used to construct installations. These can be tied on any suitable natural anchor point:

(1) Tree Wrap Anchor. The tree wrap is an anchor used to relieve tension on the actual knot itself. This system requires more rope when tied around a large anchor point.

(a) Tie a Figure 8 Loop on the standing end of the rope and wrap the 4-6 times around an anchor point.

(b) Attach a locking carabiner through the knot's loop and clip it onto the running end of the rope.

TREE WRAP

NOTE: If the anchor point to be wrapped is larger then 18' around, then three wraps will suffice.

(2) Swammi Wrap Anchor. When tying the Swammi Wrap, ensure that the joining knot is kept on the side of the anchor point. The Swammi Wrap can be loosened under load if need be. This system can be substituted for a tree wrap when the installation's rope length is an issue.

(a) Select a suitable anchor point and tie a Swammi Wrap around it.

(b) Clip a locking carabiner into as many wraps as possible. This will serve as the attachment point for the anchor system.

(3) Figure of 8 Anchor. The Figure of 8 Anchor is a quick and efficient system. This system cannot be loosened under load. It is tied in the following manner:

(a) Tie a Figure of 8 Loop on the standing end of the rope and wrap the knot around a suitable anchor point. Attach a locking carabiner through the knot's loop and clip it onto the running end of the rope.

(b) Before tensioning the anchor system, adjust the tree wrap so that the running end of the rope runs smoothly through the carabiner towards the direction of pull. This will prevent any lateral tension.

(c) If two ropes are used, the upper rope's carabiner has its gate upwards, and the lower rope's carabiner has it's gate downward.

c. Primary and Secondary Natural Anchor Point System. The two anchor points should be in line with the direction of pull. The primary anchor is the point nearest the running end while the secondary anchor point is directly behind the primary. It is constructed in the following manner:

(1) The primary anchor knot should be an around the object clove hitch. This is chosen for ease of untying the system after tension has been placed on the rope.

(2) The secondary knot is tied around a suitable anchor point ensuring that the rope is taunt between the two anchor points.

(3) Only one anchor point is required when:

(a) Using a tree with a diameter of 12" or more, and it can handle the intended load safely.

(b) When constructing a retrievable system.

NOTE: If constructing an anchor system in which only one natural anchor of less then 12" diameter exists, it must be backed up by two pieces of artificial protection.

TRANSITION: Now that we have discussed natural anchors, are there any questions? Let's move on to artificial anchors.

4. (10 Min) **ARTIFICIAL ANCHOR SYSTEMS**. Any time we use anything other then a natural feature, we are using an artificial anchor point. Artificial anchors can be constructed in the ground, or on the rock itself. The following are artificial anchor systems:

a. Single Timber Deadman. This system is constructed in the ground and it requires considerable time and effort. The steps of it's construction are as follows:

(1) Dig a trench 6 feet long and 3 feet deep and wide enough to work in at a 90 degree angle to the direction of the pull.

SINGLE TIMBER DEADMAN

(2) Dig another trench about 12 inches wide. This trench is dug so that it intersects the main trench at a right angle in the middle. The bottom of this trench should be parallel to the direction of pull and should join the bottom of the main trench.

(3) Take an anchoring device (i.e. log, engineer stakes, bundled up branches etc.) that is strong enough to support the intended load. The anchor is placed into the main trench and covered with dirt with the exception of that part of the anchor that joins the second trench. Stakes approximately 3 feet long should be driven approximately 1-1/2 feet into

the ground between the dead-man and the slanted side of the trench to assist in holding the dead-man in place if the soil is soft.

b. Picket Hold Fast. The picket hold fast is an easier anchor to construct than the deadman, and can be used almost anywhere. The strength of the picket system depends on the pickets and the soil or snow conditions. The picket system can be used for both high and non-tension systems. Construction is as follows:

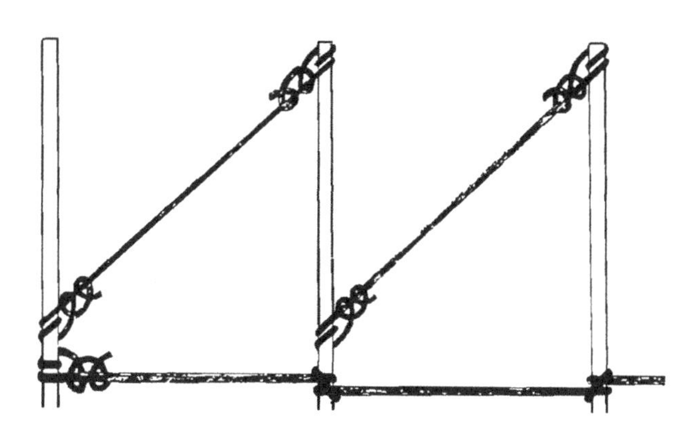

PICKET HOLD FAST

(1) Three stakes (i.e. logs, engineer stakes, etc.) are driven into the ground at a 30 degree angle against the direction of pull.

(2) The line of pickets should be driven in line with the direction of pull. The distance between the pickets can be anywhere from 3-12 feet apart depending on the terrain and soil conditions.

(3) Before tying any rope to the picket anchors, the base of the pickets must be buffed or padded.

(4) To tie a rope to the pickets, go to the furthest picket away from the cliff edge. Tie a round turn and two half hitches at the base of the picket. Take the running end of the rope and tie an over the object clove hitch to the base of the middle picket. Then with the running end of the rope tie an over the object clove hitch to the picket closest to the edge of the cliff face. Ensure that there is tension on the rope in between each picket.

(5) To tie off the pickets to themselves, go to the furthest picket from the cliff edge. And with a sling rope, tie a round turn and two half hitches at the base of that picket. With the running end of that sling rope tie a round turn and two half hitches at the top of the middle picket. Using a second sling rope tie a round turn and two half hitches to the base of the middle picket. Then with the running end of that sling rope tie a round turn and two half hitches to the top of the picket closest to the cliff edge.

c. Equalized Anchor. This system is built with a minimum of three pieces of protection. It can be tied with the standing end of a rope or by utilizing a practice coil.

(1) Construction with the Standing End of the Rope:

(a) Tie a Figure of 8 Loop in the standing end of the rope.

(b) Place a carabiner in each artificial anchor point and attach the knot in either of the outside carabiners.

(c) Clip the rope into the remaining carabiners.

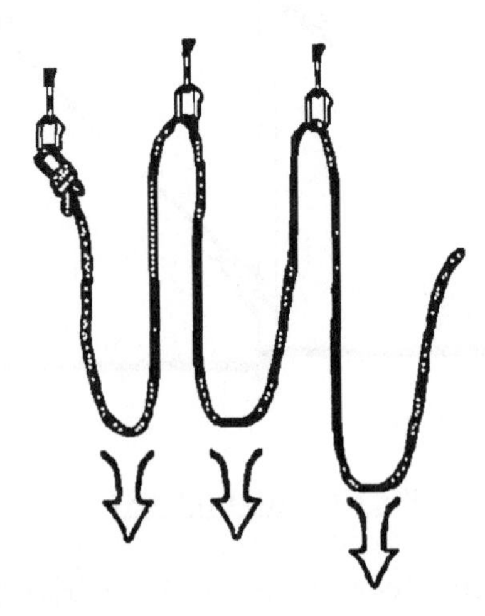

(d) A bight of rope is then pulled down after each carabiner into the anticipated direction of pull. With all three bights, tie an overhand knot around itself to include the running end of the rope.

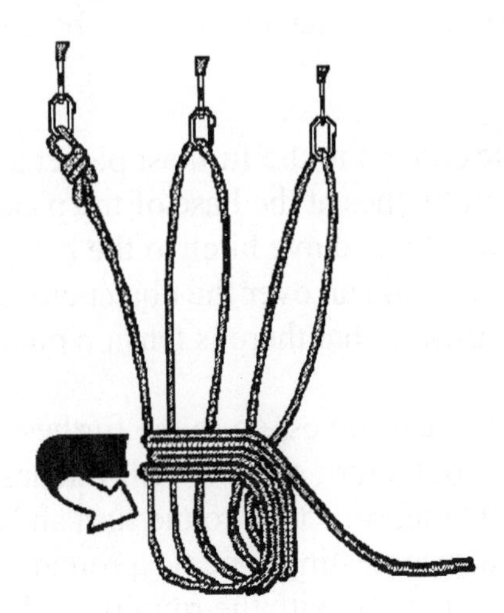

(e) With the running end of the rope coming from the overhand knot, tie a Figure of 8 Loop and attach it to the overhand knot's loop with a locking carabiner.

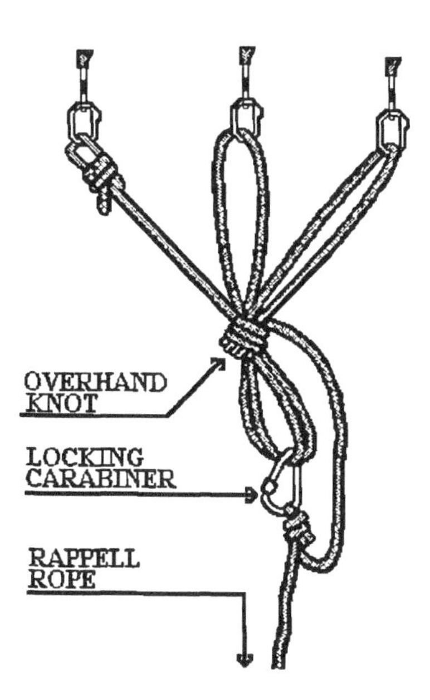

(f) Ensure that there is sufficient slack in the dead rope to prevent a possible shock loading of the system.

(2) Construction with the Cordalette Method:

(a) Create an endless loop with a practice coil and clip it into each of the artificial anchor points.

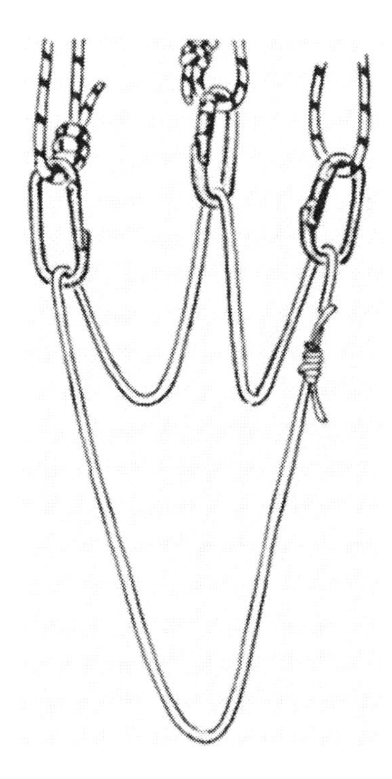

(b) A bight of rope is then pulled down between each carabiner into the intended direction of pull.

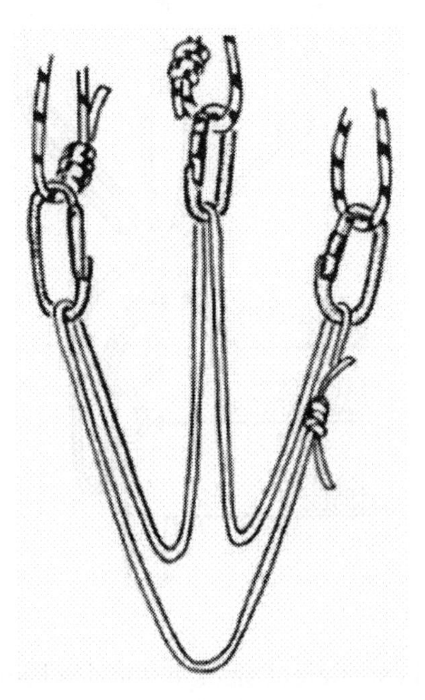

(c) With all three bights, tie an overhand knot on itself.

(d) Place a locking carabiner into the overhand knot's loop. This will serve as the attachment point for the anchor system.

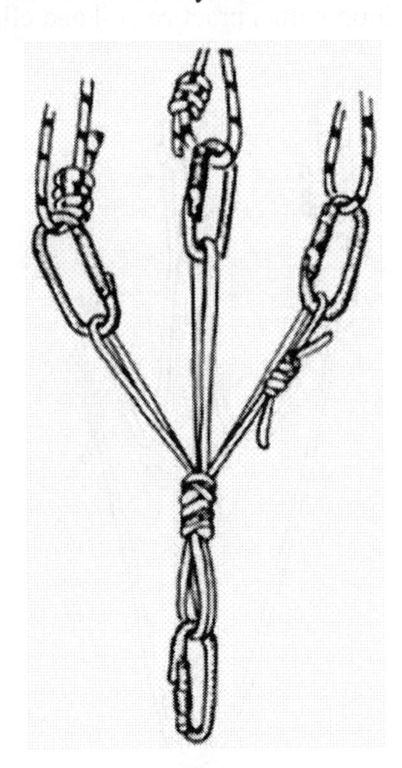

NOTE: If constructing a two-rope high-tension installation, a minimum of three artificial anchors per rope will be used.

TRANSITION: During this period of instruction we have discussed natural and artificial anchors, are there any questions? If you have none for me then I have some for you.

PRACTICE (CONC)

a. Students will establish natural and artificial anchor systems.

PROVIDE HELP (CONC)

a. The instructors will assist the students when necessary.

OPPORTUNITY FOR OUESTIONS (3 Min)

1. QUESTIONS FROM THE CLASS

2. QUESTIONS TO THE CLASS

Q. What are the two types of anchors?

A. (1) Natural.
(2) Artificial.

Q. How many pieces of protection are needed for an artificial anchor in a high tension installation, that requires two ropes?

A. A minimum of six pieces of protection.

Q. What are the different types of artificial anchors?

A. (1) Single Timber Dead-man
(2) Picket Hold fast
(3) Chock Anchor System

SUMMARY (2 Min)

a. During this class we have looked at the two types of anchors, to include natural and the artificial anchors. We have discussed the methods of tying these anchors as well as criteria for anchor points.

b. Those of you with IRF's please fill them out at this time and turn them into the instructor. We will now take a short break.

UNITED STATES MARINE CORPS
Mountain Warfare Training Center
Bridgeport, California 93517-5001

SML
SMO
02/11/02

LESSON PLAN

STREAM CROSSING

INTRODUCTION (5 Min)

1. **GAIN ATTENTION**. Anywhere in the world a mountain stream swollen by melting snow runoff poses a formidable obstacle to movement and can also be a hazard to your Marines.

2. **OVERVIEW**. The purpose of this period of instruction is to introduce the student to stream crossing by discussing the site selection considerations, safety precautions, preparations, and team and individual methods. This lesson relates to movement in mountainous terrain.

INSTRUCTOR NOTE: Have students read learning objectives.

3. **INTRODUCE LEARNING OBJECTIVES**

   a. TERMINAL LEARNING OBJECTIVE. In a summer mountainous environment, cross a mountain stream in accordance with the references.

   b. ENABLING LEARNING OBJECTIVES. (MLC)

      (1) Without aid of the references, list in writing the considerations in site selection for a stream crossing, in accordance with the reference.

      (2) Without the aid of the reference, list in writing the safety precautions taken while conducting a stream crossing, in accordance with the reference.

      (3) In a summer mountainous environment, execute the individual preparations taken prior to crossing a stream, in accordance with the reference.

      (4) In a summer mountainous environment and given a staff, cross a mountain stream using the staff method, in accordance with the reference.

(5) In a summer mountainous environment, cross a mountain stream using team methods, in accordance with the reference.

c. ENABLING LEARNING OBJECTIVES. (SMO)

(1) With aid of the references, state orally the considerations in site selection for a stream crossing, in accordance with the reference.

(2) With the aid of the reference, state orally the safety precautions taken while conducting a stream crossing, in accordance with the reference.

(3) In a summer mountainous environment, execute the individual preparations taken prior to crossing a stream, in accordance with the reference.

(4) In a summer mountainous environment and given a staff, cross a mountain stream using the staff method, in accordance with the reference.

(5) In a summer mountainous environment, cross a mountain stream using team methods, in accordance with the reference.

4. **METHOD/MEDIA**. The material in this lesson will presented by lecture and demonstration. You will practice what you have learned in upcoming field training exercises. Those of you with IRF's please fill them out at the conclusion of this lesson.

5. **EVALUATION**.

a. MLC - You will be tested later in the course by written and performance evaluations on this period of instruction.

b. SMO - You will be tested by verbal and performance evaluations.

TRANSITION: Are there any questions on the purpose, learning objectives, how the class will be taught, or how you will be evaluated? Before we can get into the actual techniques of stream crossing, we must first choose a site to perform the stream crossing.

BODY (55 Min)

1. (10 Min) **SITE SELECTION**

a. Mountain streams and rivers are military obstacles and therefore are danger areas for units crossing them. In order to reduce the time in the vicinity of the danger area, a recon team should precede the main body and select the best crossing site. The site selection for a stream crossing should include these eight considerations:

(1) Look for logjams, rocks or fallen trees that will provide a dry crossing if possible.

(2) If a dry crossing is not possible, select a crossing point at a wide and shallow point where the current is slower.

(3) Avoid sharp bends. They can be deep with a strong current on the outside of the bend.

(4) Look for a firm, smooth bottom. This is because large rocks and boulders provide poor footing and cause a great deal of turbulence in the water.

(5) It maybe easier to cross several small channels of water rather than one large one.

(6) Do not cross just above rapids, falls, or logjams, taking a fall or slipping could have serious consequences.

(7) Cross in the early morning. The water level will be lower since there has been less daylight for the snow to melt. Also, on sunny days, you will have more time to dry clothing and equipment.

(8) There should be a suitable spot downstream for safety swimmers.

<u>TRANSITION</u>: Now that we've discussed selecting our site, are there any questions? Let's discuss the safety precautions that must be taken.

SMO CUE: TC 3

2. (5 Min) **<u>SAFETY PRECAUTIONS</u>**. The following two safety precautions must be taken while conducting stream crossings.

   a. There must be a safety line at a minimum of a 45-degree angle downstream across the stream. This is for anyone who slips and is swept downstream to grab so that he is capable to stopping himself

   b. There must be safety swimmers downstream. These are strong swimmers who are positioned downstream to help anyone who is swept downstream. The safety swimmers will use throw bags.

NOTE: The safety line, as well as any other lines that must be taken across the stream, will be taken across by using the lead swimmer method taught in *ONE-ROPE BRIDGE.*

<u>TRANSITION</u>: Now that we have talked about safety precautions, are there any questions? Once the safety lines are in place and the safety swimmers are in position, individual preparations must be made to cross the stream.

3. (10 Min) **<u>INDIVIDUAL CROSSING PREPARATIONS</u>**. Prior to beginning to cross a stream, there are certain preparations that each individual should take. The six preparations are as follows:

a. Wear pack with shoulder straps fastened snugly. Waterproof the pack for buoyancy if possible.

b. Weapons will be slung diagonally over the shoulder with the weapon itself being between the pack and the individual's back.

c. Button all pockets and remove blousing garters. This prevents the water from creating added drag against the individual, which it would do if it could flow into open pockets.

d. Wear boots to protect the feet, but remove socks and insoles to keep them dry.

e. Wear the minimum amount of clothing. This reduces the amount of clothing that must be dried after the crossing.

f. Do not wear helmets in swift moving currents, due to the possibility of the current forcing the helmet/head under water.

NOTE: If you are in a tactical situation, the actual situation will dictate which of the above precautions will be taken.

TRANSITION: Now that we have discussed all the actions taken prior to crossing, are there any questions? Let's discuss the actual techniques used to cross. First we will discuss the individual methods.

4. (10 Min) **INDIVIDUAL CROSSING METHODS**. There are three individual methods that may be used.

   a. Staff Method. A strong staff or pole about 6 feet long is used as a crossing aid. It should be strong enough to support the Marines weight and trimmed clean of any branches. Placing both hands on the pole, the Marine should place the staff just upstream of his intended path. He should use the staff as the third leg of a tripod and should move only one leg or the staff at a time. He should face upstream using the staff to retain his balance. The staff is also used as a probe to discover bottom irregularities that could trip the Marine. The Marine should drag his feet instead of picking them up.

STAFF CROSSING METHOD

b. Swimming. This is an obvious method, if your Marines are good swimmers. This is not always the case, so usually this method is not a preferred one.

(1) In fast, shallow water, the Marine should angle across on his back with his feet downstream and his head up. He should use his hands to tread water and his feet to fend off obstructions.

(2) In fast, deep water, the Marine should angle across the stream on his stomach with his head upstream, to establish a proper ferry angle.

c. Belayed Method. In chest deep or water with a strong current, a rope can assist greatly. A rope will be secured from bank to bank, with the far anchor slightly downstream from the near anchor. The rope will be anchored off so that it lays at a minimum of 45 degrees. The Marine attaches himself to the rope by using a sling rope as safety line, tying a bowline around his waist and a figure-of-eight loop with a steel locking carabiner inserted. He then attaches the carabiner to the crossing line and crosses using the current to assist him.

5. (10 Min) **TEAM CROSSING METHODS**

a. Line Abreast Method. Small units (squad to platoon) can cross in moderate currents up to chest deep, by lining arms in a line abreast or chain method. The largest man of the chain is placed on the upstream side of the group. The group will enter the stream parallel to the flow of the stream. The middleman of the chain will control the group's movement and give the command when to step.

LINE ABREAST METHOD

b. Line Astern Method. Three or more men can line up facing the current. The upstream man, who should be the largest man in the group, breaks the current while the downstream men hold him steady. The upstream man may use a staff, similar to the individual staff method, to steady himself. All men side step at the same time with one man calling the cadence.

LINE ASTERN METHOD

c. Huddle Method. Between three and eight men can face inboard as in a football huddle. They will wrap their arms around each other's shoulders and cross the stream in this formation. The upstream man should change position as they cross because the entire formation will rotate. This prevents one man from becoming exhausted in the upstream position.

HUDDLE METHOD

TRANSITION: Now that we have discussed individual and team methods, are there any questions at this time? If you have none for me, then I have some for you.

PRACTICE (CONC)

a. Students will conduct a stream crossing.

PROVIDE HELP (CONC)

a. The instructors will assist the students when necessary.

OPPORTUNITY FOR QUESTIONS (3 Min)

1. QUESTIONS FROM THE CLASS

2. QUESTIONS TO THE CLASS

Q. What are the two safety precautions that must always be taken for stream crossings?

A. (1) A safety line will be used downstream.
(2) Safety swimmers will be stationed downstream.

Q. What are the three team methods of stream crossing?

A. (1) Line astern method
(2) Belayed method.
(3) Huddle method

Q. What are the five individual preparations that should be taken prior to a stream crossing?

A. (1) Wear pack with shoulder straps fastened snugly; do not fasten the waist strap.
(2) Weapons slung diagonally over the shoulder with the weapon between the pack and the individual's back.
(3) Button all pockets and remove blousing garters
(4) Wear boots to protect the feet but remove socks and insoles.
(5) Wear minimum amount of clothing.

SUMMARY (2 Min)

a. We have just discussed crossing streams with special emphasis on safety considerations, site selection criteria, individual preparations, and methods.

b. Those of you with IRF's please fill them out at this time and turn them in to the instructor. We will now take a short break.

TACTICAL STREAM CROSSING SOP

TM 1:

Actions upon reaching the stream:

1. Conduct recon of crossing site
2. Establish near side security
3. Determine whether or not a snow bridge will have to be constructed
4. Provide guide to bring remainder of the unit forward

Actions during crossing the stream:

1. Maintain security during the crossing
2. Will be the last team to cross the stream

Actions after crossing the stream:

1. Establish security as directed
2 Wait for the order to move out

TM 2:

Actions upon reaching the stream:

1. Set up ORP with TM 3
2. If necessary, move forward to construct a snow bridge for the crossing
3. TM 1 will guide you to the site

Actions during the crossing:

1. First team to cross the steam,
2. Maintain accountability of all Marines crossing the stream

Actions after crossing:

1. Recon far side of crossing site
2. Establish 360 security on far side
3. Give signal for rest to unit to cross
4. Place other teams in 360 as they cross
5. Wait for order to move out

TM 3:

Actions upon reaching the crossing site:

1. Establish 360 with TM 3
2. Wait for guide from TM 1

Actions during the crossing:

1. Second team to cross

Actions after crossing:

1. Establish security as directed by TM 2
2. Wait for order to move out

Actions if compromised:

1. Actions if compromised prior to the crossing- TL will make the decision to attack or attempt the crossing, depending on the size of the enemy, this will be situational dependent.
2. Actions if compromised during the crossing- TL will make the decision to either cross the remainder of his unit or bring the Marines on the far side back across to engage the enemy, again this will be situational dependent.
3. Actions if compromised after the crossing- TL will make the decision to engage the enemy or withdraw, possibly back across the stream, once again this is situational dependent.

NOTE: In most cases, the stream crossing may be used as a water resupply point.

UNITES STATES MARINE CORPS
Mountain Warfare Training Center
Bridgeport, California 93517-5001

SML
SMO
ACC
02/11/02

LESSON PLAN

MECHANICAL ADVANTAGE SYSTEMS FOR LIFTING AND TIGHTENING

INTRODUCTION

(5 Min)

1. **GAIN ATTENTION.** Image yourself as a patrol leader conducting a patrol in a mountainous environment. You have a thirteen man patrol, who all have sling ropes, carabiners, and the squad has two 165' static ropes. Your patrol comes up to a swift moving stream, which cannot be crossed by foot and is 100' wide. The rope will reach across, but sags in the water endangering Marines by drowning. The rope must be tightened to keep the Marines out of the water and aid in the crossing. This class will demonstrate and apply the physics of tightening a rope through mechanical advantage.

2. **OVERVIEW.** The purpose of this period of instruction is to introduce the student to the basics of mechanical advantage and the different types of pulley systems used. This period of instruction relates to all high tension systems.

3. **METHOD/MEDIA.** The material in this lesson will be presented by the lecture and demonstration method. You will practice what you have learned in the one rope bridge class practical application. Those of you with IRF's please fill them out at the end of this period of instruction.

4. **EVALUATION.** This period of instruction has no learning objectives and students will be evaluated throughout their training. The procedure of mechanical advantage will be tested on the one Rope Bridge.

TRANSITION: Are there any questions over the purpose, learning objectives, how the class will be taught, or how you will be evaluated? Let's discuss the theory behind the mechanical advantage system.

BODY (50 min)

1. (10 Min) **BASIC THEORY OF THE MECHANICAL ADVANTAGE SYSTEM.** How can a man weighing 200 pounds lift a load three times above his weight with ease?

   a. Consider first the heavy block in the diagram below suspended from two ropes. The upward force on the block is the tension in the ropes, and the sum of the two tensions must equal the weight of the block. If the whole system is symmetrical, each rope is under tension equal to half the weight of the block.

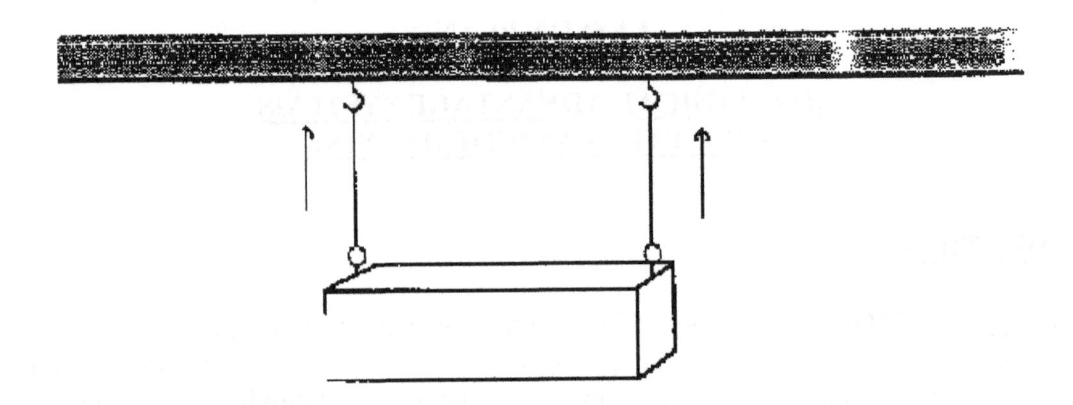

   b. Now look at the diagram below where the block has been attached to a pulley. There is now only one rope, which passes through the pulley. The tension in the rope is the same throughout; if it were different on one side than on the other, the pulley would turn until the tension on the two sides equalized. The tension in the rope is still only half the weight of the block.

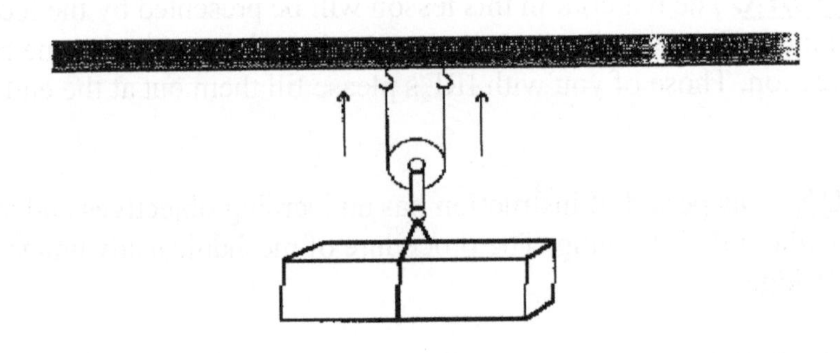

   c. The Work Principle

      (1) Webster's definition of work states that; work is an activity in which one exerts strength or faculties to do or perform something. It is a sustained physical or mental

effort to overcome obstacles and achieve an objective or result. What kind of advantage can we achieve using Pulley systems to reduce the amount of work we do?

(2) While pulleys are useful, they do not give something for nothing. Ignoring the problem of friction (force which opposes the movement of one surface sliding or rolling over another with which it is in contact), the input and output forces are in inverse ratio to the respective distance. To solve this we use this formula:

$$\frac{\textbf{EFFORT}}{\textbf{LOAD}} = \frac{\textbf{LOAD DISTANCE}}{\textbf{EFFORT DISTANCE}}$$

EXAMPLE: If you have a log that weighs 100 lbs. And you need to lift it two feet, using a 2:1 ratio system, the formula is:

$$\frac{50}{100} = \frac{2 \text{ feet}}{4 \text{ feet}}$$

(3) The frictionless pulley does not alter the product of force and distance. There is another limitation on the definition of work. Only the force in the direction of motion counts. What does this mean? The angle on which you direct the pull of your system will determine the actual advantage that it will receive. *REFER TO NATURAL AND ARTIFICIAL ANCHORS*

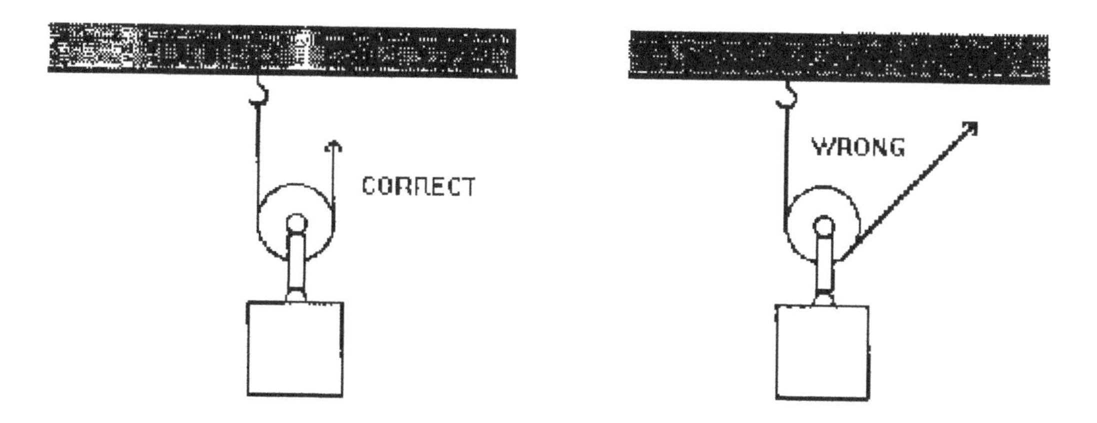

TRANSITION: Are there any questions over the basic theory? Let's now discuss the construction of various ratio systems.

2. (30 Min) **CONSTRUCTION OF THE RATIO SYSTEMS**.

a. Construction of a 1: 1 ratio system

(1) Attach the pulley to a suitable anchor point.

(2) Anchor one end of the rope to the load.

(3) Run the other end of the rope through the pulley.

(4) To equalize the load, pull the rope until there is tension on the system. If the load weighs 100 lbs., theoretically it should only take 100 lbs. of force to lift the load, but due to friction it may require 110 lbs. of force to equalize the load. To lift the load one foot, you will have to pull the rope one foot.

(5) This is referred to as a directional pulley or redirect, and gives no mechanical advantage at all.

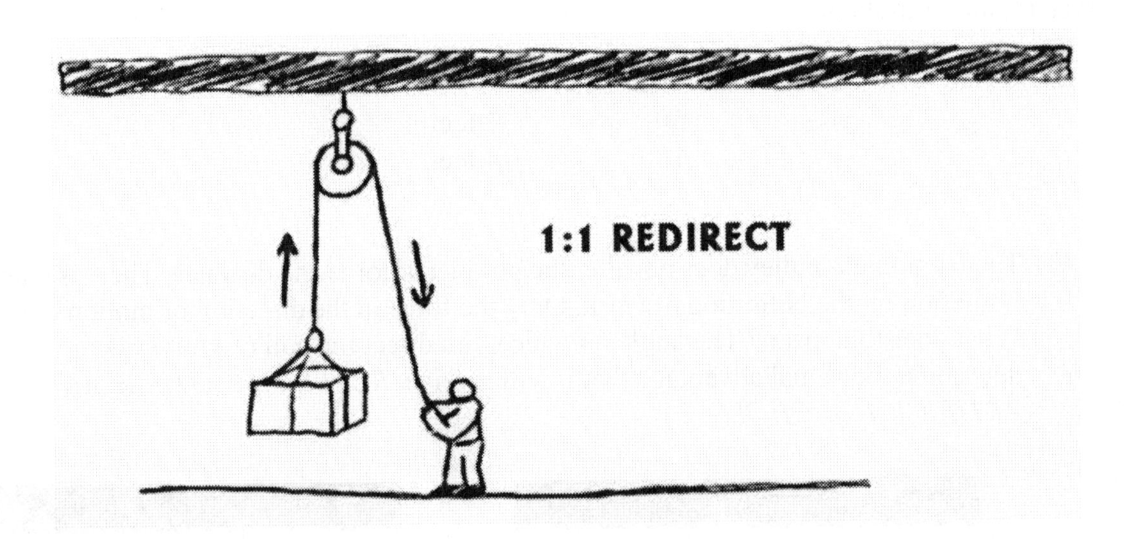

b. Construction of the 2:1 ratio system (C-pulley).

(1) Attach one end of the rope to a suitable anchor.

(2) Attach a pulley to the load.

(3) Run the other end of the rope through the pulley.

(4) Now you can raise the load with the amount of force that is equal to half of its weight.

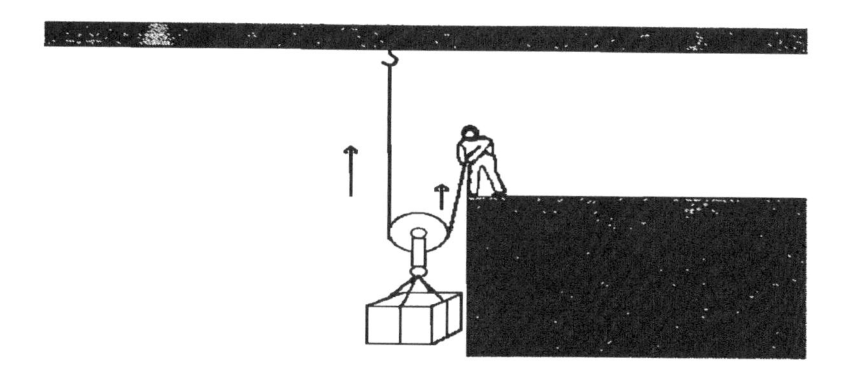

2:1 RATIO SYSTEM

(5) To lift the load two feet with this system, you must pull four feet of rope through the pulley.

(6) When each rope is equalized, divide by two. So if the load weighs 100 lbs., each line has 50 lbs. of supported weight.

c. Construction of the 3:1 ratio system (Z-Pulley)

(1) For tightening a rope installation:

(a) Anchor one end of the rope to a suitable far anchor using the tree wrap method.

(b) Tie a swami wrap around the near anchor and clip a steel-locking carabiner with the gate up and large axis facing the far anchor. This carabiner is referred to as the Main Anchor Carabiner (MAC).

(c) Take the running end of the rope from the far anchor and clip it into the MAC. Tie a stopper knot (Auto block), with a short prusik on the rope and attach it back into the MAC and lock it down.

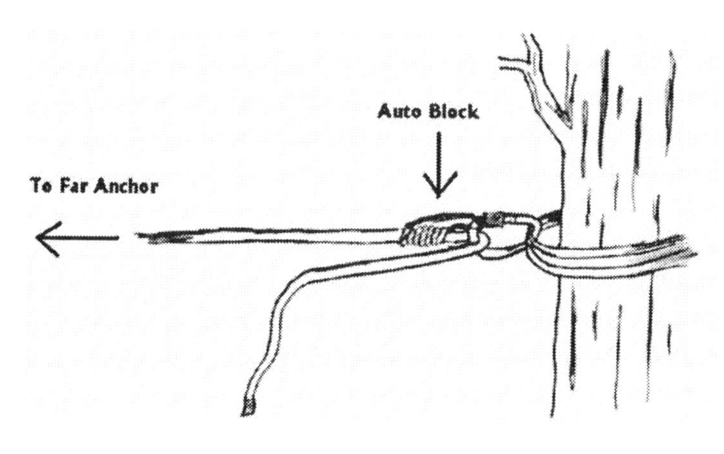

AUTO BLOCK

(d) Using another short prusik, come out from the swami wrap a few feet and tie a stopper knot (French prusik) on the rope and clip a steel locking carabiner into it with the direction of pull towards the near anchor. Clip the running end of the rope from the Auto block into the carabiner and lock it down.

(e) Now pull the running end of the rope to tighten the rope installation.

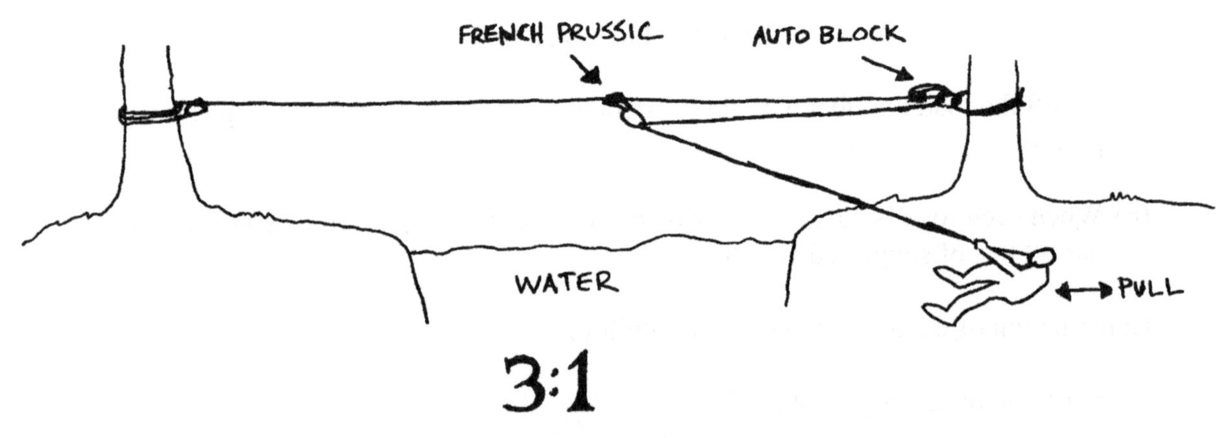

TIGHTENING A 3:1 MECH ADVANTAGE

d. Construction of the 9:1 ratio system (Z-Z Pulley)

(1) For tightening a rope installation:

(a) Anchor one end of the rope to a suitable far anchor using the tree wrap method.

(b) Tie a swami wrap around the near anchor and clip a steel locking carabiner through all the wraps with the gate up and large axis facing the far anchor. This carabiner is referred to as the Main Anchor Carabiner (MAC).

(c) Take the running end of the rope from the far anchor and clip it into the MAC. Tie a stopper knot (Auto block), with a short prusik, on the rope and attach it back into the MAC and lock it down.

(d) Using another short prusik, come out from the swami wrap a few feet and tie a stopper knot (French prusik) on the rope and clip a steel locking carabiner into it with the direction of pull towards the near anchor. Clip the running end of the rope from the Auto block into the carabiner and lock it down.

(e) Take the running end of the rope back to the swami wrap; attach another steel locking carabiner into the MAC with the large axis facing down and out to the far anchor. Clip the rope into this carabiner and lock it down.

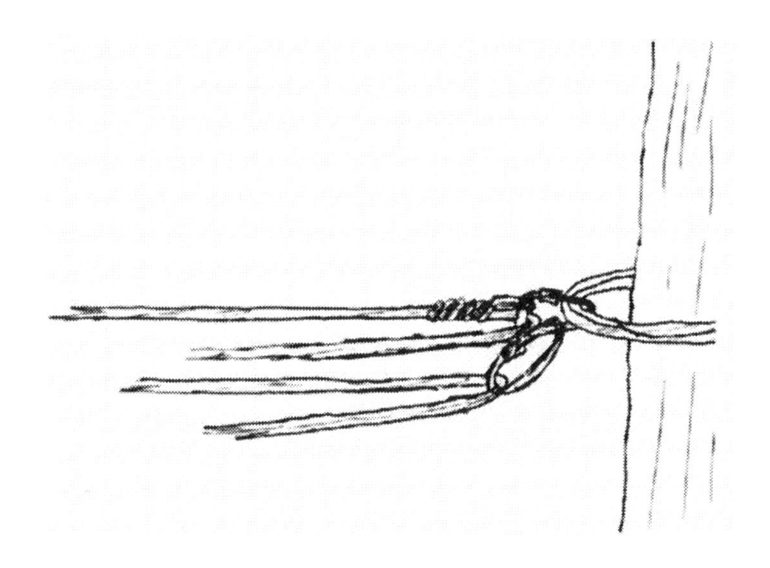

CARABINER INTO THE MAC

(f) Holding the rope that runs between the 2nd stopper knot and the carabiner hanging from the MAC, go back to a few inches before the French prusik and tie a 3rd stopper knot (French prusik) on that line of rope with a steel locking carabiner attached to it. Take the running end of the rope and clip it into the carabiner and lock it down.

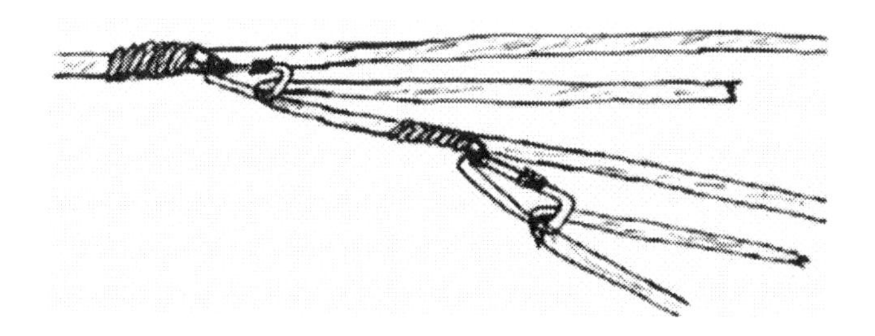

CLIPPING INTO THE PRUSIKS

(g) Now pull the running end of the rope to tighten the rope installation.

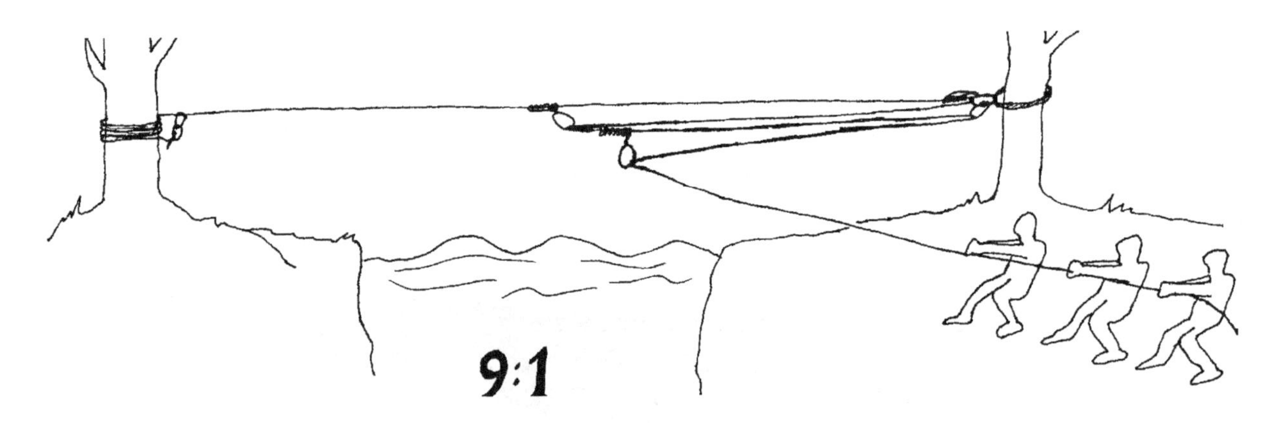

TIGHTENING A 9:1 MECH ADVANTAGE

NOTE: Care must be taken when increasing the ratio system. Breakage and damage of the ropes, carabiners and pulleys is very possible if the force end of the rope is greater than the load end of the rope. **No more than two men will tension the installation at any time if using pulleys.**

TRANSITION: Are there any questions over the construction of the systems? Because we are Marines, we are used to carrying a great deal of weight. But when we are in the mountains we should take every opportunity to lighten our loads and reduce the amount of work we need to do. If you have no questions for me, then I have some for you.

PRACTICE (CONC)

a. Students will use a mechanical advantage system to tighten a rope installation. Practical application of this will be on the one Rope Bridge.

PROVIDE HELP (CONC)

a. The instructors will assist the students when constructing the one Rope Bridge.

OPPORTUNITY FOR QUESTIONS (3 Min)

1. QUESTIONS FROM THE CLASS

2. QUESTIONS TO THE CLASS

Q. What are the two practical uses for the mechanical advantage system?

A. (1) As a tightening system.
(2) As a device to raise personnel or equipment up steep or vertical terrain.

SUMMARY (2 Min)

a. During this period of instruction we have discussed the terms and definitions, the practical uses, the equipment required for the construction of the 2:1 (C-Pulley), the 3:1 (Z-Pulley) and the 9:1 (Z-Z Pulley) ratio systems.

b. Those of you with IRF's please fill them out at this time and turn them in to the instructor. We will now take a short break.

UNITED STATES MARINE CORPS
Mountain Warfare Training Center
Bridgeport, California 93517-5001

SML
SMO
ACC
02/11/02

LESSON PLAN

ONE-ROPE BRIDGE

INTRODUCTION (5 Min)

1. **GAIN ATTENTION.** At times in mountainous terrain, it will become necessary to cross-rivers or streams. If the obstacle is a river, it can normally be crossed by utilizing various fording techniques. If it is a ravine, you may have to rappel down one side and scramble up the other. The most expeditious technique that can be used for crossing such an obstacle, particularly where a large body of men are involved, is by the use of a man portable easily erected and dismantled rope bridge.

2. **OVERVIEW.** The purpose of this period of instruction is to provide the students with the necessary skills required to cross-streams using a one-rope bridge. This material will be covered by discussing site selection, how to estimate the distance to be crossed, construction, crossing techniques, rescue techniques, and retrieving the bridge. This lesson relates to mountain movement.

INSTRUCTOR NOTE: Have students read learning objectives.

3. **INTRODUCE LEARNING OBJECTIVES**

    1. TERMINAL LEARNING OBJECTIVES. In a summer mountainous environment, conduct bridging, in accordance with the references.

    2. ENABLING LEARNING OBJECTIVES (SML) and (ACC)

        (1) Without the aid of references, list in writing the criteria for site selection for a one-rope bridge, in accordance with the reference(s).

        (2) Without the aid of references, given a MAC Kit, and a simulated or actual obstacle, construct a one rope bridge, in a specified time limit and in accordance with the references.

(3) In a summer mountainous environment, cross a one-rope bridge using the rappel seat method, in accordance with the reference(s).

(4) In a summer mountainous environment, execute a rescue on a one-rope bridge, in accordance with the reference(s).

(5) In a summer mountainous environment, retrieve a one-rope bridge, in accordance with the reference(s).

(6) Without the aid of references, list in writing the rescue techniques, in accordance with the reference(s).

3. ENABLING LEARNING OBJECTIVES (SMO)

(1) With the aid of references, list orally the criteria for site selection for a one-rope bridge, in accordance with the reference(s).

(2) With the aid of references, given a MAC Kit, and a simulated or actual obstacle, construct a one rope bridge, in a specified time limit and in accordance with the references.

(3) In a summer mountainous environment, cross a one-rope bridge using the rappel seat method, in accordance with the reference(s).

(4) In a summer mountainous environment, execute a rescue on a one-rope bridge, in accordance with the reference(s).

(5) In a summer mountainous environment, retrieve a one-rope bridge, in accordance with the reference(s).

4. **METHOD/MEDIA.** The material in this lesson will be presented by lecture and demonstration. You will practice what you have learned in upcoming field training exercises. Those of you with IRF's please fill them out at the end of this period of instruction.

5. **EVALUATION**

a. MLC- You will be tested later in the course by written and performance evaluations on this period of instruction.

b. SMO - You will be tested by a verbal exam and by a performance examination.

TRANSITION: Are there any questions over the purpose, learning objectives, how the class will be taught, or how you will be evaluated? During this period of instruction we will discuss site selection, construction, retrieving, and crossing the one rope bridge. Let's move on to the first objective, site selection.

BODY (65 min)

1. (5 Min) **SITE SELECTION**

   a. The two criteria for site selection for a one-rope bridge are:

      (1) There must be suitable anchors on both sides of the stream.

      (2) The anchors must offer good loading and unloading platforms.

   b. Other considerations involved are:

      (1) The site chosen for the initial crossing does not have to be at the location for the construction of the bridge, just as long as the rope can be taken to the selected site the crossing.

      (2) The site chosen for the lead swimmer to cross should be as free as possible from obstacles in the water, such as large boulders, stumps or logs as discussed in *STREAM CROSSING*.

      (3) The anchors must be close enough for the 150 foot coil to reach both near side and far side anchors. Keep in mind that it will take approximately 1/3 of the 150 foot rope for tightening and anchoring of the bridge.

2. (5 Min) **DISTANCE ESTIMATION**. The follow methods can be used to determine the distance between anchor points:

   a. Azimuth Method. Shoot an azimuth to a point on the far side of the intended obstacle to cross. Then move LEFT or RIGHT (perpendicular to the azimuth) until you get a 15 degree offset of your previous azimuth. Next, measure the distance that you paced in feet from your first azimuth to your second azimuth, and multiply that distance by three. This total will give you the approximate distance across the obstacle in feet.

   b. Unit Average Method. Get three Marines to judge (best guess) the distance across the intended obstacle. Add up the total accumulated distance, then divide by three. This will give an estimation of the distance across the obstacle.

Example:		
	Marine 1	240 ft
	Marine 2	230 ft
	Marine 3	250 ft
		720 ft

Divided by 3 = 240 ft

3. (2 Min) **ORGANIZATION**. The organization of construction is broken down into four groups. They are as follow:

   a. The Bridging Team - consisting of the bridge NCO and another Marine.

   b. The Safety Line Team - consisting of the lead swimmer and his belay man.

   c. The Mule Team - consisting 3 Marines for a 9:1, and 6 Marines for a 3:1.

   d. The Security Team – consisting of the rest of the party.

4. (15 Min) **CONSTRUCTION**. This bridge is ideal for squad and platoon sized units, for its quick construction and minimal amount of equipment required.

   a. Once the site for the bridge has been designated by the bridge NCO, the Safety Line Team will move up stream from this site to enter the water. Their first step is to flake out the rope. Once this is accomplished, the lead swimmer's will take a bight of the rope and tie a figure 8 loop at the end that will be going to the far side, ensuring the knot has a 18"-24" loop. The lead swimmer's upstream arm will go in it, with the belay man tending his rope the lead swimmer will cross the never using the flow to assist him. (Ferry angle as taught in STREAM CROSSING)

   b. At the same time as the lead swimmer is crossing, the bridging team will be preparing the near side anchor. The bridging team will tie a swammi wrap around the anchor, using a sling rope or practice coil, and ensuring the square knot is behind or on the side of the anchor. Once the anchor is secure, flake out the bridging line in order to send it across.

   c. Once the lead swimmer is across he will move to the site where the far side anchor will be established and secure his rope. The belay man will then move down to the near side anchor, and as close to the waters edge as possible. He will attach the bridging line to the safety by tying a middle of the rope in a figure 8 knot to the safety line and an end of the rope figure 8 knot to the bridging line connecting the two ropes with an 85 carabiner. The lead swimmer will then pull the rope across. The belay man with the help of another Marine will belay the bridging line across keeping tension on the rope to keep the rope out of the water.

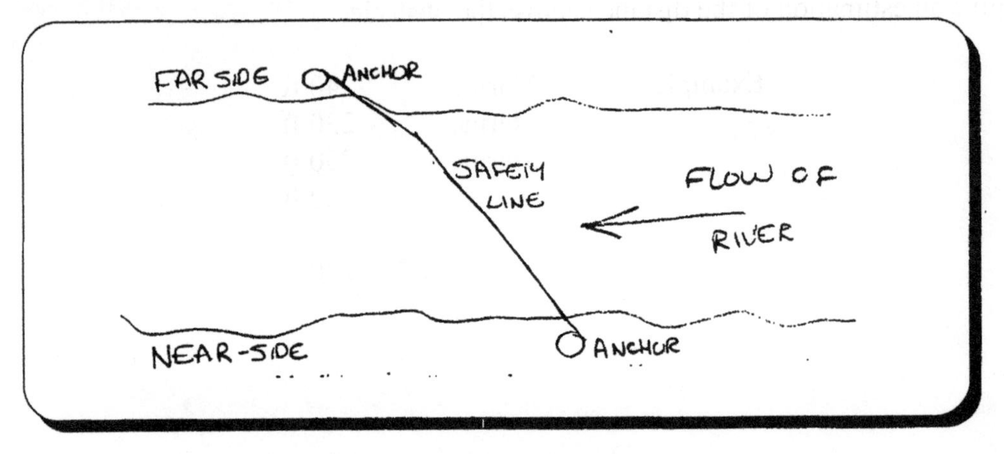

d. After the bridging line is across the lead swimmer will detach the bridging line from the safety line and secure it to the far side anchor using a tree wrap, wrapping from right to left, ensuring there are 3-5 wraps.

e. Once the bridging line is secure. The lead swimmer and the belay man will secure the safety line on both banks, ensuring the rope is creating a ferry angle (minimum 45 degrees). This will be the safety line for the Marines crossing the bridge.

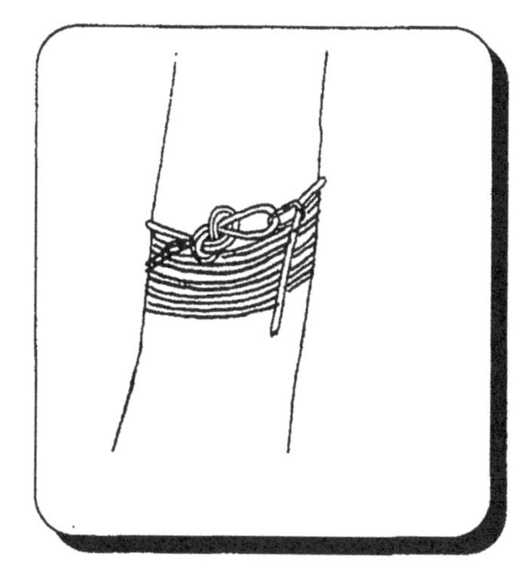

NOTE: If a fall from a rope bridge would result in death or serious injury, then utilize a safety line that will be anchored parallel with the bridge and hand tightened. The Marine crossing will clip into both ropes prior crossing.

f. On the near side bank:

1. The bridging team will need the following gear to build the mechanical advantage:

   (a) Four steel locking carabiners & 3 -Three foot prusiks

2. Once the bridging line is secured on the far side, the bridging team will take one 85 and clip it into the near side anchor (swammi wrap). Then they will take the bridging line and clip into that 85, then using one three foot long prusik (16" loop) the bridging line will be secured to the 85 by tying a French Prusik, which will act as a braking knot. The bridging team will now pull the rope taut, and begin construction of the mechanical advantage.

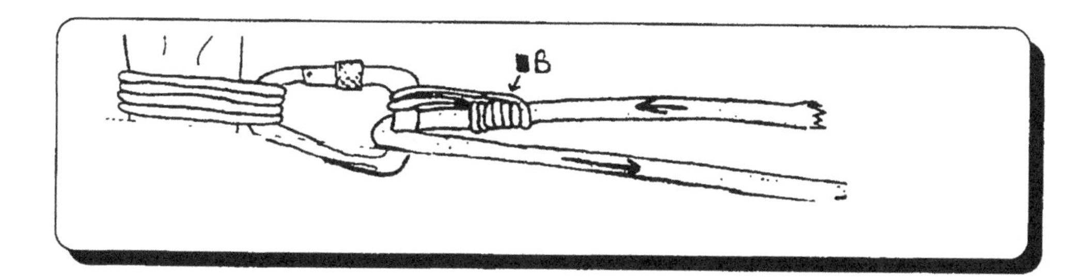

3. The first step is to take one three foot prusik (16" loop) and tie a French prusik on the bridging line as far away from the anchor as possible. Then take one 85 and clip it to the loops of the French prusik.

4. Taking a bight from the running end of the rope clip it into the 85 hanging from the tails of the French prusik.

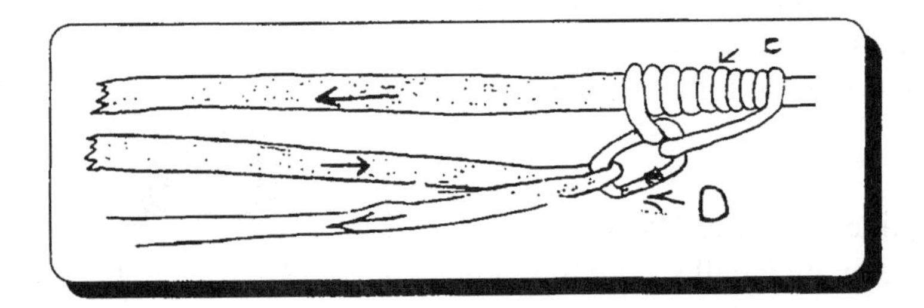

5. Then bring a bight back from the clipped in rope to the anchor, and taking one 85 clip it into the 85 created in step (2), take the bight of rope and clip into this 85.

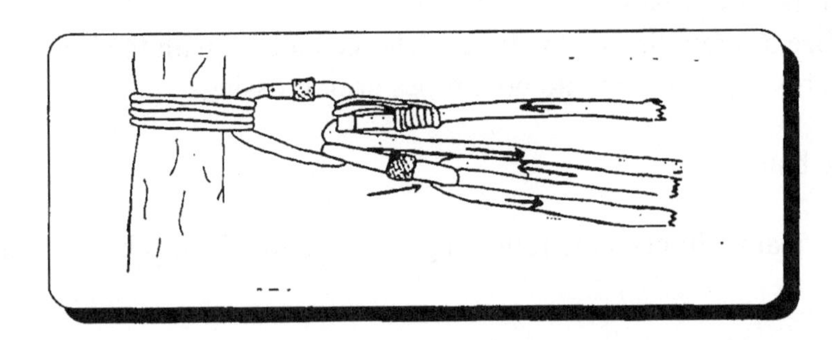

6. Take one three foot prusik and tie a French prusik onto the bottom end of the bight clipped into the 85 created in step (3), take one 85 and clip it into the tail loops of the French prusik.

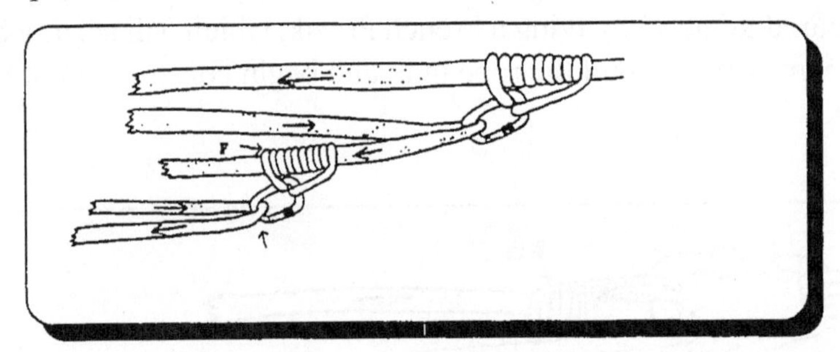

7. From the anchor, take a bight of rope from the 85 created in step (6) and clip it into the 85 created in step (6).

8. You have made approximately a 9:1 mechanical advantage.

NOTE: If the manpower is available the squad/platoon has the option of building a 3:1 mechanical advantage vise a 9: 1, the mule team will just have to add another body having no more then 3 bodies on a 9:1 and 6 bodies on a 3:1

g. Tensioning of the Bridge

1. Now that the bridge is built, the bridge NCO can call up the mule team to tension the bridge.

2. This procedure is facilitated by the mechanical advantage system that was just put into place. The braking knot (French prusik) is used to hold tension on the bridging line while the bridge is being tensioned.

3. The mule team will begin to pull on the running end of the rope coming out of the mechanical advantage. They will pull the rope straight back, trying to keep it in line with the bridge as best as possible.

4. The mule team will tension the rope as much as possible, the bridge NCO will monitor the system. Once the mule team cannot pull anymore tension, the bridge NCO will then have them hold the rope in place, and will reset the braking knot. Once the braking knot is set the bridge NCO will ask for slack from the mule team and cycle the system out. This process will continue until the bridge is tight.

5. On the last cycle the bridge NCO will set the brake knot and with the help of another Marine they will make a bight out of the running end and bring it around the tree while keeping tension.

6. Take a bight and make a complete round turn on the body of the 85 created in step (2). Last, the bridge NCO will tie two half hitches encompassing all the ropes just behind the anchor.

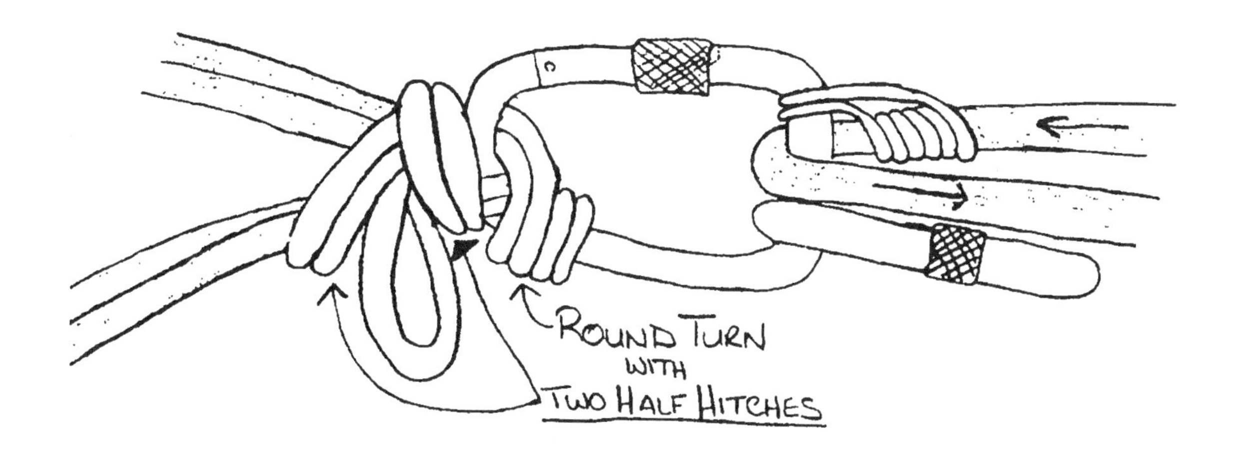

7. At this stage the bridge is tight and secured, the bridge NCO will now call up the remainder of the squad/platoon to cross. The bridge NCO will monitor the crossing of all Marines, and will be the last one to cross.

TRANSITION: Now that we have discussed the construction of the bridge, are there any questions over the material? Next we will discuss the crossing of the bridge.

5. (10 Min) **CROSSING.** The method used to cross is known as a horizontal traverse. This traverse can be accomplished in the following ways.

   a. Rappel Seat Method

      (1) The Marine ties himself into a rappel seat and inserts a large steel locking carabiner with the gate facing down and away.

      (2) The Marine faces the bridge with his right hand towards the near anchor to snap into the bridge, once the 85 is locked onto the bridge rope, a helper flips the 85 over so that the locking nut screws down. A carrier rope with a girth hitch may be added to assist the Marine in mounting the rope.

      (3) The Marine hangs below the bridge from his rappel seat with his head pointing in the direction of the far anchor and allows his legs to hang free.

      (4) Pulling with his arms makes progress.

      (5) This method is the safest and therefore the preferred method.

      (6) If the Marine must take a pack across, he may wear it with the waistband secured: However, the preferred method is to have another carabiner (aluminum non-locking is sufficient) attached to the pack frame at the top and attach this to the bridge behind the Marine putting his legs through the shoulder straps and pulling the pack across with him.

(7) One man at a time will cross, although one can load and another can unload concurrently.

(8) Weapons will be worn across the shoulder; muzzle down with a tight sling securely attached to the weapon.

b. Pulley Method. This method is used when the one rope bridge is long, uphill, or speed is vital and the Marines crossing it have a lot of heavy personal equipment; M24OG's, radios, etc.

(1) Equipment required. One pulley, four steel locking carabiners, and a hauling line twice the length of the obstacle.

(2) Setting up the system. Construction of the first suspension point is done by attaching a pulley to the one rope bridge, one 85 carabiner is attached to the pulley, gate down, and a second 85 is attached to the first 85 with the gate facing the near side bank. Into this

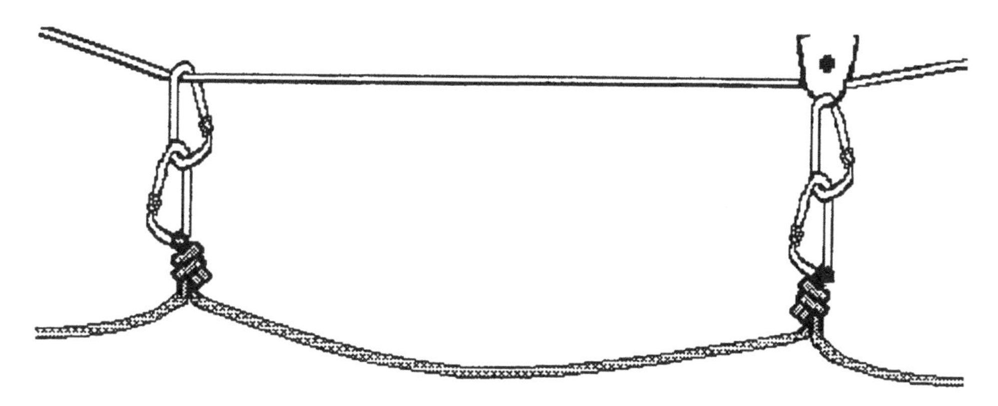

second 85, attach the hauling rope with a figure-8 loop. The figure-8 is placed halfway along the hauling rope and attached to the 85, which is then locked. Four feet down the line another figure-8 on a bight is placed into the second suspension point.

(3) Connection. The Marine clips his 85 into the lower 85, his equipment is clipped into the second 85. The mule team starts to pull.

c. Helmets will be worn for all methods of crossing the bridge. Gloves are optional.

NOTE: There are additional methods of crossing a one rope bridge such as the Commando crawl, Monkey crawl and hand over hand techniques, however, they are not taught at MWTC due to safety reasons.

TRANSITION: Are there any questions over crossing? Now we will cover rescue techniques.

6. (8 Min) **RESCUE TECHNIQUES.** If an individual is unable to complete the crossing on the one rope bridge using the rappel seat method of crossing (i.e. from a fall or exhaustion), a rescue will have to be made in the following manner:

a. Reach. First try to reach the victim by using an object such as a pole, your hand, if the victim is close enough, etc ...

b. Throw. If reaching the victim is impossible, try throwing a rope to the victim and have him attach the rope to himself, preferably his seat, and pull him back to the desired side of the installation.

c. Tow. If the victim is unable to catch or reach a rope being thrown to him or the victim is unconscious, tie a figure-of-eight loop into the middle of the safety line and then connect the rope to the rope bridge with a steel locking carabiner. The mule team will pull the carabiner up against the victim's seat and begin towing the victim to the desired side of the installation. The rope used to tow the victim should be twice the length of the span of the rope bridge (If necessary, ropes can be tied together to accommodate the span). This will allow towing from either side without having to throw or have a Marine carry the safety line back across the bridge each time a towing rescue is performed.

d. Go. If all else fails, the last option will be to go after the victim. The rescuer will move out onto the rope bridge with a safety line attached to himself at approximately eight feet from the end of the safety line. In the end of the safety line tie a figure-of-eight loop and insert a steel locking carabiner in the end of it. Once the rescuer has made contact with the victim. He will attach the steel locking carabiner to either the victim's seat (preferably the victim's carabiner that is attached to the rope bridge), or to the bridge itself. To ensuring that the carabiner is placed so that it pulls against the victim's carabiner. The mule team now starts to pull both the victim and the rescuer to the desired side of the installation.

e. Cut. If while crossing the rope over water the individual goes underwater and no other rescue technique can be employed, the rope will be cut.

TRANSITION: Are there any questions over rescue techniques? Now we need to know how to retrieve our ropes.

7. (10 Min) **RETRIEVING THE ONE ROPE BRIDGE**

a. Before the bridge NCO sends the last Marine to cross, they must make the bridge retrievable.

b. The first step is to break down the mechanical advantage, ensuring the braking knot is set before doing so. The system will be broken down until the bridging line is attached only to the brake knot.

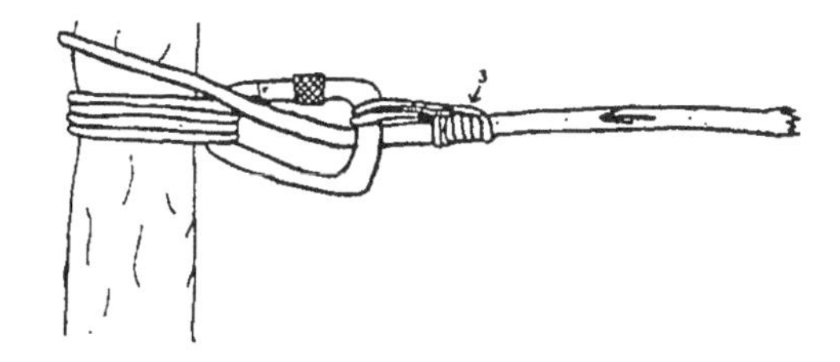

c. They will now take the rope around to the back of the anchor, and tie a slip figure 8, with the loop of the slip 8, take it and attach it to the bridging line just in back of the braking knot with a 85.

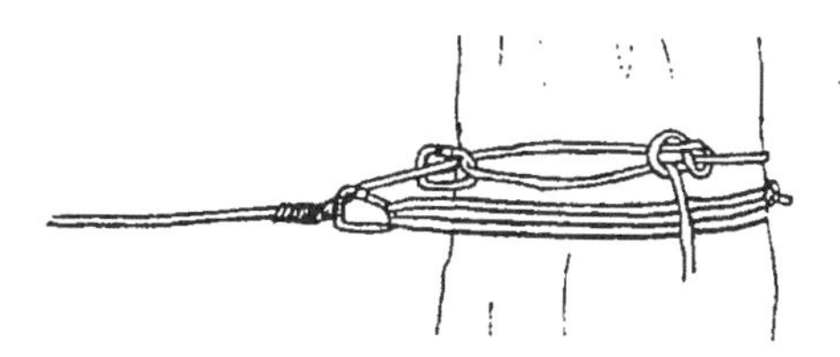

d. Now pull the running end coming out of the slip 8 until it tightens, and bites on the rope. Once the rope bites, tie a thumb knot behind the slip 8 as close to the slip figure 8 as possible.

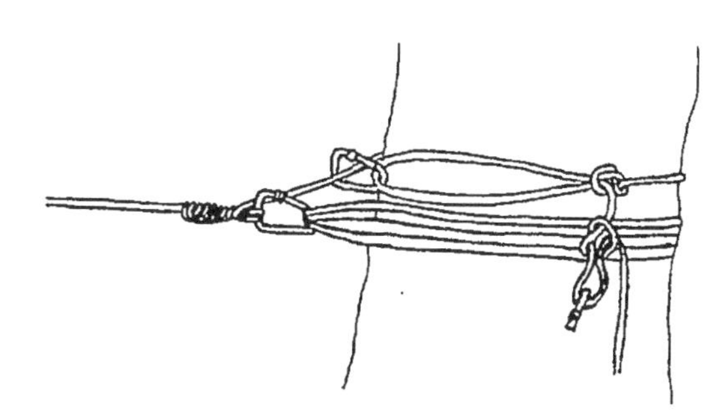

e. The bridge NCO, with the help of another Marine, will now untie the swammi wrap, until all of the tension is on the bridging line.

f. The safety line will now be brought up, and attached to the end of the bridging line with an overhand knot.

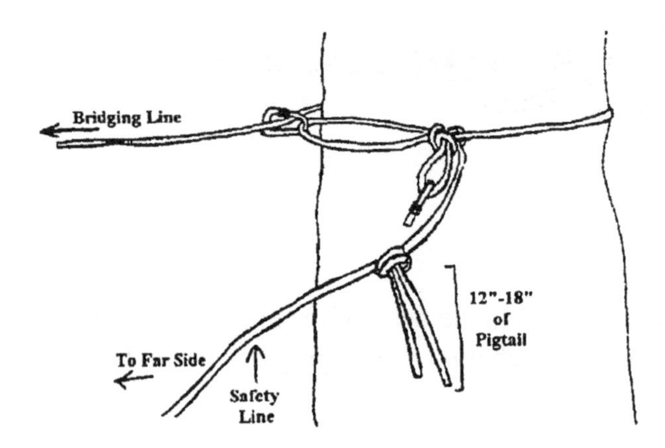

g. The bridge NCO and the other. Marine will now cross the bridge.

NOTE: It is possible that the bridge may lose tension after breaking the bridge down. The packs of the last two Marines across can be sent over with other Marines, prior to the slip figure being tied.

h. After both Marines have crossed, the safely line will be brought up from down stream.

i. The tree wrap will now be taken off. Once the tree wrap is off, all knots and carabiners will be taken out.

j. The safety line will now be pulled, the bridge is retrieved.

TRANSITION: When in mountainous terrain it is important to be aware of the fact that you may encounter a great variety of terrain in a very short time. Employing a rope bridge is a very worthwhile skill. Are there any questions over the retrieving the bridge or any thing covered in this class. If you have none for me, then I have some for you.

PRACTICE (CONC)

a. Students will construct a rope bridge, cross it and retrieve it.

PROVIDE HELP (CONC)

a. The instructors will assist the students when necessary.

OPPORTUNITY FOR QUESTIONS (3 Min)

1. QUESTIONS FROM THE CLASS

2. QUESTIONS TO THE CLASS

Q. What are the two criteria for site selection for one-rope bridge?

A. (1) Suitable anchors on both sides of the stream.

(2) Good loading and unloading platforms.

Q. What are the two different methods used to cross a one-rope bridge?

A. (1) Rappel-seat method.

(2) Pulley method.

SUMMARY (2 Min)

a. We have discussed the site selection, construction, crossing and retrieval of a one-rope bridge.

b. Those of you with IRF's please fill them out at this time and turn them in to the instructor We will now take a short break.

UNITED STATES MARINE CORPS
Mountain Warfare Training Center
Bridgeport, California 935-17-5001

SMO
SML
ACC
02/11/02

LESSON PLAN

A-FRAMES

INTRODUCTION (5 Min)

1. **GAIN ATTENTION**. In mountain operations it is often necessary to move inexperienced troops and equipment over steep, rocky terrain. A fixed rope installation is one method to move personnel up or down these obstacles. However, this method is not always sufficient for moving their equipment. So, the construction of a vertical hauling lines system may be necessary.

2. **OVERVIEW**. The purpose of this period of instruction is to provide the student with the necessary information to establish A-frames for use with evacuating patients and moving gear up and down steep cliffs. This will be accomplished by discussing the definitions, purposes, criteria of site selection, and usage of these devices. This lesson relates to vertical hauling lines and suspension traverse.

INSTRUCTOR NOTE: Have students read learning objectives.

3. **INTRODUCE LEARNING OBJECTIVES**

   a. TERMINAL LEARNING OBJECTIVES. In a summer mountainous environment and given the necessary equipment, construct an A-Frame, in accordance with the references.

   b. ENABLING LEARNING OBJECTIVES. (SML) and (ACC)

      (1) Without the aid of references, state in writing the definition of an A-Frame, in accordance with the reference.

      (2) Given A-Frame poles, 6 sling ropes and an anchoring rope, construct an A-Frame, in accordance with the reference.

c. ENABLING LEARNING OBJECTIVES (SMO)

(1) With the aid of references, state orally the definition of an A-frame, in accordance with the references.

(2) Given A-Frame poles, 6 sling ropes and an anchoring rope, construct an A-Frame, in accordance with the reference.

4. **METHOD/MEDIA**. The material in this lesson will be presented by lecture and demonstration. You will practice what you have learned during upcoming field training exercises. Those of you with IRF's please fill them out at the end of this period of instruction and pass them forward.

5. **EVALUATION**

a. MLC - You will be evaluated later in this course by written and performance examination.

b. ACC - You will be tested later in the course by written on this period of instruction and by performance evaluations during the cliff assault.

c. SMO - You will be tested by a verbal examination and also by a practical application method.

TRANSITION: Are there any questions over the purpose, learning objectives, how the class will be taught, or how you will be evaluated? We will start by discussing the A-frame.

BODY (60 Min)

1. (15 Min) **A-FRAME**. Natural objects such as trees, bushes and boulders can be utilized to gain height in a rope installation. There are different artificial devices that can be used for the same purpose, one example being the A-Frame.

a. Definition. An A-Frame is an artificial device used to gain height in a rope installation. Although an A-Frame can be used with various types of rope installations its most common use is in conjunction with a vertical hauling line.

b. Materials. The materials required for construction of an A-Frame are:

(1) Two-poles approximately 8 feet long and not less than 3 inches in diameter. The actual size of the poles will depend on the type of load; a very heavy item would require a stout pole and a taller item would require a longer pole to provide sufficient clearance for the load.

(2) To construct an A-Frame, 4-6 sling ropes are required. This number may vary depending on the diameter of the poles and the length of the sling ropes being used.

c. Nomenclature.

(1) Apex. The point near the top of the A-Frame where the poles cross each other.

(2) Butt ends. The bottom ends of the poles used in construction of the A-Frame. The butt end is larger in diameter than the end at the apex.

TRANSITION: Now that we have discussed what an A-Frame is, are there any questions? Now we are ready to learn how to actually construct an A-frame.

2. (35 Min) **CONSTRUCTING AN A-FRAME**.

a. Place two sturdy poles of approximately the same length side by side, ensuring that the butt ends are flush together.

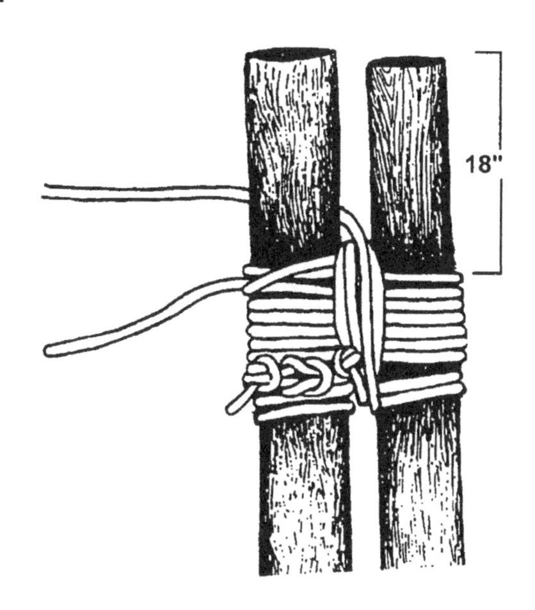

b. Take a sling rope and come down 18 inches from the top of the shortest pole leaving an 18 inch pigtail extended. Next, tie a clove hitch ensuring that the 18 inch pigtail is toward the top of the pole and the locking bars of the clove hitch are to the outside.

c. Wrap the sling rope 6-8 times horizontally around both poles, hitch down toward the butt ends.

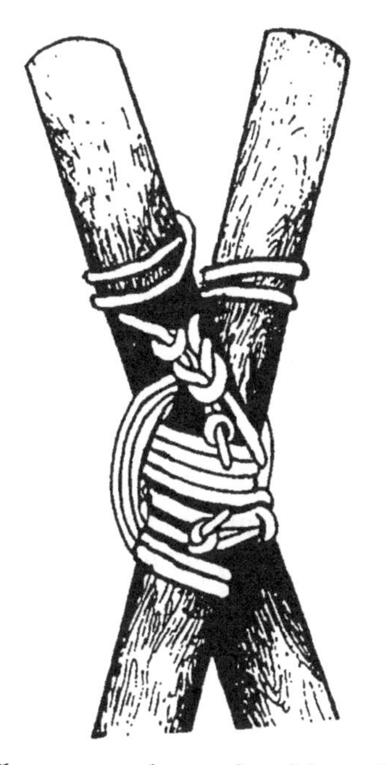

(1) It will be necessary to join another sling rope to the sling rope that you started with. The knot used to join the sling ropes is a square knot finished with two overhand knots.

(2) The joining of the sling ropes must be done on the horizontal wraps and the square knot must be on the side of one of the poles so it will not interfere with the vertical wraps.

d. Wrap 4-6 times vertically between the poles and around the horizontal wraps.

e. Tie off the sling by using the 18 inch pigtail that extends from the clove hitch; tie it off by using a square knot. The square knot should be so tight that the overhands will be tied on the pigtails themselves.

The square knot should not be tied on the inside of the apex.

f. Spreader Rope (Bar)

(1) Tie a sling rope or ropes between the poles at the bottom of the A-Frame with a clove hitch with locking bar facing out and two half hitches on both butt ends.

(2) If more than one sling rope is needed, join the sling ropes with a square knot finished with two overhand knots. Make sure the ropes are joined in the middle of the A-Frame.

NOTE: During testing two sling ropes will be used for the spreader bar.

(3) Adjustment can be made to either side when needed.

g. Anchoring the Butt Ends of the A-Frame

(1) Place the butt ends of the A-Frame poles into natural or manmade pockets.

(2) Use natural or artificial anchors to prevent the butt ends from moving. In order to keep the butt ends in place, additional anchors may be necessary along with anchoring the A-Frame at the middle and bottom. This is done with sling ropes in order to keep the A-Frame stationary. A clove hitch and two half hitches or a round turn and two half hitches are tied to the A-Frame and normal anchor knots are used on the anchors.

h. Anchoring and tying-off the A-Frame.

(1) The top part of the A-frame (the Apex) is normally anchored off. There are a few different techniques used in anchoring off the A-frame. The installation being used will dictate how it is to be anchored. (This will be discussed later in *VERTICAL HAULING LINES and SUSPENSION AND TRAVERSE)*

i. Task organization for the A-Frame. To construct the A-Frame quicker and more efficiently, separate tasks can be assigned to different individuals. The following tasks should be assigned:

(1) One individual will tie the sling ropes around the poles to form the apex.

(2) While the apex is being tied another individual should be constructing the anchor. After the spreader bar is completed, this individual will anchor down the A-Frame.

(3) Once the poles are spread, a third Marine should tie the spreader bar. Ensure the spreader bar is not tied over the anchor line.

TRANSITION: Are there any questions over the construction of the A-Frame? If there are no questions, we'll go over how to gain height using a tree branch.

PRACTICE (CONC)

a. Students will construct an A-frame.

PROVIDE HELP (CONC)

a. The instructors will assist the students when necessary.

OPPORTUNITY FOR QUESTIONS (3 Min)

1. QUESTIONS FROM THE CLASS

2. QUESTIONS TO THE CLASS

Q. What is the definition of an A-Frame?

A. It is an artificial device used to gain height in rope installations.

Q. What are the materials required to construct an A-Frame?

A. (1) Two poles approximately 8 feet long and not less than 3 inches in diameter.
(2) 4-6 sling ropes.

SUMMARY (2 Min)

a. During this period of instruction we have discussed the materials needed to construct an A-Frame and the actual techniques involved in the construction of an A-Frame.

b. Those of you with IRF's please fill them out at this time and turn them in to the instructor. We will now take a short break.

UNITED STATES MARINE CORPS
Mountain Warfare Training Center
Bridgeport, California 93517-5001

SML
SMO
ACC
02/11/02

LESSON PLAN

VERTICAL HAULING LINES

INTRODUCTION

(5 Min)

1. **GAIN ATTENTION.** In mountain operations it is often necessary to move inexperienced troops and equipment over steep, rocky terrain. A fixed installation is one method to move personnel up or down these obstacles. However, this method is not always sufficient for moving their equipment. So, the construction of a vertical hauling line system may be necessary.

2. **OVERVIEW.** The purpose of this period of instruction is to provide the student with the necessary information to establish vertical hauling lines for use in evacuating patients and moving gear up and down steep cliffs. This will be accomplished by discussing the definitions, purposes, criteria of site, and usage of this system. This lesson relates to cliff assault.

INSTRUCTOR NOTE: Have students read learning objectives.

3. **INTRODUCE LEARNING OBJECTIVES**

   a. TERMINAL LEARNING OBJECTIVES. In a mountainous environment, establish a vertical hauling line, in accordance with the references.

   b. ENABLING LEARNING OBJECTIVES.

      (1) Without the aid of references, describe in writing the purpose of a vertical hauling line, in accordance with the references.

      (2) Without the aid of references, list in writing the criteria involved in site selection for a vertical hauling line system, in accordance with the references.

      (3) In a summer mountainous environment, construct an endless rope for use in a vertical hauling line system, in accordance with the references.

4. **METHOD/MEDIA.** The material in this lesson will be presented by lecture and

demonstration. You will practice what you have learned during upcoming field training exercises. Those of you with IRF's please fill them out at the end of this period of instruction

5. **EVALUATION.**

a. MLC - You will be evaluated later in this course by written and performance examination.

TRANSITION: Are there any questions over the purpose, learning objectives, how the class will be taught, or how you will be evaluated? We will start by discussing the vertical hauling line.

BODY (45 Min)

1. (5 Min) **VERTICAL HAULING LINE**

a. Purpose. The purpose of a vertical hauling line is to move equipment and troops over steep, rocky terrain.

b. Materials. The materials needed for a vertical hauling line are:

(1) Two static ropes. (An additional rope is required if using a knotted hand line)

(2) Minimum of six sling ropes.

(3) Four steel locking carabiners.

(4) An appropriate belay device. (four non-locks if using a crab brake)

(5) Two rescue pulleys. (Optional - these are preferred but locking carabiners may be used.)

(6) Two A-Frame poles, strong enough to support the load and tall enough to provide adequate clearance for the load.

TRANSITION: We have covered what the vertical hauling is, are there any questions? Without a proper site, building any installation in the mountains could be useless, as well as dangerous.

2. (2 Min) **SITE SELECTION FOR A VERTICAL HAULING LINE.** Picking the proper site is critical to the safe and efficient operation of a vertical hauling line. The site must meet the following four criteria:

a. It must have a suitable top anchor.

b. It must have good loading and unloading platforms.

c. There must be sufficient clearance for the load at all points.

d. If using an A-Frame, you must be able to anchor the butt ends of the A-Frame.

TRANSITION: Now that we have discussed what is required for site selection, are there any questions? Let's take a look at the construction of a vertical hauling line using an A-Frame.

3. (15 Min) **CONSTRUCTION OF A VERTICAL HAULING LINE USING AN A-FRAME**

a. Construct an A-Frame that can support the load to be moved and that provides adequate clearance.

b. To anchor the A-Frame, use the following procedures. One static rope will be used.

(1) Double the rope and find the middle.

(2) Lay the middle of the static rope over the apex of the A-Frame, leaving an 18 inch bight over the apex. This is known as the anchor bight.

(3) Tie clove hitches above the lashing on each side of the apex, ensuring that the clove hitch locking bars are facing each other and are next to the lashing. The 18 inch bight is left dangling from the top of the apex.

(4) Anchor the A-Frame, using a round turn and, two half hitches. Angle the A-Frame 30 degrees from the vertical and tie off using two half hitches. The rest of the rope is coiled neatly and placed out of the way.

(5) Construct a second 18 inch bight from the top of the apex using a sling rope. This is known as the safety bight. Attach the sling rope to each side of the apex of the A-Frame above the first clove hitches the same manner as the anchor bight with the locking bars to the outside. Adjust the spacing between the anchor bight and safety bight to 1-3 inches (2-4 fingers). Secure the pigtails of the sling rope with a square

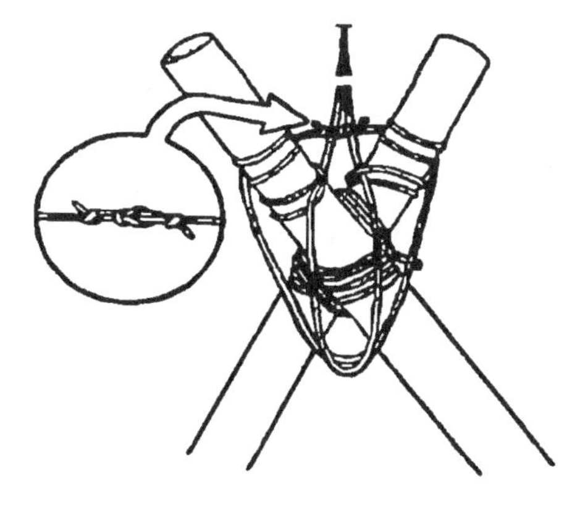

SAFETY BIGHT

knot and two overhands at the rear of the apex of the A-Frame.

(6) Two steel locking carabiners with gates opposite and opposed are inserted into both bights hanging down from the apex.

c. Hauling Line. A hauling line is constructed next. Hauling lines can be constructed in two different ways. Either by using the end of the rope or securing it to itself making an endless rope.

(1) For lowering and raising heavy loads and personnel; the end of the rope is used. (This method is less efficient but much safer than the endless rope) It is constructed by running one end of the rope through the carabiners/pulleys on the safety/anchor bight and tying a figure 8. The other end is then incorporated with an anchor (i.e. tree wrap) and an appropriate lowering/raising system discussed later.

(2) If the hauling line is to be controlled from the bottom of the cliff, it is converted into an endless rope by tying the two ends together using a square knot finished with two overhand knots. Next, two directional figure-of-eight knots are tied on opposite sides of the rope, one at each loading/unloading platform.

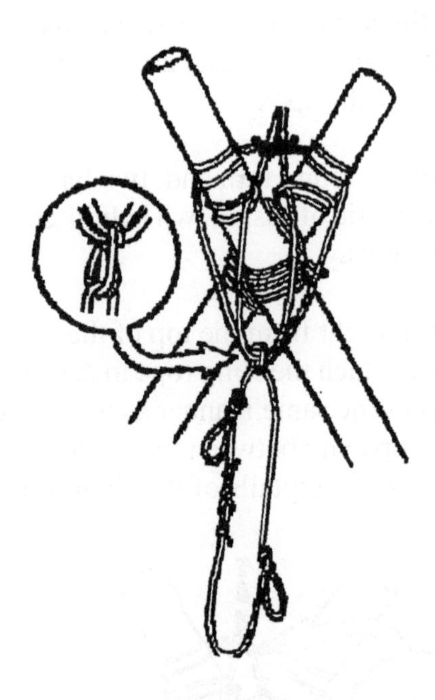

ENDLESS LOOP

(3) After completing the hauling line, personnel at the base of the cliff should pull on the line to test it and the A-Frame.

d. Additional Considerations

(1) Eliminate excessive friction. Use pulleys when possible.

(2) To help keep the line taut and the load from banging into the cliff, tie a swami wrap on a tree with a carabiner and pulley on it. Run the endless rope through the pulley and

tighten the joining knot on the endless rope. This will minimize the mule team size and make it easier to move the loads up and down.

(3) Remove any obstacles and any loose objects, which could be dislodged by men or equipment.

(4) Station two men at the unloading platform to operate the A-Frame during use. These operators are tied in with a safety line.

(5) Use multiple anchors when establishing an A-Frame/vertical hauling line. This will prevent total system failure should one particular anchor fail.

(6) A knotted handline may be used to aid individuals up a vertical hauling line and is constructed as follows: Overhand knots are be placed approximately 10-12 inches apart, ensuring that there is approximately 20 feet without knots at one end to be used to anchor. To anchor the handline, tie a round turn and a bowline finished with an overhand, around the anchor and throw the knotted handline over the apex of the A-Frame. The knotted handline is used as a simple fixed rope by any personnel ascending the vertical hauling line.

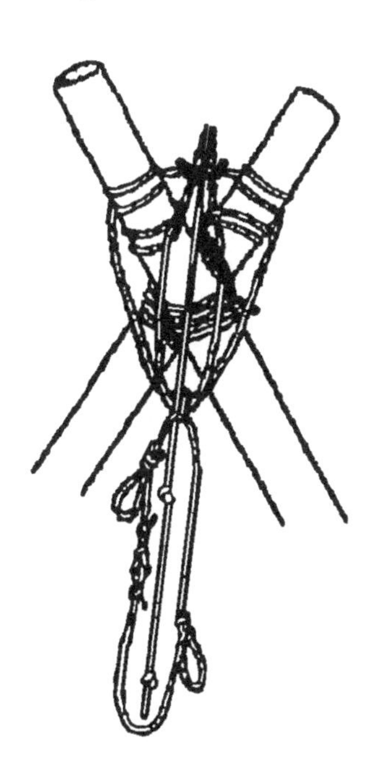

KNOTTED HANDLINE

TRANSITION: Are there any questions over the A-frame construction? We will now discuss how to gain height using a tree.

4. (5 Min) **USING A TREE TO GAIN HEIGHT**. When using a tree to gain height in a vertical hauling line installation; there are some considerations that need to be applied. The following factors are to be considered:

a. Select a branch that has not been chopped, burned or is rotten.

b. The branch must be at least 6" in diameter or of suitable thickness to support the intended load.

c. The branch should be at a sufficient height to offer adequate clearance for the load.

d. A loose swammi wrap will be tied around the intended branch.

e. Two steel locking carabiners with gates opposite and opposed are inserted into the swammi wrap.

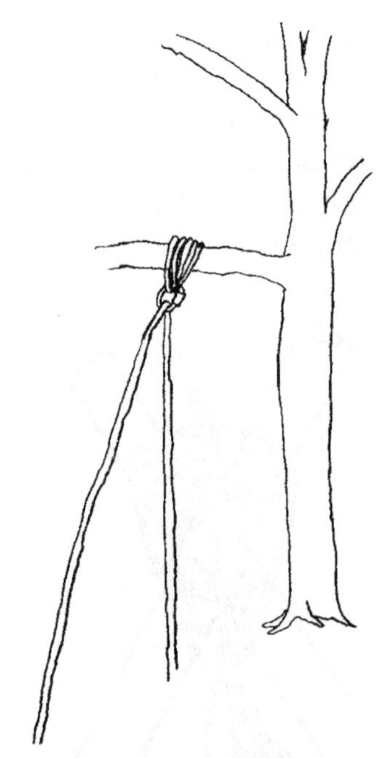

USING A TREE TO GAIN HEIGHT

f. The desired rope installation will be attached through the carabiners.

TRANSITION: Are there any questions over the construction using a tree? If not, we will now learn how to operate a vertical hauling line.

5. (3 Min) **OPERATION OF THE VERTICAL HAULING LINE**

a. Personnel or equipment are secured to the hauling line and raised or lowered by a team of Marines pulling on the rope, or belaying it down.

(1) If equipment and personnel are only being lowered, the hauling line can be used from the top with the same belay used with the *SUSPENSION TRAVERSE.*

(2) If equipment is being raised from the bottom, an endless rope is the most efficient

hauling line.

(3) Marines and equipment being raised or lowered will load and unload through the center of the A-Frame, not to the sides.

(4) Mule Team. If equipment and personnel can be raised from the top of the cliff, a mule team can be used there. To establish a mule team, the hauling line is run through a pulley or carabiner (preferably a pulley) at the anchor point behind the system. The line is then led away at about a 90-degree angle to a cleared area. A safety line should be attached to the hauling line to prevent the load from falling if the hauling line is dropped by the mule team. This is done with an autoblock/French prusik on the carabiner or pulley on the anchor. A group of six men, (mule team) assemble on this rope on the side of the rope away from the cliff or outside the bight of the rope if on the bottom. They will haul personnel and supplies up the cliff by grasping the rope and simply walking away with it. One man monitors the auto block to ensure it does not run through the 85 carabiner (He will not wear gloves). If the mule team looses control of the line the safety man will let go of the autoblock. The mule team responds to a single commander by using the following verbal/hand and arm signals.

6. (5 Min) **CONSTRUCTION AND OPERATION OF A BRAKING/BELAY DEVICE**

a. Because the A-Frame/vertical handing line can be a very strong, sturdy system, we often put heavy loads on it. When working with heavy loads, a braking/belay device that generates a great deal of friction is required. The most commonly used device for this purpose is the Carabiner Brake or Crab. The equipment needed is as follows:

(1) Four steel locking 82's or aluminum non-lock carabiners.

(2) One steel locking carabiner (large D).

b. Construction.

(1) Insert one large steel carabiner into the anchor with the gate up.

(2) Insert two small carabiners with gates facing to the right into the one large steel locking carabiners.

(3) Pull up a bight of rope and insert it through the two small carabiners.

(4) Clip two more small carabiners with the gate facing down and to the right into and through the first two small carabiners.

NOTE: All directions apply while facing the anchor.

c. Operation.

(1) More than one rope may be fed through the double crab.

(2) As the rope flows through the crab brake, make sure that the direction of the rope (if coming in contact with the gates of the locking carabiners) is such that the rope is tightening the gates.

(3) The flow of the rope through the crab brake should be constant and smooth.

(4) Regardless of the type of belay device, two Marines will be required in lowering. One to operate the belay device (i.e. crab brake) and one to monitor the safety prusik.

NOTE: The double carabiner brake/double crab is a one way brake used for lowering only and cannot be reversed under a load.

TRANSITION: Now that we have covered the operation, are there any questions? If there are none for me, then I have some for you.

PRACTICE (CONC)

a. Students will construct an A-frame and vertical hauling line and operate a vertical hauling line.

PROVIDE HELP (CONC)

a. The instructors will assist the students when necessary.

OPPORTUNITY FOR QUESTIONS (3 Min)

1. QUESTIONS FROM THE CLASS

2. QUESTIONS TO THE CLASS

Q. What is the purpose of a vertical hauling line?

A. The purpose of a vertical hauling line is to move equipment and troops over steep, rocky terrain.

Q. What are the four criteria for site-selection for a vertical hauling line?

A. (1) Suitable to anchor.
(2) Good loading and unloading platform.
(3) Sufficient clearance for the load.
(4) Butt ends of the A-Frame must be anchored.

SUMMARY (2 Min)

a. During this period of instruction we have discussed the materials needed to construct a vertical hauling line with an A-frame and the actual techniques involved in the construction of an A-frame and a vertical hauling line.

b. Those of you with IRF's please fill them out at this time and turn them in to the instructor. We will now take a short break.

UNITED STATES MARINE CORPS
Mountain Warfare Training Center
Bridgeport, California 93517-5001

SML
SMO
ACC
02/11/02

LESSON PLAN

SUSPENSION TRAVERSE

INTRODUCTION (5 Min)

1. **GAIN ATTENTION.** If we had to move personnel and equipment over ravines, rivers, and chasms and up or down cliffs, we would have a major difficulty with the task. One of the simplest methods of making the task easier is to use a suspension traverse. This is basically a line of non stretch rope placed from one point to another, which men and equipment can be moved up or down any obstacle, with the only limiting factors being the length of the rope. This class will provide your unit with the knowledge to build a suspension traverse.

2. **OVERVEIW.** The purpose of this period of instruction is to introduce the students to the method of erecting, and operating a suspension traverse. This lesson relates to cliff assaults.

INSTRUCTOR NOTE: Have students read learning objectives.

3. **INTRODUCE LEARNING OBJECTIVES**

   a. TERMINAL LEARNING OBJECTIVE. In a summer mountainous environment, conduct a suspension traverse as a member of a squad, in accordance with the reference.

   b. ENABLING LEARNING OBJECTIVES. (MLC) and (ACC)

      (1) Without the aid of references, list in writing the considerations when selecting a site to establish a suspension traverse, in accordance with the reference.

      (2) Without the aid of references, construct a suspension traverse utilizing an A-frame, in a specified time limit and in accordance with the references.

      (3) In a summer mountainous environment, construct a belay line, in accordance with the references.

(4) Without the aid of references and given a list of the hand and arm signals, demonstrate the hand and arm signals for operating a suspension traverse, in accordance with the references.

c. ENABLING LEARNING OBJECTIVES. (SMO)

(1) With the aid of references, state orally the considerations when selecting a site to establish a suspension traverse, in accordance with the reference.

(2) In a summer mountainous environment, construct a belay line, in accordance with the references.

(3) With the aid of references and given a hand and arm signal, state orally the proper commands, in accordance with the references.

4. **METHOD/MEDIA.** The material in this lesson will be presented by lecture and demonstration. You will practice what you have learned in upcoming field exercises. Those of you with IRF's please fill them out at the conclusion of this period of instruction.

5. **EVALUATION**

a. MLC - You will be tested later in the course by written and performance evaluations on this period of instruction.

b. SMO - You will be tested by a verbal and performance test.

TRANSITION: Are there any questions over the purpose, learning objectives, how the class will be taught, or how you will be evaluated? Let us now discuss the definition of a Suspension Traverse.

BODY (70 Min)

1. (5 Min) **DEFINITION**. A suspension traverse is a high- tension rope installation established at a suitable angle (not less than 30 degrees and not more than 65 degrees approximately) which allows a suspended load (no more than 250 lb. is recommended) to be moved over cliffs, ravines and rivers.

NOTE: The limit of 250 lb. has been recommended not because the suspension traverse would fail, but it is about the limit that can be safely manhandled next to a cliff edge.

TRANSITION: Are there any questions over the definition? Let us discuss the selection of a site for a suspension traverse.

2. (5 Min) **SITE SELECTION**. The three considerations for site selection for a suspension traverse are:

a. Suitable upper and lower anchors must be available.

b. Good loading and unloading platforms.

c. Sufficient clearance for the load.

TRANSITION: Now that we have covered where to establish the suspension traverse, are there any questions? Let's look at the construction.

3. (10 Min) **CONSTRUCTION OF A SUSPENSION TRAVERSE**.

a. Top to Bottom.

(1) When constructing the suspension traverse from top to bottom, a suitable high-tension anchor system will be placed around the top anchor point.

(2) The rope will be deployed and the construction/mule team will rappel down in order to construct the bottom anchor.

(3) Select a sound bottom anchor point and construct a suitable high-tension anchor system.

(4) The rope is pulled to the bottom anchor and clipped through the anchor system's 85.

(5) With an 18" Prusik Cord, tie a French Prusik to the rope and secure it also to the anchor system's 85. This will serve as the brake.

(6) On the rope running from the far anchor point, tie a friction knot and attach an 85 through its loops. Clip the running end of the rope to the 85.

(7) Attach a second 85 into the anchor system and clip the running end of the rope into it.

(8) Where the rope comes out from the friction knot's 85, tie a second friction knot complete with an 85 and clip the running end of the rope through it.

(9) You have now made a 9:1 mechanical advantage system, with a brake attached.

NOTE: If pulleys are available, place them in the 85 carabiner on the friction knot.

b. Tensioning the installation.

(1) Initially the traverse is tensioned by one man. This is accomplished by pulling on the 9:1, the French Prusik is used to hold the tensioned line when the friction knots are moved.

(2) When the installation has been tensioned as much as possible, it is secured by taking the standing end around the anchor point and pulling a large bight through the anchor system's 85.

(3) Still maintaining tension on the installation, tie the bight off with a round turn and two half hitches.

SAFETY NOTE: NO MORE THAN TWO MEN WILL TENSION THE INSTALLATION AT ANY TIME IF USING PULLEYS.

NOTE: When two ropes are used, care should be taken to anchor the lines as close together as possible, in such a manner that the ropes do not cross each other.

c. Bottom to Top.

(1) When constructing a suspension traverse from bottom to top, the lead climber will attach the traverse's ropes to the back of his harness.

(2) Once the lead climber reaches the top safely, the traverse's ropes will be anchored off using a suitable high-tension anchor system. The lead climber will then bring up the number 2 climber.

(3) The installation is constructed and tightened from the bottom.

TRANSITION: Now that we have discussed the construction, are there any questions? If you have to gain height at the top and there is no suitable tree, branch, or rock to place the static line across then you will have to construct and use an A-frame (see A-Frame outline).

4. (5 Min) **USE OF AN A-FRAME**.

a. Bringing up A-frame poles. If you have to gain artificial height at the top and there is no suitable tree branch/rock etc., then the A-frame logs, sling ropes and carabiners need to be moved to the top of the cliff. This can be accomplished in the following manner:

(1) The A-frame logs are tied to the ropes utilizing the Timber Hitch (See ROPE MANAGEMENT). The extra sling ropes and carabiners are clipped into the rope.

(2) The climbers will haul the logs to the top and construct their portion of the installation.

NOTE: The A-frame is constructed in the same manner as taught in *A-FRAMES*. The suspension traverse rope is located in the upper V (apex) of the A-frame. The A-frame can be placed under the static rope before or after the rope is tensioned. The latter way is better on a long suspension traverse as it helps tension the rope further.

b. Securing the A-Frame.

(1) An over the object clove hitch is tied in the middle of a sling rope and placed over one pole at the apex.

(2) A prusik is tied around the suspension traverse lines with the ends of the sling rope on both sides of the A-frame. This is done before erecting the A-frame.

(3) The butt ends must be anchored using a practice coil.

(4) The ropes should be buffed wherever necessary.

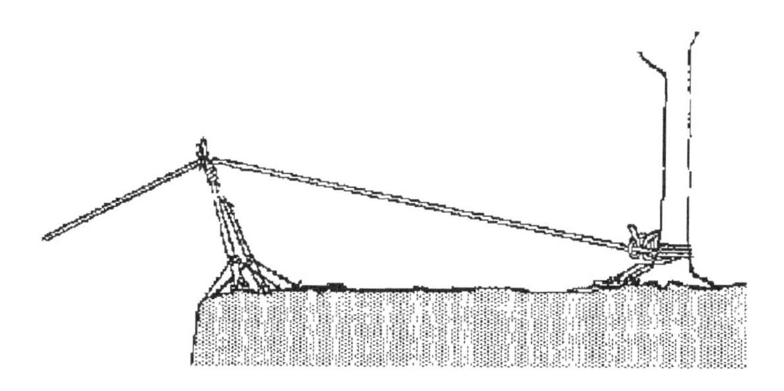

SECURING THE A-FRAME

NOTE: As you can see from the figure above, the A-frame should be inclined forward on the edge of the cliff face, this is done to offset the strain on the A-frame to stop it from collapsing forwards or backwards. The approximate angle is between 45-60 degrees.

TRANSITION: Now that we have discussed use of an A-Frame, are there any questions? Let us now discuss the use of a tree to gain height.

5. (5 Min) **USING A TREE TO GAIN HEIGHT**. When using a tree to gain height in a suspension traverse rope installation; there are some considerations that need to be applied. The following factors are to be considered:

   a. Select a branch that has not been chopped, burned or is rotten.

   b. The branch must be at least 6" in diameter or of suitable thickness to support the intended load.

   c. The branch should be at a sufficient height to offer adequate clearance for the load.

   d. A suitable anchor point must be located behind the intended branch and in relation to the direction of pull of the load.

USING A TREE TO GAIN HEIGHT

e. To secure the rope in place over the branch, use a sling rope and tie a clove hitch around the branch. Secure the pigtail of the sling rope to the rope installation by tying end of the line prusiks. This will prevent the rope from rubbing side to side and also function as the third leg.

TRANSITION: Are there any questions regarding the use of a tree to gain height? If not, let's talk about the use of personnel.

6. (5 Min) **USING PERSONNEL TO GAIN HEIGHT**. If there is no reasonable object available to gain height, personnel may have to manhandle the equipment over the edge. The following factors are to be considered:

a. The edge of the obstacle must be buffed to prevent unnecessary abrasion to the ropes.

b. The weight of the load must not exceed 100 lbs.

c. The personnel handling the load must be secured to the cliff with a safety line.

TRANSITION: Are there any questions over the use of personnel? Let's discuss the technique used to attach loads to the static line.

7. (10 Min) **CONSTRUCTION OF THE BELAY LINE**. The belay rope is used to attach personnel and equipment to the bridging line and belay system. It is tied as follows:

a. Ensure that the belay line is established at the top of the bridge with an adequate anchor, braking system, and a safety prusik.

b. Tie a figure eight loop at the end of the belay line, and attach it with a carabiner to the load.

c. Twelve to eighteen inches from the end of the belay line tie another figure eight loop and attach this loop to the bridging line with a carabiner.

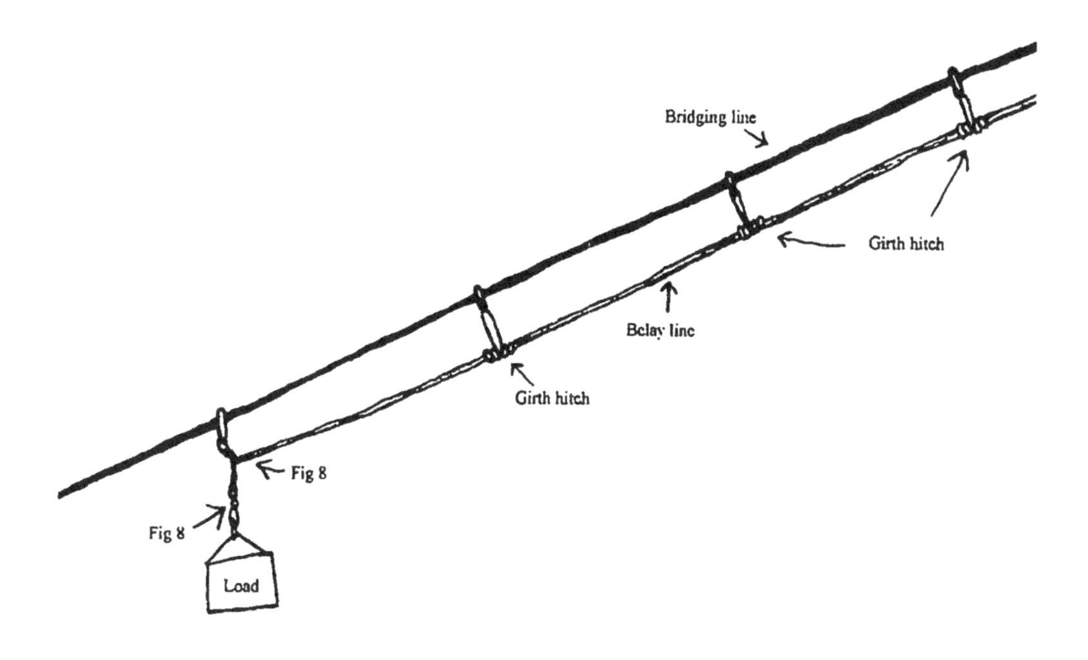

SUSPENSION TRAVERSE BELAY LINE

d. Using web runners and aluminum carabiners girth hitch the web runners to the belay line, and clip into the bridging line at intervals of approximately 15 feet. This is to ensure that the belay line doesn't sag excessively and tangle in trees or on rocks.

TRANSITION: Are there any questions over the construction of the belay line? When lowering gear, a belay system must be used. The Marine assigned as the belay man plays a very important role. Let us now discuss the operation of the suspension traverse.

8. (10 Min) **OPERATION OF A SUSPENSION TRAVERSE**. A suspension traverse is used for the transportation of men and equipment up and down a vertical obstacle. The operating procedure is as follows:

a. Point NCOIC. The Point NCOIC has overall responsibilities for the operation of the suspension traverse. The Point NCOIC duties include:

(1) Supervision of the construction of the installation.

(2) Control of the load being lowered and raised.

(3) Deploys wed runners on the belay line.

(4) Ensure that the brakes are employed properly.

(5) In command of the mule team.

b. Lowering. Whenever a load is lowered down a suspension traverse, a belayer and a belay line are used to control the load. Ideally the belay man should observe the load being lowered. If this isn't possible, the belay man will receive his commands from the Point NCOIC.

(1) Carabiner Brake.

(a) To make the single carabiner brake, two small carabiners are hooked to the anchor point, opposite and opposed. Two more carabiners are locked, gates to the right, to the first set. The last sets of carabiners are placed across the body of the second set, gates down and right.

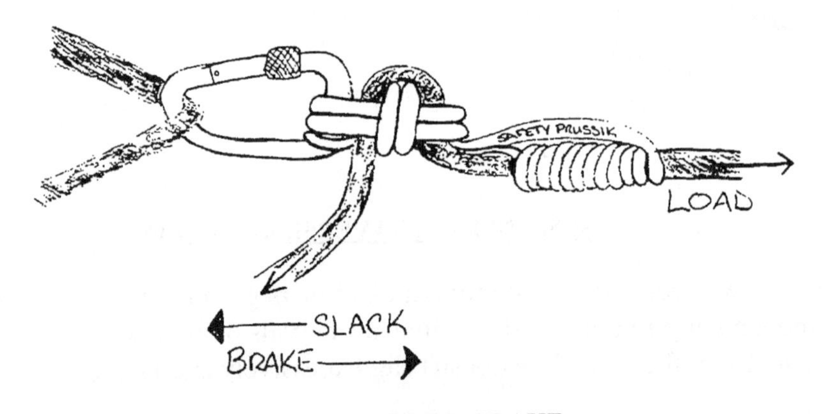

CARABINER BRAKE

(b) If the gates were against the ropes, the rope action would in time unlock the carabiners, with dangerous consequences (if using locking ovals).

(c) To place tension or brake the rope, the belay man simply pulls the rope against the carabiners, towards the load, which creates more friction on the rope.

(d) The carabiner brake is a good method for lowering a load due to lots of friction and good control, but the carabiner brake does not make for a good raising system because of the large amount of friction. This will make unnecessary work for the mule team.

NOTE: If the loads require more or less friction, you can add or take away the carabiners, which are placed across the body.

SMO CUE: TC 8

(2) Figure 8. The Figure 8 rappel device is the easiest method for belaying a load.

(a) Thread a bight of the belay line through the large "O" ring and pull it over the small "O".

(b) Clip a locking carabiner through the small "O" ring and attach it to an anchor point.

(c) Braking is conducted in the same fashion as in rappelling.

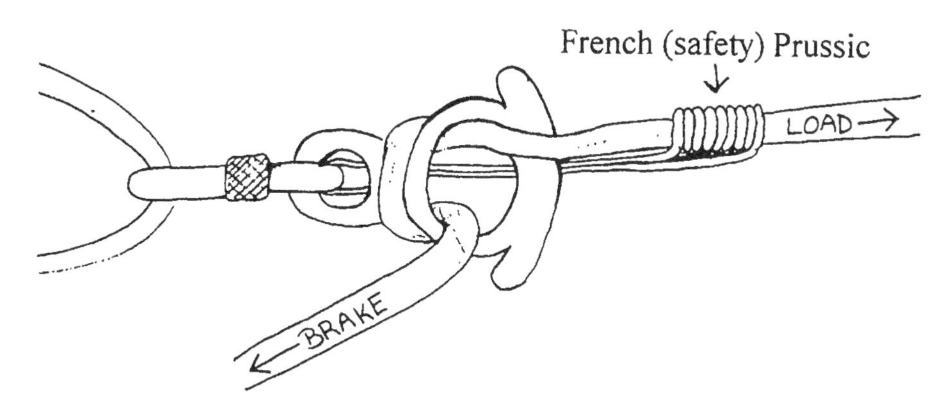

OPERATION OF THE FIGURE 8

(3) Munter Hitch. The Munter Hitch is a good brake because of the minimal gear required. Use with heavy loads.

(4) Stitch Plate. The stitch plate is a user-friendly device that offers quick attachment and removal. Unlike some other belay techniques, the stitch plate has no rope on rope friction. This system can be used for all load sizes.

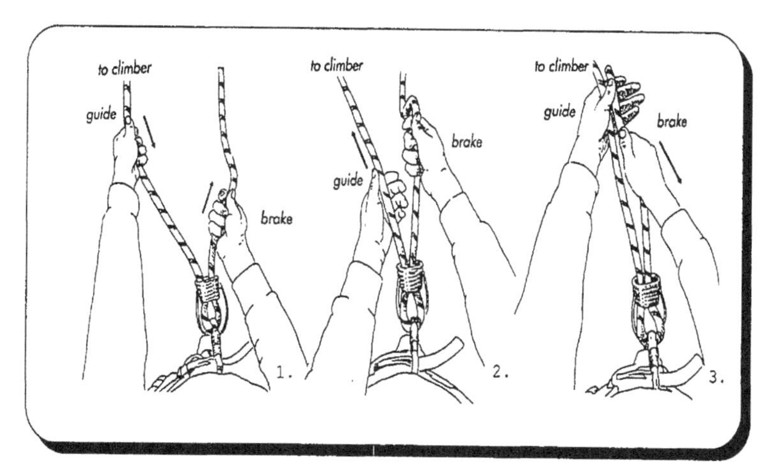

OPERATION OF THE STITCH PLATE

(5) Safety Brake. For added security, an end of the line friction knot is attached to the belay line in the front of the braking device. The safety brake will have it's own anchor point. One man is tasked with the operation of the safety brake.

c. Raising the load. If a large amount of heavy weapons, ammunition, or other logistical equipment is to be moved over an obstacle, a suspension traverse is the most expedient method we can use. Because part of the weight of the supplies to be lifted will be carried by the static line. This is not true in a vertical hauling line where all the weight is suspended vertically. The main method of raising supplies is by utilizing a "Mule Team". This is a group of men 6 or more who will do the lifting.

(1) Mule Team.

(a) To establish a mule team, the belay/haul line is run through a pulley or a steel locking carabiner at the top of the system, just as for lowering.

NOTE: This is why choosing a belay device that is suitable for the load and can be removed quickly is so important.

(b) If the cliff head does not afford an open area for the mule team to operate directly back, the haul/belay line can be re-directed. The line is then led away at an angle to a cleared area.

(c) Six or more men assemble on the rope, on the side away from the cliff. They will haul supplies and personnel up the cliff by grasping the rope and simply walking with it. This uses leg muscles and can be maintained for long periods of time. The mule team responds to a single commander by using the following verbal/hand and arm signals:

COMMAND	MEANING	SIGNAL
*Pick up the rope.	Mule Team picks up the rope.	None
Take the strain.	Take up the slack in the rope.	Arm up, palm out.
Walk away.	Walk away with the rope lifting the load.	Arm motion away from the installation.
Check.	Stop in place, holding the load.	Arm up, fist clinched.
Walk back.	Walk toward the installation holding the load.	Arm motion towards the installation.

* The rope should be held unless specifically ordered to lay it down. Also the use of the prusik braking system on the belay rope is necessary, in the event that the mule team lose control of the rope.

* As with all installations requiring commands, some form of tactical commands need to be established. Those can be established by unit SOP.

TRANSITION: Now that we have discussed the commands, are there any questions? We will now discuss load considerations.

9. (5 Min) **GENERAL LOAD CONSIDERATIONS**. The type of equipment that needs to be lowered or raised will range from weapons to ammunition, personnel to litters, all of which need to be handled differently.

   a. Weapons. Normally the types of weapons that would have to be raised or lowered are the crew served weapons organic to a rifle battalion, i.e. 81mm Mortar, 50 caliber machine gun, dragon etc. The basic method of securing these weapons is accomplished by using two sling ropes and tying a clove hitch to the front and rear of the weapon. A figure of eight is then tied into the ends and attached to the suspension traverse. The belay line is now attached to the load.

   b. Ammunition/Equipment. The articles to be moved are to be secured through bundling or banding them together. A rope is then taken and tied around the equipment in a package wrap fashion. The rope is wrapped one time around the bundle, then when the rope is brought back together; the ends will be crossed and wrapped around the remainder of the bundle (90 degrees to the previous wrap). This is then secured by using a square knot with two over hands and attached to the carrier rope by a steel locking carabiner.

   c. Litters. The two standard litter types used are the collapsible and the stokes. These litters are to be rigged for a medevac as per MOUNTAIN CASUALTY EVACUATION. The litter to be used will be attached to the Suspension Traverse as follows:

      (1) To attach the belay line to the litter you tie a figure of eight loop at the end of the belay line, then attach it to the lower body pre-rigs. Then tie a second figure of eight loop approximately four to five feet from the first figure eight loop, and attach it to the upper body pre-rigs.

NOTE: Make sure that when you are doing this, that you are at a safe distance from the edge of the cliff head, and the brakeman is on belay.

      (2) At this stage it is worth checking and adjusting the length of the pre-rigs, to ensure that the casualty's head is higher than his feet. Once all that is done and checked, the belay line is taken tight and the litter is attached to the suspension traverse, the lower body carabiner is attached first, followed by the upper body carabiner. This is awkward to do. The helpers must have a safety line on while moving around at the cliff head.

      (3) Before you start to lower, a last minute safety check is essential to ensure that carabiners are locked, knots are secure, casualty is secure, that the brake man is on the brake and the safety man is manning the prusik. Now you can commence lowering or raising the casualty.

   d. Personnel. Personnel will wear a rappel seat clipped into the belay line.

TRANSITION: Are there any questions about load considerations? If there are none for me, then I have some questions for you.

PRACTICE (CONC)

a. Students will construct and operate a suspension traverse.

PROVIDE HELP (CONC)

a. The instructors will assist the students when necessary.

OPPORTUNITY FOR QUESTIONS (3 Min)

1. QUESTIONS FROM THE CLASS

2. QUESTIONS TO THE CLASS

Q. What are the three considerations in selecting a site for a suspension traverse?

A. (1) Good upper and lower anchors.
(2) Good loading and unloading platforms.
(3) Sufficient clearance for the load.

Q. What are the commands given between the belay man and the individual going down a suspension traverse?

A. (1) Last name, on suspension traverse!
(2) Last name, on belay!
(3) Slack.
(4) Tension.
(5) Last name, off suspension traverse!
(6) Last name, off belay!

Q. What are the five hand and arm signals used when using a mule team?

A. (1) Pickup the rope.
(2) Take the strain.
(3) Walk away.
(4) Check.
(5) Walk toward.

SUMMARY (2 Min)

a. During this period of instruction, we have covered the correct selection of a site and the equipment needed for a suspension traverse, techniques for construction, lowering and raising loads, commands used for these operations, and general considerations.

b. Those of you with IRF's please fill them out at this time. We will now take a short break

UNITED STATES MARINE CORPS
Mountain Warfare Training Center
Bridgeport, California 93517-5001

SML
SMO
ACC
02/11/02

LESSON PLAN

RAPPELLING

INTRODUCTION

(5 Min)

1. **GAIN ATTENTION**. Rappelling is a method for descending vertical or near vertical cliffs. It will enable your unit to move quickly over rugged terrain to swiftly accomplish its mission. Learning how to establish a rappel site and then applying it is an essential skill that all Marine units should possess while operating in a mountainous environment.

2. **OVERVIEW**. The purpose of this period of instruction is to introduce the students to rappelling by discussing the dangers, site selection considerations, commands involved in rappelling, and the different types of rappels. This lesson relates to cliff assault.

3. **INTRODUCE LEARNING OBJECTIVES**.

   a. TERMINAL LEARNING OBJECTIVES. In a summer mountainous environment and given a designated cliff face, conduct rappelling operations, in accordance with the references.

   b. ENABLING LEARNING OBJECTIVES. (SML) and (ACC)

      (1) Without the aid of references, list in writing the criteria involved in site selection for a rappel site, in accordance with the references.

      (2) Without the aid of references, list in writing the duties of the rappel point NCOIC, in accordance with the references.

      (3) Without the aid of references, describe in writing the types of rappels, in accordance with the references.

      (4) In a summer mountainous environment, rappel down a moderate slope using a hasty rappel, in accordance with the references.

(5) In a summer mountainous environment, execute a seat-shoulder rappel, in accordance with the references.

(6) In a summer mountainous environment, rappel down a cliff face at night using a seat-hip rappel, in accordance with the references.

(7) Without the aid of references, list in writing the duties of the first man down a rappel lane, in accordance with the references.

(8) Without the aid of references, execute the commands used between a rappeller and his belay man when conducting a rappel, in accordance with the references.

(9) In a summer mountainous environment, tie off on a cliff face while conducting a seat-hip rappel, in accordance with the references.

(10) In a summer mountainous environment, retrieve a rappel rope, in accordance with the references.

c. ENABLING LEARNING OBJECTIVES. (SMO)

(1) With the aid of references, state orally the criteria involved in site selection for a rappel site, in accordance with the references.

(2) With the aid of references, state orally the duties of the rappel point NCOIC, in accordance with the references.

(3) With the aid of references, describe orally the different types of rappels, in accordance with the references.

(4) In a summer mountainous environment, rappel down a moderate slope using a hasty rappel, in accordance with the references.

(5) In a summer mountainous environment, execute a seat-shoulder rappel, in accordance with the references.

(6) In a summer mountainous environment, rappel down a cliff face at night using a seat-hip rappel, in accordance with the references.

(7) With the aid of references, state orally the duties of the first man down a rappel rope, in accordance with the references.

(8) With the aid of references, execute the commands used between a rappeller and his belay man when conducting a rappel, in accordance with the references.

(9) In a summer mountainous environment, tie off on a cliff face while conducting a seat-hip rappel, in accordance with the references.

(10) In a summer mountainous environment, retrieve a rappel rope, in accordance with the references.

4. **METHOD/MEDIA**. The material in this lesson will be presented by lecture and demonstration. You will practice what you have learned in upcoming field training exercises. Those of you with IRF's please fill them out at the end of the period of instruction.

5. **EVALUATION**.

a. MLC - You will be tested later in the course by written and performance evaluations on this period of instruction.

b. ACC - You will be tested later in the course by written on this period of instruction and by performance evaluations during the cliff assault.

c. SMO - You will be tested by a verbal and performance evaluation on T- 11.

TRANSITION: Are there any questions over the purpose, learning objectives, how the class will be taught, or how you will be evaluated? First, let's discuss why rappelling is dangerous and site selection.

BODY (90 Min)

1. (2 Min) **INHERENT DANGER OF RAPPELLING**. Rappelling is inherently dangerous because rappellers rely totally on the equipment.

a. To ensure a safe training evolution, two ropes will be utilized at MWTC.

b. All the rappels taught at MWTC can be completed with a single rope if the situation arises.

c. If utilizing a one-rope system with the carabiner wrap, the rope will be attached to the carabiner with two wraps vice one.

TRANSITION: Are there any questions over the inherent dangers of rappelling? If not, let's discuss the criteria's involved in site selection.

2. (5 Min) **SITE SELECTION**. When selecting a rappel site consider these three factors.

a. There must be a good anchor. As previously taught in *NATURAL* and *ARTIFICIAL ANCHORS*, natural anchors are preferred.

b. The rappel route down should be as free of obstacles (i.e., vegetation, debris) as possible.

c. There must be suitable loading and unloading platforms.

NOTE: The evaluation of a site for the above factors should be made by the rappel point NCOIC, who should be the most experienced rappeller in the unit.

TRANSITION: Are there any questions over site selection? Once an area has been found that meets these requirements, a rappel site may now be established. Let us now talk about duties of the rappel point NCOIC.

3. (10 Min) **DUTIES OF THE RAPPEL POINT NCOIC**. Once a rappelling site has been selected, one person will be appointed to each rappel lane as a rappel point NCOIC. These individuals should have experience as rappellers. The rappel point NCOIC has ten duties and responsibilities, which are:

   a. Ensures that the anchor points are sound and that the knots are properly tied.

   b. Ensures that loose rock and debris is cleared from the loading platform.

   c. Allows only one man on the loading platform at a time and ensures that the rappel point is run in an orderly manner.

   d. Ensures that each man is properly prepared for the particular rappel; i.e. gloves on, sleeves down, helmet secured, rappel seat tied correctly and secured properly.

   e. Attaches the rappeller to the rope and ensures the rappeller knows the proper braking position for that particular rappel.

   f. Ensures that the proper commands or signals are used.

   g. Dispatches each man down the rope.

   h. The rappel point NCOIC will be the last man down the rope.

   i. The rappel point NCOIC will ensure that the ropes are inspected after every 50 rappels.

   j. The rappel point NCOIC will maintain a rope log.

TRANSITION: Before we go any further are there any questions over the duties of the rappel point NCOIC? We will now discuss the different types of rappels taught here at MWTC.

4. (5 Min) **TYPES/USE**. The three types of rappels and when they are preferred to be used are:

   a. Hasty Rappel. It is used when carrying loads down moderate slopes.

   b. Seat-Shoulder Rappel. It is used for heavily laden troops over vertical to near vertical cliff faces.

   c. Seat-Hip Rappel. It is used when carrying loads over vertical to near vertical faces.

TRANSITION: Are there any questions on the different types of rappels? Next we will discuss how each of these rappels are used.

5. (5 Min) **HASTY RAPPEL**. The hasty rappel is the easiest type of rappel to prepare for. It requires no equipment other than a rope and gloves.

   a. Conduct. A hasty rappel is conducted in the following manner:

      (1) Sleeves will be rolled down and gloves will be put on.

      (2) Face slightly sideways.

      (3) Place the rappel rope across your back, grasping it with both hands, palms forward, and arms extended.

      (4) The hand nearest the anchor is the guide hand. The hand farthest from the anchor is the brake hand.

      (5) Lean out at a moderate angle to the slope.

      (6) Descend down the hill facing half sideways, taking small steps and continually looking downhill while leading with the brake hand.

      (7) Feet should not cross and the downhill foot should lead at all times.

   b. Braking. The steps for braking during a hasty rappel are as follows:

      (1) Bring the lower (brake) hand across the front of the chest to brake.

      (2) At the same time, turn to face up toward the anchor point.

HASTY RAPPEL

TRANSITION: Are there any questions about the technique of the hasty rappel? Now let us discuss the seat-shoulder rappel.

6. (5 Min) **SEAT-SHOULDER RAPPEL**. The seat-shoulder rappel relies on friction as the main effort of controlling the descent. It is very efficient for men with heavy packs because it provides support for heavy loads on the back.

a. Conduct. A seat-shoulder rappel is conducted in the following manner:

(1) Put on your rappel-seat, roll down your sleeves and put on your gloves.

NOTE: To avoid causing a possible injury, it is advised that the rappel seat be constructed at the rear of the rappel site loading platform. If the rappel seat is worn for too long, the rappel seat could loosen up enough to cause you to slip out of your seat while rappelling.

(2) The steel locking carabiner is placed on the rappel-seat so that the gate opens down and away, to prevent the gate from opening once the wraps are placed into the carabiner.

(3) Step-up to the rope with your left shoulder facing the anchor.

(4) The rappel rope is attached to the rappeller's hard point carabiner as follows:

(a) Snap the rope into the locking carabiner.

(b) Taking slack from the standing (anchor) end of the rope, make one wrap with the rope around the body of the carabiner and back through the gate.

(c) Ensure that the locking nut of the carabiner is fastened to lock the carabiner closed.

NOTE: If you are using only onc rope to rappel with due to the tactical situation or equipment availability, the procedures are the same, EXCEPT the individual will make two wraps around the body of the carabiner instead of one.

(5) Take the rope across your chest, over your left shoulder, diagonally across your pack and down to the right (brake) hand.

(6) Descend by walking down the cliff using the braking procedure to control the rate of descent. Look under your brake arm for possible obstacles to avoid.

b. Braking. The steps for braking during a seat-shoulder rappel are as follows:

(1) Lean back.

(2) Face directly uphill while bringing the brake hand across the chest.

TRANSITION: Are there any questions concerning the seat-shoulder rappel? If not, let us discuss the seat-hip rappel.

7. (10 Min) **SEAT-HIP RAPPEL**. The seat-hip rappel is the most commonly used rappel.

a. Conduct. A seat-hip rappel is conducted in the following manner:

(1) Construct the rappel seat; roll down sleeves and put gloves on.

(2) The steel locking carabiner is placed on the rappel-seat so that the gate opens up and away.

(3) Step up to the rope with your left shoulder facing the anchor.

(4) The rappel rope is snapped into the carabiner as follows:

(a) Snap the rope into the locking carabiner.

(b) Taking slack from the standing (anchor) end of the rope, make one wrap with the rope around the body of the carabiner and through the gate again.

(c) Ensure that the locking nut of the carabiner is fastened to lock the carabiner closed.

SEAT-HIP RAPPEL

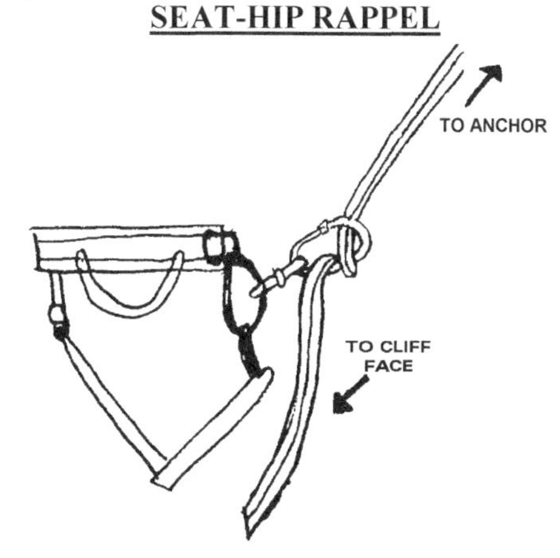

NOTE: If you are using only one rope to rappel with due to the tactical situation or equipment availability, the procedures are the same, EXCEPT the individual will make two wraps around the body of the carabiner instead of one.

(5) The rappeller will grasp the running end of the rope with the brake (right) hand, palm down and turned slightly inboard, near the hip.

RAPPELLER WITH BAKE HAND ON ROPE

b. Braking. The steps in braking for a seat-hip rappel are as follows:

(1) Grasp the rope tightly with the brake hand.

(2) Take the brake hand and place it in the small of the back. This will create enough friction to stop all momentum.

NOTE: At no time will you bound or jump while you are descending. You "walk down" the cliff face using the proper braking procedure to control your rate of descent.

TRANSITION: Now that we have discussed the hasty rappel, seat- shoulder rappel and the seat-hip rappel, are there any questions? Let's now talk about the safety factors of the first man down.

8. (5 Min) **SAFETY OF THE FIRST MAN DOWN**. To ensure the safety of the first man down, the rappeller will:

a. Before deploying the ropes, tie the two ends of the ropes together with an overhand knot.

b. Tie a friction knot on the standing end of the rappel rope with a Prusik cord.

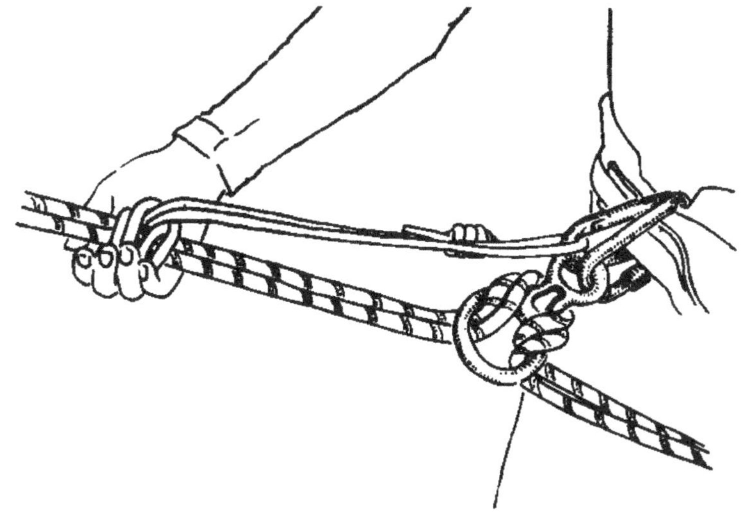

SAFETY PRUSIK

c. Attach the rappel ropes to the harness's hard point with a locking carabiner.

d. Attach the Prusik cord to the harness's hard point with a locking carabiner. This cord is also referred to as a safety Prusik.

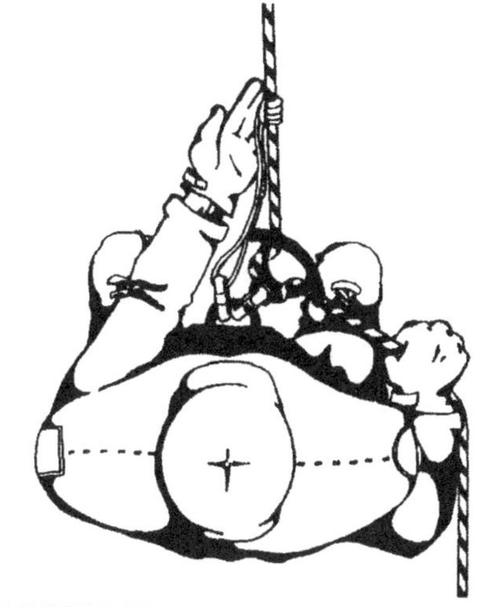

RAPPELLER WITH SAFETY PRUSIK

NOTE: If the Safety Prusik is too short, it can be extended with a web runner.

TRANSITION: We've just discussed the safety of the first man down, are there any questions? Now let's talk about the duties of the first man down.

9. (10 Min) **DUTIES OF THE FIRST MAN DOWN**. The first man down the rope has specific duties. They are as follows:

   a. Selects a smooth route down for the ropes.

   b. Clears the route of loose rocks and debris.

c. The first man down will untie the overhand knot and straighten the ropes out once he reaches the bottom.

d. The first man down belays the next man down the rope. There are two methods of belaying the next man down. The method used will be decided upon the restriction of the unloading platform.

(1) Confined unloading platform:

(a) The belay man will stand facing the cliff face, arms up with palms facing upward and over lapping. The rope will pass through the opening of the over lapping hand's thumb and index fingers.

RAPPELLER AND BELAY MAN

(b) To stop a fallen rappeller, the belay man will grab the rope with closed fists and pull straight down bringing the forearms parallel to the deck.

BELAY MAN

(2) Open unloading platform:

(a) The belay man should stand facing the cliff with the ropes under both arms and behind his back.

(b) To stop a fallen rappeller, the belay man will move away from the cliff face while holding the ropes firmly as in a standing hip belay.

e. Take charge of personnel as they arrive at the bottom to include appointing a belay man.

TRANSITION: Now that we have discussed the duties of the first man down, are there any questions? Next we will discuss the individual rappelling commands.

10. (5 Min) **RAPPELLING COMMANDS.** In order to conduct rappelling operations safely, it is essential that everyone understands the sequence of events. The following voice commands or rope tugs will be utilized:

VOICE COMMANDS	GIVEN BY	MEANING
"Lane # on Rappel"	Rappeller	I am ready to begin rappelling
"Lane # on Belay"	Belayer	I am on belay and you may begin to rappel
"Lane # off Rappel"	Rappeller	I have completed the rappel and am off the rope
"Lane # off Belay"	Belayer	I have completed the belaying of the rappeller

ROPE TUGS	GIVEN BY	MEANING
Three Tugs	Rappel Point NCOIC	The rappeller is ready to begin rappelling
Three Tugs	Belayer	I am on belay and the rappeller may begin to rappel
Three Tugs	Belayer	The rappeller is off the rappel rope

TRANSITION: Are there any questions about the rappelling commands or rope tugs? If not, let us discuss the possibility of tying off.

(10 Min) **TYING-OFF**. Occasionally, it may be necessary to stop during a rappel before reaching the bottom of a cliff. The following sequence is used:

a. The commands used in tying-off are as listed below:

COMMAND	GIVEN BY	MEANING
"Lane # Tying-off"	Rappeller	I am ready to tie-off, give me some slack.
"Lane # Tying-off"	Belayer	I have given you enough slack and you may tie-off.
"Lane # On rappel"	Rappeller	I have completed tying-off and I am ready to resume rappelling.
"Lane # On belay"	Belayer	I am on belay, you may rappel.

b. The procedure used when tying-off is as follows:

(1) The rappeller gives the command, "Lane #, tying-off."

(2) The belayer gives the rappeller slack and gives the command "Lane # Tying-off". He continues to hold the rope in the belay position, remaining alert and watching the rappeller.

(3) The rappeller quickly brings his brake hand to the twelve o'clock position, so that the running end and standing end are parallel.

(4) The rappeller grasps all ropes with his guide hand as close to the carabiner as possible.

(5) The rappeller steps over the rope so that the running end is going between his legs.

(6) The rappeller releases the rope with his brake hand then reaches under the running end, over the standing end. He then takes up a bight from the running end about two feet long and pulls it over the standing end and under the running end, forming a half hitch.

(7) He pulls the half hitch tightly against the guide hand.

(8) He works the half hitch down snugly against the carabiner while maintaining contact with the guide hand as long as possible.

NOTE: A loose half hitch could bind into the carabiner causing difficulties in clearing the knot out. Make sure that the first half hitch is dressed down tightly before moving it against the carabiner.

(9) Place another half hitch above the one already tied.

(10) To untie, reverse the steps. Remove the safety half hitch, then shrink the first half hitch down to a small loop by grabbing the running end of both ropes and pulling them straight out to the left. Place your right hand in the middle of the two bights until they are snug on all four fingers. Then remove one finger and make the bights snug on three fingers. Repeat this process until it's down to one finger and both the bights are equal.

(11) Now, grasp the running end with both hands and smartly jerk the running end of the ropes upward to pop the small loop out from the first half hitch. From this position, keep the guide hand around both of the ropes next to the carabiner; step back over the rope so that the running end is to your right side. Grab just the running end with your brake hand and quickly set the brake behind the small of your back then, readjust you guide hand onto the standing end. You now should be in the seat-hip rappel position.

NOTE: The wraps may bind up some after untying, where no further movement down is possible. To alleviate this, keep the brake on and force your body weight down to pop the wraps in the carabiner back to their intended position and then continue on with the rappel.

(12) The rappeller gives the command "Lane #, on rappel."

(13) The belayer takes up the slack and gives the command "Lane #, on belay."

(14) The rappeller resumes to rappel down.

TRANSITION: Are there any questions over tying off? We will now discuss how to perform different types of rescues during rappelling operations.

11. (5 Min) **RESCUE TECHNIQUES**. When conducting rappelling operations, the possibility of a rappeller getting caught on the rope due to either his clothing or equipment can occur. There are two types of rescues that can be performed to free the rappeller from this situation.

a. The Self-Rescue Technique.

(1) After realizing that you are caught up on the rope, check with your belay man to ensure that he has a solid brake set.

(2) Using a safety Prusik, tie a friction knot onto the rappel rope approximately an arms length above the area that is fouled. Anchor the other end of your safety Prusik to your rappel seats hard point. This is the same technique used as discussed earlier with the first man down the rope.

(3) Work the friction knot up until there is no tension on the rappel device. This will give you the necessary slack to free the malfunction.

(4) Once you have corrected the problem, continue to rappel down using the first man down method.

b. The Buddy Rescue Technique. This method is used when the rappeller is unable to correct the problem by himself.

(1) A rescuer will rappel down on another rope to the disabled rappeller and tie-off.

(2) The rescuer will then establish a safety Prusik onto the victim as taught in the self-rescue technique.

(3) After clearing the malfunction, continue with rappelling operations.

c. Rappelling Casualty Rescue. There are three different methods in which to get an injured rappeller to the bottom.

(1) The first method is to allow the belay man to lower the casualty by slowly releasing the tension on the rappel rope.

(2) The second method is used for critical injuries or when the belay man cannot properly control the casualty's descent.

(a) The rescuer will rappel down using another rope. Once the rescuer gets to the casualty, the rescuer will have his belay man brake him off, enabling the rescuer to use both hands.

(b) If necessary, the rescuer will perform the basic first aid needed. Once the casualty is ready to be lowered, the rescuer will call down to both of the belay men to simultaneously lower both men. The rescuer will hold onto the casualty the entire way down so that he doesn't bounce off the rock face.

(3) The third method is called a tandem rappel. Its used to rescue a casualty who has sustained an injury serious enough that he cannot operate the rappel device himself and requires the assistance of a second rappeller. This type of rescue can be conducted from the top of the cliff face or while on rappel, depending on where the injury occurred.

(a) If the injury occurs while on top of the cliff face, the following steps should be taken:

1. Take either a long sling rope or Prusik cord and tie a figure-eight knot offset so that one length of the cord is longer then the other.

2. Tie a figure-eight knot at each end.

3. Take a rappel device and attach it to the offset figure-eight loop with a carabiner and attach it to the rappel rope.

4. Next, attach the casualty's hard point to the short end of the sling rope/Prusik cord and the rescuer to the other end.

5. The rescuer will then utilize a safety Prusik in the same method as the first man down technique, except that the friction knot will be tied below the rappelling device and controlled with the brake hand.

6. The rescuer will then maneuver himself under the casualty to provide assistance and support.

7. At this point you are ready to rappel both rescuer and casualty at the same time.

NOTE: Your rappel device should be far enough in front of you so that the casualty will not be able to reach it.

(b) If the injury occurs during the rappel, the following steps are taken:

1. The rescuer will preset the same system on a separate rappel rope and rappel down to the casualty.

2. The rescuer will then take the short end of the sling rope and attach it to the casualty's hard point.

3. With a knife, the rescuer will cut the casualty's rappel rope away.

4. Rappel down to the bottom of the cliff face.

NOTE: THIS IS THE ONLY TIME THAT A KNIFE WILL BE USED DURING A RAPPEL. EXTREME CAUTION SHOULD BE USED. IT IS NOT TO BE DONE IN TRAINING, BUT ONLY DURING AN ACTUAL RESCUE AND AS A LAST RESORT.

TRANSITION: Are there any question over rescue techniques? Let's now discuss how to make our rappel installation retrievable.

12. (10 Min) **RETRIEVABLE RAPPELS**. Once a unit has rappelled down a vertical obstacle, it may be necessary to retrieve the rope(s). Depending on the height of the obstacle either one or two ropes will used to construct the rappel lane.

a. One-Rope Retrievable Rappel:

(1) Find the middle of the rope and place it directly behind a suitable anchor point.

(2) Join the pigtails of the rope with an overhand/figure 8 knot and deploy the rope down the obstacle.

(3) On one side of the rope in front of the anchor point, tie an over-the-object clove hitch onto a locking carabiner.

(4) On the other side of the rope, tie a figure 8 loop and attach it to the same carabiner.

(5) The first man down will utilize a safety Prusik and untie the overhand knot after he reaches the bottom.

(6) All others conduct proper rappel procedures.

(7) The last man down will disconnect the carabiner from the rope, and ensuring that the middle of the rope is directly behind the anchor point, rappel down the rope. If point NCO is an assault climber, use the stitch plate/ATC-type belay device to straighten the ropes.

(8) The rope is then retrieved by pulling on either end of the rope.

b. Two-Rope Retrievable Rappel. This type of system is identical to the one rope retrievable rappel except for a few considerations.

(1) The reason for using two ropes vice one rope is that the height of the rappel is greater.

(2) When two ropes are used they should be joined together using a square knot. This knot will be placed out of the system when securing the carabiner to the ropes. This will prevent total failure of the system if the knot should fail.

(3) All else remains the same as the one rope retrievable rappel except:

(a) The last man down will disconnect the carabiner from the rope and will move the joining knot as close to the vertical obstacle's edge as possible. This will prevent the knot from possibly getting caught up while retrieving.

(b) The last man down will then place a carabiner on the rope below the knot. This will enable him to know which line to pull for retrieval. If the last man down is an assault climber, use the stitch plate/ATC-type belay device to straighten the ropes.

TRANSITION: Now that we have covered retrievable rappels, are there any questions? As Marines we are required to carry a variety of weapons and equipment into combat. Let us discuss considerations for our equipment so that it doesn't enable our rappelling efforts.

13. (3 Min) **EQUIPMENT**. Equipment should be worn in accordance with unit SOP. The unit has the responsibility to determine which methods it feels are most beneficial to the mission. Weapons should be worn across shoulder; muzzle down and away from the brake hand with a tight sling securely attached to the weapon. Weapons should also be dummy corded to the individuals.

TRANSITION: Are there any questions over equipment at this time? If you have none for me, then I have some for you.

PRACTICE (CONC)

a. Students will execute all rappel considerations outlined.

PROVIDE HELP (CONC)

a. The instructors will assist the students when necessary.

OPPORTUNITY FOR QUESTIONS (3 Min)

1. QUESTIONS FROM THE CLASS

2. QUESTIONS TO THE CLASS

Q. What are the three criteria involved in site selection for a rappel site?

A. (1) Good anchors.
(2) The route down the cliff face should be as free of obstacles as possible.
(3) There must be suitable loading and unloading platforms.

Q. What are the three types of rappels?

A. (1) Hasty Rappel
(2) Seat-Shoulder Rappel
(3) Seat-Hip Rappel

Q. When would the hasty rappel be the desired type of rappel to use?

A. When carrying loads down a moderate slope.

SUMMARY (2 Min)

a. During this period of instruction we have covered site selection, how to establish a rappel site and how it should be run, which included duties of the rappel point NCOIC and duties of the first man down a rope. We've also covered how to establish a retrievable rappel and the three types of rappels, as well as how to tie-off.

b. Those of you with IRF's please fill them out at this time and turn them in to the instructor. We will now take a short break.

UNITED STATES MARINE CORPS
Mountain Warfare Training Center
Bridgeport, California 93517-5001

SML
SMO
ACC
02/11/02

LESSON PLAN

BALANCE CLIMBING

INTRODUCTION

(5 Min)

1. **GAIN ATTENTION**. In almost every mountain operation, you will be required to negotiate rocky slopes and faces. "While not all will require technical skills, some of them will and you will be unable to accomplish your mission unless you have mastered the basic skills of balance climbing.

2. **OVERVIEW**. The Purpose of this period of instruction is to introduce the student to the techniques used in balance climbing. This will be accomplished by discussing safety requirements, individual preparations, commands, types of holds, and considerations for body position and movement. This lesson relates to top roping and cliff assaults. (SMO)

INSTRUCTOR NOTE: Have students read learning objectives.

3. **INTRODUCE LEARNING OBJECTIVES**

   a. TERMINAL LEARNING OBJECTIVE. In a summer mountainous environment, execute balance climbing, in accordance with the references.

   b. ENABLING LEARNING OBJECTIVES. (MLC) and (ACC)

      (1) Without the aid of references, describe in writing the safety requirements for balance climbing, in accordance with the references.

      (2) Without the aid of references, list in writing the individual preparations for a balance climb, in accordance with the references.

      (3) In a summer mountainous environment, execute the duties of a spotter for a balance climb, in accordance with the references.

      (4) In a summer mountainous environment, execute the commands used between the climber and the spotter during a balance climb, in accordance with the references.

(5) In a summer mountainous environment, execute each type of hold used in balance climbing, in accordance with the references.

(6) Without the aid of references and given the acronym "CASHWORTH", describe in writing the considerations for proper body position while climbing, in accordance with the references.

c. ENABLING LEARNING OBJECTIVES. (SMO)

(1) With the aid of references, state orally the safety requirements for balance climbing, in accordance with the references.

(2) With the aid of references, state orally the individual preparations for a balance climb, in accordance with the references.

(3) In a summer mountainous environment, execute the duties of a spotter for a balance climb, in accordance with the references.

(4) In a summer mountainous environment, execute the commands used between the climber and the spotter during a balance climb, in accordance with the references.

(5) In a summer mountainous environment, execute each type of hold used in balance climbing, in accordance with the references.

(6) With the aid of references and given the acronym "CASHWORTH", describe orally the considerations for proper body position while climbing, in accordance with the references.

4. **METHOD/MEDIA**. The material in this lesson will be presented by lecture and demonstration. You will practice what you have learned in upcoming field training exercises. Those of you with IRF's please fill them out at the conclusion of this period of instruction.

5. **EVALUATION**.

a. MLC - You will be tested later in the course by written and performance evaluations on this period of instruction.

b. ACC - You will be tested later in the course by written and performance evaluations on this period of instruction.

c. SMO - you will be tested by an oral and performance examination.

TRANSITION: Are there any questions over the purpose, learning objectives, how the class will be taught, or how you will be evaluated? Civilians call what you are about to learn "Bouldering". The Marine Corps refers to it as balance climbing. Regardless of the term used, the techniques you will be learning are the basis upon which all your other climbing training will rest. By practicing the techniques you are about to learn, you will acquire the skills necessary to operate successfully in any area where you might encounter precipitous terrain. The first thing we need to know in balance climbing are the safety precautions.

BODY (60 Min)

1. (5 Min) **SAFETY PRECAUTIONS**. There are two safety precautions that always apply to balance climbing. They are as follows:

   a. Never climb more than 10 feet above the ground. By this it is meant that the climber's feet are never more than 10 feet above the ground.

   b. A spotter is required for all balance climbs.

SPOTTER WATCHING THE CLIMBER

TRANSITION: Now that we have discussed the safety precautions, are there any questions? The next thing we will discuss are what an individual has to do to prepare himself for a balance climb.

2. (5 Min) **INDIVIDUAL PREPARATIONS**. Prior to beginning a balance climb there are seven things that the climber must do to prepare himself. They are as follows:

   a. Helmet with a serviceable chinstrap must be worn.

   b. Sleeves rolled down to give hand and arm freedom of movement. Blouse tucked in to your trousers. In case of fall, it may catch on a rock and cause you to flip over sideways.

   c. All watches, rings, and jewelry must be removed before climbing.

d. Gloves will not be worn, as they can slip, and also give a false feel for the rock.

e. Unblouse trousers, if they restrict movement.

f. Soles of boots clean and dry.

g. Select route where vegetation is minimal. Never use vegetation for hand or foot holds.

TRANSITION: Are there any questions over the individual preparations? As I mentioned, you never balance climb without a spotter. Who is this "spotter and what are his duties?

3. (5 Min) **DUTIES OF THE SPOTTER**. The spotter is the balance climber's partner who, rather than climbing himself, acts as the safety man for the climber during the climb. The five duties of the spotter are as follows:

   a. Positions himself directly behind the climber before the climb starts.

   b. Maintains his position facing the cliff, directly below the climber and approximately 3-4 feet away from the base of the cliff, for the duration of the climb. He will move diagonally as necessary to remain below the climber.

   c. The spotter will stand with his feet shoulder width apart and arms ready to stop the climber if he falls.

   d. If the climber falls, the spotter will not "catch" him; he will prevent the climber from falling further down the hill. He will do this by pushing the climber towards the base of the cliff, thereby preventing him from tumbling backwards.

   e. At no time will the spotter allow anyone to come between himself and the face of the cliff while a balance climb is taking place. He will require anyone who wants to pass by his position to go behind him.

TRANSITION: Are there any questions over the duties of the spotter? Now we're almost ready to go out and start climbing. However, before that, let's learn the commands used between the climber and spotter.

4. (5 Min) **SPOTTING AND CLIMBING COMMANDS**. The following are the commands used by both the spotter and climber.

COMMAND	GIVEN BY	MEANING
"Last Name, climbing."	Climber	I am ready to climb.
"Climb Climber."	Spotter	I am ready to spot you.
"Last Name, off Climb."	Climber	I am off the climb.
"ROCK!"	Climber or Spotter.	A rock has been knocked off the rock face and is falling.

"FALLING!"	Climber	I am Falling.

a. If the command "ROCK" is given, all personnel in the vicinity will take the following action:

(1) If close to the cliff face, move against the cliff face with your face against the cliff face and your hands between you and the cliff face.

(2) If not close to the cliff face, look up to locate the rock and avoid it.

5. (5 Min) **ACTIONS IF FALLING**. If, while making a balance climb, the climber feels himself slipping and beginning to fall, he will take the following action:

a. Sound the command "falling".

b. Push himself away from the rock face.

c. Maintain proper body position as follows:

(1) Head up.

(2) Hands out toward the rock.

(3) Body relaxed.

(4) Feet kept below the body, slightly apart.

(5) Ensure you face the cliff face as you fall.

TRANSITION: Now that we have discussed the safety precautions and considerations, are there any questions? Let's move on to the actual climbing.

6. (5 Min) **TYPES OF HOLDS**. There are five basic holds that are used in balance climbing. They are as follows:

a. Push Holds.

(1) Most effective when hands are kept low.

(2) Often used in combination with a pull hold.

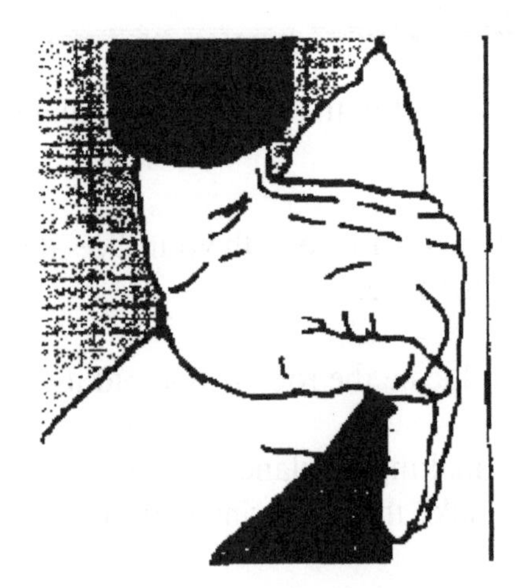
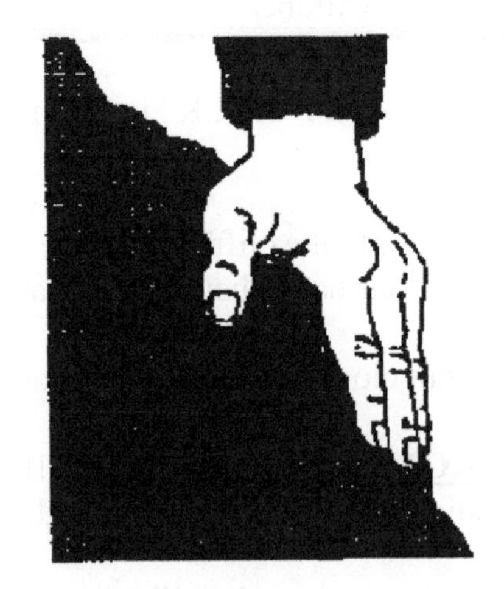

TYPES OF PUSH HOLDS

b. Pull Holds.

(1) The easiest hold to use and, consequently, often overused.

(2) Can be effective on small projections.

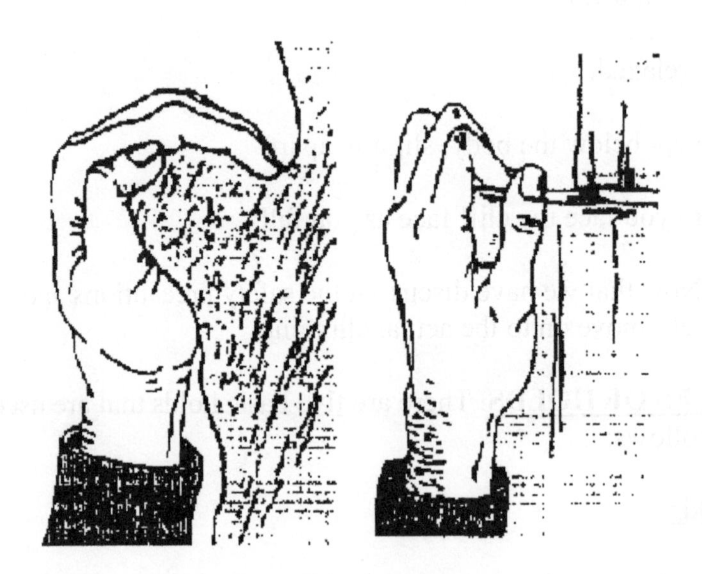

TYPES OF PULL HOLDS

c. Foot Holds

(1) Feet should be positioned with the inside of the foot to the rock.

(2) Use full sole contact as much as possible.

(3) Avoid crossing your feet. If you must cross your feet use a change step. A change step is a method of substituting one foot for the other foot on the same foothold.

(4) Making maximum use of footholds, climbing with your feet, is an effective means of conserving your body strength, since your leg muscles are stronger than your arm muscles.

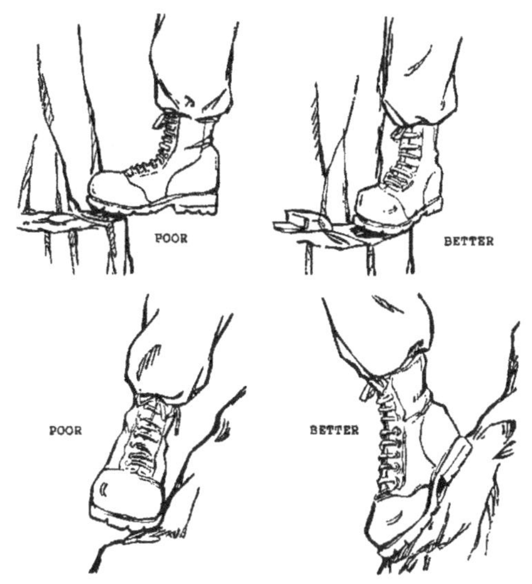

USE OF FOOT HOLDS

d. Friction Holds. A friction hold is anytime you are relying on the friction of your foot or hand against the face of the rock for traction, rather than pushing/pulling against a projection on the face of the rock.

(1) It is a type of hold that feels very insecure to an inexperienced climber.

(2) The effectiveness of this type of hold is dependent upon many things i.e., type, condition and angle of the rock face, type of boot soles, confidence, etc.

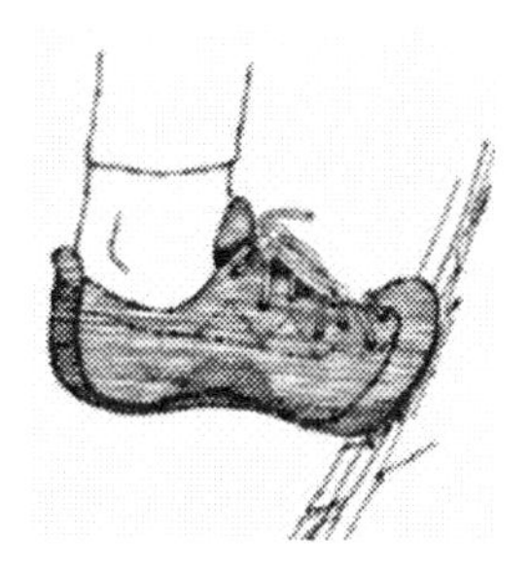

FRICTION USED IN SMEARING

e. Jam Holds. This type of hold involves jamming/wedging any part of your body or your entire body into a crack/opening in the rock.

(1) An important consideration is that you do not jam such that you cannot free that portion of your body after you complete the move. This sounds like a ridiculous

statement; however, you must remember that after you complete your move, you may be withdrawing the portion of your body that you used from a different angle than you inserted it.

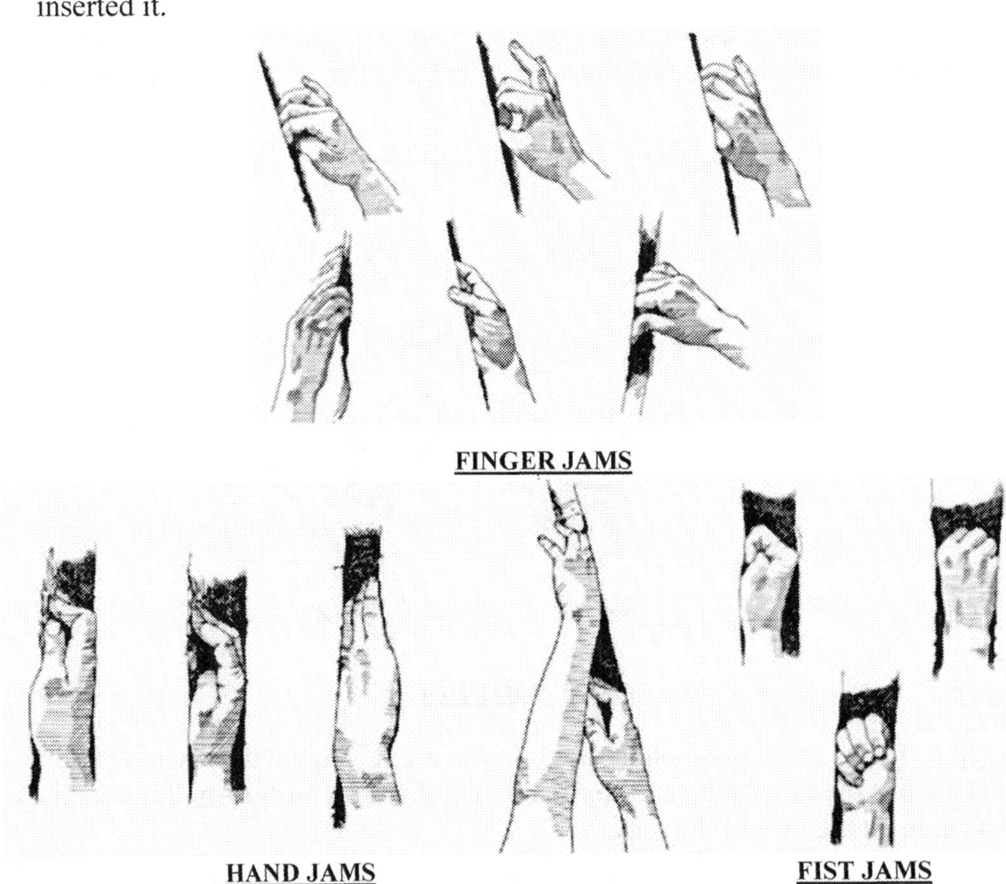

FINGER JAMS

HAND JAMS **FIST JAMS**

7. (5 Min) **COMBINATION HOLDS.** The five types of holds just mentioned above are not just used individually. They are most often used in combinations with each other. Some examples are:

   a. Chimney Climbing. This is when you insert your entire body into a crack in the rock and by using both sides of the opening, and possibly all five types of basic holds, move up' the crack.

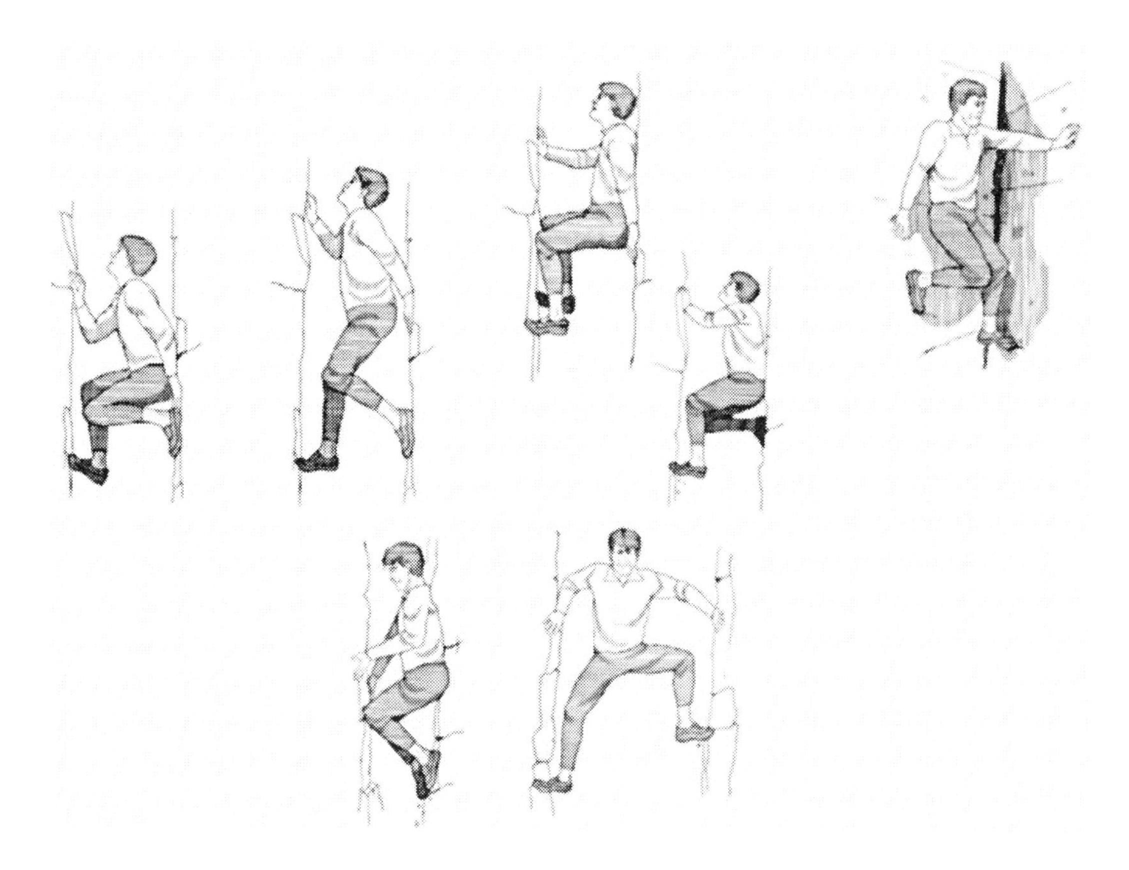

CHIMNEY CLIMBING TECHNIQUES

b. Lie-back. This is a combination of both pull holds with your hands and friction holds with your feet.

LIE BACK TECHNIQUE

c. Push-Pull. As the name implies, this is when you use a push hold and a pull hold together.

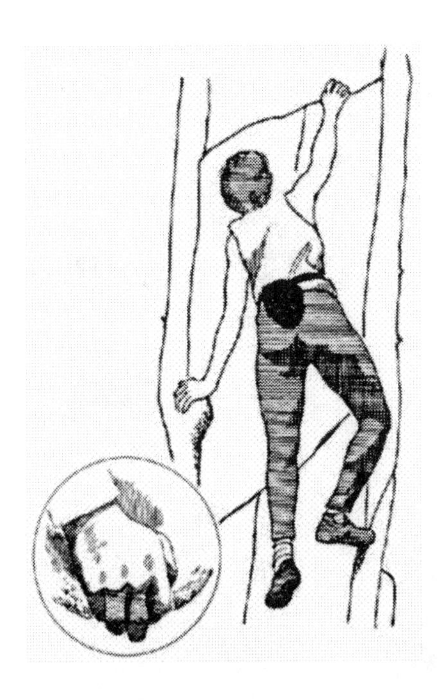

PUSH-PULL

d. Mantling. This is a technique where you continue to climb without moving your hands off a projection by pulling yourself up until your hands are at chest level and then invert your hands and push on the same projection.

MANTELING

MANTELING WITH FACE HOLD

e. Cross-Pressure in Cracks. This is a technique of putting both hands in the same crack and pulling your hands apart to hold/raise yourself.

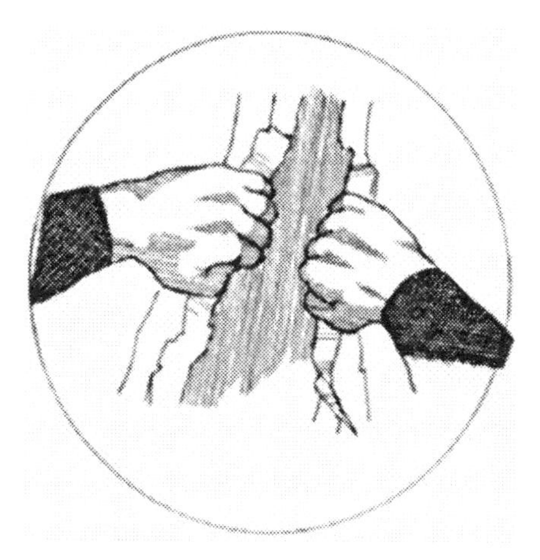

CROSS-PRESSURE METHOD

f. Inverted. Pull or push.

INVERTED TECHNIQUE USING COUNTERFORCE PRESSURE BETWEEN FEET AND HANDS

g. Pinch. As the name implies this is a grip used on tiny little nubbins.

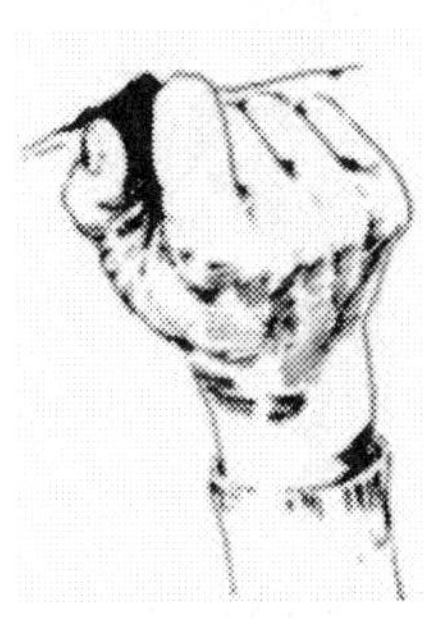

PINCH

h. Stemming. The spreading of arms or legs to maintain a proper body position. (i.e. usually used in a book or chimney.)

STEMMING

TRANSITION: Are there any questions over the types of holds? There are some general guidelines that aid the climber that we should be aware of

8. (5 Min) **GENERAL USE OF HOLDS**. How you use an individual hold is dependent on your experience level, or sometimes, your imagination. Here are some general guidelines.

   a. Most handholds can be used as foot holds as you move up the rock.

   b. Use all holds possible in order to conserve energy.

   c. Even small projections may be used as holds.

   d. Do not make use of your knees or elbows due to the reason that it is skin on bone and a slip could occur if pressure is exerted on them. Knees and elbows can be used with the extension of a limb jam.

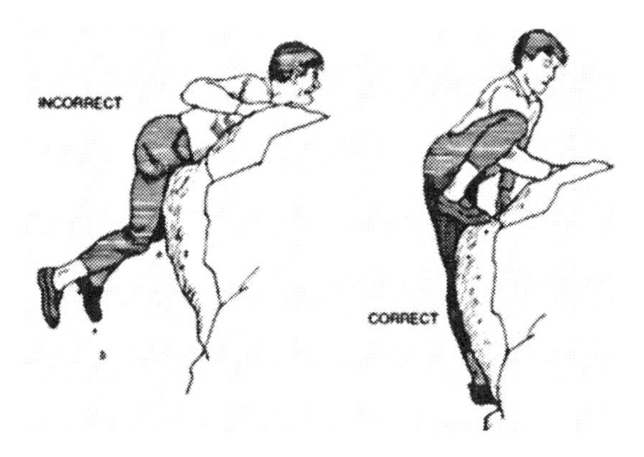

PROPER AND IMPROPER

9. (5 Min) **MOVEMENT ON SLAB**. Movement on slab is based on friction holds.

   a. Use any and all irregularities in the slope to gain additional friction.

   b. Traversing requires both hands and feet.

TRAVERSING

   c. Descending steep slab may require turning inboard to face the slab and backing down.

   d. The biggest mistake in slab climbing is leaning into the rock. Maintain maximum friction by keeping weight centered.

SLAB CLIMBING

TRANSITION: Are there any questions on the general use of holds or movement on slab? To help us maintain proper body position, we will use the acronym "CASHWORTH".

10. (5 Min) **BODY POSITION**

a. The climber should climb with his body in balance by keeping his weight centered over and between his feet. Don't hug the rock. Don't over extend and become "spread-eagled". While climbing, keep in mind the acronym "CASHWORTH" for proper body position and movement.

C - Conserve energy.

A - Always test holds.

S - Stand upright, on flexed joints.

H - Hands kept low; handholds should be waist to shoulder high.

W - Watch your feet.

O - On three points of contact.

R - Rhythmic movement.

T - Think ahead.

H - Heels kept low, lower than the toes.

TRANSITION: Now that we have covered cashworth, are there any questions? Following these principles will aid you in a less tiring and stressful climb. Above all remain calm. If there are no questions for me, then I have some for you.

PRACTICE (CONC)

a. Students will execute balance climbing.

PROVIDE HELP (CONC)

a. The instructors will assist the students when necessary.

OPPORTUNITY FOR QUESTIONS (3 Min)

1. QUESTIONS FROM THE CLASS

2. QUESTIONS TO THE CLASS

Q. What are the two safety requirements for balance climbing?

A. (1) Never climb higher than 10 feet above the ground.
(2) A spotter is required for all balance climbs.

Q. Name five types of holds?

A. (1) Push.
(2) Pull.
(3) Foot.
(4) Friction.
(5) Jam.

Q. What is the meaning of the acronym "CASHWORTH"?

A. C - Conserve energy.
A - Always test holds.
S - Stand upright on flexed joints.
H - Hands kept low.
W - Watch your feet.
O - On three points of contact.
R - Rhythmic movement.
T - Think ahead.
H - Heels kept low.

SUMMARY (2 Min)

a. During this period of instruction, we have talked about what balance climbing is, the safety precautions that must be used in balance climbing, the proper body position and the acronym "CASHWORTH". We have also discussed different types of holds and the duties of both the climber and the spotter.

b. Those of you with IRF's please fill them out at this time and turn them into the instructor. We will now take a short break.

UNITED STATES MARINE CORPS
Mountain Warfare Training Center
Bridgeport California, 93517-5001

SML
ACC
SMO
02/11/02

LESSON PLAN

TOP ROPING

INTRODUCTION

(5 Min)

1. **GAIN ATTENTION**. On June 6,1944, the allied forces performed one of the greatest feats in the history of warfare, an opposed landing on hostile enemy territory. Part of the plan for this landing called for elements of the United States Army to scale the cliffs above the landing beaches. This was accomplished by the use of ladders, which were transported to the cliff head on the landing craft, no one had to lug them around. But in a real mountainous environment you probably will not have this luxury. Fortunately we are accomplishing the same mission with a coil of climbing rope, and I submit to you, would you rather carry a sixty foot ladder around or a 150 foot coil of rope.

2. **OVERVIEW**. The purpose of this period of instruction is to introduce the student to top roping. This will be accomplished by discussing the preparations for climbing, belaying, and commands. This lesson relates to cliff assaults.

INSTRUCTOR NOTE: Have students read learning objectives.

3. **INTRODUCE LEARNING OBJECTIVES**.

   a. TERMINAL LEARNING OBJECTIVE. In a summer mountainous environment, and given a rock face and a belay man, conduct top roping with cartridge belt and rifle, in accordance with the references.

   b. ENABLING LEARNING OBJECTIVE. (SML) and (ACC)

      (1) In a summer mountainous environment, establish a belay stance from the top of a cliff for use in a top rope climb, in accordance with the references.

      (2) In a summer mountainous environment, belay a climber using a Munter Hitch, in accordance with the references.

(3) In a summer mountainous environment, demonstrate the use of climbing commands while conducting a top rope, in accordance with the references.

c. ENABLING LEARNING OBJECTIVES. (SMO)

(1) In a summer mountainous environment, belay a climber using a Munter Hitch, in accordance with the references.

(2) In a summer mountainous environment, state orally the climbing commands used while conducting a top rope, in accordance with the references.

4. **METHOD/MEDIA**. The material in this lesson will be presented by lecture and demonstration method. You will practice what you have learned in upcoming field exercises. Those of you with IRF's please fill them out at the end of this period of instruction.

5. **EVALUATION**

a. MLC - You will be tested later in the course by written and performance evaluations.

b. ACC - You will be tested by a performance exam.

c. SMO - You will be tested by a performance examination.

TRANSITION: Are there any questions over the purpose, learning objectives, how the class will be taught, or how you will be evaluated? First, let's discuss the preparations for establishing a belay stance from the top.

BODY (55 Min)

1. (15 Min) **ESTABLISHING A BELAY STANCE FROM THE TOP**

a. The belayer will establish a sitting belay stance on the cliff head by:

(1) Constructing a suitable anchor with the standing end of the rope.

(2) With the direction of pull away from the anchor, tie a directional figure 8 loop near enough to the cliff's edge so that the climber can be observed, if possible.

(3) Tying a swammi Wrap around oneself, clip a locking carabiner through all the rear wraps and clip a large locking carabiner through all the front wraps.

(4) Secure the rear locking carabiner into the directional figure of 8 loop.

(5) Secure the running end of the rope to the large locking carabiner utilizing a suitable friction belay (i.e., munter hitch, stitch plate, etc.).

NOTE: Gloves will not be worn while belaying a climber during a top rope.

TRANSITION: Are there any questions over establishing a belay stance from the top? If not, let's discuss securing the climber to the rope.

2. (5 Min) **SECURING THE CLIMBER TO THE ROPE**

   a. The climber will tie into the end of the top rope by:

      (1) Constructing a Swammi wrap around oneself.

      (2) Tying a retrace figure 8 loop through all of the Swammi's wraps or;

      (3) A figure of 8 loop clipped into a locking carabiner secured through all the Swammi's wraps.

   b. Alternative methods.

      (1) A bowline on a coil as discussed in ROPE MANAGEMENT or;

      (2) Tie into the hard point of the harness as discussed in SIT HARNESS.

TRANSITION: Are there any questions over the securing of the climber to the rope? Let's talk about establishing a belay stance from the bottom.

3. (15 Min) **ESTABLISHING A BELAY STANCE FROM THE BOTTOM**.

   a. Belaying from the bottom is commonly referred to as a "Yo-Yo" or "Sling Shot" belay. This system can be constructed for either a direct or indirect belay.

      (1) Construct a suitable top anchor utilizing either a pulley or one steel locking or two non-locking carabiners as the attachment point for the anchor.

      (2) The rope will travel up from the bottom and pass through the attachment point then return to the bottom.

      (3) The climber will be secured to one end of the rope and the belayer from the other end.

BELAY STANCE FROM THE BOTTOM

TRANSITION: Are there any questions concerning the establishment of a bottom belay stance for a top rope? Let's discuss the climber's responsibilities.

4. (5 Min) **CLIMBER'S RESPONSIBILITIES**

   a. The climber ensures that the belayer is anchored and on belay, by use of commands or prearranged signals prior to beginning the climb.

   b. The climber will not out climb the belayer; this will cause slack in the rope between the belayer and the climber.

   c. Avoid placing excess pressure/weight on belay man.

   d. Weapons will be worn across shoulder, muzzle down and to the left, with a tight sling securely attached to the weapon.

TRANSITION: Are there any questions over the climber's responsibilities? Let's discuss the commands for the climber and his belay man.

5. (5 Min) **CLIMBING COMMANDS AND SIGNALS.** In order to conduct top rope operations safely, it is essential that everyone understands the sequence of events. The following voice commands or rope tugs will be utilized:

VOICE COMMANDS	GIVEN BY	MEANING
"Lane #, Up Rope"	Climber	Belayer needs to take in the slack.
"That's Me"	Climber	Excess slack has been taken up between us.
"Lane #, On Belay"	Belayer	I am on belay.
"Lane #, On Climb"	Climber	I am ready to climb.
"Climb Climber"	Belayer	You may begin to climb.
"Lane #, Slack!"	Climber	Pay out rope
"Lane #, Giving Slack"	Belayer	I am paying out rope
"Lane #, Tension!"	Climber	Take in rope
"Lane #, Applying Tension"	Belayer	I am taking in rope

ROPE TUGS	GIVEN BY	MEANING
Three Tugs	Climber	Belayer needs to take in the slack.
Three Tugs	Belayer	I have taken up the slack and you may begin to climb.
Three Tugs	Climber	Climbing.

TRANSITION: Without the proper usage of commands, injuries can occur. Are there any questions over the commands? If you have none for me, then I have some for you.

PRACTICE (CONC)

a. Students will conduct and establish a belay for top roping.

PROVIDE HELP (CONC)

a. The instructors will assist the students when necessary.

OPPORTUNITY FOR QUESTIONS

1. QUESTIONS FROM THE CLASS (3 Min)

2. QUESTIONS TO THE CLASS

Q. What are the climber's responsibilities?

A. (1) Ensure the belayer is anchored and on belay prior to beginning his climb.
(2) Don't out climb his belayer.
(3) Avoid placing excess pressure/weight on the belayer.
(4) Ensure equipment is properly secured.

Q. What is the climbing voice command when all excess rope has been taken up between the climber and belay man?

A. "That's me"

SUMMARY (2 Min)

a. During this period of instruction we have discussed how to establish a top and bottom belay stance, the responsibilities of the climber, climbing commands, and preparations prior to top roping.

b. Those of you with IRF's please fill them out at this time and turn them in to the instructor. We will now take a short break.

UNITED STATES MARINE CORPS
Mountain Warfare Training Center
Bridgeport, California 93517-5001

SML
ACC
02/11/02

LESSON PLAN

SIT HARNESS

INTRODUCTION (5 Min)

1. **GAIN ATTENTION.** The number of climbing sit harnesses available in the world today is quite staggering. Here at the MWTC, in co-operation with Yates Inc., we have designed a climbing sit harness that can be used for a multitude of mountaineering tasks, including two party climbing on rock or ice, rescues, rappelling, big wall climbing, and the other interests. It is one of the first harnesses made in the USA that has passed the UIAA test and is rated to 5950 lb. Further, it is a one size fits all harness.

2. **OVERVIEW.** The Purpose of this period of instruction is to familiarize the students with the MWTC sit harness, its component parts, how to care for it, how to put it on and adjust it for a comfortable fit, and how to tie into it.

INSTRUCTOR NOTE: Have students read learning objectives.

3. **INTRODUCE LEARNING OBJECTIVES**

   a. TERMINAL LEARNING OBJECTIVE

      (1) Without the aid of references, properly wear and maintain the MWTC sit harness, in accordance with the references.

   b. ENABLING LEARNING OBJECTIVES

      (1) Without the aid of references and given a diagram, name in writing the parts of a sit harness, in accordance with the references.

      (2) Without the aid of references, wear a sit harness, in accordance with the references.

4. **METHOD/MEDIA.** The material in this lesson will be presented by lecture and demonstration. You will practice what you have learned during upcoming field training exercises. Those of you with IRF's please fill them out at the conclusion of this period of instruction.

5. **EVALUATION**

a. SML - You will be tested by written and performance evaluation.

b. ACC - You will be tested by a written exam and performance evaluation throughout the course.

TRANSITION: Are there any questions over the purpose, learning objectives, how the class will be taught, or how you will be evaluated? During this period of instruction we will cover the nomenclature of the SIT harness, the wearing of the harness, tying into a rope, and care and maintenance. Let's first go over nomenclature of the sit harness.

BODY (35 Min)

1. (5 Min) **NOMENCLATURE**

a. Waist belt.

b. Leg loops (adjustable).

c. Buttocks straps (adjustable).

d. Fastex buckle - 2.

e. Doughnut.

f. Equipment loops.

g. D-ring (older models).

h. Waist belt tie-in point.

i. Crotch strap.

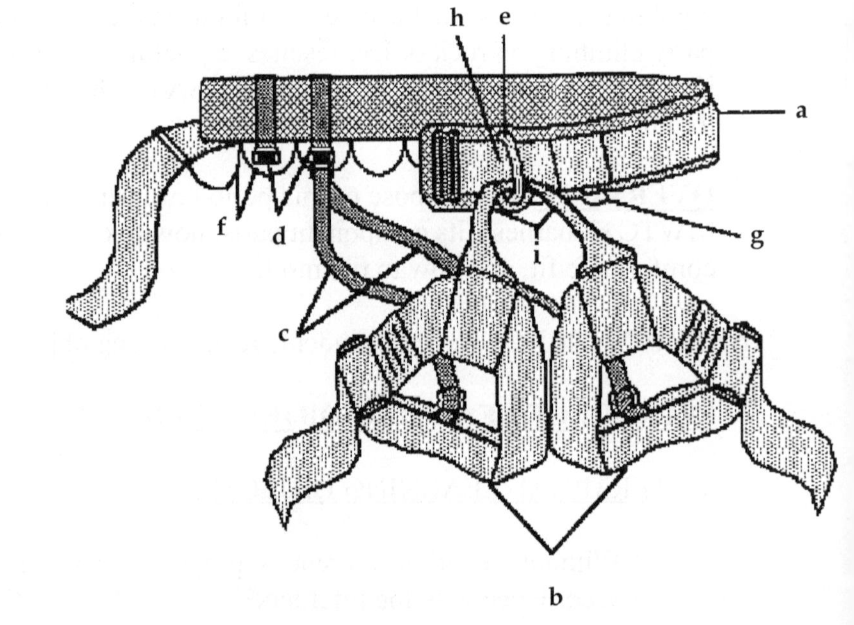

NOTE: The harness is made out of various size mountaineering tape and should be maintained in the same manner as a climbing rope is, refer to ROPE MANAGEMENT.

TRANSITION: Are there any questions over the nomenclature? Let's discuss the wearing of the harness.

2. (5 Min) **WEARING OF THE SIT HARNESS**

a. First, disconnect the fastex buckle at the rear of the harness.

b. Hold the harness in front of you, put your feet through the leg loops ensuring that the buckles on the leg loops are outboard, on your thighs.

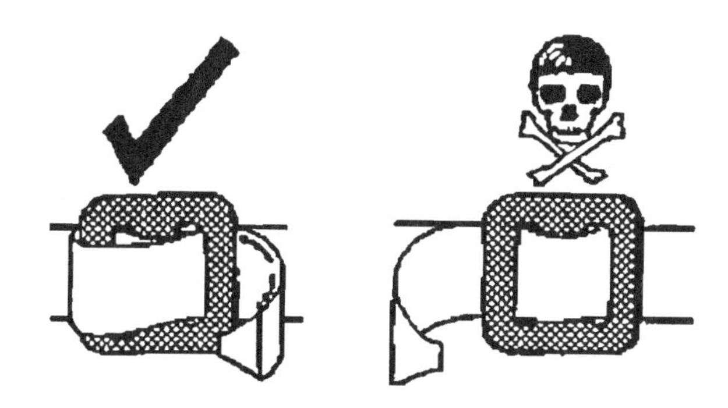

c. Fasten the waist belt into the buckle, ensuring that it is a tight but comfortable fit. **You must ensure that the waist belt is threaded back through the buckle, this action locks the waist belt to the buckle. Failure to do this will cause the waist belt to slip through the buckle when it is under load.**

d. Adjust the leg loops so that they are high on your thigh, once you have them adjusted, get your buddy to clip the fastex buckle male to the most comfortable female fastex. Then adjust the buttock straps so as the leg loops are held up.

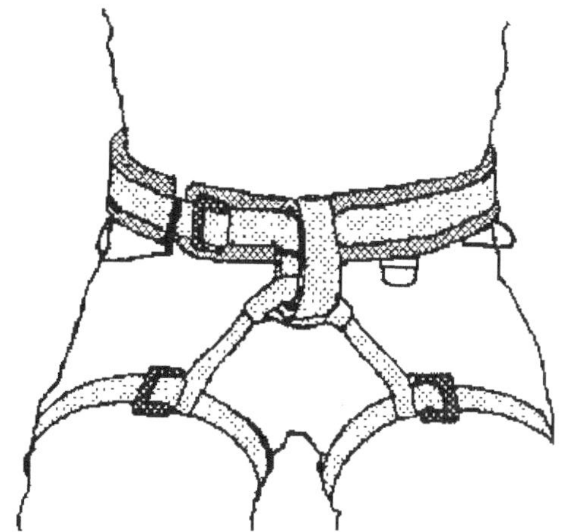

e. If all the above are done correctly, the harness should now be a comfortable but snug fit, after you have fitted it for the first time there is no need to go through the same procedure each time you put it on. Simply hold the harness in front of you, step into it and attach the waist belt to the buckle in the approved manner.

TRANSITION: Are there any questions over the wearing of the sit harness? Let's talk about tying in the rope to the sit harness.

3. (5 Min) **TYING IN THE ROPE (END OF ROPE)**

a. Pass the end of the climbing rope up through the crotch strap, then through the doughnut and through the waist belt tie-in point.

b. Tie the rope using a retraceable figure eight, adjusting the knot to get it as close as possible to the body.

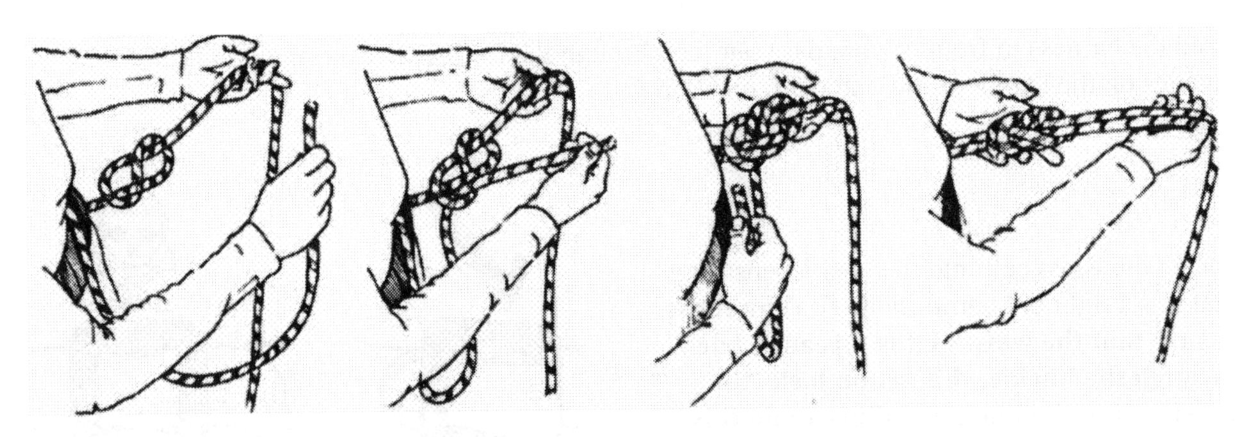

RETRACE FIGURE 8 IN SIT HARNESS

4. (5 Min) **TYING IN THE ROPE (MIDDLE OF THE ROPE)**

   a. Take up a bight of rope and tie a figure eight loop, then take a steel locking carabiner and attach the carabiner to the harness by securing the crotch strap, doughnut, and the waist belt tie-in loop. Now attach the figure eight loop to the carabiner.

5. (5 Min) **CARE AND MAINTENANCE**

   a. Avoid contact with chemicals, as this will damage the nylon.

   b. Regularly inspect for signs of abrasions and normal wear. Pay particular attention to wear points such as the tie-in loops, buckles, and sewn joints.

   c. Keep away from heat such as open flames, cigarettes, etc.

   d. If soiled by grit and sea water, wash in lukewarm water with pure soap and allow to dry in a warm room away from direct heat.

   e. Two to three years of life can be expected during normal climbing use.

   f. It is recommended that a harness that has experienced a serious fall should be discarded.

   g. Under no circumstances will you ever tie into the equipment rack of any sit harness as a belay, anchor point, etc.

NOTE: The time to discard a harness is on a climber's own discretion.

TRANSITION: Now that we have discussed tying in and care, are there any questions? If you have none for me, then I have some for you.

PRACTICE (CONC)

   a. The students will properly use the sit harness.

PROVIDE HELP (CONC)

a. The instructors will assist students when necessary.

OPPORTUNITY FOR QUESTIONS (3 Min)

1. QUESTIONS FROM THE CLASS

2. QUESTIONS TO THE CLASS

Q. What is the nomenclature of the sit harness?

A. (1) Waist belt.
(2) Leg loops (adjustable).
(3) Buttocks straps (adjustable).
(4) Fastex buckle - 2.
(5) Doughnut.
(6) Equipment loops.
(7) D-ring.
(8) Waist belt tie-in point.
(9) Crotch strap.

SUMMARY (2 Min)

a. What we have just covered is the nomenclature, wearing, care and employment of the MWTC sit harness.

b. Those of you with IRF's please fill them out and turn them in to the instructor. We will now take a short break.

UNITED STATES MARINE CORPS
Mountain Warfare Training Center
Bridgeport, California 93517-5001

SML
ACC
02/11/02

LESSON PLAN

PLACING PROTECTION

INTRODUCTION (5 Min)

1. **GAIN ATTENTION**. As a lead climber, the success of the climb relies not only on your ability to climb but also on your ability to protect yourself. No matter how well your belay man is at belaying or how good your equipment is, if the protection is improperly placed, the consequences of a fall could be death or serious injury.

2. **OVERVIEW**. The purpose of this period of instruction is to introduce the student to the basics of pro placement to include placement of chocks, SLCD's, hexes, properly clipping the rope into this protection and how to rack protection. This lesson relates to party climbing.

INSTRUCTOR NOTE: Have students read learning objectives.

3. **INTRODUCE LEARNING OBJECTIVES**.

   a. TERMINAL LEARNING OBJECTIVE.

      (1) In a summer mountainous environment, and given the proper equipment, place rock climbing protection, in accordance with the references.

4. **METHOD/MEDIA**. The material in this lesson will be presented by lecture and demonstration. You will practice what you learned during upcoming field training exercises. Those of you with IRF's please fill them out at the end of this lesson.

5. **EVALUATION**.

   a. MLC - You will be tested by a performance evaluation during the passout climb.

   b. ACC - You will be tested by a performance evaluation during the passout climb.

TRANSITION: Are there any questions over how the class will be taught or how you will be evaluated? Our first topic will be placing protection and anchors.

BODY (50 Min)

1. (5 Min) **GENERAL**. Placing protection is the skill of establishing points of protection along the route by using natural features or by lodging artificial devices in the rock. The leader clips the rope through each piece of protection, while the belayer at the bottom of the pitch waits in position to hold the leader if he should fall. This technique makes lead climbing safely possible. Since carabiners are used in all protection placements, the novice should learn some general principles as for their use:

   a. The gates of the carabiners clipped into the pro should be down and out, or the gates should be down and facing to whichever side minimizes the risk of the gate being forced open by a rock edge or nubbin during a fall.

   b. Rope should be run from the climber through the carabiner without twists and kinks.

   c. When using a runner, be sure the runner is not twisted before clipping into the carabiner; otherwise, the carabiner may end up facing the wrong way.

2. (10 Min) **NATURAL FEATURES AS PROTECTION**. Natural features offer some of the very best protection but should always be tested for stability as discussed in *NATURAL AND ARTIFICIAL ANCHORS.*

   a. Trees

      (1) Trees provide the most common and obvious natural protection. To attach the runner to a tree, wrap the runner around the tree and clip a carabiner into both looped ends.

      (2) The runner should usually be as low on the tree trunk as possible, although it may sometimes be desirable to put it higher up or even on a branch to avoid creating a sharp bend in the climbing rope or to provide a higher point of protection.

   b. Scrubs and Bushes

      (1) Locate the central root and construct the anchor as near to the base as possible to avoid leverage.

   c. Rocks and Boulders

      (1) Extreme caution should be used when utilizing rocks and boulders because of the danger it poses if it should come loose. Always ensure that the intended stone is well embedded or is too heavy to move.

      (2) Place the runner at the base of the stone to prevent possible leverage.

d. Spikes and Flakes

(1) Spikes and flakes often provide good placement. A runner is attached by placing it over the spike / flake as near the base as possible. Sometimes, to prevent the action of the climbing rope from pulling the runner off, it can be attached to the spike / flake with a girth hitch.

(2) When possible, avoid placing the runner where it may be pulled against a sharp edge of the rock in a fall. When this is unavoidable, padding sharp edges with another runner or other soft material may be helpful.

e. Threads and Chockstones

(1) Runners can be can be threaded through the holes or placed around the chockstone and clip a carabiner into both looped ends.

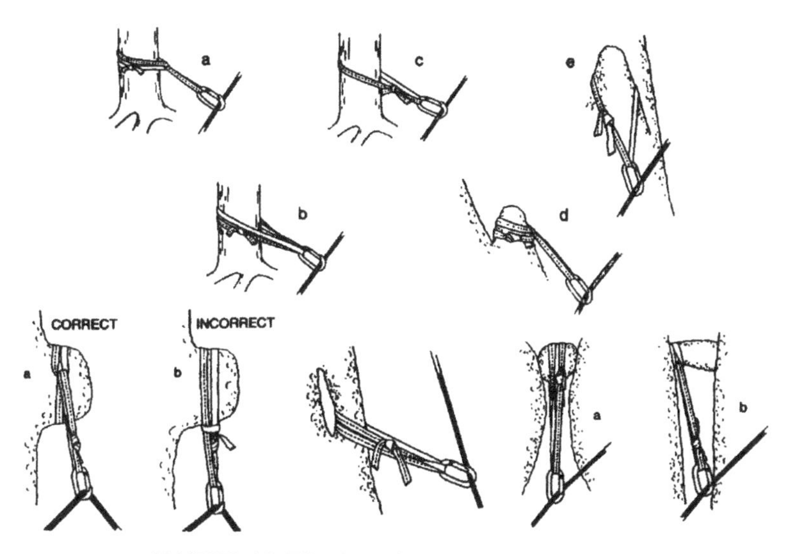

NATURAL FEATURES AS PROTECTION

3. (15 Min) **ARTIFICIAL PROTECTION**. In the absence of natural protection, climbers will use artificial protection devices known as chocks, spring-loaded cams and bolts.

a. Chocks. The principle in placing chocks is deceptively simple; find a crack with a constriction at some point and place a chock of the appropriate size above the constriction and pull down on the loop to set the chock. Chocks can either be passive, meaning it sits in a constricting crack until it takes a load then wedges itself into the crack. Or it can be camming meaning that the chock is placed in a crack in such a way that when it takes a load it tries to rotate. The rotation causes the chock to cam and lock into place. However simple in theory, placing chocks require a good eye and some experience. There are basically two different types of chocks used at MWTC, wedges and hexcentrics.

(1) Wedges. This is a passive chock that is also known as a stopper, wire or nut. A wedge chock has a wired loop attached to it and is tapered down from top to bottom so that it can fit into the constriction of a crack. It is constructed to have a wide side, the

strongest side, and a narrow side. The goal is to get the greatest possible contact between the chock and the rock.

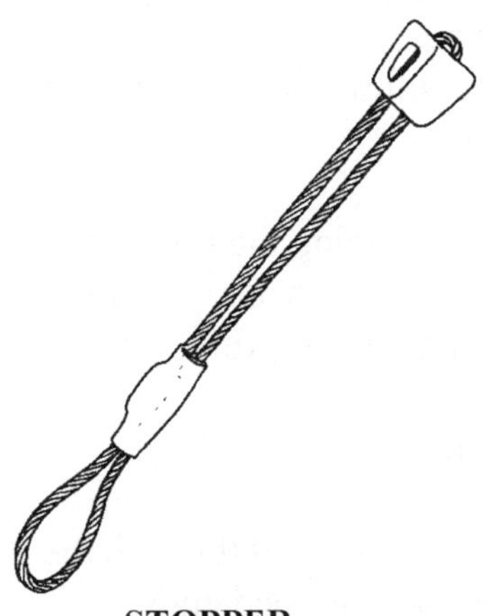

STOPPER

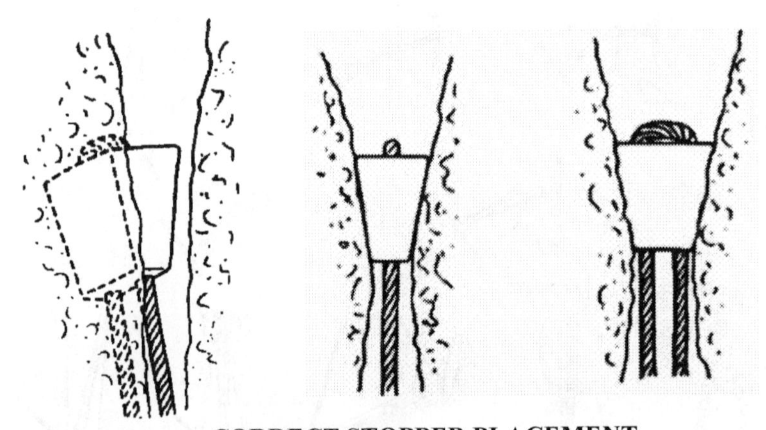

CORRECT STOPPER PLACEMENT

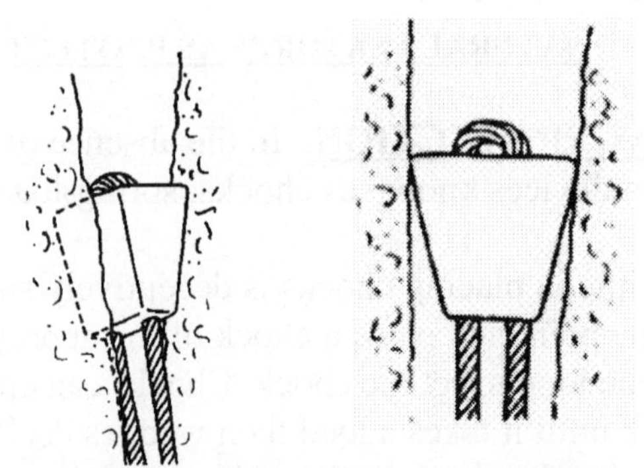

INCORRECT STOPPER PLACEMENT

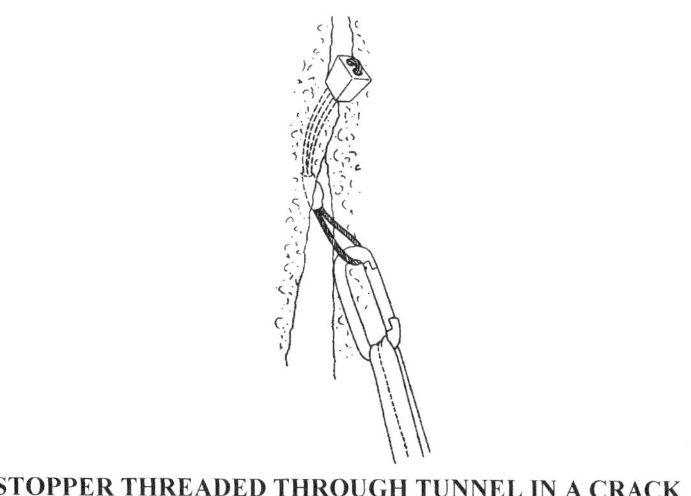

STOPPER THREADED THROUGH TUNNEL IN A CRACK

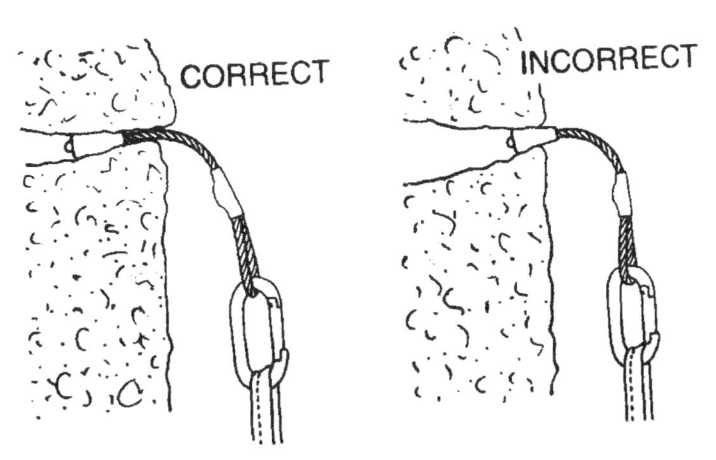

STOPPER PLACEMENT ON A LIP

(2) Hexcentrics. These chocks can be used in either the passive or camming mode and are also known as hexes. Hexcentric chocks have a wired or Type I cordage loop attached to it and have six different sized sides that are tapered, allowing it to be placed into a constricting crack.

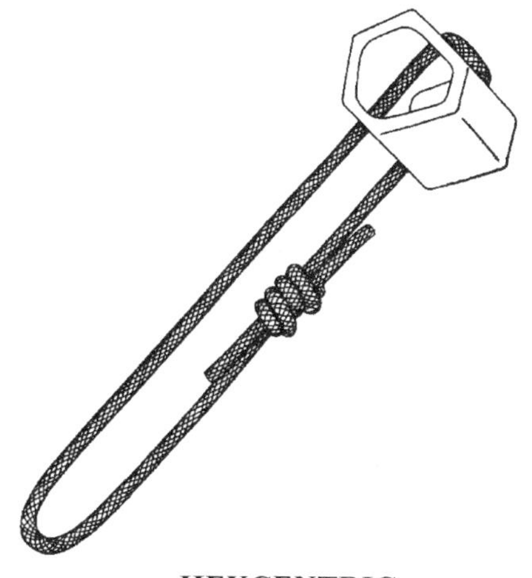

HEXCENTRIC

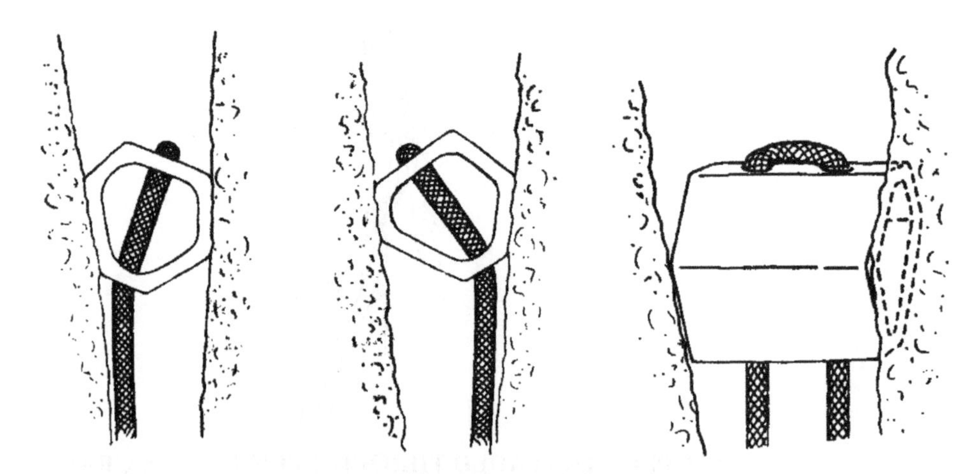

HEXCENTRIC PLACEMENT

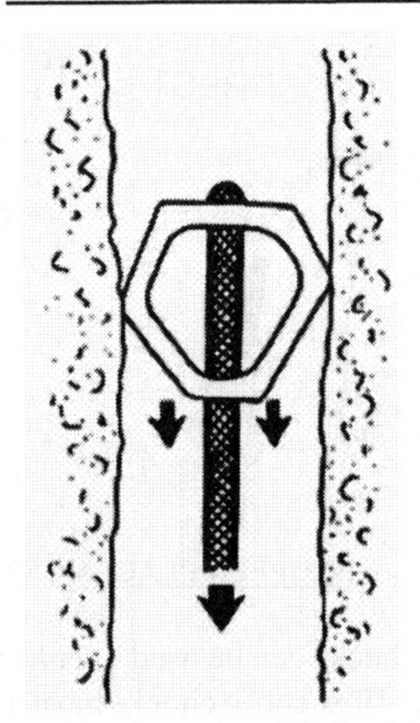

INCORRECT HEXCENTRIC PLACEMENT

(3) The following factors should be considered when placing chocks:

(a) When possible, avoid cracks that have crumbly or deteriorating rock on one side. Many cracks that look good could have a small, loose flake on one side; often well disguised with grass and dirt. Some very tempting cracks are in fact formed by a detached flake against a large mass of rock.

(b) When attempting to place a chock, always look for a likely constriction in the crack first, then select a chock that will fit, rather than selecting a chock and looking for somewhere to put it.

(c) When a place is found, choose a chock that will have as much surface area as possible in contact with the rock. A chock with just one side resting on a small crystal is likely to be unsound and unsafe. A chock that sticks partially out of a crack is usually poor protection. If the crack is too shallow to get the chock all the way in, use a smaller chock or find a deeper portion of the crack.

(d) When placing a #1 or #2 stopper, always back that piece up with a secondary piece of protection as soon as possible.

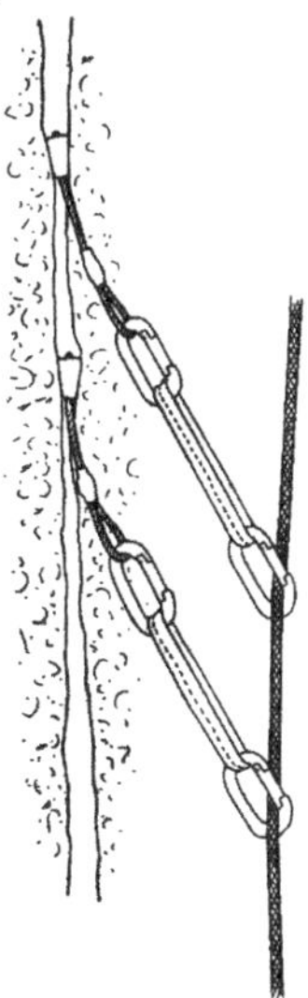

BACKING UP A #1 OR #2 STOPPER

(e) Outward flaring cracks are a problem. A few types of chocks are specifically designed to fit in flaring cracks as long as the angle of the flare is not too great.

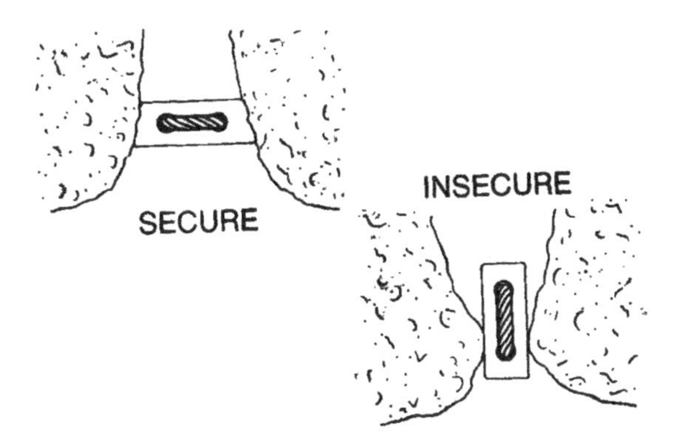

INCORRECT PLACEMENT IN FLARING CRACK

(f) Stacking Chokes. Parallel-sided cracks can produce a problem. Chocks of various sizes can be stacked in many ingenious ways, but only one method will be described here. If two wedge shaped chocks are placed in contact, one upside-down, their surfaces will be approximately parallel. If the pair barely fits into a parallel-sided crack, they are set firmly and the rope is clipped into the one placed right side up (loop to the direction of pull). The upper chock should be clipped to the lower one in some way to keep it from being lost if the placement should pop out.

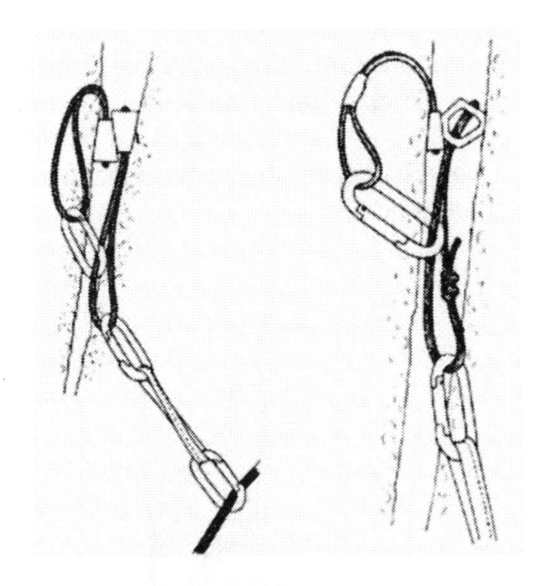

STACKED CHOCKS

(g) Horizontal cracks can take simple placements if the interior of the crack is wider than the edges at some point. If not, a placement of stacked chocks may hold. In either case if the chock is on a rope loop which runs over a sharp lower edge, as often happens, it cannot be depended on to hold a hard fall, for the loop may be cut If it is on a wire it is usually safe.

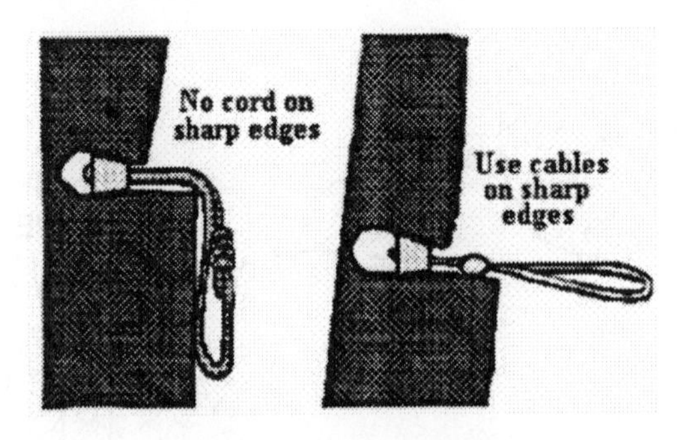

CHOCKS ON EDGES

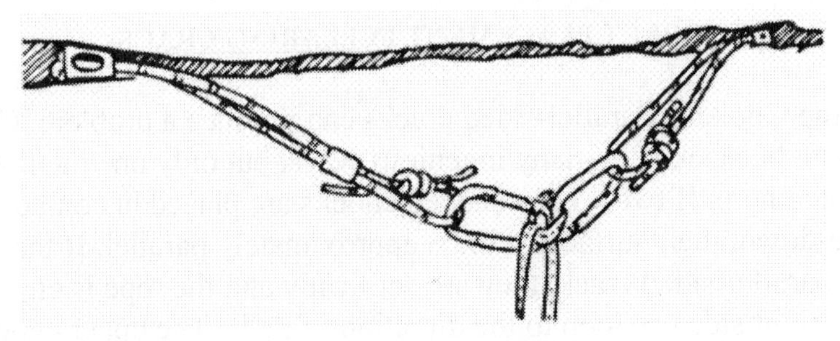

STOPPER PLACEMENT IN A HORIZONTAL CRACK

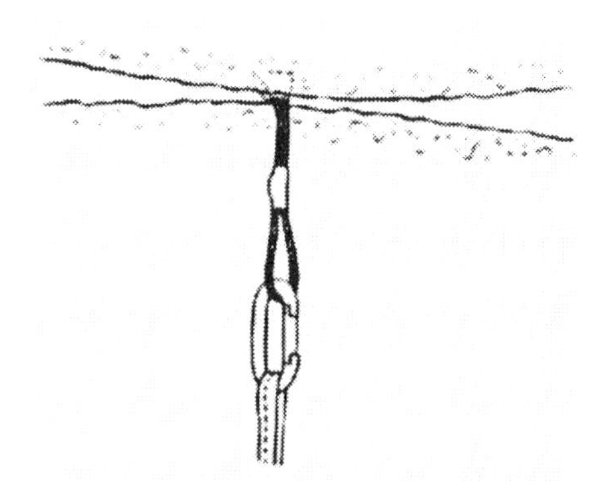

STOPPER PLACEMENT BEHIND HORIZONTAL CRACK

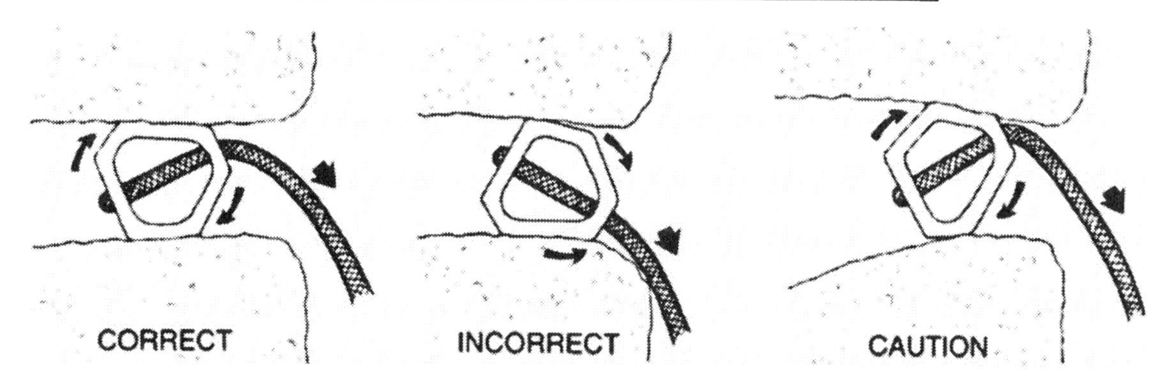

HEXCENTRIC PLACEMENT IN A HORIZONTAL CRACK

(h) When placing a chock with a rope loop, the side of the loop with the knot should face out when there is a choice, in order to ensure that at least one-half of the loop fits into the crack. When the loop will not fit into the crack at all, a wired chock is usually a better choice. Once a chock is placed, the climber can either clip the climbing rope into it directly or extend the chock loop with a runner. The loop of a wired chock is always extended with a quick draw to prevent "walking" the pro out of the rock. Ensure that enough quick draws are carried to use on placements.

b. Spring-Loaded Camming Devices (SLCD). This is a mechanical device known as quadcams or camalots. The four cams in a spring-loaded camming device, also known as lobes, are connected to a trigger mechanism. Pull the trigger and the cams retract, narrowing the profile of the device so it can fit into a particular crack. Release the trigger and the cams again rotate outward expanding the profile until the cams grip the sides of the crack. SLCD's provide reliable placements in cracks where ordinary chocks are difficult or impossible to place; such as parallel-sided cracks, flaring cracks, and cracks under roofs. They can be placed quickly and make it possible to do some extremely difficult pitches that are otherwise virtually impossible to protect.

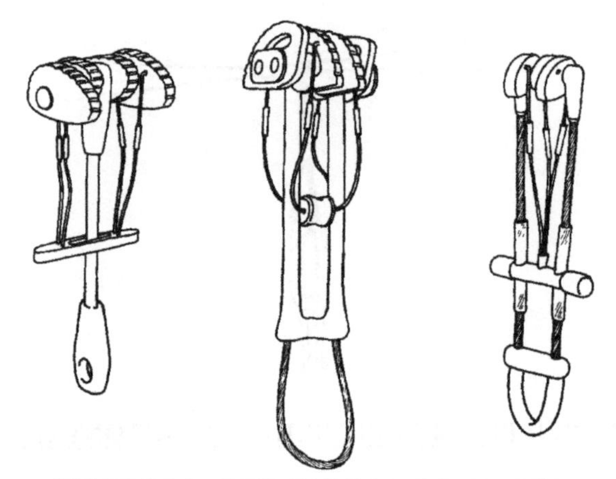

SPRING LOADED CAMMING DEVICES

(1) The following factors should be considered when placing SLCD:

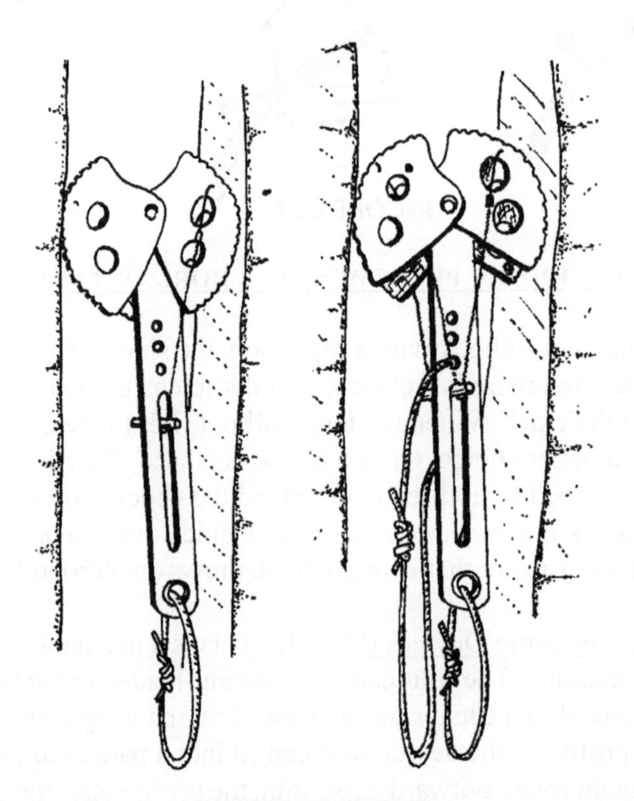

CAM PLACEMENT

(a) Care should be taken to ensure all individual cams of the SLCD have a good purchase on the rock. If one or more cams have no purchase, this is considered improper placement, and an SLCD can easily be pulled out of the rock when a fall is taken.

(b) If the trigger must be pulled all the way back to place the SLCD into a crack, the cams will be over-retracted (over-cammed) preventing the SLCD from performing properly. It may also be impossible to remove. Use a smaller one or another type of pro.

(c) Avoid under camming, where the cams are over expanded causing the cams to rock on its tips; also known as "tipped" placement.

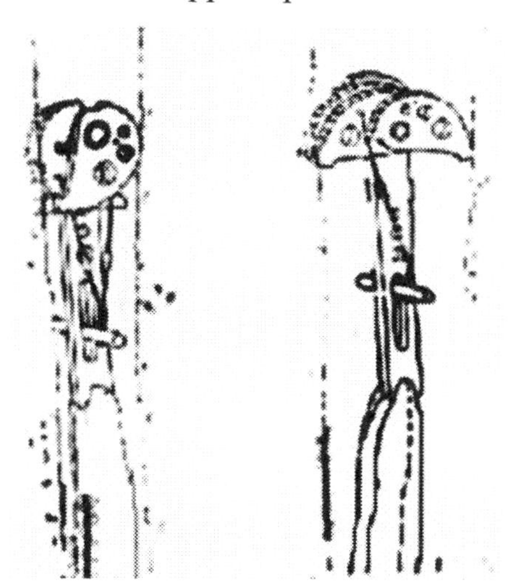

INCORRECT CAM PLACEMENT

(d) Do not put the SLCD any deeper into the crack than it has to be for a good placement. If placed too deeply, it may be impossible to reach the trigger to remove it.

(e) When placed behind flakes or in deep cracks, SLCD's have been known to "walk up" into the crack out of reach as a result of rope action. A SLCD should be extended with a runner in such placements to reduce the possibility of the pro walking.

(f) Always align a SLCD's stem and cams in the direction of pull.

c. Fixed Bolts

(1) Fixed bolts are often found in civilian rock climbing areas and are not uncommon on alpine rock climbs. Bolts are most commonly used for belay anchors, but also provide protection on otherwise unprotected stretches of rock.

(2) Bolts generally provide the best and certainly the most convenient protection. Unfortunately, they are not always sound and should be checked. Bolts can be checked by examining the rock around them for evidence of crumbling or cratering, and they should be tested by clipping into them with a separate carabiner and jerking on it before clipping in the rope. Never hammer on a bolt to test or improve it, since this will permanently weaken it.

(3) When clipping into a bolt, it is best to do so with a carabiner-runner-carabiner combination, known as a “quickdraw”, which will reduce rope drag.

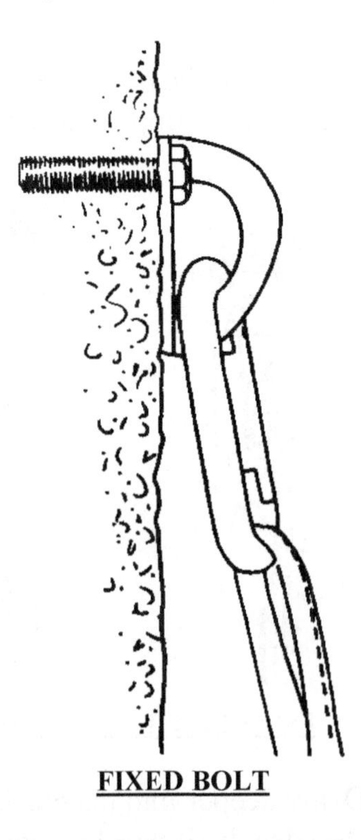

FIXED BOLT

TRANSITION: Now that we have discussed placing protection, are there any questions? Next, we will discuss clipping in.

4. (5 Min) **ATTACHING THE ROPE TO THE PROTECTION**. After placing a piece of protection, the climber must secure his climbing rope to it. The carabiner is the basic tool used to connect the climbing rope to the protection. As easy as it sounds, there are factors to consider when clipping the rope into the carabiner:

   a. Ensure that the gate of the carabiner is facing away from the rock, especially near books and edges. If necessary, extend the protection with a runner.

   b. Always place the carabiner so that the gate is facing down and away. Depending on the climber’s route of travel will decide which direction the gate is facing:

      (1) When traversing left of the protection, the gate will face toward the right.

      (2) When traversing right of the protection, the gate will face toward the left.

      (3) When climbing straight up, the gate direction will not matter but ensure that it is facing away from sharp edges.

   c. To clip the carabiner, grab the rope from your retrace figure of eight and pull up a bight. If a bigger bight is necessary, hold the rope between your teeth and pull up a bigger bight.

d. Always clip the carabiner from the rear to the front. This will prevent the rope from backtracking over the gate causing it to unclip during a fall. This is also known as "The Death Clip".

e. The carabiner should allow the rope to run smoothly without twists or kinks. Always inspect the run of the rope through the carabiners to ensure against "Z" clips.

f. When this is done correctly, the rope going to your retrace figure of eight should be coming out of the front of the carabiner. The rope going to the belayer will be between the carabiner and the rock.

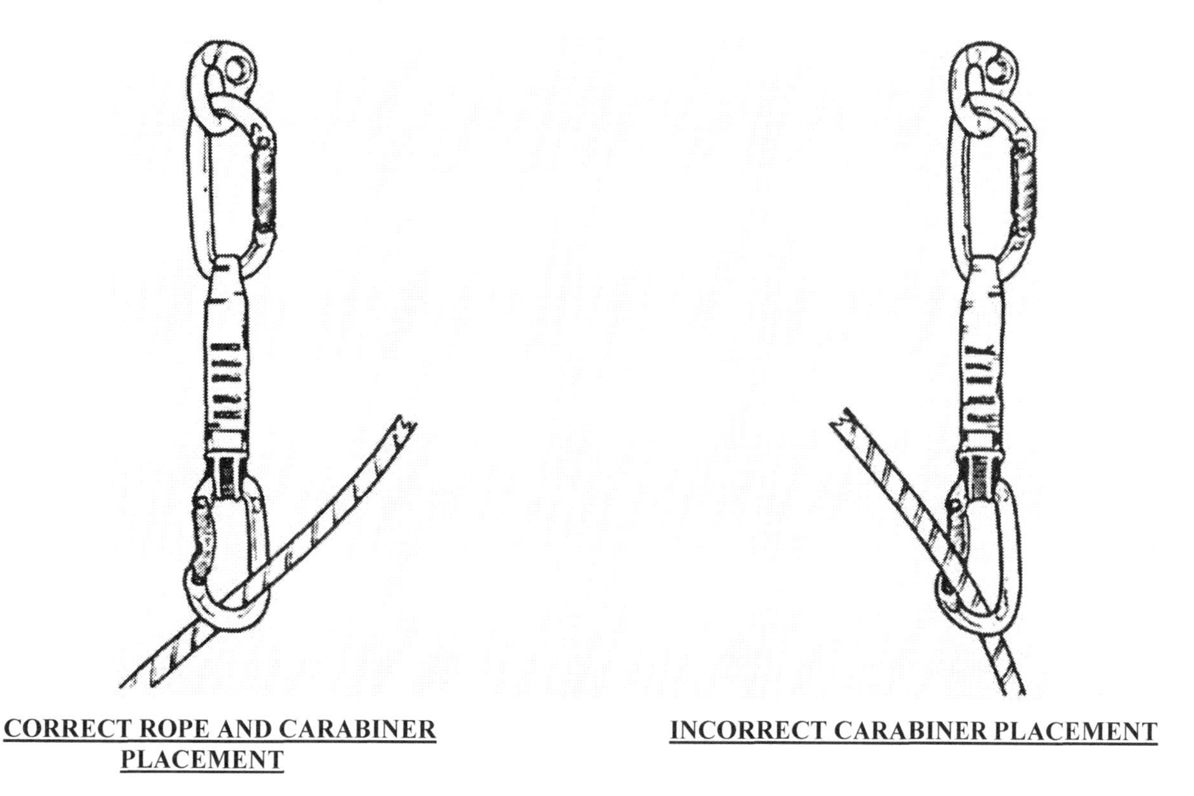

CORRECT ROPE AND CARABINER PLACEMENT

INCORRECT CARABINER PLACEMENT

TRANSITION: Are there any questions over clipping in? Next we will discuss the racking of equipment.

5. (5 Min) **RACKING EQUIPMENT**. A climber's selection of gear for protection is called his rack. A rack consists of a varying numbers of chocks, SLCD's, carabiners and quick draws. Pro should be carried in a logical sequence of size and type so that the lead climber can quickly find the correct type for the crack he wishes to fit. Pro can be carried on an equipment sling across the body, attached to the equipment loops on the harness, or a combination of both. The amount of pro carried is dependent upon the type of climb to be attempted. The following are methods for racking:

INSTRUCTOR NOTE: Have students take their full racks for the first couple days of climbing until they learn what they need to take.

a. Rack several pieces of pro of the same size on one carabiner, arranging these on the rack in order of size. This has the advantage of having extra pro on hand if you don't get the placement right the first time.

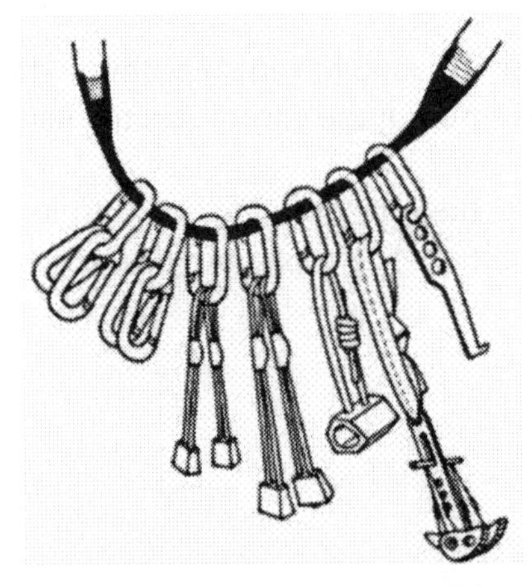

PROTECTION GROUPED TOGETHER

b. Long slings are worn by doubling the loop over and clipping both ends with a carabiner.

c. Quick draws are preset carabiner and runner combinations that allow quick and easy extension of a piece of protection. It is helpful to have the gates of the carabiners opposite of each other so that the bottom carabiner is down and away from the rock. This makes clipping in much easier.

d. Other item such as a stitch plate, rappel gloves, chock pick, etc., need to be placed at the back of the harness.

RACKED EQUIPMENT ON CLIMBER

TRANSITION: Now that we have covered racking equipment, are there any questions? Are there any questions over the amount of gear? If you have none for me, then I have some for you.

PRACTICE (CONC)

a. Students will properly place climbing protection.

PROVIDE HELP (CONC)

a. The instructors will assist the students when necessary.

OPPORTUNITY FOR QUESTIONS (3 Min)

1. QUESTIONS FROM THE CLASS

2. QUESTIONS TO THE CLASS

Q. If the bottom carabiner in a quickdraw is facing left which way should the rope be clipped in?

A. From the left.

Q. How can parallel sided cracks be overcome in a climb using chocks?

A. Stacking.

SUMMARY (2 Min)

a. During this class we have covered the various techniques for placing protection, how to clip into quickdraws and how to rack your gear.

b. Those of you with IRF's please fill them in at this time. We will now take a short break.

UNITED STATES MARINE CORPS
Mountain Warfare Training Center
Bridgeport, California 93517-5001

SML
ACC
02/11/02

LESSON PLAN

BELAYING FOR PARTY CLIMBING

INTRODUCTION

(5 Min)

1. **GAIN ATTENTION**. In two party climbing it is essential that the belayer is secured to an anchor. If his partner fell, it is possible that the belayer could be dragged up the cliff face or pulled off. On a multi-pitch climb, (a climb which involves a number of belay positions), the climbers must have a set method of securing themselves to the rock face.

2. **OVERVIEW**. The purpose of this period of instruction is to introduce the students to belaying for party climbing, by demonstrating how to belay and how to tie into all the anchors that are available. This lesson relates to placing protection and two party climbing.

3. **INTRODUCE LEARNING OBJECTIVES**

a. TERMINAL LEARNING OBJECTIVES

(1) In a summer mountainous environment and given the necessary equipment, conduct belaying techniques for party climbing, in accordance with the references.

b. ENABLING LEARNING OBJECTIVES

(1) Without the aid of references, describe in writing the methods of belays, in accordance with the references.

(2) Without the aid of references, describe in writing the types of belays, in accordance with the references.

(3) Without the aid of references, describe in writing the minimum amount of natural anchors required to create a belay anchor, in accordance with references.

(4) Without the aid of references, describe in writing the minimum amount of artificial protection required to create a belay anchor, in accordance with the references.
(5) In a summer mountainous environment, establish a climbing anchor using natural anchor points, in accordance with the references.

(6) In a summer mountainous environment, establish an artificial anchor system, in accordance with the references.

4. **METHOD/MEDIA**. The material in this lesson will be presented by lecture and demonstration. You will practice what you have learned in upcoming field training exercises. Those of you with IRF's please fill them out at the conclusion of this period of instruction.

5. **EVALUATION**

a. SML - You will be tested by a written examination and a performance evaluation.

b. ACC - You will be tested by a written examination and a performance evaluation.

TRANSITION: Are there any questions over the purpose, learning objectives, how the class will be taught, or how you will be evaluated? Let's take a look at belaying concepts

BODY (90 Min)

1. (5 Min) **THE CONCEPT OF BELAYING**. Belaying is the procedure by which the belayer, also known as the #2 climber, manages the rope that is tied to the lead climber, also known as the #1. The following procedures take place during the belay sequence of a climb:

a. The belayer establishes an anchor system (a strong attachment point to the mountain) and takes a stance (bracing against the terrain to resist a hard pull on the rope).

b. Rope is paid out as the climber advances, keeping a minimum of slack between the roped team so that any fall will be stopped as short as possible.

c. If the climber falls, the belayer will apply the brake with the use of a belay device.

d. Upon reaching the top, the lead climber will establish a belay stance to top rope up the number 2.

NOTE: Alertness and appreciation for the importance of the belayer's role is critical. A leader belayed by a novice without the knowledge and training in belay techniques would be wise to climb as if there were no belay at all.

TRANSITION: Now that we have discussed the concept of belaying, are there any questions? Let's discuss methods and types of belays.

2. (5 Min) **TYPES OF BELAY ANCHORS**. There are two basic types of belay anchors: Indirect and Direct.

a. Indirect Belay. The term stems from the fact that the belayer's position is between the anchor point and the climber, therefore absorbing some of the Impact Force in the event of a fall. So the force of the fall goes indirectly to the belay anchor point through you.

b. Direct Belay. The direct belay exerts the Impact Force of a falling climber directly to the belay anchor point. The belayer is not in the system but still remains in control of the belay device.

DIRECT AND INDIRECT BELAYS

TRANSITION: Now that we have discussed the types of belays, are there any questions? Let's discuss the elements of a belay chain.

3. (5 Min) **ELEMENTS OF A BELAY CHAIN**. There are three principle elements of a belay chain used in rock climbing: the Anchor, the Belayer and the Climber (ABC).

   a. The Anchor. The anchor is a term used to describe the method by which the belayer attaches himself (with the rope or additional equipment) to the mountain so that he cannot be pulled off his belay stance. The anchor can be either an indirect or a direct belay using either natural or artificial points, or a combination of both.

      (1) When using natural anchors during climbing operations, a minimum of two natural anchor points will be utilized.

      (2) When using artificial anchors during climbing operations, a minimum of three artificial anchor points will be utilized.

(3) If constructing an anchor system in which only one natural anchor can be located for use, it must be backed up by two pieces of artificial protection.

(4) Anchors on the bottom should normally be behind the belay man and at or below his waist, terrain permitting. Anchors on the top should be at waist height or above, terrain permitting.

b. The Belayer. The term belayer is used to describe the static climber's mission of providing security with a rope to the lead climber in case of a fall.

(1) The belayer should position himself as near to the climbing route as possible to prevent the "Zipper Effect". This is covered in *ALTERNATIVE BELAYS AND ANCHORS.*

(2) When using the indirect belay stance, the belayer should secure himself snuggly between the anchor and the climber to absorb some of the Impact Force and to prevent any possible dragging, which could possibly jar the brake hand off the rope.

(3) When using the direct belay stance, the belayer should ensure that the anchors could support a Factor Two fall. This is covered in *PARTY CLIMBING.*

(4) When belaying from above, ideally the belayer should try to establish a sitting belay stance.

c. The Climber. Besides climbing the route, the climber is responsible for:

(1) Places protection into the rock through which the rope is passed through so that in case of a fall, the length of the fall is reduced.

(2) Establishes the top anchor belay to bring up the number 2.

TRANSITION: Now that we have discussed the elements of a belay chain, are there any questions? Let us discuss the establishment of a belay.

4. (5 Min) **BELAYER'S RESPONSIBILITIES**. Before establishing a belay stance the belayer must:

a. Ensure that the anchors are sound and in conjunction with the direction of pull.

b. Locate the exact position of where he will be belaying the lead climber. This is known as the belay stance.

c. When possible, the belay stance should be slightly offset from the lead climber's intended route to avoid possible hazards such as falling rocks, equipment, etc…

d. Ensure that the climbing rope is back stacked near the belay stance.

e. With his respected end of the climbing rope, tie a retrace figure of eight into the hard point of his harness.

TRANSITION: Now that we have discussed the responsibilities of the belayer before a climb, let us talk about establishing the belay.

5. (30 Min) **ESTABLISHING THE BELAY**. A properly designed and secured belay stance is essential if the risk of a serious injury is to be minimized while conducting climbing operations. Whether using natural or artificial anchor points will constitute the design of the belay stance:

a. Establishing a Natural Anchor Belay Stance Using the Rope. This method is desired when the anchor points are near by. Minimum amount of equipment is necessary to build this system, but it may require the use of a lot of rope.

(1) Approximately three feet from the retrace figure of eight loop on the harness, tie a figure of eight loop.

(2) Attach a large steel locking carabiner onto the retrace figure of eight. This will now be referred to as the Main Anchor Carabiner (MAC).

(3) Clip the figure of eight loop into the MAC. This will now be referred to as the remote. The remote allows the belayer to escape the anchor system if necessary. This will be covered in *RESCUE FOR PARTY CLIMBING.*

(4) From the remote, take a bight of rope and place it around the furthest natural anchor point then back to the belay stance.

(5) Ensuring that the rope is taut, tie an over the object clove hitch with the bight of rope and clip it into the MAC.

(6) From the clove hitch on the MAC, take a bight of rope and place it around the second natural anchor point then back to the belay stance.

(7) Again ensuring that the rope is taut, tie an over the object clove hitch and clip it into the MAC and lock down the carabiner.

(8) The climber is now in a secure belay stance. He will sound off with the command "Off Climb".

(9) He will now clip another steel locking carabiner into the MAC. This will serve as the belaying carabiner to attach the belaying device to.

(10) With the live rope near the lead climber's retrace figure of eight, attach it to the belay device ensuring that the belaying carabiner is locked down when complete.

(11) The #2 is now ready to belay the lead climber.

b. Establishing a Natural Anchor Belay Stance Using Slings / Runners. Using slings / runners will shorten the distance to the anchors points when they are further away. More equipment will be necessary to build this system, but it may require less rope to construct.

(1) Place a sling / runner around the furthest natural anchor point with a non locking carabiner.

(2) Attach a MAC onto the retrace figure of eight.

(3) With the rope from the retrace figure of eight loop on the harness, attach it to the furthest natural anchor point ensuring that there will be slack in the rope. This will enable him to escape the system if needed.

(4) With the rope from the natural anchor point, return to the designated belay stance.

(5) Ensuring that the rope is taut, tie an over the object clove hitch and clip it into the MAC.

(6) From the clove hitch on the MAC, take the rope and clip it straight through the second natural anchor point and return to the belay stance.

(7) Again ensuring that the rope is taut, tie an over the object clove hitch and clip it into the MAC and lock down the carabiner.

(8) The climber is now in a secure belay stance. He will sound off with the command "Off Climb".

(9) He will now clip another steel locking carabiner into the MAC. This will serve as the belaying carabiner to attach the belaying device to.

(10) With the live rope near the lead climber's retrace figure of eight, attach it to the belay device ensuring that the belaying carabiner is locked down when complete.

(11) The #2 is now ready to belay the lead climber.

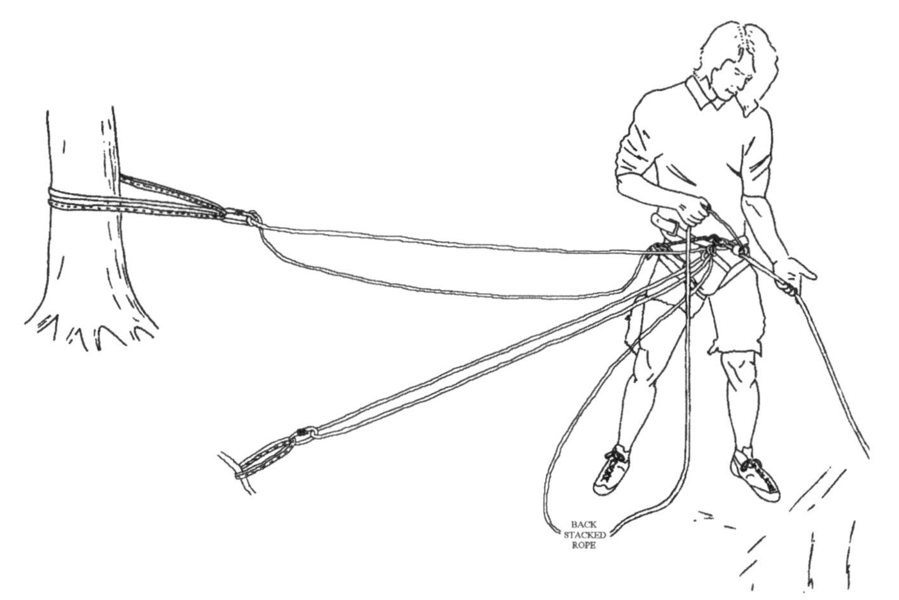

BELAYING WITH TWO NATURAL ANCHORS USING SLINGS AND RUNNERS

c. Establishing an Artificial Anchor Belay Stance. When no natural anchor points are available, the #2 will establish the belay stance utilizing at least three pieces of artificial protection. All three pieces of protection will be placed in the direction of pull.

NOTE: Special considerations for anchors established on ledges and during amphibious assaults will be discussed in *ALTERNATIVE BELAYS AND ANCHORS.*

(1) Place the three pieces of artificial protection into the rock, keeping in mind the anticipated direction of pull if the lead climber fell. Ideally the # 2 would place the three pieces behind him, below the waist and spaced out evenly.

(2) Attach a MAC onto the retrace figure of eight.

(3) With the rope from the retrace figure of eight loop on the harness, attach it to the furthest artificial anchor point with an over the object clove hitch ensuring that there will be slack in the rope that returns to the climber. This will enable him to escape the system if needed.

NOTE: Runners and slings are not required but can be used for extension purposes.

(4) With the rope from the artificial anchor point, return to the designated belay stance.

(5) Ensuring that the rope is taut, tie an over the object clove hitch and clip it into the MAC.

(6) From the clove hitch on the MAC, take the rope and clip it straight through the second artificial anchor point and return to the belay stance.

(7) Again ensuring that the rope is taut, tie an over the object clove hitch and clip it into the MAC and lock down the carabiner.

(8) Attach a non locking carabiner into the MAC.

(9) From the second clove hitch on the MAC, take the rope straight through the third piece of artificial protection and attach it to the non locking carabiner on the MAC with an over the object clove hitch.

(10) The climber is now in a secure belay stance. He will sound off with the command "Off Climb".

(11) He will now clip another steel locking carabiner into the MAC. This will serve as the belaying carabiner to attach the belaying device to.

(12) With the live rope near the lead climber's retrace figure of eight, attach it to the belay device ensuring that the belaying carabiner is locked down when complete.

(13) The #2 is now ready to belay the lead climber.

NOTE: Whenever using artificial protection, slings / runners are not required but can be used for extension purposes. Always keep in mind not to disturb the direction of pull when using slings / runners.

d. Establishing an Anchor Belay Stance using both Natural and Artificial Anchors. If constructing an anchor system in which only one natural anchor can be located for use, it must be backed up by two pieces of artificial protection. The method of constructing this system will be the same as constructing an artificial anchor belay stance.

TRANSITION: Now that we have discussed how to establish a belay stance, are there any questions? Let us now talk about methods of belaying a fall.

6. (5 Min) **METHODS OF BELAYING A FALL**. There are two methods of belaying a lead climber's fall: Static and Dynamic.

a. Static Belay.

(1) A static belay is a method, which does not allow the rope to run through the belay device, therefore stopping the falling climber quickly. The belayer brakes immediately after the fall occurs, therefore preventing any unnecessary slack developing between himself and the climber.

(2) This technique is used when the belay anchors points, and the running belays are sound. It is also used to stop a falling climber from hitting any projection (a ledge or rock outcrop) that is below him when he falls. This is the most common belay used on rock.

b. Dynamic Belay.

(1) A dynamic belay is a method, which deliberately allows some of the rope to run through the belay device, thus slowly bringing the falling climber to a halt. The belayer will gradually apply braking pressure to the rope to reduce the Impact Force on the belay anchor points and the running belays.

(2) This technique is used when the belay anchor points and the intermediate points of protection are not very sound. It is mostly used when ice climbing, although the technique is widely used when climbing unstable rock.

TRANSITION: Now that we have talked about the methods to belay a fallen climber, let us discuss the disassembly of the belay stance.

7. (5 Min) **DISASSEMBLY OF THE BELAY STANCE**. Upon hearing the command "Off Climb" from the lead climber, the #2 will:

a. Remove the live rope from the belay device and wait until the next command of "Up Rope".

b. The lead climber will pull up the rope until it becomes taut against the MAC of the #2.

c. The #2 will give the command "That's Me". This informs the lead climber to secure the rope to his belay device.

d. After securing the climbing rope to his belay device, the lead climber will give the command "On Belay".

e. The #2 will now disassemble the belay stance in reverse order of the way it was constructed. Ensuring that lead climber continues to take the slack out of the rope after each anchor removed.

f. After all the anchors have been removed, the #2 will ensure that all the gear is stowed properly and that the nut pick is readily available.

g. The #2 will now give the command "Climbing".

h. The lead climber will answer back with the command "Climb Climber".

i. The #2 will begin to climb. At this time, the roles are switched. The belayer will become the climber and the lead climber will become the belayer.

TRANSITION: Now that we have discussed the disassembly of the belay stance, are there any questions? Let us now talk about retrieving the protection.

8. (5 Min) **RETRIEVING THE PROTECTION**. While the #2 is climbing, he will retrieve all the pieces of protection along the route. The following are the actions of the #2 during his ascent:

   a. When the he reaches a piece of protection, he will sound off with the command "Point", this informs the belayer to set the brake with the belay device.

   b. The belayer will answer back with the command "Point", to ensure the climber that he has set the brake.

   c. The #2 will remove the protection and stow it away on his rack. If the piece of protection becomes difficult to remove, he will utilize the nut pick.

   d. After stowing the protection, he will sound off with the command "Climbing", this informs the belayer that he has removed the protection and that it has been properly stowed.

   e. The belayer will answer back with the command "Climb, Climber", this informs the climber that he is ready to begin belaying procedures.

   f. The #2 will repeat these actions until he tops off.

   g. When the #2 reaches the belay position, he will either:

      (1) Make himself secure by moving 10 feet back for the edge of the climb or attach himself to an anchor point.

      (2) Stop and exchange/reorganize climbing gear with the lead climber for next pitch, and then continue climbing.

TRANSITION: Now that we have discussed retrieving the protection, are there any questions? Multi-pitch climbs require the belayer to establish a belay stance during the climb. Let us now talk about the belay method known as hanging belays.

9. (10 Min) **ESTABLISHING A HANGING BELAY STANCE**. Multi-pitch climbs require the climber to establish a belay stance midway through the climb. Since it takes approximately 15 feet of rope to construct a hanging belay, constant awareness of the amount of rope used during the climb must be acknowledged. This can be accomplished through communications between the climber and the belayer. Constructing the hanging belay stance is basically the same as an artificial anchor belay stance with a few considerations.

   a. When the belayer has approximately 25 feet of live rope left to pay out to the climber, he will sound off "Twenty Five Feet". This informs the climber to begin looking for a belay stance.

b. The climber will begin to locate a good position to establish the hanging belay stance understanding that he will use approximately 10 to 15 feet of rope to construct the belay stance.

c. Once a position has been located he will sound off with the command "Point", this informs the belayer to set the brake with the belay device.

d. The belayer will answer back with the command "Point", to ensure the climber that he has set the brake.

e. The climber will begin to construct his belay stance by placing a piece of protection in the downward direction of pull at approximately chest level to the climber or higher, but within arms reach.

f. With the rope from the retrace figure of eight loop on the harness, attach it to the piece of protection with an over the object clove hitch.

g. The climber will attach a MAC onto the retrace figure of eight.

h. Ensuring that the rope is taut from the piece of protection, tie an around the object clove hitch and clip it into the MAC.

i. The climber will place a second and third piece of protection, again in the downward direction of pull at approximately chest level to the climber or higher, but within arms reach.

j. From the clove hitch on the MAC, the climber will take the rope and clip it straight through the furthest piece of protection and again ensure that the rope is taut, attach it on to his MAC with an around the object clove hitch and lock down the carabiner.

k. The climber will now attach a non-locking carabiner into the MAC.

l. From the second clove hitch on the MAC, the climber will take the rope straight through the third piece of protection.

m. Again ensuring that the rope is taut, attach it to the non-locking carabiner on the MAC with an over the object clove hitch.

n. The climber now will place his fourth piece of protection in the upward direction of pull at approximately waist level to the climber or lower, but within arms reach.

NOTE: The principle behind this is so that when the #2 climber passes the belay stance and becomes the lead climber, this fourth piece of protection will prevent the belayer from being pulled upward if the leader falls.

o. From the clove hitch on the non-locking carabiner, take the rope and attach it to the forth piece of protection with an over the object clove hitch.

p. The climber is now in a secure belay stance. He will sound off with the command "Off Climb".

q. He will now clip another steel locking carabiner into the MAC. This will serve as the belaying carabiner to attach the belaying device to.

r. The climber will pull up the rope, ensuring to back stack it carefully to prevent unnecessary entanglement.

s. Once the climber receives the command "That's Me", he will attach the live rope to the belay device ensuring that the belaying carabiner is locked down when complete.

t. The climber is now ready to belay his partner and will sound off with the command "On Belay".

TRANSITION: Now that we discussed hanging belay, are there any questions? Let us now discuss the changing over of gear during a climb.

10. (5 Min) **CHANGING OVER GEAR.** The most efficient method of multi-pitch climbing is for a pair of climbers to alternate leading pitches. At each belay stance the climbers will have to change over and reorganize the gear for the next pitch. When the #2 climber reaches the belay stance, the #1 climber will tie off the belay device and connect a web runner from his donut to the #2 climber's donut. Now the #2 climber is effectively secured so that the gear can be changed over.

TRANSITION: Are there any questions over changing over the gear? With the basics you have learned you should have little trouble in understanding the mechanics of an effective belay. If you have no questions for me, then I have some for you.

PRACTICE (CONC)

a. The students will practice establishing anchors.

PROVIDE HELP (CONC)

a. The instructors will assist the students when necessary.

OPPORTUNITY FOR QUESTIONS (3 Min)

1. QUESTIONS FROM THE CLASS

2. QUESTION TO THE CLASS

Q. What is the minimum number of anchors used in a natural anchor?

A. Two.

Q. What is the minimum number of anchors used in an artificial belay anchor?

A. Three.

SUMMARY (2 Min)

a. We have discussed belaying considerations, types of belays, methods of belays, components of a belay chain, the methods of anchoring, and how to tie in to the anchors.

b. Those of you with IRF's please fill them out and turn them in to the instructor. We will now take a short break.

UNITED STATES MARINIE CORPS
Mountain Warfare Training Center
Bridgeport, California 93517-5001

SML
ACC
02/19/02

LESSON PLAN

PARTY CLIMBING

INTRODUCTION (5 Min)

1. **GAIN ATTENTION**. Two party climbing is the means by which trained military mountaineers ascend vertical to near vertical rock features, without the benefit of a top-rope. In the military, party climbing is used as a means of ascending a cliff face to set up ropes and other associated installations from the top of a rock feature, to prepare the way for a unit to undertake a cliff assault, or as a means of crossing an obstacle.

2. **OVERVIEW**. The purpose of this period of instruction is to introduce the student to the method of two party climbing, as well as an understanding of fall factors. This period of instruction brings together placing protection, anchor belays for party climbing, and the balance climbing class. This lesson relates to cliff assault.

3. **INTRODUCE LEARNING OBJECTIVES**

   a. TERMINAL LEARNING OBJECTIVE.

      (1) In a summer mountainous environment and given the necessary equipment, two party climb a rock face, in accordance with the references.

   b. ENABLING LEARNING OBJECTIVES

      (1) In a summer mountainous environment and given the necessary equipment and a designated cliff face, lead a party climb, in accordance with the references.

      (2) In a summer mountainous environment and given the necessary equipment and a designated cliff face, belay a party climb, in accordance with the references.

      (3) In a summer mountainous environment, execute climbing commands for party climbing, in accordance with references.

      (4) Without the aid of references, list in writing the dangers to avoid for route selection in party climbing, in accordance with the references.

(5) Without the aid of references, list in writing the responsibilities of a lead climber prior to a climb, in accordance with the references.

(6) Without the aid of references, list in writing the times you would place protection, in accordance with the references.

4. **METHOD/MEDIA.** The material in this lesson will be presented by lecture and demonstration. You will practice what you have learned in upcoming field evolutions. Those of you with IRF's please fill them out at the end of this lesson and return them to the instructor.

5. **EVALUATION**

a. SML - You will be tested by a written exam and a performance evaluation.

b. ACC - You will be tested by a written exam and a performance evaluation.

TRANSITION: Are there any questions over the purpose, learning objectives, how the class will be taught, or how you will be evaluated? Before a climb can start we need a belay.

BODY (75 Min)

1. (5 Min) **THE BELAY.** Where a belayer chooses to establish his belay is an important consideration. It is possible that the belayer may sit or stand for hours on end so a position of comfort should be chosen. There are two basic stances, sitting and standing. Normally sitting is used on the top anchor and standing on the bottom.

TRANSITION: Now that we have talked a little about the belay, let's now look at some dangers to avoid when party climbing.

2. (5 Min) **DANGERS TO BE AVOIDED WHEN SELECTING A ROUTE**

a. Wet or Icy Rock. These impediments can make an otherwise easy route almost impassable.

b. Rocks overgrown with moss, lichen, or grass. These areas can be very treacherous when wet or dry.

c. Tufts of grass and small bushes growing from loosely packed soil. These normally appear firm, but can give way suddenly when they are pulled or stepped on.

d. Gullies that are subject to rock fall. If you have to use a gully that has evidence of rock fall in it, then try to stay to the sides.

e. The most common danger is the overestimation of your own ability.

TRANSITION: Once at a cliff face, it is imperative that we keep in mind our responsibilities in order to maintain organization.

3. (5 Min) **LEAD CLIMBERS RESPONSIBILITIES**. Before starting a climb the lead climber must:

   a. Pre-select probable route.

   b. Ensure he has the proper equipment to complete the route.

   c. Ensure that the climber and the #2 are tied into their respective ends of the rope.

   d. Ensure that the #2 has selected the proper anchors for the belay anchor system.

   e. The lead climber may have to construct a gear rack to carry his equipment on. It is constructed as follows;

      (1) Take a length of 1 inch tubular nylon webbing and tie the ends together using a water/tape knot forming a loop so it fits over your head and shoulder and runs diagonally across your chest. If tubular nylon webbing is not available, then utilize one of your runners, if the route will permit. A sling rope could also be used as a gear rack.

TRANSITION: Now that we have covered dangers to be avoided when selecting a route, are there any questions? As we climb, and we have all of this gear on our body, we need to know when and where to place our pro.

4. (10 Min) **RULES FOR PLACING PRO.** In the placing protection class the method of placing protection was covered; however, there are four rules for placing pro. These are as follows:

   a. First runner rule. A good piece of protection should be placed as high as possible just before leaving the ground to prevent hitting the ground in a fall. Once that first piece of pro is at waist level, a second piece should be placed as high as possible. A third piece should be placed following the same guidelines as the second piece (this will reduce the chances of the leader bottoming out).

   b. Every 10 to 15 feet. This is done to prevent the possibility of taking a long fall. On a short route, 30 to 50 feet, protection should be placed every six to eight feet to prevent bottoming out during a fall.

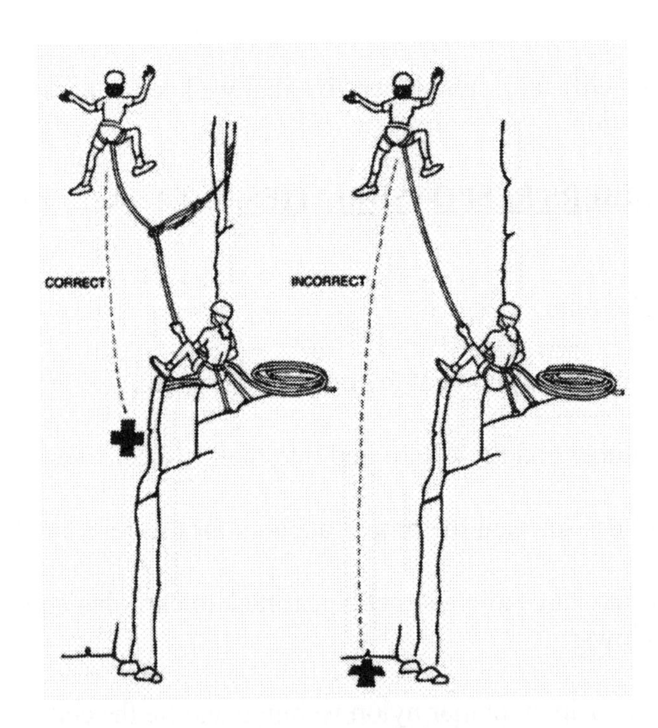

BOTTOMING OUT DUE TO NO PROTECTION

c. Before and after a hard move. A leader should place pro before and after a hard move (crux) because a fall is more likely. If you are not satisfied with that piece of pro, back it up with another one. If a ledge is encountered during a climb, the first runner rule will apply again.

PLACING PROTECTION BEFORE AND AFTER A HARD MOVE

d. When the climber feels that protection needs to be placed. When in doubt - stitch the route.

e. When placing a number 1 or number 2 stopper, always backup these pieces of protection with a secondary piece as soon as possible.

5. (5 Min) **PREVENTING ROPE DRAG.** Rope drag causes all sorts of problems. It can hold a climber back, throw him off balance, pull his pro out, and can make it hard for the leader to pull enough rope up to clip the next piece of protection. Also, rope drag can affect how well a belayer responds to a fall by reducing the ability to provide a dynamic belay. Keeping the rope in a straight line from the belayer to the climber is the best way to reduce rope drag.

   a. Pro Placement.

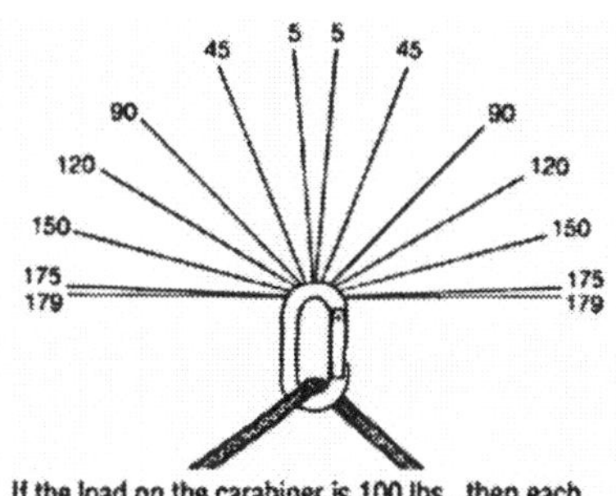

If the load on the carabiner is 100 lbs., then each anchor will be subjected to X pounds of force depending on the angle:

ANGLE BETWEEN PROTECTION (DEGREES)	LOAD ON PROTECTION (LBS.)
179	5727
175	1147
150	200
120	100
90	71
45	54
5	51

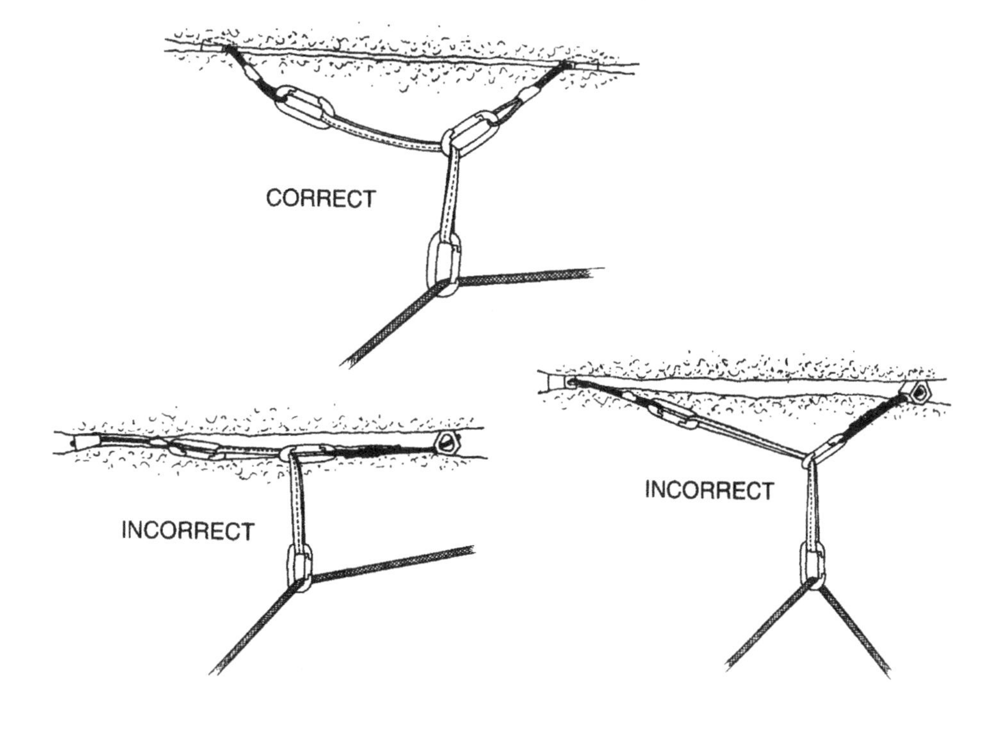

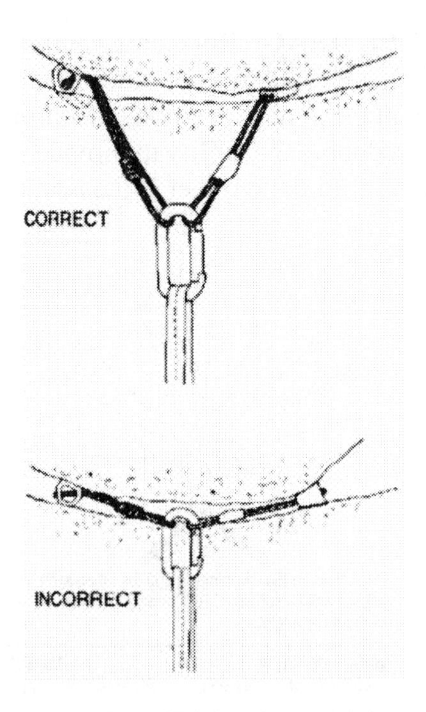

b. Runners / Quick draws. Protection should be placed so that the rope follows as straight of a line as possible. If the protection placements do not follow a straight line up the pitch, and the rope is clipped directly to these placements, it will zigzag up the cliff, causing severe rope drag. If the protection cannot be placed in a direct line, runners or quick draws can be used to extend the protection. This will allow the rope to hang straight and run more freely through the protection system.

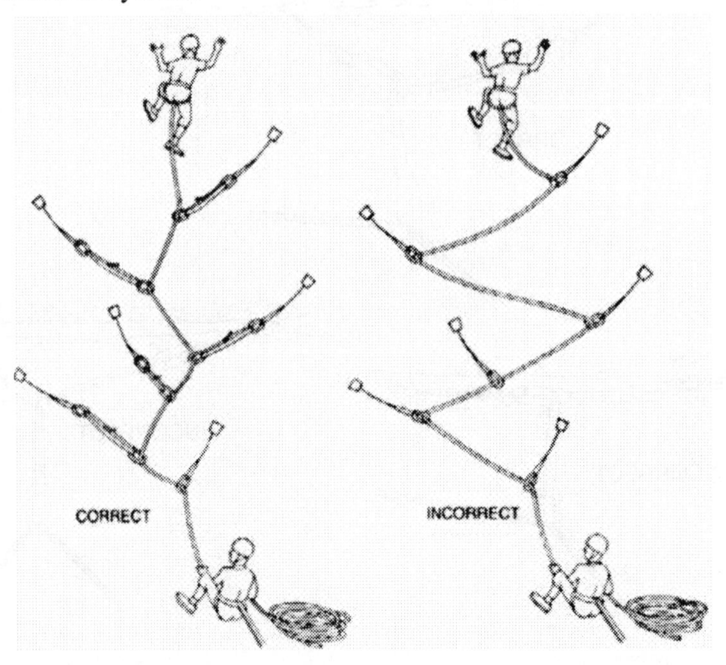

USE OF RUNNERS TO PREVENT ZIG-ZAGGING OF ROPE

(1) If you use an extra long runner you can create another problem. The extension may keep the rope in a straight line, but it may also add dangerous extra feet to the length of a fall. In such a case, it is sometimes better to accept some rope drag in order to get better security in case of a fall.

(2) If the protection placements happen to be in a straight line, the rope will run straight and there will be less rope drag even if it is clipped directly to the protection. However, you must be aware that rope movement can and will jiggle a chock out of position. So a quick draw can be used to isolate the pro from rope movement.

(3) Quick draws should always be placed on wire protection and SLCD's without pre-sewn runners that are used for running belays. There is no requirement to place quick draws on wire runners used for a belay stance, unless you need some extension.

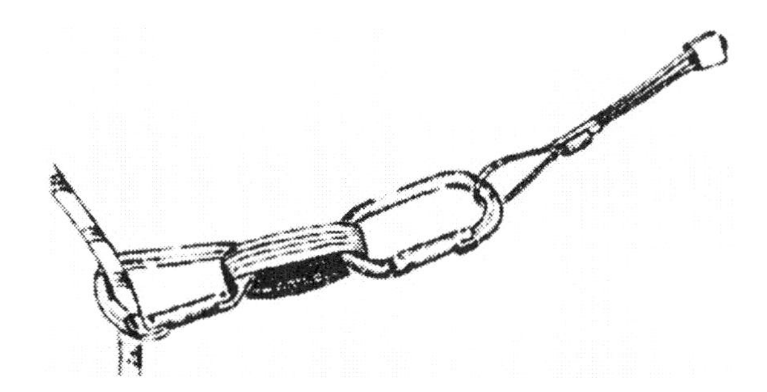

QUICK DRAW ON A PIECE ON WIRE PROTECTION

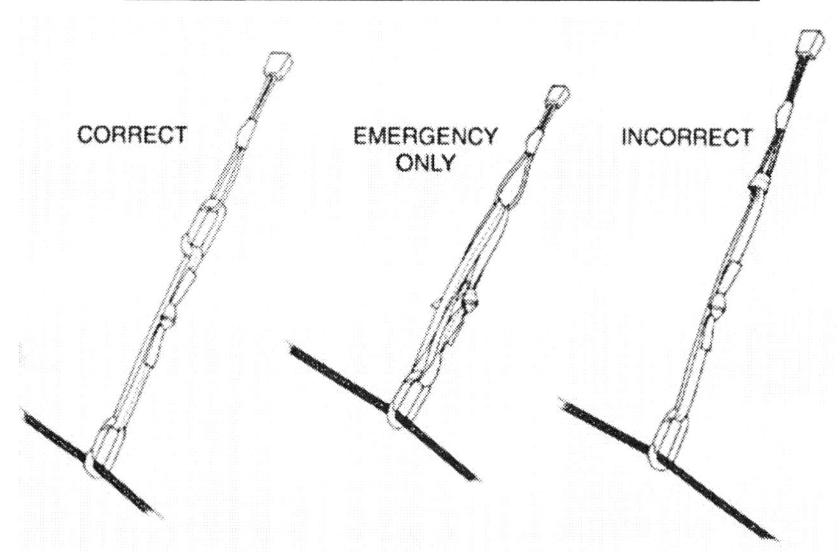

QUICKDRAWS ON PROTECTION

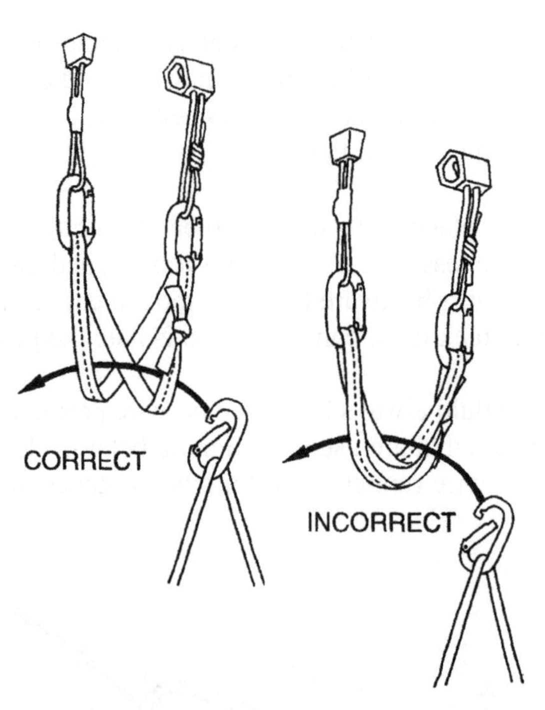

PROPER PLACEMENT OF CARABINER THROUGH A RUNNER ON MULTIPLE PIECES OF PRO

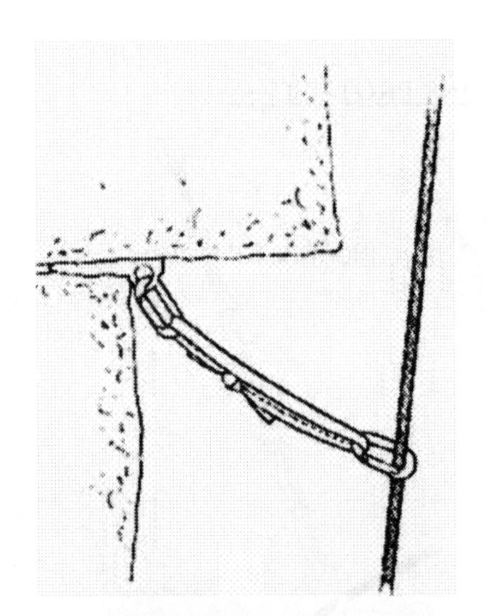

PROPER ROPE PLACEMENT AND RUNNER LENGTH

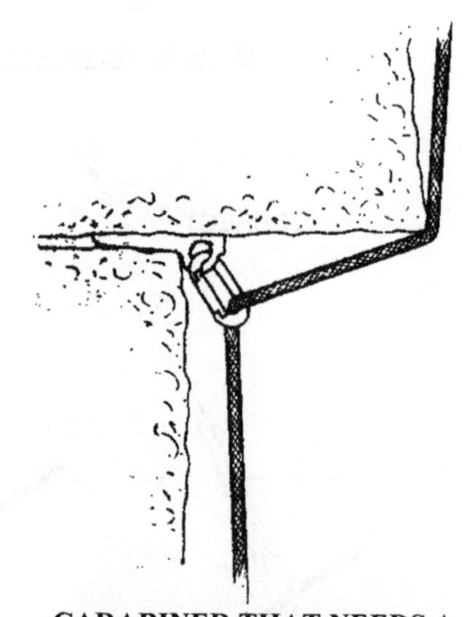

CARABINER THAT NEEDS A RUNNER AND CARABINER TO REDUCE ROPE FRICTION

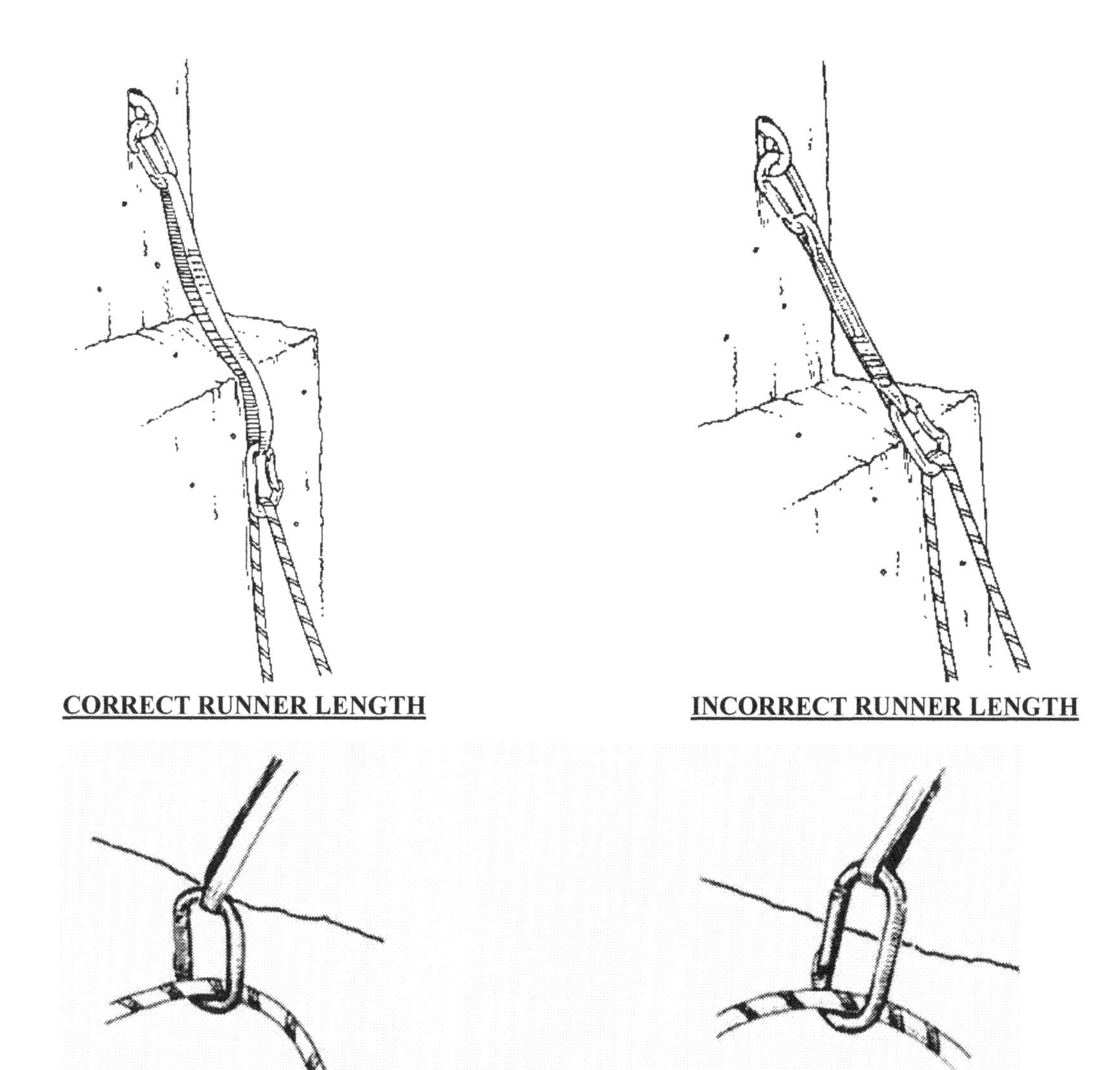

CORRECT RUNNER LENGTH **INCORRECT RUNNER LENGTH**

CORRECT PLACEMENT OF CARABINER ON AN EDGE **INCORRECT POSITION OF CARABINER ON AN EDGE**

(4) Quick draws do not need to be placed on corded hexcentrics or chocks unless you need extension.

6. (10 Min) **CONSIDERATIONS FOR PARTY CLIMBING**

a. A two man climbing team is faster than a three-man team is.

b. The strongest climber of a team always takes the hardest pitch.

c. Climbers will tie into their harnesses with a retrace figure-of-eight; they will not untie until they are off the climb, 10 ft back away from the cliff edge, or secured to an anchor point.

d. As you climb you must use the correct climbing commands to prevent confusion.

e. Where possible, the leader should use natural anchors for protection.

f. The leader must not climb to the length of the rope before selecting a good belay stance, ideally 15 feet of rope is required for a belay anchor.

(1) When the lead climber reaches a good position, he will set up a belay position and bring up the #2 climber.

g. If a traverse is encountered it is protected as shown. The leader is not only protecting himself but also the #2 man as well.

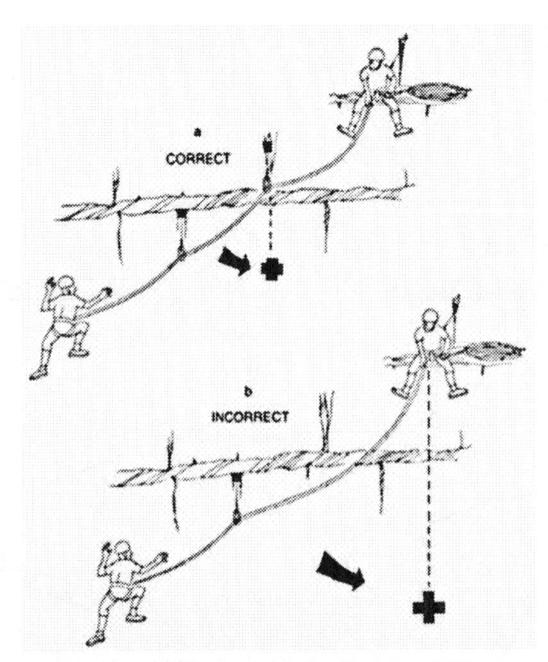

PRO PLACEMENT FOR TRAVERSING

NOTE: Protecting a traverse: A: correct, placing protection after a hard move and a traverse can reduce the potential pendulum fall for the second climber. B: incorrect, if the second climber falls on the traverse with inadequate protection, he faces a long pendulum fall.

TRANSITION: Now that we have discussed rules for placing pro, preventing rope drag and considerations for party climbing, are there any questions? Next we'll talk about the climbing commands used when party climbing.

7. (10 Min) **CLIMBING COMMANDS.** During a climb you will often find yourself in a position where you cannot see your climbing partner. This accompanied by the effects of wind, weather conditions or the distance between each other, often makes communication very difficult. For this reason we use a very strict set of commands in order to communicate with as little confusion as possible.

a. Verbal Commands

Command	Given By	Meaning
“On Belay”	Belayer	I have a solid anchor, and am ready for you to climb.
“Climbing”	Climber	I am ready to begin the climb.
“Climb”	Belayer	Go ahead.
“Point”	Climber	I have reached a point where I am going to place protection, watch me.
“Point”	Belayer	I am watching, giving or taking rope as needed.
“25 feet”	Belayer	The climber has 25 feet of rope left, and must find a belay stance soon.
“FALLING”	Climber	The climber is about to fall.
“Slack”	Climber	Pay out some rope.
“Slack”	Belayer	Allows climber to take rope that is needed.
“Tension”	Climber	Telling belayer to take up excess rope.
“Tension”	Belayer	I am taking up excess rope.
“Off climb”	Climber	I am at the next belay and I am anchored.
“Off Belay”	Belayer	I have taken off the belay.

NOTE: At this point the belayer becomes the number 2 climber and the lead climber is the belay man.

Command	Given By	Meaning
"Up Rope"	#2 (Climber)	Take up excess rope between us
"That's me"	#2 (Climber)	The Climber is ready to be put on belay and the rope is taut on the climber and not snagged
"On Belay"	No 1 (Belayer)	I am in the belay stance and I am ready for you to climb

NOTE: From this point on the commands are the same as before.

b. Non-verbal Commands. When undertaking a tactical cliff assault the use of verbal climbing commands are not of much use. To that end the climbing pair utilize a method of sharp tugs on the rope to communicate with each other. These are as follows:

Command	Meaning
1 Tug	Give me slack
2 Tug	Give me tension.
3 Long Tugs	I am secured on the belay/the rope is secure.

.

NOTE: Practice of the above Tug commands is essential, as during the normal course of a climb the rope will move around in the belayer’s hand and can deceive the belay man.

TRANSITION: Now that we have discussed climbing commands, are there any questions? Without adequate communications, conducting a two-party climb can be extremely dangerous to both parties. With an understanding of belays, it is important to realize that during a multi-pitch climb the leader can fall beyond his belay man without hitting the ground, therefore it is important to learn about the forces exerted in such a fall.

8. (10 Min) **<u>FALL FORCES</u>.** The general standard of climbing equipment is continually being improved. In the event of a fall by a lead climber, an enormous amount of force is applied on the rope. As the rope stretches, part of the energy converted into heat by the friction between the rope fibers. There are two parts in determining the force of a fall, Impact Force and Fall Factor.

   a. <u>Impact Force</u>. (Shock Force) This is the amount of force the belay man has to exert on a falling climber through the rope, anchor and belay device to stop his fall. The amount of impact force needed to stop his fall is determined by the fall factor.

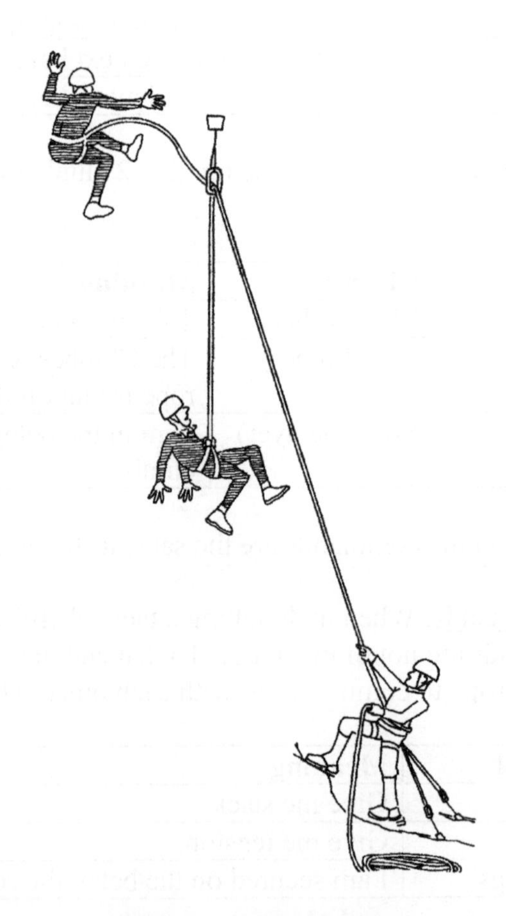

<u>IMPACT (SHOCK) FORCE</u>

   b. <u>Fall Factor</u>. This is the amount of force generated by a leader fall onto the rope. A fall factor is purely a mathematical equation, which its numeric result ranges between 0 at the lowest and 2 at the greatest. Fall factor is simply the length of the fall divided by the rope run out. The equation looks like this:

      (1) Fall Factor = $\frac{\text{Length of Fall}}{\text{Rope Run Out}}$

Example #1

10 meter run out above a belay on a rock face results in a 20 meter fall. (10 x 2)

$\frac{20}{10}$ = 2 Fall Factors, the highest Fall Factor.

Example #2

10 meter run out above a belay on a rock face, but with a piece of protection placed at 5 meters resulting in a 10 meter fall. (5 x 2)

$\frac{10}{10}$= 1 Fall Factor

Example #3

10 meter run out above a belay on a rock face, but with a piece of protection placed at 7.5 meters resulting in a 5 meter fall. (10 – 7.5)= 2.5 x 2 = 5

$\frac{5}{10}$= 0.5 Fall Factor

FALL FACTOR EXAMPLES

c. The following table shows the relationship of Fall Factor and Impact Force for a 180-pound climber:

RELATIONSHIP OF FALL FACTOR AND IMPACT FORCES FOR 180 LB CLIMBER			
FALL FACTOR	IMPACT FORCE	FALL FACTOR	IMPACT FORCE
0.0	360 lbs	1.0	1676 lbs
0.1	683 lbs	1.2	1817 lbs
0.2	868 lbs	1.4	1947 lbs
0.4	1137 lbs	1.6	2067 lbs
0.6	1347 lbs	1.8	2181 lbs
0.8	1521 lbs	2.0	2288 lbs

NOTE: The less a climber weighs, the less Impact Force the belay has to sustain. The more the climber weighs, the more force is exerted

d. There are several points that should be considered during a climb to minimize the seriousness of a fall:

(1) The belayer must always be in a position to withstand an impact force of 2300 lbs.

(2) Protection should be placed more frequently during the first half of the climb.

(3) It is better to belay after a series of cruxes than before them. Belays after cruxes minimize Fall Factors in potentially dangerous situations.

(4) A seat harness should always be used to distribute the forces of a fall.

(5) Any fall involving a Fall Factor in excess of 1.0 should be regarded as serious and the rope should be retired from lead climbing duties.

TRANSITION: We have discussed how to party climb and the considerations for Fall Forces, now let's talk about tactical considerations for military operations.

9. (5 Min) **CONSIDERATIONS FOR NIGHT PARTY CLIMBING AND GEAR SILENCING**

a. Try to get eyes on the cliff and complete a cliff sketch before nightfall in order to study the terrain and probable route.

b. All considerations for party climbing apply at night.

c. Non-verbal commands should be applied.

d. Get rid of all belay devices to reduce noise, use the Munter hitch for belaying.

e. Place tape around carabiners to reduce noise, but ensure it does not interfere with gate operation.

f. If using Hexcentrics, tape the sides and ends of size 6-11 hexes to muffle the echoing effect when they bang together.

g. Use 550 cord attached to the nut pick to extend it from your chest pocket.

h. NVG's should be used on the bottom and top of the cliff to help with accountability.

TRANSITION: Up to this point, we have talked about the dangers to avoid when selecting a route, lead climbers responsibilities, rules for placing protection, climbing commands, night party climbing, gear silencing and Fall forces. Are there any questions? If you have none for me, then I have some for you.

PRACTICE (CONC)

a. Students will set an anchor, belay and lead a party climb.

PROVIDE HELP (CONC)

a. The instructors will assist the students when necessary.

OPPORTUNITY FOR QUESTIONS (3 Min)

1. QUESTION FROM THE CLASS

2. QUESTIONS TO THE CLASS

Q. What dangers should be avoided when selecting a route?

A. (1) Wet or icy rock.
(2) Rocks overgrown with moss, lichen or grass.
(3) Tufts of grass/small bushes growing from loosely packed soil.
(4) Gullies that are subject to rock fall.
(5) The most common danger is overestimation of your ability.

SUMMARY (2 Min)

a. During this period of instruction we have discussed the dangers to avoid in party climbing, four responsibilities of a lead climber, climbing commands, when you place pro and Fall Forces.

b. Those of you with IRF's please fill them out and pass them to the instructor. We will now take a short break.

UNITED STATES MARINE CORPS
Mountain Warfare Training Center
Bridgeport, California 93517-5001

SML
ACC
02/19/02

LESSON PLAN

MILITARY AID CLIMBING

INTRODUCTION

(5 Min)

1. **GAIN ATTENTION.** Occasionally, a military climber may find himself up against a rock barrier, which has insufficient handholds, or foot holds to climb or that the technical grade is above what he can climb. One way or another, the climber has to get to the top of the cliff in order to complete the mission. Utilizing aid techniques can get a climber through a tough section where he would normally be stuck and unable to continue.

2. **OVERVIEW.** The purpose of this period of instruction is to introduce to the student military direct aid, by discussing and demonstrating aid climbing, as well as the equipment and techniques used. This lesson relates to party climbing.

3. **INTRODUCE LEARNING OBJECTIVES**

   a. TERMINAL LEARNING OBJECTIVE. In a summer mountainous environment, conduct military aid climbing, in accordance with the references.

4. **METHOD/MEDIA.** The material in this lesson will be presented by lecture and demonstration method. You will practice what you have learned in upcoming field training exercises. Those of you with IRF's please fill them out at the end of this period of instruction.

5. **EVALUATION.** This period of instruction is a lesson purpose class.

TRANSITION: Are there any questions over the purpose or how the class will be taught? Before discussing how to military aid climb, we will talk about what aid climbing actually is.

BODY

(50 Min)

1. (5 Min) **INTRODUCTION TO AID CLIMBING**. A free climber depends entirely on his footwork, ability, skills and physical strength to move up the rock face. The equipment used while free climbing is for safety; i.e., protection. Equipment is not used to ascend the rock directly. If you were to place a piece of pro and hang on it, or use it to reach a higher hold,

you would not be free climbing in the "pure" sense of the word. Any use of equipment to directly ascend a rock face is called aid.

TRANSITION: Next we will discuss situations which may require direct aid.

2. (5 Min) **REASONS FOR AID.** There are many situations, which a military climber may choose to aid, let's name a few.

   a. Above your ability. You conducted the cliff recon and picked a route, which appeared to be within your ability. Now you're 2 pitches up and stuck.

   b. Fatigue or Injury. The route may be a simple 5.6, but due to it being sustained 5.6 the entire pitch and not just one 5.6 move, you transition to aid due to fatigue. You take a lead fall on the second pitch and injure your arm or leg. As long as you have one good arm and leg you can continue on, utilizing aid techniques.

   c. Weather. You arrive at the cliff site and it begins to rain or snow. Aid can be accomplished with gloves on, even when wearing ski march or ice boots.

   d. Save Energy for Mission or Rescues. As an assault climber, conducting a cliff assault is just a means of crossing an obstacle, in route to the objective. Once your Company or Platoon is up the cliff, you still need the energy to continue on with the mission. In the event of a TRAP mission or high angle rescue, the real work begins once the casualty is down off the cliff, and must be carried to the extract zone.

TRANSITION: Now that we have discussed reasons direct aid may be applied, are there any questions? Let's talk about some of the gear we'll be using.

3. (15 Min) **GEAR FOR MILITARY AID CLIMBING.** Some specialized gear is required for military aid climbing however; this gear can be constructed from the equipment a climbing team already possesses. The three minimum pieces of aid equipment are the Etrier, Daisy Chain and Fifi Hook.

   a. Etrier. Also known as Aiders. These are ladder like slings that allow climbers to step up from one placement to the next when they are clipped into a piece of pro. Etriers become your hand and footholds on otherwise unclimbable rock. The etriers can be constructed from an 18 foot tape or 21 foot prussic cord (tied in a loop with a square knot). At the top of the web/cord, tie a two-inch loop with an overhand knot. 10 to 12 inches down from the overhand pinch the two strands of web/cord together and pull up one side until it is offset enough for a comfortable step. Tie another overhand knot and repeat the process for the entire length of the web/cord. It is important to alternate which side is offset so that all the steps are not on one side. An 18-foot tape should provide 6 or 7 steps.

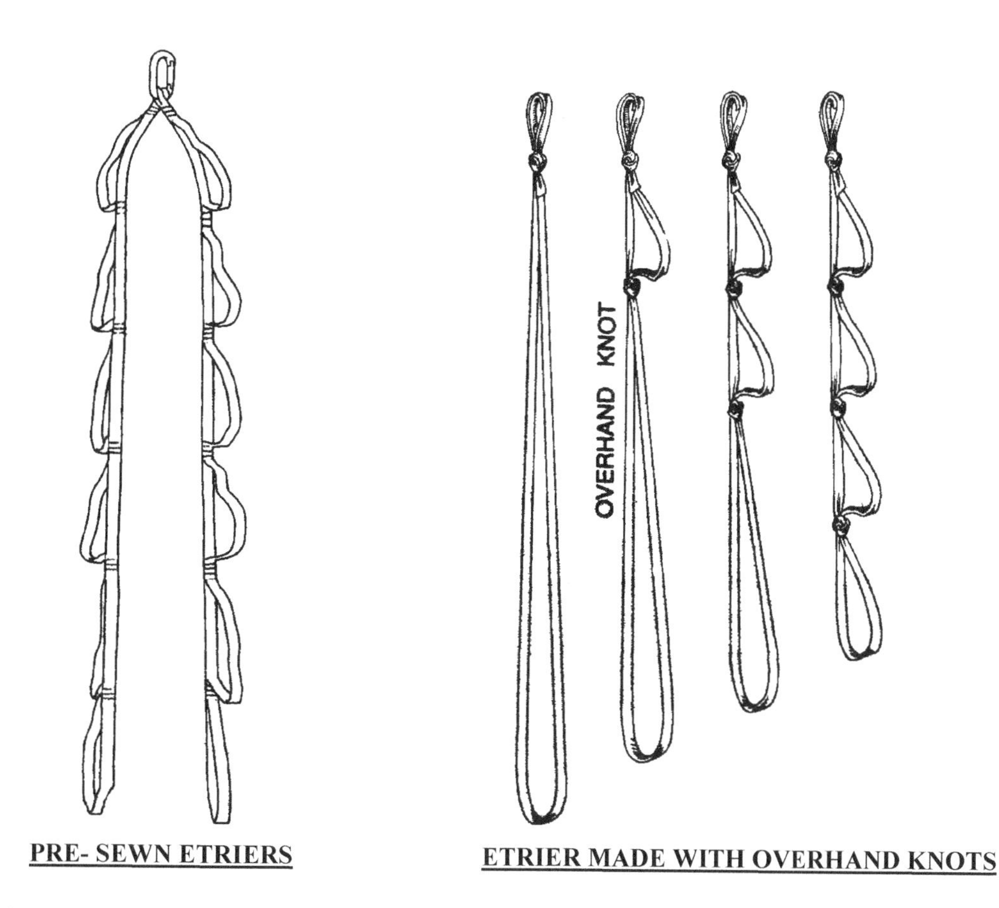

PRE- SEWN ETRIERS

ETRIER MADE WITH OVERHAND KNOTS

b. Daisy Chain. A length of webbing or cordage with clip in loops along its entire length used to connect a climber to a piece of pro. A daisy chain can be made from a 48-inch web runner. It is constructed in a similar manner as an etrier however; all the loops should be only 3 to 6 inches apart and on the same side. Keep in mind that the daisy chain should not extend beyond arm's reach.

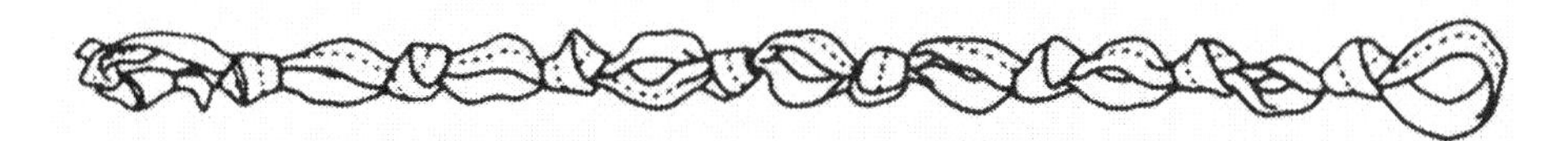

DAISY CHAIN

c. Fifi Hook. The fifi hook also called the "Arm Saver", is used by the lead climber to clip in short from his harness to the next piece of pro after ascending the etrier. This allows the climber to rest his arms while assessing the next placement and sorting through his rack. The actual fifi hook is not part of the MAC Kit but can be made by girth hitching a small runner to the hardpoint of your harness and clipping a carabiner to it.

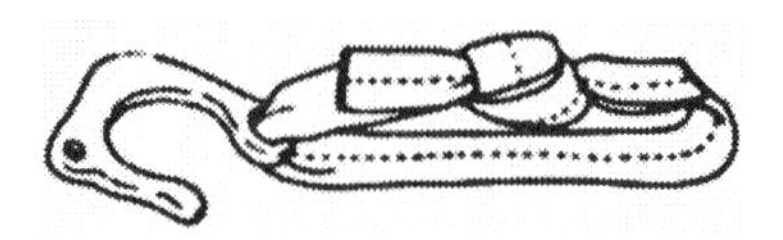

FIFI HOOK

TRANSITION: Let's take a minute to look at the Aid Rack.

4. (5 Min) **THE RACK.** If you determined from the recon that an entire pitch or more of aid is necessary for the climb, you must augment the standard rack as necessary from the MAC Kit. In direct aid, you must find a placement every arms length, unlike free climbing where you have 8 to 10 feet or more to look for a good placement. This presents two problems, a massive amount of equipment necessary and getting used to moving upward on marginal placements.

   a. Amount of Equipment. The amount of pro necessary to complete a 50 meter pitch may be as much as 40 placements, and out of the 40, many may be of the same size. For this reason you must tailor your rack to the climb. Recon the route, focusing on the crack width, then draw the appropriate amount of chocks and cams suitable for the crack. You must consider the possibility of having to leave two carabiners per piece of pro, so now you are looking at 80 free carabiners. This does not include the carabiners used to clip your pro to the rack. To save some equipment you can bypass putting a runner on every piece of protection. Clip in two linked carabiners to the pro, the second carabiner acts as a runner where you clip the rope.

   b. Slings. Due to the weight involved with this much equipment it may be necessary to draw two padded racks. Sling them over both right and left shoulders and clip equipment to both racks and the climbing harness.

TRANSITION: Now that we have discussed the gear required, are there any questions? Let's talk about the techniques used in military aid climbing.

5. (10 Min) **TECHNIQUES USED IN MILITARY AID CLIMBING.**

   a. Some of the simplest forms of aid climbing are known as "Hang Dogging" or "French Free" in the civilian climbing community. Hang Dogging involves clipping into a piece of protection, having the belayer take in all the slack and locking the belay device off so the climber can rest.

HANG DOGGING

b. French Free involves using protection to pull up or to stand up on. The most basic method is using aid for one move.

(1) Set in a piece of pro and clip an etrier into it.

(2) Step into the etrier and slowly transfer weight onto it in order to test the solidity of the piece of pro.

(3) If it holds, conduct a bounce test with one foot in the etrier.

(4) Step up higher in the etrier and bypass the difficult move.

(5) If the problem is lack of a handhold, a short web runner can be clipped into the pro and used as a handle.

FRENCH FREE

c. Aiding through multiple moves is more difficult and time consuming.

(1) The first aid movement is to place pro as high as possible and clip in two linked carabiners.

AID CLIMBER USING ETRIER AND PLACING PRO

(2) Next, clip the daisy chain into the upper most carabiner of the pro just placed. All slack needs to be adjusted out of the daisy chain. Moving the carabiner down the daisy chain to the next clip in point does this, then slowly apply body weight to it. At this point you are hanging from the daisy chain, if you can reach up to the pro, your adjustment is correct.

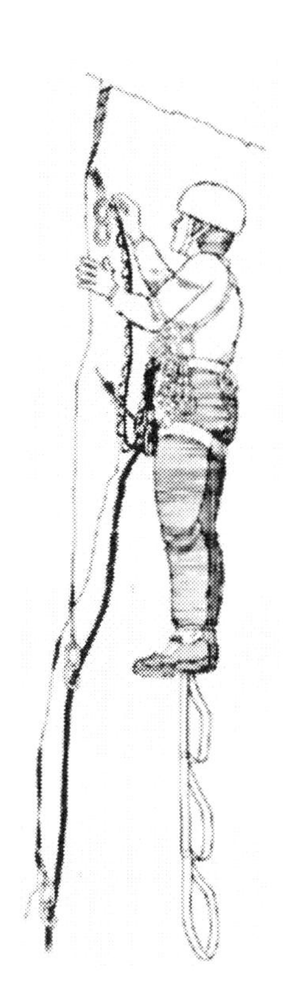

CLIPPING DAISY CHAIN TO PROTECTION

(3) Now conduct a hanging bounce test. This is to ensure that the piece is solid prior to committing to it.

HANGING BOUNCE TEST

(4) If the piece is solid, clip the etrier to it, upper most carabiner, and ascend the etrier.

(5) Once your waist is level with the piece of pro, clip in the fifi hook to the upper most carabiner, sit back and rest. By clipping everything into the uppermost carabiner, the lower carabiner is free for the rope.

(6) Now, clip the rope, into the lower, or free carabiner on the pro that you are hanging from. Note, if you clip the rope into the pro at an arms length, like when free climbing, you then ascend the etrier and the piece blows, you just increased the length of your fall.

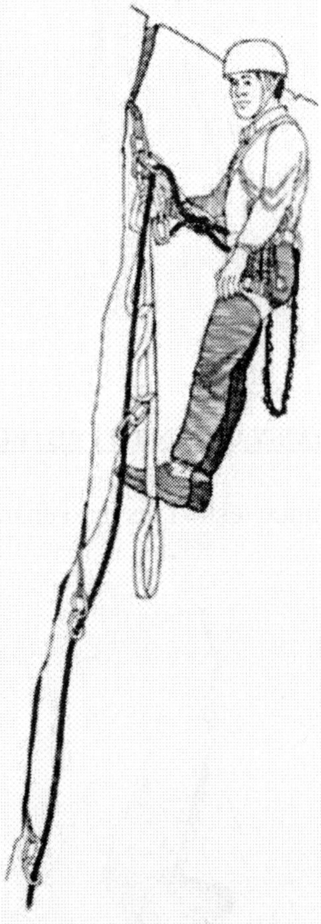

RESTING ON A FIFI HOOK

(7) The sequence begins again; however, now that you are off the ground and hanging from the fifi hook, you place the pro, two carabiners, daisy chain and test. If the new piece blows, you should only fall a couple inches until your fifi hook on the previous piece stops you. For this reason, only unclip your fifi hook after the bounce test and just before ascending the etrier.

NOTE: Increased stability on a climb can be provided by using two etriers with a Daisy Chain attached to each however this requires drawing the extra equipment in advance.

d. Due to the extreme amount of time involved in direct aid, the leader, once at the belay, should fix the rope to the anchor. The #2 climber should ascend the rope by use of mechanical ascenders, cleaning the pro while on a rope ascender. If the climbers do not have a set of rope ascenders, then it will be necessary for the #2 climber to aid the pitch and clean as he goes while being top roped. The only difference being, once the fifi hook has been released from the last piece of pro, the climber will hang from the daisy chain, or clip in the fifi hook to one of the steps of the etrier, hang and remove the previous piece of pro.

TRANSITION: Knowing the proper sequence and performing it correctly is critical. Are there any questions over techniques or the sequence?

PRACTICE (CONC)

a. Students will practice what was taught in upcoming field exercises.

PROVIDE HELP (CONC)

a. Instructors will provide help as necessary.

OPPORTUNITY FOR QUESTIONS (3 Min)

1. QUESTIONS FROM THE CLASS

2. QUESTIONS TO THE CLASS

Q. What is aid climbing?

A. Using equipment to directly ascend a rock face.

Q. What are three specialized pieces of gear used in aid climbing?

A. An etrier, daisy chain, and fifi hook.

SUMMARY (2 Min)

a. During this period of instruction we have covered what military aid climbing is, the specialized gear required and how to conduct aid climbing. Remember that military aid climbing utilizes the equipment a leader already possesses as part of his climbing rack.

b. Those of you with IRF's please fill them out and turn them in. We will now take a short break.

UNITED STATES MARINE CORPS
Mountain Warfare Training Center
Bridgeport, California 93517-5001

SML
ACC
02/19/02

LESSON PLAN

ALTERNATIVE BELAYS AND ANCHORS

INTRODUCTION (5 Min)

1. **GAIN ATTENTION**. At some stages during a climb there might be a need to construct belays or anchors that are unique to a certain situation. In some cases it may not be possible to construct a three point artificial or a two point natural anchor due to time constraints or lack of good protection available. Alternative belays and anchors can take on many guises. Be it rock spikes, flakes, boulders, trees, chocks wedged in cracks, and pitons. Most alternative belays and anchors are direct and dynamic.

2. **OVERVIEW**. The purpose of this period of instruction is to provide the students who, after passing the climbing phase, have further knowledge and skills in the use of alternative belays and anchors. This lesson relates to *PARTY CLIMBING.*

3. **METHOD/MEDIA**. The material in this lesson will be presented by lecture and demonstration.

4. **EVALUATION**. This period of instruction is a lesson purpose class.

TRANSITION: Are there any questions over the purpose or how the class will be taught? As mentioned earlier, certain situations will dictate that you may have to use some sort of alternative belay or anchor. We will first discuss the belays.

BODY (35 Min)

1. (15 Min) **BELAY**. Belay is a method to stop a fall by applying friction to the rope and by resisting the forward pull of the fall. Besides the use of traditional devices in belaying operations, other techniques can be utilized. The following are a few expedient methods of belaying:

a. Body Belay Method. The body belay is used as a convenient and quick belay when ascending and descending moderate terrain with experienced troops. Since friction burns are a real danger, the arms must be covered and gloves should be worn. This method can be used in both the standing or sitting stance. The following techniques apply for the body belay:

(1) The belayer should position himself so that his legs are braced straight into the direction of pull.

(2) Place the rope around the belayer's back so that the rope rests on top of the hips while he grasps the rope tightly with both hands.

(3) To pay out rope, slide the running end of the rope (nearest the climber) forward with a guide hand and clasps both strands of the rope above the brake hand.

(4) Allow the brake hand to slide back down to its original position and repeat the process.

(5) To brake the climber, the brake hand wraps the rope around the waist. A twist of rope can be taken around the brake arm to increase friction.

b. Redirect Belay Method. When belays are established away from the base of a climb, the rope runs at a low angle from the belayer to the first piece of pro on the rock. From here the rope changes direction and goes abruptly upward. If the leader should fall, the rope goes taut and tries to run in a straight line from the belayer to the top piece of pro. This effect puts great strain on the bottom piece of pro. If it pulls out, the line of pro could be yanked out one after another from the bottom up. This is known as the "Zipper Effect". The zipper effect can be prevented by moving the belay stance within 10 feet of the base of the climb or by creating opposition protection.

ZIPPER EFFECT

(1) Place one chock in the crack so it can take a downward pull.

(2) Below that emplacement, place another chock with an upward pull.

(3) Clip a carabiner into each chock wire / cord.

(4) Using a large runner, tie the two carabiners together using clove hitches so that there is tension between the two chocks, which will help hold them in place.

(5) Create a loop in the slack of the runner near the upward direction of pull chock by tying an overhand knot and attach a carabiner into the loop.

(6) Clip the climbing rope into this carabiner in the normal fashion.

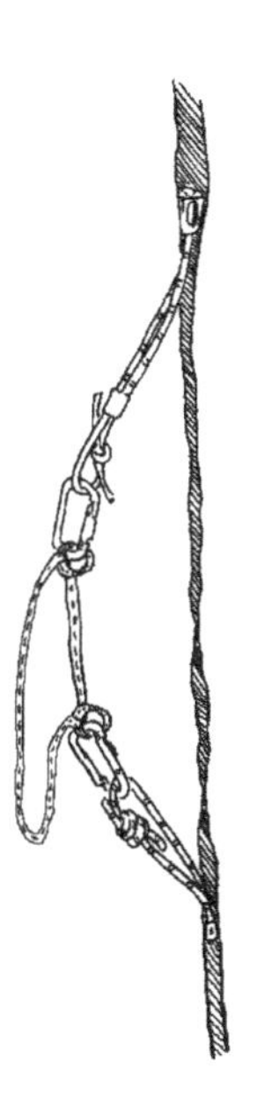

OPPOSITION PROTECTION WITHOUT THE OVERHAND KNOT

1) Place one chock in the crack so it can take a downward pull.

2) Below that emplacement, place another chock with an upward pull.

3) Clip a carabiner into each chock wire / cord.

4) Clip one end of a large runner into the bottom piece of protection.

5) Run the other end up to the second piece of protection and tie a clove hitch through the carabiner ensuring there is tension between the pieces to prevent the pulling out of either piece.

6) The running end of the runner may now be attached to the rope with a carabiner as a normal piece of protection.

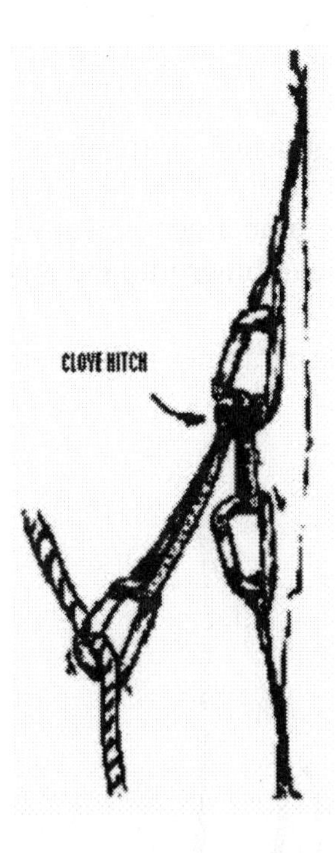

ALTERNATIVE REDIRECT

c. Expedient Belay Devices. Interlinked carabiners, figure eight descenders and even pitons can be used as expedient belay devices.

(1) Carabiner Brake Method. Also known as the "Crab" brake. This method is somewhat complex to set up, but it has the advantage of not requiring any special equipment.

(a) Attach one locking or two non-locking opposite and opposed carabiners to the hard point on a harness.

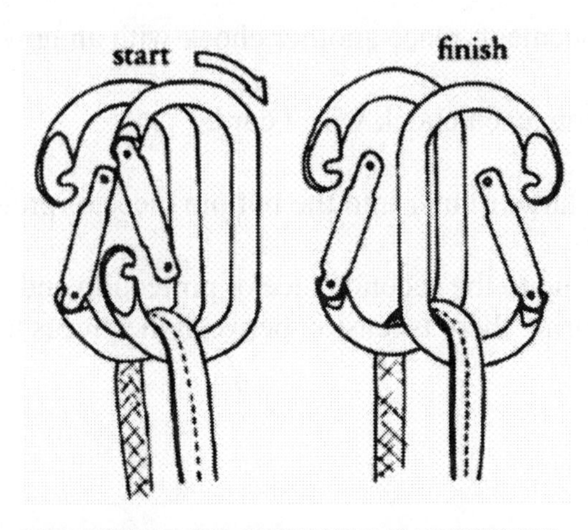

OPPOSITE AND OPPOSED CARBINERS

(b) Clip another pair of non-locking carabiners to the other carabiners already attached to the hard point.

(c) Take a bight of the rope through the outer pair of carabiners.

(d) Take another carabiner and clip it across the outer pair of carabiners beneath the bight of the rope.

(e) The rope then runs across the outer edge of this final carabiner known as the brake crab.

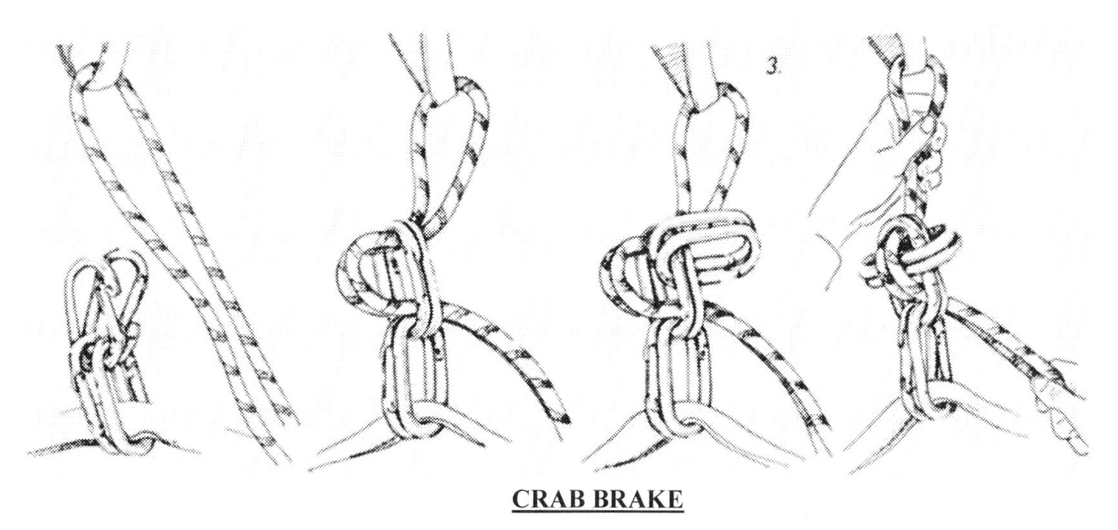

CRAB BRAKE

(f) To brake, pull the rope forward causing a bight around the rope and the carabiner.

2) Figure Eight Method. There are three ways to use the figure eight as a belay device.

(a) Use the figure eight in the normal manner as in rappelling.

(b) Pull a bight of the rope through the small hole of the figure eight and clip a carabiner through the bight. Belay in the same manner as with an ATC or stitch plate.

(c) Place the figure eight on a carabiner in the normal fashion. Pull a bight through the large hole of the figure eight and clip it through the carabiner.

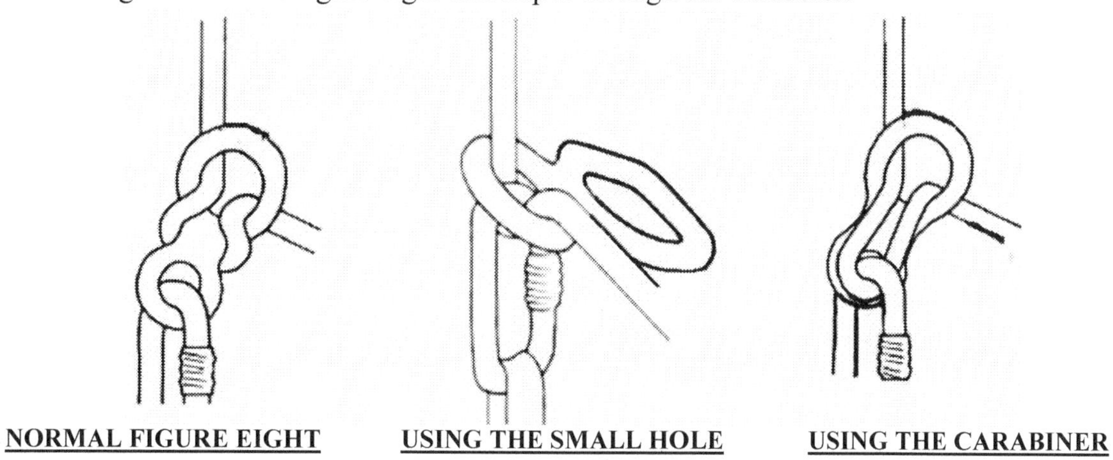

NORMAL FIGURE EIGHT **USING THE SMALL HOLE** **USING THE CARABINER**

(3) Petzl Grigri. The grigri is a specialized belay device that doesn't require any stopping force at all from the belayers hand. It works on the same principal as the safety belts in your car. With slow steady movements the rope feeds through freely. When there is a shock load (as in a fall) the grigri locks, jamming the rope with a cam. There is some tendency for the grigri to lock up when the leader makes a sudden move up. It also works badly or not at all with wet or icy ropes; that together with its weight and bulk, makes it largely unsuitable for mountaineering but quite useful in climbing gyms and in rock-climbing areas.

(a) Open the grigri and run the rope around the cam. Pay attention to the routing and ensure the load (leader) is on the end of the cam closest to the carabiner hole.

(b) Close the grigri and hook the carabiner from your harness (MAC) through the hole in the bottom of the grigri.

(c) Check the diagram and ensure the rope is run the correct way.

(d) Perform a function check by feeding the rope through the grigri with both hands. Then take the load rope (rope from the climbers harness) and give it a jerk. The grigri should lock and stop all movement.

(e) The left hand is the guide and the right controls the slack. Should a climber take a fall and you need to release the load or lower the climber, the left hand operates the load release lever. Note that you must maintain control of the slack rope with the right hand to help control the speed of the descent.

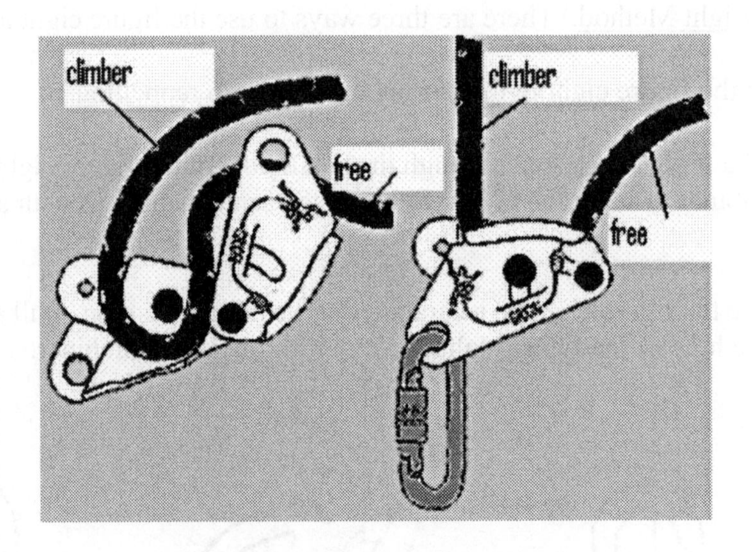

LOADING THE PETZL GRIGRI

(4) Piton Carabiner Method. Not all pitons can be used for this technique due to some pitons being designed with sharp edges. Shallow angle pitons work the best.

(a) Place a locking carabiner into the hard point on a harness.

(b) Clip another carabiner with a piton attached through its eyelet to the other carabiner already attached to the hard point.

(c) Take a bight of the rope through the outer carabiner.

(d) Place the piton beneath the bight so its pointed end rests on the opposite side of the carabiner it is attached to.

(e) The rope then runs across the top of the piton.

(f) To brake, pull the rope forward causing a bight around the rope and the piton.

TRANSITION: Now that we discussed alternative belays, let us talk about alternative anchors.

2. (10 Min) **ANCHORS**. Anchors are the ultimate security for any belay, the anchor should be able to hold the longest possible fall. There are many different types of anchors for different situations. The following are a few different methods of establishing an anchor system:

   a. The Cordelette Method. This method is good because sometimes the extra 10 or 15 feet of rope may be required to get to a better belay stance. It requires a 16-foot Type II cordage that is constructed into an endless loop.

      (1) Place three pieces of protection with carabiners attached.

      (2) Clip the Type II cordage into all three pieces of protection ensuring that the joining knot is located near one of the carabiners so that it will not be involved in the system.

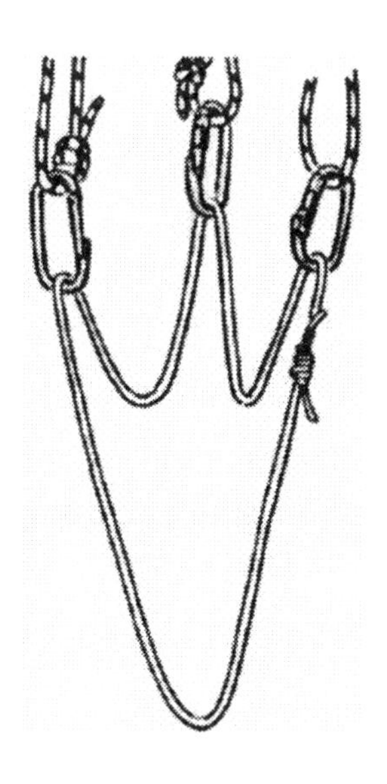

(3) A bight will be pulled down between each of the pieces of pro so that there are three equal bights pointing in the anticipated direction of pull.

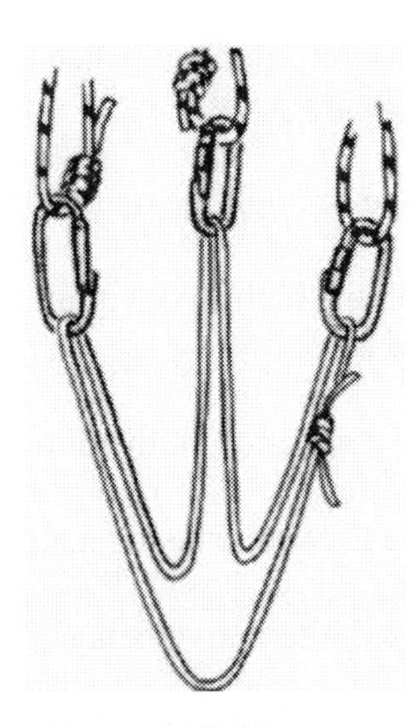

(4) Tie an overhand knot at the end of the bights creating three separate equalized loops.

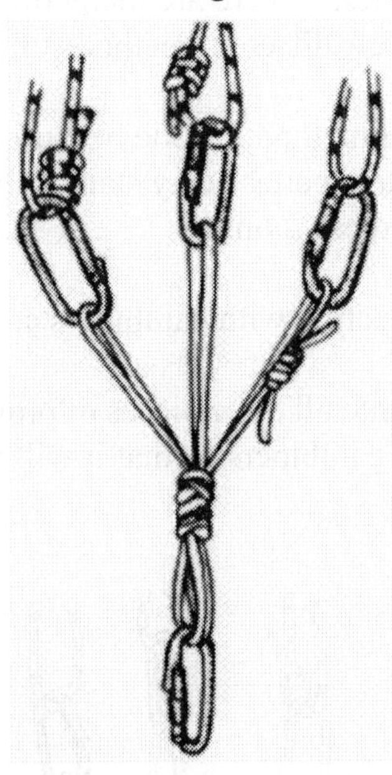

(5) Clip a carabiner these loops. This carabiner will now serve as the main anchor point carabiner to the constructed system.

b. Multi-Point Equalizing Anchor. This method is good because it self equalizes if any pieces of pro happen to fail. It requires a long sling runner, which is clipped into three pieces of protection as with the cordelette method except:

(1) An additional carabiner that will serve as the main anchor point carabiner is clipped on the bottom of the hanging runner.

(2) The runner is grasped between each anchor, twisted 180 degrees, and clipping each resulting loop into the main anchor point carabiner. Such twists guarantee that the main carabiner is clipped into, and not around the runner, so that it will stay attached to the runner even if two of the three points fail.

(3) The smaller the runner the smaller the drop if one anchor fails. Therefore, if the anchors are wildly separated bring them together with other slings before equalizing.

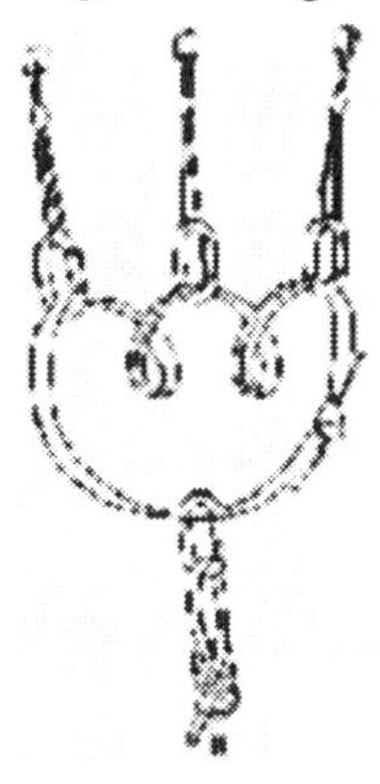

MULTI-POINT EQUALIZING ANCHOR

c. Special Considerations for Belay Stance Anchors. To safe guard the #2 from possible hazards i.e. belaying from a ledge, waves from the sea during amphibious assaults, etc., two pieces of protection will be placed in the direction of pull while the third will be place in the opposite direction of pull.

(1) Place the three pieces of artificial protection keeping in mind the anticipated direction of pull if the lead climber fell.

(2) Ideally the # 2 would place two of the pieces behind him with the direction of pull up. These two anchors should be the first pieces clipped when constructing the anchor.

(3) The third piece will be placed in front of the belay stance with the direction of pull down. This piece of protection will prevent the belayer from being pulled off his stance unintentionally.

(4) The belay stance will be constructed in the manner as taught in *BELAYING FOR PARTY CLIMBING.*

TRANSITION: During this period of instruction we have talked about various alternatives to belaying and anchoring. Remember safety is paramount and only use alternative methods if all other methods have been exhausted.

PRACTICE (CONC)

a. Students will practice establishing Alternative belays.

PROVIDE HELP (CONC)

a. Instructors will assist the students when necessary.

OPPORTUNITY FOR QUESTIONS (3 Min)

1. QUESTIONS FROM THE CLASS

2. QUESTIONS TO THE CLASS

Q. What are three alternative methods to belay?

A. (1) Body Belay
(2) Redirect Belay
(3) Expedient Belay Devices

SUMMARY (2 Min)

a. During this period of instruction we have discussed the various alternative methods of belays and anchors.
b. Those of you with IRF's please fill them out and turn them in to the instructor. We will now take a short break.

UNITED STATES MARINE CORPS
Mountain Warfare Training Center
Bridgeport, California 93517-5001

SML
ACC
02/19/02

LESSON PLAN

RESCUE TECHNIQUES FOR PARTY CLIMBING

INTRODUCTION (5 Min)

1. **GAIN ATTENTION.** Climbing and mountaineering by its nature is an occupation which involves a certain amount of risk, many of these risks can be eliminated or reduced by sound training, common sense, and by wearing and using the right equipment for the task at hand. However, accidents may occur even though you have taken all prudent precautions to protect yourself and your partner. You may very well find yourself having to rescue your partner or even have to be rescued yourself. It is therefore important that you should have a sound working knowledge of the basic self help and rescue techniques used.

2. **OVERVIEW.** The purpose of this period of instruction is to familiarize the student with the methods in which to rescue himself or his climbing partner in the event one sustains an injury serious enough that they can no longer continue to climb. This lesson relates to *PARTY CLIMBING.*

3. **METHOD/MEDIA.** The material in this lesson will be presented by lecture and demonstration. You will practice what you have learned immediately following this period of instruction.

4. **EVALUATION.** This period of instruction is a lesson purpose class.

TRANSITION: Are there any questions over the purpose or how the class will be taught? First we will talk about what we will need to know to determine a rescue.

BODY (55 Min)

1. (5 Min) **ASSESSING THE SITUATION.** A party climb rescue can become time consuming, and precious minutes are involved when dealing with injuries. Whenever an injury occurs while climbing, whether it involved the #1 or #2 climber, you should first assess the situation and determine the following:

    a. Is the climber conscious or unconscious?

b. How severe is the injury involved?

c. What's the location of the injured climber? (i.e. less then half the distance of the rope or more then half the distance of the rope)

d. Do you have enough gear to conduct the rescue?

e. Can the other climbing teams in the area assist you in the rescue?

f. How familiar are the climbers with the rescue techniques?

NOTE: The answers to the above questions will determine which method should be used to rescue the climber.

TRANSITION: Now that we have discussed the factors involved when to conduct a rescue, are there any questions? Now we will discuss how to do it.

2. (5 Min) **ESCAPING FROM THE SYSTEM**. This is a phrase used to describe the technique of releasing one's self from the belay, while ensuring that the climber you are responsible for is safe and secure. The reasons for escaping are varied and too numerous to mention; however, once you have established the reason to escape, you should work logically, safely, and try to keep it as simple as possible. The procedures are as follows:

   a. Lock the climber off.

   b. Once the belay device is locked off pass a bight through the belay device carabiner.

   c. Tie the belay device off around the load rope with two-half hitches.

   d. Untie your retraced figure of eight knot and pull the end of the rope out through your harness.

TRANSITION: Now that we discussed how to escape the system, are there any questions? Now let's talk about rescuing an injured lead climber.

3. (20 Min) **RESCUE AN INJURED LEAD CLIMBER.** When conducting two party climbing, there is a greater chance that the lead climber will suffer injuries from a fall vice the #2 climber who is being top roped. After assessing the situation, a plan must be devised as soon as possible. The following are different methods which could be utilized:

   a. Lowering the Injured Climber. The method of lowering will depend upon two factors; how much rope was used and if the injured climber is conscious or not.

      (1) Less than Half the Rope Distance. If less than half the rope has been used during the climb, it may be feasible to just lower the victim. Normally, the cause of the injury was due to a fall, which seated the protection firmly. If the injured climber is conscious, he

will back the lowering protection point up with another piece of protection and equalize it.

(2) More than Half the Rope Distance. If more than half the rope has been used during the climb and the injured climber is conscious, he will conduct the same actions as above and the following:

(a) Clean the route on his way down.

(b) Descended past the half way mark and locate a feasible stance to tie himself off to the rock using an equalized anchor system.

(c) Pull some slack up from the bottom and tie it off to a hard point.

(d) Untie the retrace figure eight from his harness and pull the rope through the lowering point.

(e) Once the rope has cleared the lowering point, he will attach the end of the rope back into his harness with a retrace figure eight.

(f) Untie the slack rope from the hard point and have the belayer take in all the slack.

(g) Taking a bite from his retrace figure eight, he will clip it into the equalized anchor system.

(h) Ensuring that the belayer has taken in all the slack and that the brake is applied, he will detach himself from the equalized anchor system.

(i) The belayer will then lower the victim.

b. Assistance from other Climbing Teams. Other climbing teams in the area can provide assistance if needed. There are two options for their assistance, which depends upon their location, either on the top or the bottom.

(1) A climbing team on the top can set up a rappel lane and conduct a tandem rappel as taught in *MOUNTAIN CASUALTY EVACUATION.*

(2) A climbing team on the bottom can perform the rescue by performing the following actions:

(a) Conduct a two party climb up to the victim and establish a hanging belay as described in *BELAYING FOR PARTY CLIMBING.*

(b) The leader will pull up all of the rope and tie a figure eight loop in the end of the rope. This loop will be clipped into the victim's donut with a locking carabiner.

(c) The leader will establish a belay to the victim and tie the belay off.

(d) The victim's retrace figure eight can be untied or cut if necessary.

(e) The leader can now untie the belay and lower the victim to the ground.

NOTE: A climbing rope will not be cut unless it is a true emergency and extreme precaution must be taken so that only the victim's rope is cut.

TRANSITION: Are there any questions over rescuing an injured lead climber. In some circumstances you may have to rescue the No. 2. Normally he will be either injured from falling rock or unconscious.

4. (20 Min) **RESCUE AN INJURED NO 2**. There are several ways to assist an injured No. 2 climber we will cover just a few.

   a. Assisted Hoist. The assisted hoist is most frequently used in situations where your second is unable to climb a particular part of a climb or he may have fallen off to one side and is unable to get back onto that climb. It is not necessary for the rescuer to escape from the system to set up the hoist, however in some circumstances it may well be easier to set up and work if you get yourself out of the system, but remember to secure yourself once you are out.

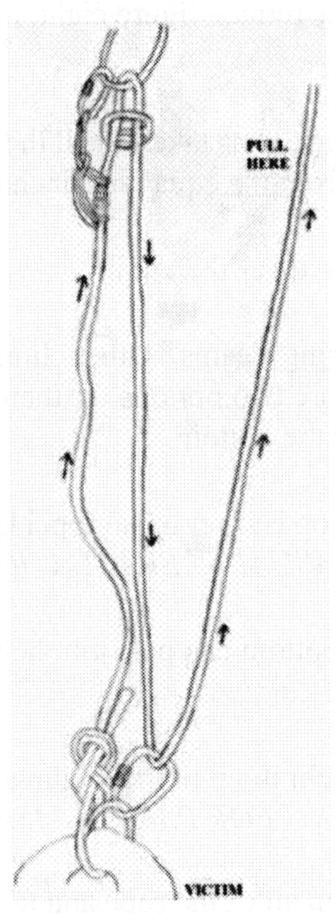

ASSISTED HOIST

NOTE: This system can only be set up and used if you are within 1/3 of the ropes length of each other.

(1) Tie off the belay device. Attach an Autoblock knot on the load rope, then secure it back to the MAC, and slide it down until all the slack is taken out.

NOTE: Make sure that the Autoblock is not too long, no more than 1 foot from the belay device, otherwise, it will slide out of arms reach when loaded.

(2) Take a bight in the loose end of the rope and clip a carabiner into it.

(3) Lower the bight and carabiner down to victim and tell the victim to clip it into the strong point of their harness, making sure that the rope is not twisted.

(4) Untie belay device and gently lower victim's weight onto the autoblock. The victim will pull the center downward moving rope.

(5) Both rescuer and victim pull at the same time, the victim walks up the rock face to assist the hoist. Should either require a rest, you simply lower the victim's weight onto the autoblock.

b. Unassisted Hoist. There may be circumstances on a multi pitch climb when you might need to hoist an injured or unconscious victim rather than try to lower them down to a safe stance. If you should be unfortunate enough to be in this type of situation, then a mechanical advantage will save a lot of energy and backache.

(1) Three-to-one-Hoist. One of the simplest of the mechanical advantages to set up, but remember to work logically and safely.

(a) First tie off your belay device and escape from the system and anchor yourself.

(b) Attach an autoblock to the loaded rope and secure it back to the MAC, slide it down until slack is taken up.

(c) Untie belay device and gently release the loaded rope onto autoblock.

(d) The loaded rope is then secured by retying the belay device.

(e) Take a short prusik loop and put a French prusik onto the loaded rope as far down as you can reach, clip in a carabiner.

(f) Untie the belay device and take the slack rope from the carabiner at the anchor point and run it down and back through the French prusik carabiner previously placed on the loaded rope.

(g) The victim's weight is now hanging on the autoblock that was put on after you escaped from the system. It may be necessary to shorten the autoblock closer to the anchor point for greater effectiveness; this can be done when the weight has been taken off the autoblock during hoisting.

(h) To hoist the victim, pull on the slack end of the rope. If the victim is conscious, he can help by walking.

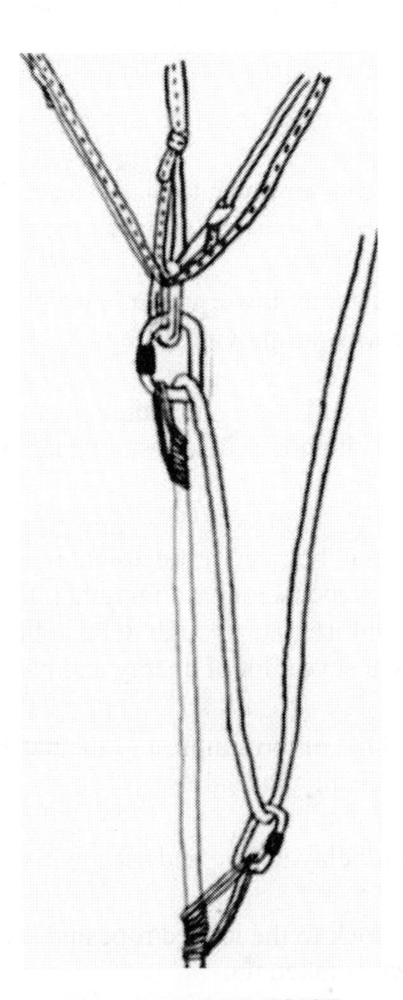

THREE TO ONE HOIST

(i) As you pull, the French prusik will come up to meet the autoblock, at this point lower the victims weight onto the autoblock and then slide the French prusik back down the loaded rope as far as possible. This procedure is repeated until the victim is where you require him. If there is enough space you can attach the pulling rope to your harness through your belay device, and use your body weight instead of you arms.

NOTE: This system can be easily upgraded to a 9:1 to enhance leverage. If available, pulleys can be used to prevent unnecessary friction.

TRANSITION: Are there any questions over the rescue of the #2? Saving the victims life is important, but remember to be safe so that you do not become a victim as well.

PRACTICE (CONC)

a. Students will practice rescuing both the injured No 1 and No 2.

PROVIDE HELP (CONC)

a. The instructors will assist the students when necessary.

OPPORTUNITY FOR QUESTIONS (3 Min)

1. QUESTIONS FROM THE CLASS

2. QUESTIONS TO THE CLASS

Q. When using the assisted hoist what is the furthest distance that you can be away from the victim?

A. 1/3 the distance of the rope.

Q. What situations would determine the need to conduct a party climb rescue?

A. (1) Is the climber conscious or unconscious?
(2) How severe is the injury involved?
(3) Do you have enough gear to conduct the rescue?
(4) What's the location of the injured climber? (i.e. less then half the distance of the rope or more then half the distance of the rope)
(5) Can the other climbing teams in the area assist you in the rescue?
(6) How familiar are the climbers with the rescue techniques?

SUMMARY (2 Min)

a. During this period of instruction we have discussed and demonstrated those sills necessary to escape from a belay and rescue a fallen climber, including hoists and improved hoists.

b. Those of you with IRF's please fill them out and turn them in to the instructor. We will now take a short break.

UNITED STATES MARINE CORPS
Mountain Warfare Training Center
Bridgeport, California 93517-5001

SML
SMO
ACC
02/19/02

LESSON PLAN

STEEP EARTH CLIMBING

INTRODUCTION (5 Min)

1. **GAIN ATTENTION.** As a mountain leader you must know how to negotiate all types of terrain in the mountains. Steep earth climbing may be done by a trained mountaineering team when climbing steep earth slopes is required.

2. **OVERVIEW.** The purpose of this period of instruction is to introduce the student to steep earth climbing, including the equipment used, procedures used, and organization.

3. **METHOD/IMEDIA**. The material in this lesson will be presented by lecture and demonstration. You will practice what you have learned in upcoming field exercises. Those of you with IRF's please fill them out at the conclusion of this period of instruction.

4. **EVALUATION.** This is a lesson purpose class, you will not be formally evaluated on this period of instruction.

TRANSITION: Are there any questions over the purpose or how the class will be taught. This brings us to the first part of our class, the equipment needed.

BODY (40 Min)

1. (5 Min) **EQUIPMENT NEEDED**.

   a. Each climber will need the following equipment.

      (1) Climbing harness

      (2) Mountain boots with a half or full shank if possible.

(3) Helmet

(4) Gloves

(5) One 165 foot climbing rope per team

(6) One earth axe (short ice axe)

(7) One alpine hammer, Northwall style ice hammer, or a heavy wall type hammer

(8) Five finger/point gripfast

(9) Twelve point crampons.

(10) Sling rope

(11) Snow stake

(12) Rebar (1,2, and 3 foot lengths)

b. The #2 climber will have a five point hand grapnel with at least 20 feet of knotted 7mm cord and a 165 foot knotted climbing rope.

c. Earth Ax Construction. The earth ax is a short ice ax which can be used to cut steps or hand holds. The pick can be used to gain holds in earth or rock. This should be "dummy – corded" to the climber.

TRANSITION: Are there any questions over the equipment needed? The next thing we will discuss is the climbing procedures.

2. (10 Min) **CLIMBING PROCEDURES**

a. Route Reconnaissance. The route is visually inspected for steepness, soil composition, rock outcroppings, ice and snow patches, and availability of anchor's. Based on this reconnaissance, the climbers construct a climbing rack best suited to the proposed climbing route.

b. Normal belay procedures are used. The belayer establishes the bottom of the climb using available anchors supplemented by a 5 point gripfast. He will tie into the end of the rope and into the gripfast.

c. Once the belay is established, the lead climber ties in and begins his climb. The climber will cut steps with the adze of his earth ax for foot holds. As soon as possible after beginning the climb, the lead attempts to place an intermediate anchor. Depending on the composition of the face being climbed, this anchor may be an ice or rock piton, a chock, camming device, or a specialized steep earth anchor. The climber must remember that any protection placed in steep earth has questionable holding strength. The climber uses his ax for a hand hold and uses his free hand for balance or uses an alpine hammer or ice hammer for his second hand tool.

d. The lead climber digs a belay platform at the end of his rope if required. He then plants his gripfast, ties into it, and sits on it for additional security. If possible, he establishes an anchor using pitons, chocks, camming devices, etc. The lead climber then belays up the #2 climber.

e. The #2 climber climbs towards the belay stance, collecting all unused protection as discussed in *PARTY CLIMBING.*

f. This procedure continues until one climber reaches the top, establishes a belay stance with an appropriate anchor, and top ropes his companion up.

TRANSITION: Now that we covered climbing procedures, are there any questions? Let's now talk about the use of the grapnel.

3. (5 Min) **USE OF THE GRAPNEL**

a. A five point grapnel can be useful in steep earth climbing. With its twenty feet of knotted cord, it provides handholds where there may otherwise be none. Short, sheer faces and overhangs have fewer holds, thus providing ideal situations for grapnel use.

b. To use the grapnel, the climber unwinds his grapnel line, secures it to himself, then throws it above himself over a ledge, cliff edge, or other near horizontal feature. Care must be taken to throw it to one side or the other so that if it doesn't hold it won't fall on the climber or his belayer. After the grapnel has landed, the climber pulls it slowly until it is securely caught.

c. The climber now climbs up the difficult section, assisted by the grapnel line, keeping the pull steadily down. Changing to a palms down grip at the top will help to keep the grapnel in place.

d. Once at the top, the climber checks the security of the grapnel and changes its position if necessary. He then establishes a belay and belays the second climber up. The second climber uses the grapnel line as necessary.

TRANSITION: Are there any questions over the use of the grapnel? Let's now discuss what the troops on the top and bottom will do.

4. (5 Min) **DUTIES OF PERSONNEL ON THE BOTTOM**

a. 180 degrees security is set up.

b. Troops at the bottom should not be directly under the climbing site because of the loose rock and dirt that could fall upon them causing an injury.

5. (5 Min) **DUTIES ONCE ON TOP**

a. Security

(1) The first man up makes a hasty recon of the area before belaying up the other climber.

(2) Once the lead climber has conducted a hasty recon, he starts setting up the knotted hand line at a suitable climbing site for a fixed rope while the #2 climber provides security.

b. The site selected should be as follows:

(1) Good natural anchors, these may have to be multiple small shrubs or bush type anchors.

(2) Artificial anchors, i.e. gripfast, pickets, deadmen, and chocks.

(3) Make sure the rope reaches the bottom.

(4) Minimum of loose rocks and dirt.

c. Knotted hand line should be pre-knotted to save time and brought up by the #2 climber.

d. 180 degree security is set up as Marines reach the top.

e. Assault leader and automatic weapons teams will go up first.

f. The assistant leader and automatic weapon team will be the last to come up as the 180 degree security on the bottom gets smaller.

g. The ropes are pulled up and coiled once everyone is on top.

h. All evidence such as anchors, indentations or rope marks on the edge of the dirt are removed.

i. Area is policed of any other evidence.

TRANSITION: Now that we have covered the duties of the Marines on the top and bottom, are there any questions? Steep earth is a vertical obstacle. To effectively move troops over the obstacle, the same principles can be used on a cliff assault. If there are no questions for me, then I have some for you.

PRACTICE (CONC)

a. The students will climb steep earth.

PROVIDE HELP (CONC)

a. The instructors will assist the students when necessary.

OPPORTUNITY FOR QUESTIONS (3 Min)

1. QUESTIONS FROM THE CLASS

2. QUESTIONS TO THE CLASS

Q. What equipment is carried by the lead climber?

A. Helmet, boots, climbing harness, crampons, gloves, one earth ax, one alpine hammer/ice hammer, gripfast, 165 foot climbing rope, climbing rack, and plastic lens glasses/goggles.

Q. What equipment is carried by the #2 climber?

A. Same as the lead climber except the #2 climber carries the grapnel.

SUMMARY (2 Min)

a. During this period of instruction we have covered how to climb steep earth terrain, the equipment needed, the roles of each team member, and the duties at the bottom and the top of the cliff face.

b. Those of you with IRF's please fill them out at this time and turn them in to the instructor. We will now take a short break.

UNITED STATES MARINE CORPS
Mountain Warfare Training Center
Bridgeport, California 93517-5001

SML
SMO
02/19/02

LESSON PLAN

TREE CLIMBING TECHNIQUES

INTRODUCTION (5 Min)

1. **GAIN ATTENTION.** During many of today's conflicts the use of aircraft is becoming increasingly important. Oftentimes pilots and their aircraft are on the leading edge of the American sword. As such they're occasionally disabled and find themselves having to bail out in a hostile or non-permissive environment. When this happens American forces will be mobilized and launch what we call a TRAP recovery force. Tactical recovery of aircraft and/or pilots is a specialized mission that may require the skills of an assault climber.

2. **OVERVIEW.** The purpose of this period of instruction is to introduce the student to the skills needed to accomplish a rescue of a pilot from a tree.

3. **METHOD/MEDIA.** The material in this lesson will be presented by lecture and demonstration. You will practice what you have learned immediately after this period of instruction. Those of you with IRF's please fill them out at the conclusion of this period of instruction.

4. **INTRODUCE LEARNING OBJECTIVES.**

   a. **TERMINAL LEARNING OBJECTIVES.** In a wooded environment and given the proper equipment, recover a pilot from a tree in accordance with the reference.

   b. **ENABLING LEARNING OBJECTIVES.**

   (1) Without the aid of references, list in writing the lead climbing methods for ascending a tree in accordance with the references.

(2) Without the aid of references, describe in writing the technique used to lower a pilot from a tree in accordance with the references.

5. **EVALUATION.** You will be tested on this period of instruction by written and performance evaluation.

TRANSITION: Now that we know what to expect and what the class is about let's talk about a pilot TRAP.

BODY (60 Min)

1. (5 Min) **TRAP**. The Tactical Recovery of Aircraft and Personnel (TRAP) is an ACE mission similar to Combat Search and Recovery (CSAR). The main difference is the recovery of aircraft and the possible use of ground forces. As such, the TRAP force is responsible for expeditiously providing recovery and repatriation of friendly aircrews and personnel, from a wide range of political environments and threat levels. Additionally, equipment will either be recovered or destroyed depending on the threat and the condition of the downed aircraft.

a. For an Assault Climber the TRAP mission may require a lead climber to climb a tree and retrieve an unconscious/disabled pilot. To accomplish the mission safely ropecraft will be necessary.

TRANSITION: Now that we have an understanding what is a pilot TRAP, let us now discuss the equipment needed for a rescue.

2. (10 Min) **EQUIPMENT.** The amount and variety of the equipment needed is situational and you must be flexible. As a general rule the following equipment is sufficient for most missions:

a. Harness or sling rope

b. Static rope/165-300 ft.

c. Ten aluminum non-locking carabiners

d. Six steel locking carabiners

e. Fifteen assorted runners (longer the better)

f. Optional:

(1) One pr. of jumars

(2) Two pulleys

(3) One set of gaffs and strap/crampons can work

(4) One line launcher

(5) Pitons w/ hammer

TRANSITION: Are there any questions concerning the equipment necessary to conduct a rescue? If not, let's talk about the different types of techniques to rescue the entangled personnel.

3. (20 Min) **ASCENDING TECHNIQUES.** Prior to executing a TRAP, a method of retrieval should be tentatively planned for. There are a number of techniques to be utilized and the situation and equipment available will determine which of these you will use.

a. Lead Climbing: Using this method a lead climber will ascend the tree using one of four techniques. All four techniques will require both the number one climber and the number two climber to tie into the end of the rope. The number two will establish an anchor and belay system as discussed in *BELAYING FOR PARTY CLIMBING.*

(1) By use of gaff and straps/crampons. This is the fastest and easiest means of ascending a tree. By using the strap, the climber negates having to place protection for themselves and the use of gaffs/crampons provide quick and easy footholds for ascending even the tallest of trees.

NOTE: Marines must be "pole climber" certified by the field wiremans course to use gaffs and strap.

(2) By using the girth hitch method. This method is tiring and slow but is very safe. Start by girth hitching two runners around a tree. These runners should be long enough to create about a 3-4ft. loop for the foot. The top runner is attached to the climber's harness by use of steel locking carabiner. Grip the tree and stand up in the loop of the bottom girth hitch. Slide the top girth hitch up as high as possible and hang from it. Grab the bottom girth hitch and move it as high as possible, insert your foot and stand up again. This process will repeat itself until you have gained your desired height. If you encounter branches along the way you will have to bypass them. This is done by having a third runner in which you will attach above the tree limb. You will place a girth hitch with the third runner and attach it to your harness, slide the bottom runner up as far as possible and stand so that you can slide the top runner up and bypass the tree limb.

(3) Using the Monkey method. This method is dependent on the size and type of tree being climbed. Smaller diameter trees with lots of branches are optimal. Essentially the lead climber will just free climb the tree using branches and shimmying up the trunk with appropriate protection for the climber being accomplished by placing runners around the tree in a quad hitch or girth hitch fashion.

(4) Party Climb method. This method is similar to that of the monkey method with the exception that the lead climber will place pitons in the tree or girth hitch runners around branches for protection. The climber will clip into these pieces of protection as taught in party climbing.

b. Jumar Climbing Techniques:

(1) Jumaring: In order to Jumar or "rope walk" to the pilot it is necessary to first establish a rope above the pilot. The use of a line launcher or 550 cord around a canteen will aid in getting the rope up the tree. 550 cord will be the most likely tool utilized by Marines when using this technique. You must first attach the 550 cord to a half-full canteen or a heavy carabiner; then throw it up to the desired branch to loop it around. Once this is done you can tie your climbing rope onto the 550 cord and pull the other end of the 550 cord up and around the tree branch until you have both ends on the ground. There are four methods that you can use to jumar or rope walk up the rope to the pilot:

a. Body Thrusting. This method is very tiresome but requires little gear. First the climber will tie a tautline hitch with the end of the rope then ties a figure of eight above the tautline hitch and clips it into his harness with a locking carabiner. The climber will then pull down on the opposite rope that he is tied into, sliding the tautline hitch up along with him which will act as the friction device (a jumar here works really well and is easier to control). To work your way back down the climber will break the tautline hitch so he can slide it down and do everything in reverse to work his way down the tree.

b. Footlocking: This method is also tiresome and requires little gear. This is similar to hand over hand method (i.e.. the O course ropes). The climber will attach a Prusik knot to the two ropes hanging down from the tree and attach that to his harness then hand over hand his way up using his feet to bend and lock the rope in place to aid him in ascending the rope. To get down the climber will do everything in reverse.

c. Jumar/Texas Kick: This method is the easiest but requires the most gear. The climber will attach one end of the rope to his harness using a figure eight with carabiner clipped into hard point. Next the climber will attach a foot jumar and a chest jumar or Texas Kick Prusik (as discussed in *Glacier Travel*) to the opposite rope to which the climber is tied into. The climber will then slide the top jumar (body Prusik) as high as possible then slide the foot jumar up and stand, unweighting the top jumar so that it can be moved up higher. This process will repeat itself until the climber has reached the desired height.

d. Haul system: This is the most desired method and requires very little gear but requires manpower. Once the rope is looped around the treelimb the climber will tie into one end of the rope, then a mule team will hoist the climber up to the desired height.

NOTE: All of these techniques require the rope to run in the crotch of a tree branch and the trunk of the tree, so rope wear should be considered because of the friction that is occurring when ascending.

TRANSITION: Now that we have reached the pilot and have him securely rigged, let us discuss the procedure for lowering him safely to the ground.

4. (10 Min) **LOWERING TECHNIQUE.**

a. Once the leader has ascended the tree to the appropriate height, he must then anchor himself slightly above the pilot. This can be done by quad hitching or girth hitching a runner to the tree and hanging from it. The leader must then construct a secondary anchor to lower the pilot down on. This can be accomplished in the same manner as before with girth hitches. Once complete the climber will then do one of two things: if no second rope is available then he will untie from his rope (making sure that he is safely tied off first) and run it through the anchor for lowering the pilot. If a second rope is available it would be wise to stay tied in for safety and use a separate rope for the pilot.

b. Once the rope is fed through the anchor for lowering, a figure eight will be tied into one end and a locking carabiner will be attached. If the pilot is conscious you can lower it to him and have him connect it to the D-rings located on the front of his harness. In the event the pilot is unconscious then you must lower yourself to him and attach it yourself. If the D-rings cannot be accessed then quad hitch a runner through the shoulder portion of the pilot's harness and attach to the rope with a locking carabiner. At this point you are ready to release the pilot from his parachute.

c. Double check your system: Ensure all slack is out of the system and a belay is on. This is critical you must ensure that you and the pilot are safe before releasing his parachute. Once you've double-checked everything you need to reach down to the pilot and pull the cable loop type canopy releases located on the shoulder portion of the pilots harness. This will cause a slight shock load onto your rope once executed, so be ready. After the pilot is free slowly lower him to the ground via your belay man.

TRANSITION: Now that the pilot is safely on the ground, there are other considerations that we must take into account.

5. (5 Min) **OTHER CONSIDERATIONS.**

a. Medical assistance: A corpsman should be present to assist in providing medical attention if needed. A backboard and cervical collar should be planned for and when lowering the pilot it is best to lower him directly onto the backboard or litter. If your climbers are well versed in high angle rescues then the pilot can be put into the litter while still in the tree if his injuries require (this is specialized and not a part of this course).

b. DOD form 1833: This classified report has key information to be used in identifying a pilot. This report is called the isolated personnel report or isoprep data. The information provided is personal and specific. Often this data is verified prior to insertion. The TRAP recovery force needs them to authenticate upon location of the pilot. **NEVER** take a filled out DOD form1833 or the data with you, memorize it.

TRANSITION: Now that I have discussed tree climbing techniques, are there any questions? If there are none, then I have some for you.

PRACTICE (CONC)

a. The student will retrieve a pilot from a tree in a safe manner using the techniques taught.

PROVIDE HELP (CONC)

a. The instructors will assist the students when necessary.

OPPORTUNITY FOR QUESTIONS (3 Min)

1. QUESTIONS FROM THE CLASS

2. QUESTIONS TO THE CLASS

Q. What are the four different climbing techniques?

A. (1) Gaffs/straps
(2) Girth hitch
(3) Monkey method
(4) Party Climb

Q. What is the last thing you do prior to releasing the pilot from his parachute?

A. Recheck your system.

SUMMARY (2 Min)

a. During this period of instruction we have discussed various techniques of ascending a tree and how to safely lower a pilot to the ground.

b. Those of you with IRF's please fill them out at this time and turn them into the instructor. We will now take a short break.

UNITED STATES MARINE CORPS
Mountain Warfare Training Center
Bridgeport, California 93517-5001

SML
ACC
02/19/02

LESSON PLAN

FIXED ROPE INSTALLATIONS

INTRODUCTION (5 Min)

1. **GAIN ATTENTION**. In a mountainous battlefield, gaps or soft spots in an enemy defense are usually cliffs or steep slopes. These gaps can be easily exploited by the use of fixed rope installations. Fixed ropes also aid heavily laden troops in ascending steep to moderate slopes.

2. **OVERVIEW**. The purpose of this period of instruction is to introduce the student to fixed rope installations. This will be accomplished by discussing the types of fixed ropes, installing, maintaining, ascending, and clearing the fixed ropes. Also, we will discuss special considerations for long pitches and cable ladders. This lesson relates to cliff assault.

INSTRUCTOR NOTE: Have students read learning objectives.

3. **INTRODUCE LEARNING OBJECTIVES**

a. TERMINAL LEARNING OBJECTIVE. In a summer mountainous environment, construct a fixed rope installation, in accordance with the references.

b. ENABLING LEARNING OBJECTIVES.

(1) Without the aid of references, define in writing each type of fixed rope installation, in accordance with the references.

(2) Without the aid of references, describe in writing the conditions for positioning the climbing rope in relationship to the climber along the route of a fixed rope installation, in accordance with the references.

(3) Without the aid of references, describe in writing the criteria for maintaining a fixed rope installation, in accordance with the references.

4. **METHOD/MEDIA**. The material in this lesson will be presented by lecture and demonstration method. You will practice what you have learned in upcoming field training exercises. Those of you with IRF's please fill them out at the conclusion of this period of instruction.

5. **EVALUATION**

   a. SML - You will be tested later in the course by written and performance evaluations.

   b. ACC - You will be tested by a written examination and by a performance evaluation on your company's cliff assault.

TRANSITION: Are there any questions on the learning objectives, what we will be covering, how this lesson will be taught, or how you will be evaluated? If not, let's begin by talking about sight selection.

BODY (90 Min)

1. (5 Min) **ROUTE SELECTION**. Ropes and caving ladders assist heavily laden troops in ascending steep terrain. They should be used where a fall might result in injury. Choose a route that will allow a Marine to be ready to fight upon reaching the top. These two factors determine the maximum possible difficulty of the route:

   a. Climbing unit's experience and ability.

   b. Climbing unit's load.

TRANSITION: Now that we discussed the factors of determining route selection, are there any questions? Let's discuss cliff assault ropes.

2. (5 Min) **CLIFF ASSAULT ROPES**. There are three types of fixed rope installations that can assist Marines up steep terrain: simple fixed ropes, semi-fixed ropes and fixed ropes.

   a. Simple Fixed Ropes. The simple fixed rope is defined as being anchored at the top end of a rope. This type of rope installation is primarily designed to aid heavily laden Marines in the ascent or descent of a steep to moderately steep slope. The rope can be used for aid when climbing by pulling on the rope and walking the feet up the slope. A knotted hand line may be used for this purpose.

      (1) Advantages:

         (a) It is simple, fast, and easy to construct.

         (b) It requires no extra gear or time to attach each climber to the rope.

(2) Disadvantages:

(a) Cannot be used on near vertical to vertical terrain.

(b) If a climber lets go of the rope, he may fall down the slope.

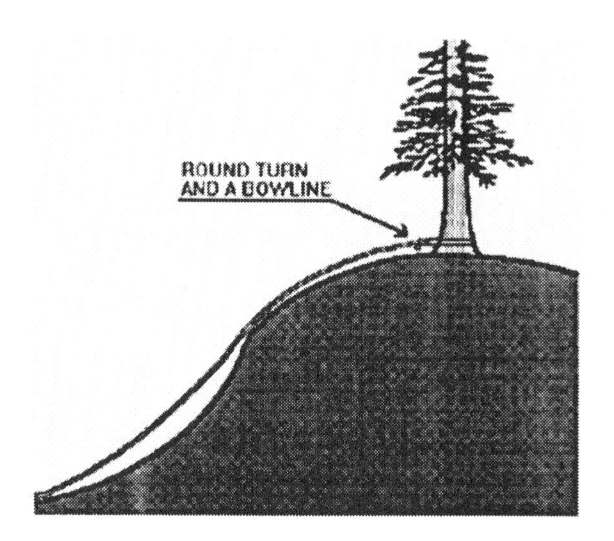

(1) Advantages:

SIMPLE FIXED ROPE INSTALLATION

b. Semi-Fixed Ropes. This type of rope installation is also designed to aid heavily laden Marines in the ascent or descent of a steep to moderately steep slope. Also known as a Fast Lane. This installation is anchored at the top and bottom but has no intermediate anchor points.

(a) More climbers can be on the rope.

(b) Least chance of injury due to safety prusik.

(2) Disadvantage:

(a) The route runs in a direct line.

c. Fixed Ropes. The fixed rope is defined as being anchored or fixed at both ends as well as with intermediate anchor points. This type of rope installation is primarily designed to protect heavily laden troops while negotiating snow / icy slopes, difficult scrambles, traverses or other slopes where balance climbing may be hazardous.

(1) Advantage:

(a) It can be used in routes that change directions.
(b) Protects the climber from a lateral fall hazard.

(2) Disadvantages:

(a) Most time consuming installation to construct.

(b) The slowest installation for climbers to negotiate.

(c) Higher chance of injury due to increased fall factor.

TRANSITION: Now that we discussed the different types of cliff assault ropes, are there any questions? Let's now discuss the construction of the cliff assault ropes.

3. (15 Min) **CONSTRUCTION OF CLIFF ASSAULT ROPES**. The difficulty of the terrain will determine what type of rope installation will be utilized.

a. Simple Fixed Rope. This is the quickest and easiest of the installations. There are two methods to establish this installation.

(1) Coiled Rope Method.

(a) A climbing team will assemble at the bottom of their assigned climbing lane.

(b) The lead climber will begin climbing up the obstacle with a coiled static rope on his body. The #2 climber will provide security.

(c) Once on top, the lead climber will locate a suitable anchor point and uncoil the rope.

(d) The lead climber will attach the static rope to the anchor point with an appropriate anchor knot.

(e) He will then deploy the rope down the obstacle.

(f) The #2 climber will ensure that there is enough rope touching the deck.

(g) The #2 climber ascends the route and ensures the rope is on the correct route.

(2) Trail Rope Method.

(a) A climbing team will assemble at the bottom of their assigned climbing lane.

(b) The number #2 will back stack a static rope while the lead climber visually inspects designated route and provide security.

(c) The lead climber will attach the static rope to his harness.

(d) The lead climber will begin climbing up the obstacle while the #2 climber provides security.

(e) The lead climber will locate a suitable anchor point and attach the static rope to it with an appropriate anchor knot.

(f) The #2 climber will ensure that there is enough rope touching the deck.

(g) The #2 climber ascends the route and ensures the rope is on the correct route.

b. Semi-Fixed Rope. Establishing this installation is the same as with the simple fixed except:

(1) The #2 will anchor and tighten the bottom of the rope before climbing up.

c. Fixed Rope. This installation is more time consuming to establish because of the intermediate anchor points but it provides extra protection against falls.

(1) The assault climber picks a route and the climbing team sets up for a two party climb.

(2) Before the lead climber begins to climb he will flake out a static rope and tie one end of the static rope into the back of his harness with a figure of eight loop.

(3) The lead climber begins two party climbing, placing protection where needed. He will also ensure that he clips the static rope into each piece of protection along with the climbing rope.

(4) The lead climber must always keep in mind that other less skilled climbers will be ascending this route. Factors to be considered while positioning the rope in relationship to the other climbers are:

(a) The rope is positioned approximately waist high.

(b) The climbers should not be forced to cross the rope at any point once it is tightened.

(5) Natural anchor points can also be used as intermediate anchor points.

ARTIFICIAL AND NATURAL ANCHOR POINTS

(6) Once on top, the lead climber will remove the static rope from his harness and anchor it off using an appropriate anchor knot. The lead climber will either tie himself off or move further than 10 feet away from the cliff face. The lead climber is now off climb and will give the appropriate signal to the #2 climber.

(7) When the #2 climber gets the "off climb" signal from his partner, he will break down the belay stance and untie from the dynamic climbing rope.

(8) At this point the lead climber will pull up the dynamic rope and coil it. Once the rope is coiled the lead climber will go report to the senior assault climber for follow on mission.

(9) The #2 climber will find the end of the static rope and tie a figure of eight knot in it. The rope will then be attached to the first piece of pro not more than waist high.

(10) With a Type II cordage, the #2 climber will tie an end of the line prusik knot onto the static rope. The running end of the cordage will be attached to his hard point utilizing a locking carabiner.

(11) The #2 climber will begin to climb the route, sliding his prusik up the rope as he climbs to protect himself in case of a fall.

(12) The static rope will be secured to each piece of protection, also referred to as intermediate anchor points. The #2 should remove any runners the lead climber placed unless needed to direct the rope. This is accomplished by tying the static rope to the piece of protection's carabiner using the slip figure eight, clove hitch, or figure of eight loop knot.

NOTE: By anchoring the rope at the intermediate anchor points, each section of rope is made independent of the others. Should one section fail, the other sections remain intact. All intermediate anchor point failures should be reported immediately to the installing unit and fixed promptly.

(13) The #2 will ensure that there is no slack between each intermediate anchor point.

(14) When the #2 climber reaches the top, all slack should be taken out at the top anchor point and reattached, ensuring that the static rope is as tight as possible.

NOTE: If the route is more then one pitch, the same procedures will be followed as mentioned above with the incorporation of a multi-pitch climb. The #2 climber will be responsible for trailing an additional static rope. When he reaches the lead climber's belay stance, their roles switch. It is the responsibility of the climber in the belay stance to adjoin the two static ropes together.

d. Assault Lane. This is an alternative method that can be used in place of a fixed rope when speed is essential for installation and recovery. This installation is established in the same manner as the fixed rope except:

(1) The intermediate anchors are not tied into the static rope.

(2) The rope can be tensioned at either the top or the bottom, preferably from the bottom.

TRANSITION: Now that we have discussed the different types of assault ropes, let us talk about how to ascend them.

4. (10 Min) **ASCENDING CLIFF ASSAULT ROPES**. Each rope installation requires different techniques for ascending. Proper commands or signals should be disseminated within a unit before ascending assault ropes. Gloves will not be worn during the ascent because this may give off a false sense of grip on the rope.

a. Ascending Simple Fixed Ropes. This is accomplished using the Hand-over-Hand Method.

(1) Straddle the rope with your legs.

(2) Grip the rope palms down, thumbs toward your body.

(3) Walk up the slope pulling with your arms while twisting the hands inward. This will create a bind between your grip and the rope.

ASCENDING A SIMPLE FIXED ROPE INSTALLATION

b. Ascending Semi-Fixed Ropes. This is accomplished using the Safety Prusik Method.

(1) With a Type II cordage, the climber will tie a prusik knot onto the static rope.

(2) The running end of the cordage will be attached to the climber.

(3) The climber will begin to climb the route, sliding his prusik up the rope as he climbs to protect himself in case of a fall.

c. Ascending Fixed Ropes and Assault Lanes. This is accomplished by using a safety line attached to the body.

(1) Tie a bowline around the chest using a sling rope ensuring that both pigtails are at least arm's lengths.

(2) At the end of each pigtail, tie a figure-of-eight loop.

(3) Place a locking or non-locking carabiner into each of the figure-of-eight loops.

(4) At the bottom of the fixed rope installation, clip both carabiners onto the rope and begin climbing. If using locking carabiners, it is not essential to lock the carabiners.

(5) Upon reaching an intermediate anchor point, unclip one of the carabiners from the rope and reattach it to the rope above the anchor point. Repeat this same procedure with the second carabiner.

NOTE: Never allow more than one person between each intermediate anchor point.

(6) This technique is repeated at each intermediate anchor until the climber tops off.

TRANSITION: Are there any questions on how to ascend an assault rope? Once the rope is up, it must be maintained to prevent a catastrophic failure of the system.

5. (5 Min) **MAINTENANCE OF A FIXED ROPE**. There are two criteria in the maintenance of fixed ropes.

   a. All assault rope installations should be buffed at points of abrasion.

   b. All assault rope installation anchors should be inspected periodically for accidental dislodging, walking of cams or improper direction of pull.

   c. All discrepancies should be reported immediately to the assault climbers for repair.

TRANSITION: Are there any questions over the maintenance of fixed ropes? Let us now talk about clearing the route.

6. (15 Min) **CLEARING THE ROUTE**. Once all the climbers have ascended the cliff it is necessary to dismantle the fixed rope to retrieve all the equipment and rope. Various methods may be used to dismantle the fixed rope depending on whether clearing the route from the top or the bottom.

   a. Rappel Method. This method is used when clearing the route from the top.

      (1) The climbing team establishes a rappel line using the climbing rope.

      (2) The climber will descend the rappel line using the first man down technique.

      (3) When he reaches an intermediate anchor point he will hang on the prusik and untie the fixed rope from the protection.

      (4) The climber will remove the protection and stow it away.

      (5) The climber will continue to clear the entire lane in this manner.

      (6) If the climber needs to ascend the cliff, he may be top roped up by his climbing partner.

   b. Safety Prusik Method. This method can be used when clearing the route from the bottom or the top.

      (1) From the bottom:

         (a) The climbing team establishes a rappel line using the climbing rope.

(b) The climber will descend to the bottom of the obstacle using the first man down technique.

(c) The climber will remove the bottom anchor and attach himself to the rope with a safety prusik.

(d) The climber will ascend the rope by sliding the safety prusik up as he climbs to protect himself in case of a fall.

(e) When he reaches an intermediate anchor point he will untie the fixed rope from the protection.

(f) The climber will remove the protection and stow it away.

(g) The climber will continue to clear the entire lane in this manner.

(2) From the top:

(a) The climber must create some slack in the lane by adjusting the top anchor point.

(b) The climber will descend the fixed rope lane using the first man down technique.

(c) When he reaches an intermediate anchor point he will hang on the prusik and untie the fixed rope from the protection.

(d) The climber will remove the protection and stow it away.

(e) The climber will continue to clear the entire lane in this manner.

(f) If the climber needs to ascend the cliff, he may be top roped up using the static rope he just unfixed.

c. Top Rope Method. This method is used when clearing the route from the bottom.

(1) The climbing team establishes a top rope.

(2) The one climber clears the route while the other climber belays him.

TRANSITION: Are there any questions over clearing the route? If not, let us discuss cable ladders.

7. (15 Min) **CABLE LADDERS**. The cable ladder is constructed with stainless steel cables and aluminum crossbars, which are also known as rungs. Each cable end has a steel ring (eyelet)

large enough to accommodate the largest carabiner in the MAC Kit. These rings allow cable ladders to be connected to each other. The cable ladder is also referred to as a caving ladder.

a. Site Selection. When selecting a ladder site consider these four factors:

(1) There must be sound anchor points located at the top and bottom of the ladder.

(2) You must be able to anchor the ladder along the route to keep it from swaying.

(3) There must be suitable loading and unloading platforms.

(4) Cable ladders provide excellent protection in rock books, chimneys and overhangs.

NOTE: Cable ladders should be set up starting from the top anchor.

b. Ladder Installation Using Natural Anchors.

(1) Ensure that there are two sound anchor points located at each end of the ladder.

(2) With one end of a sling rope (or 1" tubular nylon tape), tie an appropriate anchor knot around a sound anchor point.

(3) With the other end of the sling rope tie a round turn and two half hitches onto a carabiner attached to the eyelet of the ladder. This knot is utilized since it is easy to adjust.

(4) Do the same to the opposite eyelet, pulling tightly enough so that there is no slack between the eyelets and the anchor points.

(5) Deploy the cable ladder down the obstacle.

(6) The bottom of the cable ladder can be secured in the same manner as the top. If the ladder is longer then the obstacle, it can be secured with a sling rope by:

(a) Tying a figure of eight loop at one end of the sling rope and clipping a locking carabiner into the loop.

(b) Attach the locking carabiner to the ladder by placing it around the cable and above a rung.

(c) Tie the sling rope off to a sound anchor point using a round turn and two half hitches. This knot is used because since it is easy to adjust.

(7) If intermediate anchors are needed on the route of the ladder, make sure each side of the ladder has equal tension.

(8) The ladder can be tightened from top or bottom, if needed. Use an appropriate anchor knot at the anchor, then tie a round turn (or slip 8) on the carabiner, tighten, and tie off with 2 half hitches.

c. Ladder Installation Using Artificial Anchors.

(1) Two pieces for artificial protection will be used for each eyelet of the cable ladder.

(2) Locate the middle of a sling rope (or a 1" tubular nylon tape) and tie a figure of eight loop and clip a carabiner through the loop.

(3) Attach the carabiner to the ladder's eyelet.

(4) Secure each sling rope's pigtail to its own piece of protection with a round turn and two half hitches. This knot is utilized since it is easy to adjust.

(5) Do the same to the opposite eyelet, pulling tightly enough so that there is no slack between the eyelets and the anchor points.

(6) Deploy the cable ladder down the obstacle.

(7) The bottom of the cable ladder can be secured in the same manner as the top. If the ladder is longer then the obstacle, it can be secured in the same method as discussed earlier. The ladder can be tightened, if needed, as discussed above.

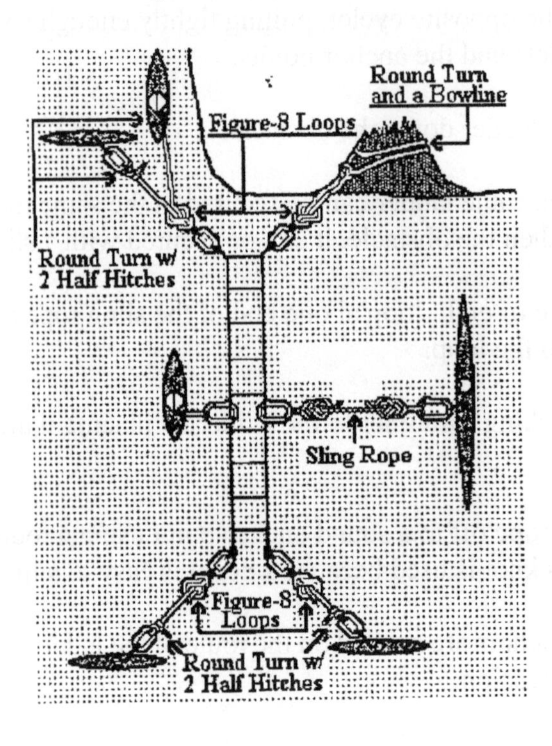

NOTE: The ladder can be tightened from either the top or the bottom so it can remain as rigid as possible.

TRANSITION: Are there any questions about the installation of cable ladders? If not, let us discuss ascending the cable ladders.

8. (10 Min) **ASCENDING THE CABLE LADDER**. This is done by one method or a combination of several. The following are techniques that can be used:

   a. Climb as any other type of ladder.

   b. Climb with one arm and one leg on each side of the ladder. Ensure that the outside foot is placed in the rung toe first while the inside boot is placed in heel first.

   c. Concentrate on using your leg muscles more then your arms.

   d. When climbing up a steep slope, climb hand over hand with the legs straddling the ladder, walking up the slope. This is similar to climbing a simple fixed rope.

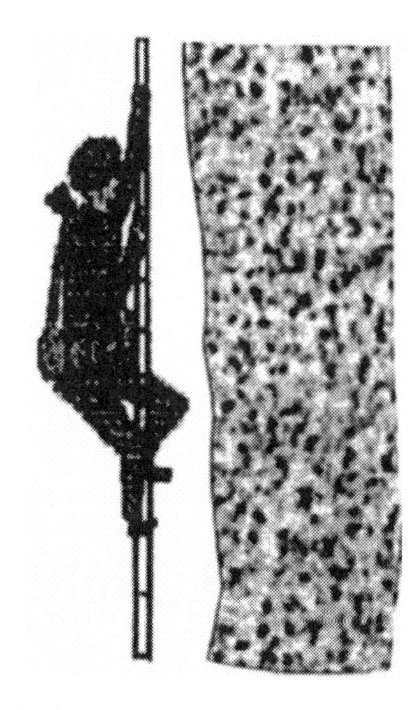

ASCENDING A CABLE LADDER

   e. Safety being a main concern, the climber must be protected as he ascends. This may be accomplished by:

   (1) The Top Rope Method.

(2) The Safety Prusik Method. By installing a semi fixed rope parallel to the ladder, the climber can use a safety prusik to ascend the cable ladder.

NOTE: The top anchor point for the method of protecting the climber will be separate from that of the top anchor point for the ladder.

TRANSITION: Now that we have discussed cable ladders, are there any questions? These are the techniques used to conduct cliff assaults. The actual conduct will be discussed in another class. If you have no questions for me, then I have some for you.

PRACTICE (CONC)

a. The students will construct a fixed rope installation.

PROVIDE HELP (CONC)

a. The instructors will assist the students when necessary.

OPPORTUNITY FOR QUESTIONS (3 Min)

1. QUESTIONS FROM THE CLASS

2. QUESTIONS TO THE CLASS

Q. What are the three types of fixed ropes?

A. (1) Simple Fixed Rope.
(2) Semi-Fixed Rope.
(3) Fixed Rope.

Q. What types of knots can be used at intermediate anchor points of a fixed rope installation?

A. Figure 8, Slip Figure 8, Clove Hitch

Q. How does the climber tie into a fixed rope installation?

A. With a round the chest bowline with a figure of eight loop at each end of the pigtail. A carabiner will be clipped into each loop. This is known as a safety line.

SUMMARY (2 Min)

a. During this class we have covered simple fixed rope, semi-fixed, fixed rope and the cable ladder installations.

b. Those of you with IRF's please fill them out at this time and turn them in to the instructor. We will now take a short break.

UNITED STATES MARINE CORPS
Mountain Warfare Training Center
Bridgeport, California 93517-5001

SML
ACC
02/20/02

LESSON PLAN

CLIFF RECONNAISSANCE

INTRODUCTION (5 Min)

1. **GAIN ATTENTION**. The invasion of Normandy didn't happen without extensive reconnaissance of the beach and cliff areas. This vital information helped the assault forces plan, prepare, and train for the steep cliffs that were awaiting them. As an assault force, Commanders cannot afford to waste precious moments on the beach waiting to gather information about the cliff and be more vulnerable than they already are.

2. **PURPOSE**. The purpose of this period of instruction is to introduce to you to cliff reconnaissance. This lesson relates to cliff assault and fixed ropes.

3. **INTRODUCE LEARNING OBJECTIVES**

   a. TERMINAL LEARNING OBJECTIVE. In a summer mountainous environment and given a designated cliff, conduct a cliff reconnaissance, in accordance with the references.

   b. ENABLING LEARNING OBJECTIVES.

      (1) Without the aid of reference, state in writing, the equipment needed to construct a cliff sketch, in accordance with the references.

      (2) Without the aid of reference, list in writing, the marginal information required on a cliff sketch, in accordance with the references.

      (3) Given a designated cliff and necessary equipment, construct a cliff sketch, in accordance with the references.

      (4) Without the aid of reference and given a diagram, list the basic symbols used on the cliff sketch, in accordance with the references.

(5) Given the proper format and needed information, prepare a cliff report, in accordance with the references.

4. **METHOD/MEDIA**. The material in this lesson will be presented by lecture. You will practice what you have learned during upcoming field training exercises.

5. **EVALUATION**. You will be tested later in this course by performance evaluation.

TRANSITION: If there are no questions, lets start by defining exactly what reconnaissance is.

BODY (55 Min)

1. (5 Min) **RECONNAISSANCE**. Conducting a cliff assault is a dangerous undertaking. Without extensive reconnaissance of the intended site, the operation will almost be doomed to fail. The more information available to the raid force commander the better the chances of success. The following units are capable of conducting cliff site reconnaissance:

   a. Force Reconnaissance Company. This is a MEF level asset whose mission s to conduct pre-assault and deep post- assault reconnaissance and surveillance in support of the Landing Force Commander. Normally there will be one platoon assigned to each MEU (SOC) deployed. This unit will normally have organic Assault Climbers or M7A qualified Mountain Leaders.

   b. Reconnaissance Company (Division). This is a Division level asset whose mission is to conduct reconnaissance and surveillance in support of the Marine Division and its subordinate elements. Normally there will be one platoon assigned to each MEU (SOC) deployed. This unit will possibly have organic Assault Climbers or M7A qualified Mountain Leaders.

   c. Surveillance and Target Acquisition Platoon (STA). This is a BLT asset, normally composed of MOS 8541 trained snipers whose mission is to conduct close reconnaissance for the Battalion Commander and deliver long range precision fire on selected targets. Normally, this element is not employed independent of BLT operations and lack much of the organic support needed for insertion into the operational area when compared to the Reconnaissance Units.

   d. Qualified Small Boat Company Scout Swimmers. They are trained to conduct cliff reconnaissance as the boat company often has the primary mission of conducting as amphibious cliff assault.

   e. SEALS. MEU (SOC) will deploy with elements from NAVSPECCWARGRP.

   f. Others. Additional units in the theater of operation may consist of Ranger Pathfinder Platoons and Special Forces A Teams.

TRANSITION: Now that we are familiar with what units are qualified to conduct site reconnaissance, let's discuss general considerations of cliff reconnaissance.

2. (10 Min) **CLIFF RECONNAISSANCE**. As covered in the Cliff Assault class, the rssault force and its climbers must be prepared to overcome the cliff by whatever means necessary. There are many factors, which drive the reconnaissance effort in support of MEU (SOC) operations. Normally, this mission can be assigned to any one of the units mentioned above at the MEU Commanders discretion. The following are offered as general considerations for conduct of a cliff reconnaissance and are not meant to dictate current unit SOP.

   a. Determine mission feasibility at the mission planning stages (METT) taking into account the enemy situation, capabilities, and probable courses of action (KOCOA, DRAW-D, SALUTE-R).

   b. Determine equipment requirements and the assault forces current capabilities. (i.e. is there pre-training or sustainment training required before mission execution.)

   c. Request an aerial reconnaissance of the area.

   d. At a minimum, conduct a detailed map reconnaissance. Request to use current MEU (SOC) reconnaissance assets if available.

   e. The unit conducting the reconnaissance should be thoroughly familiar with assault climber operations and the assaulting units capabilities.

      (1) If the unit conducting the reconnaissance does not have a qualified Mountain Leader or an assault climber they should bring one with them.

      (2) Ideally, any qualified M7A Mountain Leader or current assault climber should lead this effort. This will provide on site expertise to provide a clear picture of the obstacle to be crossed, identify possible climbing points, equipment requirements, as well as a tentative time estimate.

   f. Gather essential data.

      (1) If the unit conducting reconnaissance is not familiar with this type of mission, they must be thoroughly briefed on the specific information required by the assault climbers.

      (2) A face-to-face coordination with the recon team leader is highly recommended.

      (3) The reconnaissance prep should include, but is not limited to sketches, photographs, or any other items of significance. Don't rule out uncommon sources like tourist maps or photos from submarines. These are very good for planning and navigation.

(4) The information must be reported in a timely manner in order to prepare the assault force.

g. On site observations.

(1) Identify top and bottom anchor points.

(2) Identify top and bottom rally points.

(3) Identify probable lanes for climbing and establishment of rope installations.

(4) Identify weakness in the cliff face such as chimneys, overhangs, rotten rock, etc…

(5) Identify natural animal habitats such as dens, caves and nests. Startling the animals may warn off the enemy of a disturbance of the cliff face.

(6) Identify possible rock slide/avalanche sites.

(7) Check for the feasibility of fire support.

TRANSITION: Now that you know some of the basic considerations in conducting this mission, let's look at one of the tools of this type of reconnaissance, the cliff sketch.

3. (5 Min) **OPERATION SECURITY**. Operational security is of utmost importance. The following are considerations to be taken:

a. The reconnaissance element should not climb the intended cliff breach points. This could compromise the plan causing a disastrous loss of surprise for the assaulting unit.

b. If the recon unit is to remain in place, surveillance of the breach points and likely avenues of approach should be established.

c. The unit should facilitate the arrival of the assault force and be prepared to assist in any way possible.

4. (10 Min) **CLIFF SKETCH**. A cliff sketch is a pictorial representation of the cliff in elevation and perspective as seen from one point of observation. It will contain a horizon line and intervening features. It is rapidly made and easily read and understood. A proper cliff sketch will have enormous value to the raid force commander and his assault climbers.

a. Equipment. The following equipment will assist in constructing the cliff sketch:

(1) Compass

(2) Binoculars equipped with a mil scale
(3) Sketch pad

(4) Soft pencil

(5) Ruler

(6) Digital Camera

b. Marginal Information. The following information is placed on the sketch after indicating the reference line and before conducting the sketch:

(1) Sketchers name, rank, and unit

(2) Date of sketch

(3) Sketchers location (8 digit grid at a minimum)

(4) Direction of view (in degrees magnetic)

(5) Magnetic north arrow

(6) Scale (if used/known)

c. Construction.

(1) Study the landscape to distinguish prominent terrain features in relation to each other. This should be done in conjunction with your military map.

(2) Select a reference point. The point should be permanent and conspicuous. This is the base that the features of your sketch will be drawn from.

(3) Establish a scale. The cliff sketch is a panoramic or birds eye view of the cliff. To maintain a correct relationship between objects and features, a proportion must be established. One method of scale is the 15 inch method:

(a) Attach a 15 inch piece of cord to your ruler and hold one end in your teeth.

(b) Hold the ruler at eye level, each 3/4" increment is equal to 50 mils in width.

(c) Utilizing a scale will increase the accuracy of the sketch.

d. Basic symbols. There are seven basic symbols used on the cliff sketch, each defines a characteristic that is of importance to the assault climber.

(1) THIN CRACK

(2) THICK CRACK

(3) LEFT FACING CRACK

(4) RIGHT FACING CRACK

(5) OVERHANG

(6) LEDGE

(7) CHIMNEY

(8) SLAB

(9) RAMP

(10) CHUTE

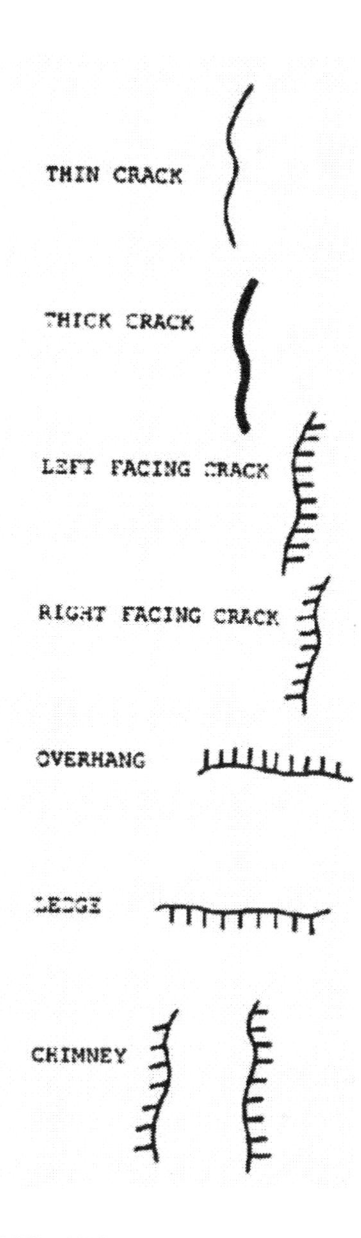

ILLUSTRATIONS OF BASIC SYMBOLS

Note: Ramp, Slab, and Chute can be written on the sketch, include the dimensions (width, depth, height) in order to know the number of lanes possible.

Example of Cliff Sketch

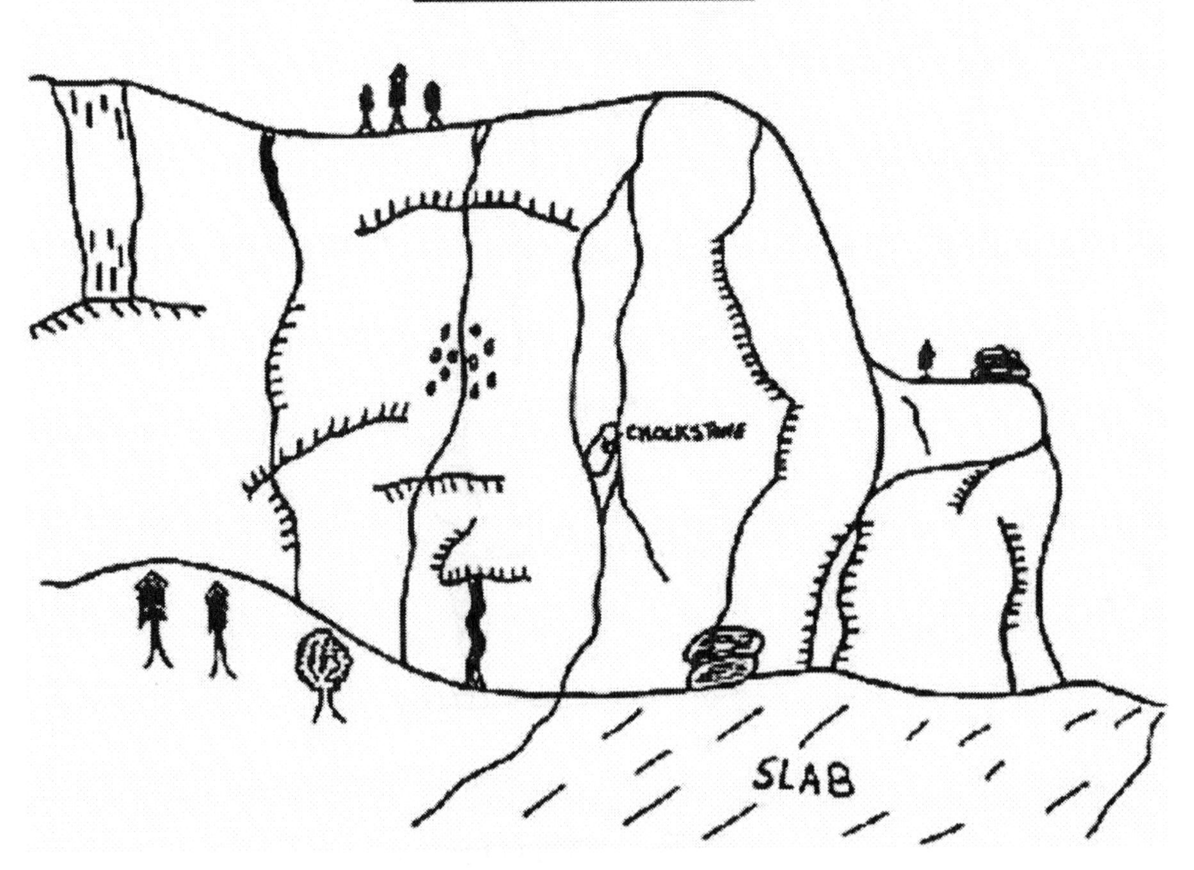

NOTE: Only details that are of military importance should be added. Details should not be added simply to fill up space or improve the appearance of the sketch.

TRANSITION: The cliff sketch is a great tool to the commander on the ground or at the objective area, but how will we get information back to the assault force on the ship before we depart on the mission. Next we will discuss the cliff report.

5. (10 Min) **CLIFF REPORT**. The cliff report contains thirteen lines, Alpha through Mike. Each of the lines will give information about a specific aspect of the cliff and the surrounding area. The following is provided in the event that a recon team is reporting directly to the assaulting unit and not their SARC.

    a. LINE ALPHA: Line Alpha indicates the units of measure to be used throughout the report.

UNIT OF MEASURE	NUMBER CODE
Meters	1
Yards	2
Feet	3

    b. LINE BRAVO: Date / time report completed.

c. LINE CHARLIE: Cliff location. Given at cliff center and expressed as a minimum 8 digit grid over a secure net or encrypted utilizing the AKAC 874.

d. LINE DELTA: Width of the cliff head, expressed in the units of measure stated in LINE ALPHA.

e. LINE ECHO: Cliff height, expressed in the units of measure stated in LINE ALPHA.

f. LINE FOXTROT: Obstacles at the base of the cliff. This line can use multiple codes.

OBSTACLE	LETTER CODE	NUMBER CODE	TYPE
Natural	A	1	Rocks
		2	Stream / River
		3	Trees / Vegetation
		4	Ditches
		5	Snow / Ice
Manmade	B	1	Buildings
		2	Fences
		3	Pylons / Wires
		4	Poles / Masts
		5	Others

g. LINE GOLF: Rock type, if known.

TYPE	NUMBER CODE
Granite	1
Basalt	2
Lava	3
Sandstone	4
Steep earth	5
Unknown	6

h. LINE HOTEL: Military classification of climbs, if determined.

CODE	DESCRIPTION
Easy	50-60 degrees, freeable, good pro placement
Moderate	60-70 degrees, good pro placement, medium -large pro
Difficult	70-80 degrees, marginal pro placements, overhangs
Severe	80-90 degrees, run-outs or unprotectable, overhangs

i. LINE INDIA: This line will identify hazards on the cliff face.

HAZARD	NUMBER CODE
Rockfall	1
Water	2
Snow / Ice	3
Vegetation	4
Other	5

j. LINE JULIET: This line will identify the number and types of tactical lanes, which can be constructed. This line may contain more than one code as applicable.

LANE TYPE	LETTER CODE
Simple fixed	A
Fixed	B
Top rope	C
Cable ladder	D
Vertical hauling line	E
Suspension traverse	F

k. LINE KILO: Is an A-frame needed? YES or NO

l. LINE LIMA: Enemy Situation. Given in the SALUTE-R format for any current enemy reports that may affect the cliff assault or are pertinent.

m. LINE MIKE: Remarks / comments. Identifies any data that is essential but not covered in the above lines.

6. (5 Min) **CLIFF REPORT EXAMPLE**. The following is an example of a cliff report. It is based on NATO report formats utilized by reconnaissance units. The advantages to this style of report are the encryption of the pertinent information and the ability to rapidly transmit the data by radio communications. The disadvantage to this style is that the receiver of the report must be thoroughly versed in the report and how to decipher the information.

a. Recon Team: Kilo One Tango this is Golf Seven Delta, stand by for CLIFFREP, over.

LINE ALPHA	3
LINE BRAVO	180800ZJAN98
LINE CHARLIE	MG67890898
LINE DELTA	250
LINE ECHO	40
LINE FOXTROT	2,4 A; 2 B
LINE GOLF	1
LINE HOTEL	Moderate
LINE INDIA	4
LINE JULIET	6 C, 3 D, 1 F
LINE KILO	Yes

LINE LIMA	None
LINE MIKE	None

b. This report describes a cliff in feet that was reconnoitered at 0800 on 18 Jan 2000. The cliff is located at grid MG67890898 and is 250 feet wide at the base. The cliff is 40 feet high and has a ditch, stream and fence in the area surrounding the base. It is made of granite and is a moderate climb. There is vegetation on the cliff face with 6 top rope lanes, 3 ladder lanes, and 1 suspension traverse site. An A-frame is needed to bring up equipment and no enemy activity was noted.

TRANSITION: Properly done the cliff report and cliff sketch can greatly help with the mission at hand. We have discussed the types of reconnaissance, cliff reconnaissance, cliff sketch, and the cliff report. Are there any questions?

PRACTICE

a. Students will practice basic cliff reconnaissance utilizing reports and cliff sketches

PROVIDE HELP

a. Instructors will assist the students when necessary.

OPPORTUNITY FOR QUESTIONS (3 Min)

1. QUESTIONS FROM THE CLASS

2. QUESTIONS TO THE CLASS

Q. What are the basic symbols used on a cliff sketch?

A. (1) Thin Crack
(2) Thick Crack
(3) Left Facing Book
(4) Right Facing Book
(5) Overhang
(6) Ledge
(7) Chimney
(8) Ramp
(9) Chute
(10) Slab

SUMMARY (2 Min)

a. During this period of instruction, we have discussed the types of reconnaissance, the cliff reconnaissance itself, the cliff sketch, and the cliff report. These areas must be highly accurate in order for a good cliff assault to be planned.

b. Those of you with IRF's please fill them out and give them to me. Everyone else take a five minute break.

UNITED STATES MARINE CORPS
Mountain Warfare Training Center
Bridgeport, California 93517-5001

SML
SMO
ACC
02/20/02

LESSON PLAN

CLIFF ASSAULT

INTRODUCTION (5 Min)

1. **GAIN ATTENTION**. For a Marine, there is almost no such thing as impassible terrain. Even cliffs can be surmounted and military operations can be conducted if the techniques are properly learned. Techniques for cliff assaults were developed and refined by the Royal Marine Commandos during World War II for amphibious raids; but the principles apply to any cliff. This class will introduce you to the principles of cliff assault. You will be applying these principles later in both day and night exercises while a member of this course.

2. **OVERVIEW**. The purpose of this period of instruction is to familiarize the students with the techniques and basic procedures of conducting a cliff assault. This lesson relates to all other lessons on mountain movement.

INSTRUCTOR NOTE: Have students read learning objectives.

3. **INTRODUCE LEARNING OBJECTIVES**

   a. TERMINAL LEARNING OBJECTIVE. In a summer mountainous environment and given a cliff face, conduct a cliff assault, in accordance with the references.

   b. ENABLING LEARNING OBJECTIVES

      (1) Without aid of the reference, list in writing the planning framework considerations when conducting a cliff assault, in accordance with the reference.

      (2) Without aid of the reference, list in writing the tactical considerations when conducting an amphibious cliff assault, in accordance with the reference.

(3) Without aid of the references, list in writing the four different techniques that may be employed to move personnel/equipment conducting a cliff assault up the cliff face, in accordance with the references.

(4) Without the aid of references, describe in writing the five phases of a cliff assault, in accordance with the reference.

4. **METHOD/MEDIA**. The material in this lesson will be presented by lecture and demonstration. You will practice what you have learned in upcoming field training exercises. Those of you with IRF's please fill them out at the conclusion of this period of instruction.

5. **EVALUATION**.

a. SML - You will be tested by a written examination and a performance evaluation.

b. ACC - You will be tested by a written exam and a performance evaluation during your company's cliff assault.

c. SMO - You will be evaluated by a performance evaluation as a company during the cliff assault.

TRANSITION: Are there any questions over the purpose, learning objectives, how the class will be taught or how you will be evaluated? This class will not just be in the classroom; you will also have the opportunity to practice it in the field, and perhaps someday in combat. With that in mind, let's look at the planning framework.

BODY (90 Min)

1. (5 Min) **PLANNING FRAMEWORK**. A cliff assault is a thoroughly planned action on a known danger area. The unit's mission is the raid beyond the cliff, not climbing the cliff itself. Beware of losing focus on the end state. Once the commander decides to execute a cliff assault, his planning is framed by the following:

a. Surprise is paramount and silence must be kept to attain surprise.

b. Speed is essential, and all ropes available must be used.

c. The cliff head must be well organized.

d. Initially, the raiding party is very vulnerable.

TRANSITION: Are there any questions over the planning framework? Let us now discuss the reconnaissance considerations.

2. (5 Min) **RECONNAISSANCE CONSIDERATIONS**. Recon teams should take an experienced assault climber team with them on insertion to ensure that:

a. Climbing points can be established on the vertical obstacles that are within the unit's ability.

b. Suitable top and bottom anchors.

c. Be able to direct assault climbers to specific routes upon arrival.

TRANSITION: Now that we have discussed the reconnaissance considerations, are there any questions? Let us now talk about the tactical considerations.

3. (5 Min) **TACTICAL CONSIDERATIONS**. If an enemy objective is in close proximity to a cliff-head, it is possible that defensive forces consider the cliff an obstacle, thus focusing their security to other more vulnerable areas. However, if the objective is outside small arms range from the cliff, the enemy will likely defend it in a 360-degree fashion. If this is the case, you have fatigued the assault unit with the climb while not achieving the desired surprise. An objective that is near the cliff assault site offers these advantages:

   a. An enemy could assume that the cliff is not crossable and therefore is a protected "wall" which he can put his back against. Therefore concentrating his defenses out board from the cliff toward more likely avenues of approach.

   b. The cliff edge to the objective distance is within our mortar range. Thus the assault unit does not have to expend the time and energy to haul mortars, crews and ammunition up the cliff to ensure fire support for their attack. Without a suitable platform, this advantage is void.

   c. The enemy security forces may not be comfortable looking over the edge of a cliff under less than ideal conditions, potentially creating a gap in security that the assault unit can exploit.

TRANSITION: Are there any questions over the tactical considerations? Let us now discuss the possibility of using deception.

4. (5 Min) **DECEPTION PLAN**. Use noise factors and diversionary attacks.

   a. Rivers or ocean waves breaking at the base of cliffs are common and mask the noise of a cliff assault very well.

   b. Weapons fire and impacts from supporting arms can also mask the noise of a cliff assault, but it can also put the enemy on alert.

   c. Shelling on or near enemy positions on a regular basis at the same time over a period of days may cause the enemy to become accustomed to the disturbance and be less vigilant during these times.

NOTE: Do not plan to use fires on or near the cliff heads because this can render the cliffs dangerous and unstable due to loose rocks and rock fall.

d. Use diversionary attacks by ground, air or indirect fire from multiple directions.

e. The cliff assault itself can be used as the diversionary attack.

TRANSITION: Are there any questions about the considerations of a deception plan? If not, let's talk about the use of communications.

5. (5 Min) **COMMUNICATIONS**. Use wire as the primary mode and radio as the alternative means to minimize radio traffic and ensure good communications.

   a. When using intra squad radios, caution must be taken because they have no crypto capability although they operate on military band only.

   b. A no Comm plan should be designed on standard signals as per the unit's SOP to include a few additional signals to deal specifically with the cliff assault.

   c. Key personnel should have a radio operator and/or runner.

TRANSITION: Are there any questions over communications? Let us now discuss the Fire Support Plan.

6. (5 Min) **FIRE SUPPORT PLAN**. Develop the Fire Support Plan using traditional deliberate attack parameters. When preparing a plan, consider the following:

   a. Use of the attacking unit's organic mortars first.

   b. Use artillery, if in range of the cliff assault.

   c. Use of the Forward Air Controller for rotary and fixed-wing aircraft.

   d. Naval Gun Fire can be utilized but be aware that it cannot cover reverse slopes.

TRANSITION: Are there any questions over the planning of fire support? There are numerous considerations to look at if the cliff assault is being conducted during an amphibious assault.

7. (5 Min) **AMPHIBIOUS CONSIDERATIONS**. The amphibious considerations are:

   a. Advance force reconnaissance operations should be employed.

   b. Hydrographic surveys/confirmatory beach reports should be conducted.

   c. All landing vehicles/crafts must be spread loaded.

d. Debarkation must be done quickly.

TRANSITION: Now that we have discussed planning, are there any questions? Once reconnaissance has confirmed the route and the plan is moving along, the planner must turn to organizing his force.

8. (5 Min) **ORGANIZATION**.

a. 1st Wave. This wave would be organized with 18-24 assault climbers (depending on size of unit), Plt / unit commanders and cliff head security (top and bottom).

BILLET	QUALIFICATIONS	RESPONSIBILITIES
Chief Assault Climber (CAC)	Senior M7A Mountain Leader Senior assault climber	Supervision of the assault climbers
#1 and #2 Assault Climbers	4 lead climbing teams from the assault climber's platoon	Lead climb routes and set in climbing points for follow on force
Unit commander	HQ elements of the unit	Complete plan through visual recon
Control Point NCO (CPNCO)	M7A Mountain Leader Experienced assault climber	Organize top of obstacle, set up control features, coordinate with MACO
Cliff Head Officer (CHO)	M7A Mountain Leader Experienced assault climber	Position security at bottom and top of vertical Obstacle, aware of all actions between beach master and control point
Beach Master (BM)	M7A Mountain Leader Experienced assault climber	Same as control point NCO but at bottom of vertical obstacle
Security Teams	Security element of the unit	Provide security at bottom and top of vertical obstacle
Lane NCO	Tactical Rope Suspension Technician or an experienced NCO	One per climbing lane, assist beach master and control point NCO in setting up control features, physically places individuals from BM position to climbing points.

NOTE: Additionally within the first wave, a company-sized unit may want to designate a vertical hauling line/suspension traverse team to establish these installations on top for heavy equipment.

b. 2nd Wave. This would constitute the remainder of the task-organized units, the assault force and the reserves. The XO is delegated in command of the second wave and will stay at the base of the vertical obstacle until the unit has negotiated the vertical face.

(1) The amount of time the unit is stationary at the vertical obstacle should be minimized. Ideally, the main second wave should move from the boat/landing craft or rally point straight into the climbing lanes (via beach master).

TRANSITION: Now that we have discussed the organization of a cliff assault, are there any questions? Let's look at some different techniques that may be used to move personnel and gear up the face of the vertical obstacle.

9. (5 Min) **ASSAULT CLIMBING TECHNIQUES**. The actual techniques used to negotiate personnel and equipment up the vertical obstacle may vary depending on a variety of factors: level of training, type of vertical obstacle to be negotiated, and/or equipment available. The following four techniques, or any combination of them may be used:

   a. Two party climb for assault climbers, all other personnel top rope.

   b. Two party climb for assault climbers, all other personnel go up fixed rope installations.

   c. Two party climb for assault climbers, all other personnel/equipment utilize vertical hauling line and/or suspension traverse.

   d. Two party climb for assault climbers, all other personnel utilize cable ladders.

TRANSITION: Are there any questions over the techniques? Now let's look at the sequence of events to be followed by each wave. Remember this is only a technique, not a principle; but it is a tested technique, and experience has shown the value of a set sequence to be followed when entering enemy territory, especially at night.

10. (25 Min) **ASSAULT SEQUENCE**. The Assault Sequence can best be understood and organized if it is broken down into five phases, with many actions and tasks taking place simultaneously within each phase.

   a. Phase One. The first wave arrives at the cliff base.

      (1) First wave moves into ORP.

      (2) CHO and leaders conduct recon of proposed climbs and establishes left and right lateral limits. Easiest climbing routes are selected for leaders.

      (3) The #2s prepare gear for climbing.

      (4) CHO and BM place flank securities into position.

      (5) Climbing teams move to designated lanes and prepare to climb.

         (a) At least one climbing team must have a radio.

         (b) All lead climbers should have NVGs, but not climb with them on.

         (c) All lead climbers should climb with minimal gear.

         (d) One or all of the lead climbers will carry a static rope. These are to be used for the rope installations.

         (e) Hardest climbs go to the best teams.

(6) Casevac plan is formulated and litters prepared.

(7) The area that the climbers ascend is not necessary the area where climbing lanes are established. First wave must be ready to move if this is the case.

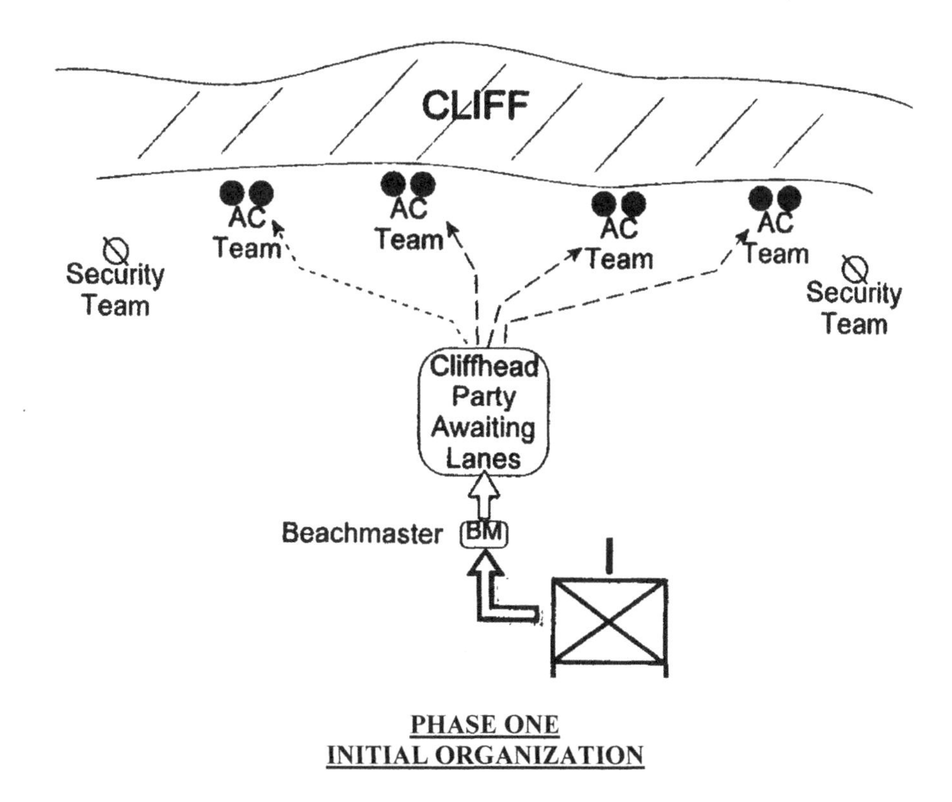

PHASE ONE
INITIAL ORGANIZATION

b. Phase Two. Begins as soon as first lead climber begins to climb.

(1) The first leader on top provides initial security.

(2) As soon as the remainder of the leaders top out, construction of the climbing lanes begin with the radio being passed off to the leader on security. The cliff-head is divided into sections for easier control.

(3) Communication must be established between the top and bottom of the cliff. If possible the CAC should be one of the leaders and have a radio.

(4) Emergency rappel lanes are established to provide for a quick withdrawal if the mission is compromised, a 4:1 ratio is required. These are usually single rope rappels.

(5) The #2 climbers are brought up and they should have the gear needed to begin construction of climbing lanes.

(6) The remainder of the first wave needed on top (CHO, CPNCO, security, etc…) begins to ascend as soon a climbing lane is established. All lanes must be cleared "hot" by the CHO before they are climbed by the second wave. Security is positioned by the CHO.

(a) CHO needs to ascend ASAP to make final decisions on location of vertical hauling lines (VHL), security, control features, etc…

(7) Assault climbers on security are relieved by designated security teams.

(8) Once on top, the Raid Force Commander (RFC) departs for the leader recon.

(9) The BM begins constructing control features at the cliff base, only when leaders have begun to construct lanes. The BM position will be dependent on the vegetation/terrain.

(10) Gear is prepared in ORP for hauling systems.

(11) Vertical hauling lines and/or suspension traverse is constructed. At least one assault climber must be left on the deck to aid in the construction/operation of the VHL.

(12) Lane NCOs move gear needed on top to VHL and it is taken up to cliff head.

(13) Control Point NCO establishes topside control features.

(14) A landline between the BM and the CPNCO is established.

(15) #2 climbers climb the ladders to check for the need of intermediate anchors.

(16) Once all of first wave is on top, Casevac Plan must be formulated to lower casualties, if necessary.

(17) ORP may be moved closer to cliff base for better silence/security.

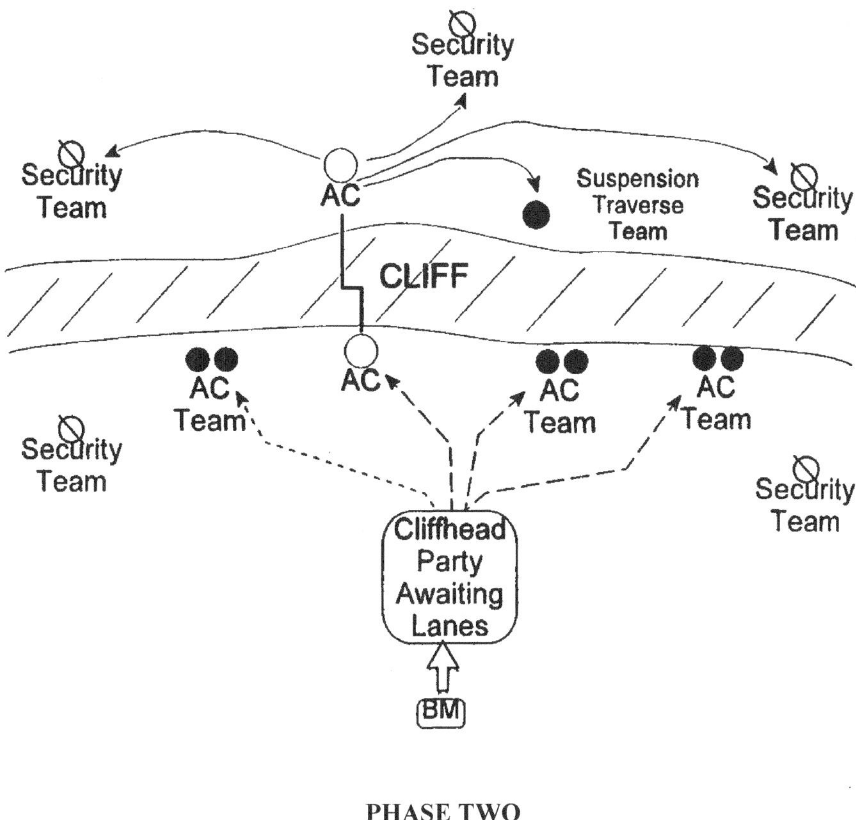

PHASE TWO
INITIAL ASCENT

c. Phase Three. Arrival of the second wave for their ascent of the cliff.

(1) Company should establish radio contact 10 minutes out.

(2) Second wave establishes an ORP, preps to climb and XO makes liaison with the BM.

(3) CHO is informed on the second waves' arrival and inquires of any changes in the numbers, injuries, changes in timeline, etc…

(4) Company should already be broken down into climbing sticks for easier control. Each Marine is assigned an alphanumeric indicator within his element.

Example:

1st Platoon Security Element	2nd Platoon Assault Element
A-1 Lcpl Tooby	B-1 Pfc Brunner
A-2 Cpl Post	B-2 Lcpl Padilla
A-3 Sgt Morron	B-3 Pvt Kesl

(5) Lane NCOs are stationed at the bottom of the lanes to assist Marines.

(6) Marines check in with BM and give their stick number. BM does final check for proper knots/equipment. The BM will then direct the Marines to specific climbing lanes.

(7) Marines with heavy loads or crew served weapons are directed to the VHL. These climbers are then directed to the climbing lanes closest to the VHL for easier recovery of their gear once on top.

(8) Once on top, all Marines follow control features to the CPNCO, and give their stick number to him. The CPNCO will then direct the Marines to their new position.

(9) Raid force begins to establish a 180-degree defensive perimeter and awaits return of the RFC. An option of staggering the raid force by element within the control feature is possible.

(10) Once all members of the second wave are on top, the XO ascends the cliff.

(11) In the event that there are missing personnel at the top, stick numbers can be checked to ensure that a Marine(s) came through the BM and the CPNCO, and that no one is still within the confines of the control features or the ORP.

(12) Upon return of the RFC, he issues final orders and briefs possible shifts in his time lines, the raid force then moves out to its objective.

(13) If the raid force is not withdrawing via the cliff, the first wave joins the second wave at the top of the cliff and continues on with the raid force. Options to the RFC are to either wait on the first wave to join the second wave and move out, or move without first wave and linkup at a later time.

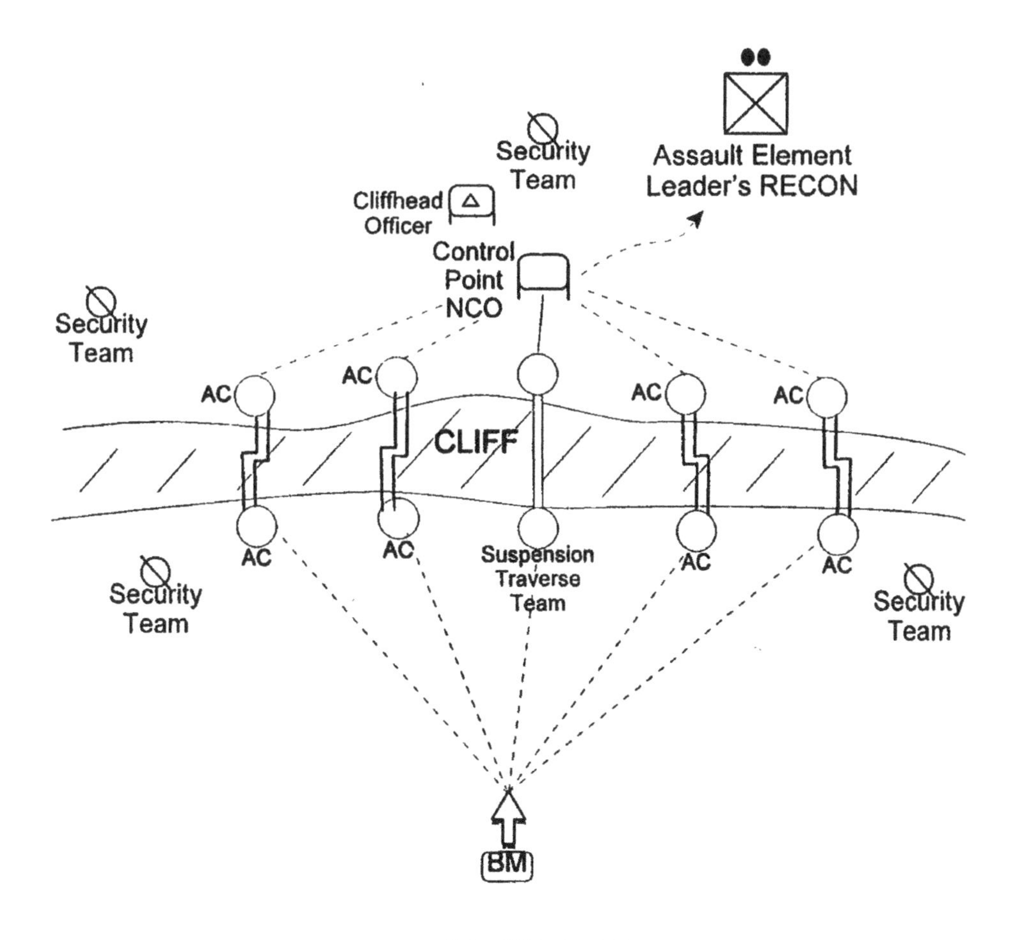

PHASE THREE
SECOND WAVE ASCENDS

d. Phase Four. Departure of the Raid Force.

(1) Approximately 15 minutes after departure of the second wave to the objective:

NOTE: The time hack is used for withdrawal if the mission is cancelled/compromised.

(a) The control features, top and bottom, are removed to prevent detection.

(b) All unnecessary equipment/personnel are moved to the cliff base.

(c) Assault climbers begin to take down climbing lanes and establish retrievable rappel lanes. Belay men must be stationed at the cliff base for the second wave.

(d) The VHL/ suspension traverse is left in place. The systems may be slacked or A-Frame collapsed to prevent detection.

(e) A central point for casualties and EPWs is established and litters are prepared for possible medical evacuations.

(f) Personnel topside form ORP and await arrival of the raid force.

(g) Personnel at the cliff base do likewise.

(h) CHO follows progress of raid force on radio and stays aware of situation (casualties, EPWs, changes in plan etc…).

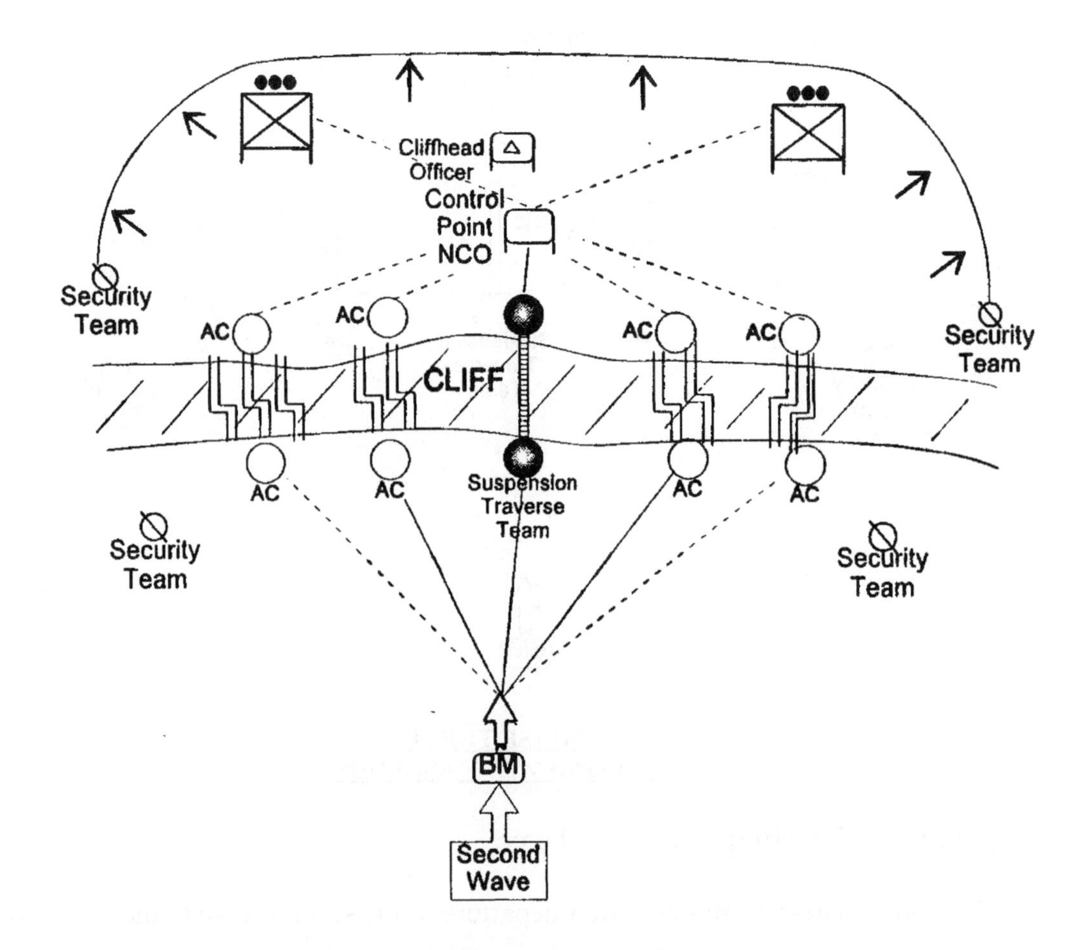

PHASE FOUR
RAID FORCE DEPARTS

e. Phase Five. Withdrawal. If the units mission dictates a withdrawal via the cliff face, the following steps should be taken:

NOTE: If the unit is compromised before the mission has been completed, then the tactical situation will dictate how the withdrawal will be accomplished.

(1) Upon contact with the raid force confirming its return:

(a) Control features are re-established at the top and bottom.

(2) Upon the re-arrival of the raid force:

(a) Raid force re-establishes 180-degree defensive perimeter. Any casualties are moved to the cliff head and evacuated.

(b) Once the order to withdrawal is given, the BM prepares to receive the descending Marines. In an amphibious operation, he would call the landing craft/boats at this time.

(c) Company XO is among the first to rappel down.

(d) A high concentration of automatic weapons and unit leaders are left on top.

(e) Squads thin their positions with the squad leader descending last. He reports his squad departure to the CPNCO.

(f) The unit commander descends after the main body departed.

(g) Once the entire second wave is down, the bulk of the first wave begins to withdrawal.

(h) Assault climbers begin to tear down the rappel lanes and VHLs/suspension traverses. All ropes are dropped to the base of the cliff and back stacked into rope bags. A-frame poles are lowered.

(i) CHO is constantly aware of the state of the cliff head.

(j) CPNCO and BM tear down their control features.

(k) Security comes in and descends the cliff via the retrievable rappel lanes.

(l) Assault climbers descend with the CAC and CHO being the last men down.

(m) The retrievable rappel ropes are retrieved. Assault climbers account for all gear and personnel.

(n) Assault climbers check in with the BM and all rejoin the unit.

TRANSITION: Now that we have discussed the five phases, are there any questions? Let us talk about some general considerations.

11. (5 Min) **GENERAL POINTS**.

a. The second wave should precede the first wave with ample enough time to establish all climbing lanes and rope systems.

b. Three climbing lanes per platoon should be established for the quickest ascent.

c. Cable ladders should be utilized to a maximum for all climbing lanes.

d. Communications between the CPNCO and the BM should be via landline. This is reliable and secure.

e. Communication between the CHO, CAC and assault climbers at the cliff base should be maintained if communication assets are available.

TRANSITION: Let us now discuss other possibilities to keep in mind.

12. (5 Min) **CONCLUSION**. The planner should be aware, or made aware of other options and scenarios that may arise concerning the ascension of a vertical cliff by a unit. Cable ladders are an excellent means of ascending a vertical cliff, but may not always be the best method. Top ropes, fixed ropes, etc… are all viable means to move up a vertical obstacle.

a. Another example is using the same route up. Instead of lead climbers climbing different routes up a cliff face, some leaders can climb up an easy route then establish their lanes on other sections of the cliff.

b. Another scenario might be in which only a security element climbs on the established cable ladders on a true vertical cliff, and then moves to secure the top of the cliff where fixed ropes can be set up. This would allow for the remainder of the company to ascend quicker and easier.

TRANSITION: Now that we have discussed the five phases of the assault, are there any questions? A cliff assault is a very daring and risky undertaking, but it can also yield some great rewards by catching the enemy completely off guard.

PRACTICE (CONC)

a. The students will conduct a cliff assault.

PROVIDE HELP (CONC)

a. The instructors will assist the students if necessary.

OPPORTUNITY FOR QUESTIONS (3 Min)

1. QUESTIONS FROM THE CLASS

2. QUESTIONS TO THE CLASS

Q. What are the four planning framework considerations that must be considered when planning a cliff assault?

A. (1) Surprise is Paramount and silence must be kept to attain it.
(2) Speed is essential and all ropes must be utilized.
(3) The cliff head must be well organized.
(4) Initially, the raiding party is very vulnerable.

Q. What are four tactical considerations if the cliff assault is to be amphibious?

A. (1) Advance force reconnaissance operations should be employed.
(2) Hydrographic surveys, confirmatory beach reports, should be conducted.
(3) All landing vehicles/crafts must be spread loaded.
(4) Debarkation must be done quickly.

Q. What are the four different techniques that may be utilized to move personnel/equipment conducting a cliff assault up the cliff face?

A. (1) Two party climb for assault climbers, all other personnel top rope.
(2) Two party climb for assault climbers, all other personnel go up fixed rope installations
(3) Two party climb for assault climbers, all other personnel/equipment utilize vertical hauling line and/or suspension traverse.
(4) Two party climb for assault climbers, all other personnel utilize rope ladders.

SUMMARY (2 Min)

a. During this period of instruction, we have covered the tactical considerations, basics of planning and the organization of a cliff assault; also, how to conduct a cliff assault, which included the five phases of an assault.

b. Those of you with IRF's please fill them out at this time and turn them in to the instructor. We will now take a short break.

UNITED STATES MARINE CORPS
Mountain Warfare Training Center
Bridgeport, California 93517-5001

SML
02/20/02

LESSON PLAN

GEOLOGY AND GLACIOLOGY

INTRODUCTION (5 Min)

1. **GAIN ATTENTION.** Did you know that 48% of the world's land area is classified as mountainous? We talk about becoming a mountain leader and fighting in a mountainous terrain, but some wonder how applicable it is. The next war will not be fought at Camp Pendleton, Camp Lejeune, or 29 Palms. Wherever we look today for potential trouble spots, we see mountains. Such as Norway, Korea, Iran, Pakistan, Afghanistan, even South America. This class is designed to lay a basic groundwork of knowledge concerning a summer mountainous environment.

2. **OVERVIEW.** The purpose of this period of instruction is to introduce the student to the basics of geology and glaciology; specifically concerning those aspects that can affect how we move and fight in the mountains. This class relates to mountain walking, cliff assaults, and all ice work.

INSTRUCTOR NOTE: Have students read learning objectives.

3. **INTRODUCE LEARNING OBJECTIVES**

   a. TERMINAL LEARNING OBJECTIVE

      (1) Without the aid of references, identify the major features of a glacier, in accordance with the references.

   b. ENABLING LEARNING OBJECTIVES

      (1) Without the aid of references and given a list of the types of rocks, describe in writing the characteristics of each type of rock, in accordance with the references.

      (2) Without the aid of references and given a list of the types of glaciers, describe in writing the characteristics of each type of glacier, in accordance with the references.

      (3) Without the aid of references, list in writing the glacial features that are impediments to travel, in accordance with the references.

(4) Without the aid of references, and given a diagram of a glacier, label the types of moraines found on a glacier, in accordance with the references.

4. **METHOD/MEDIA.** The material in this lesson will be presented by lecture and demonstration. You'll practice what you have learned in upcoming field training evolutions. Those of you with IRF's please fill them out at the conclusion of this period of instruction.

5. **EVALUATION.** You will be tested later in the course by written and performance evaluations.

TRANSITION: Are there any questions over the purpose, learning objectives, how the class will be taught, or how you will be evaluated? Let us start by getting a basic understanding of the make-up of the earth, which in turn will lead us into our discussion on mountain and glaciers.

BODY (60 min)

1. (5 Min) **THE EARTH'S COMPOSITION**

a. What do we call the study of the earth? Geology.

b. When we study the earth, we discover that it is divided into three distinct layers:

(1) Crust. The only portion of the earth that man has direct knowledge of is the crust, the outer 1% of its radius. It is from 5-40 miles thick, being thinnest under the oceans, and thickest under the mountains.

(2) Mantle. This layer, is about 1800 miles thick, separates the crust from the core. Scientists are unsure as to the exact appearance, consistency, density, or temperature of the mantle, though they have agreed that it is probably made up of solid matter. The most interesting thing about the mantle, from the standpoint of physical geography, is that despite the overall solid character, it contains layers or zones of differing strength and rigidity. The upper most layer of the mantle combines with crust to form the lithosphere, which is divided into units or plates. Located immediately beneath the lithosphere is a thick layer of weak plastic-like mantle called the asthenosphere. Many earth scientists now believe that the major source of energy for tectonic forces comes from movement within the asthenosphere produced by thermal or convection currents originating deep within the mantle.

(3) Core. The core of the earth is believed to be composed primarily of nickel and iron, and hence the material of the core is sometimes called Nife. The core material is under enormous pressure, probably as much as three to four million times the atmospheric pressure at sea level. Through the transmission of seismic waves into the structure of the earth, we discovered that the earth's core is molten. However, because increased pressure can change the melting points of elements, the innermost core of the earth appears to be solid.

TRANSITION: We've covered the earth's composition, are there any questions? Lets now cover the different rock classifications that we encounter in a mountainous environment.

2. (5 Min) **ROCK TYPES.** In general, a rock is a coherent aggregate (a whole made up of parts) of mineral particles. Although, the number of minerals making up most of the rocks of the lithosphere are limited, they are combined in so many different ways that the variety of rock types is enormous. Nevertheless, we have basically three types of rock classifications: Igneous, sedimentary, and metamorphic.

    a. Igneous. These are 'fire-formed" rocks which have solidified from a hot liquid melt, called magma. Examples are granite and basalt.

        (1) Igneous rocks rise from a depth in the earth as a molten magma. If the magma cools and solidifies before reaching the surface, the igneous rocks are termed intrusive. Intrusive rocks, such as granite, cool slowly and result in a tightly knit fabric of crystals, which form a tough hard rock, generally excellent for climbing. Intrusive rocks normally have a great many small cracks and fissures, which may be, used for hand and foot holds.

        (2) The cores of most major mountain ranges in the world are granite. In general, the older the mountains, the more granite rock has been exposed at the surface and the better the climbing.

        (3) When a magma rises too close to the earth's surface, the molten rock may either flow onto the surface and cool or be ejected in a volcanic explosion as ash and lava. These igneous rocks are termed extrusive, as they cool and solidify in the atmosphere. Extrusive rocks which are ejected by volcanic action have very little strength or cohesion and are very difficult for climbing. Extrusive rocks which cool more slowly, such as basalt, can be almost as good for climbing as granite, but such rocks rarely make up the major portion of a mountain range.

    b. Sedimentary. Rocks, which are deposited by the action of water, wind, and or ice, or chemically precipitated from water. Sandstone, shale, and coal are sedimentary rocks usually deposited by rivers and oceans, whereas limestone is precipitated from seawater.

    c. Metamorphic. These "changed" rocks were originally igneous or sedimentary rocks, which, due to temperature and/or pressure within the earth, have been altered physically and or chemically. Examples of metamorphic rocks are; slate from shale, marble from limestone, and gneiss from granite.

        (1) Sedimentary and metamorphic rocks are, in general, not as good for climbing as are igneous, as they tend to be much more friable (breakable or "rotten"). Exceptions are:

            (a) Some types of sandstone in very and (dry) regions where there is little water to weaken the rock's cementing agents.

(b) Some granite-like metamorphic rocks called gneiss. Often, too, sedimentary and metamorphic rocks contain high concentrations of clay-like minerals, which become very soft and slippery when wet.

TRANSITION: Are there any questions over rock types? Like rock types, mountains are also classified by their method of formation.

3. (5 Min) **FORMATION MECHANICS**

a. Mountains. Most mountain ranges are the result of interior stresses in the earth's interior. In order to relieve these stresses, thick sections of the crust slowly bend (fold) or fault (break). The resultant surface relief caused by folding and faulting is then often magnified by the processes of erosion.

b. Plains/Plateau. In semi-arid regions, extreme erosion of basically flat-lying rock layers can produce a typical badlands topography, which is impassable to vehicles and may require mountaineering techniques. Examples of this type of topography are the Grand Canyon area and the Black Hills and Badlands of the Dakotas.

(1) Fault Block Mountains. Fault Block Mountains are bounded on one or more sides by faults, dividing the crust into up and down thrown blocks. The Tetons of Wyoming and the Sierra Nevada's are some examples.

(2) Folded Mountains. Folded mountains, such as the Appalachians, have numerous faults but the principal structures are large scale folds, which are again modified by erosion.

(3) Domal Mountains. Domal mountains are usually the result of the upward movement of magma and the supplement folding of the rock layers overhead. Erosion may strip away the overlying layers, exposing the central igneous core. Examples are Stone Mountain in Georgia and the Ozark Mountains.

(4) Volcanic Mountains. Volcanic mountains are an exception to the rule that folding/faulting are needed to form mountains. The mountains of the Hawaiian Islands and Japan are good examples of this type.

TRANSITION: Now that we have covered formation mechanics, are there any questions? It should be remembered that the mentioned classifications are idealized, and the major mountain ranges usually consist of a mixture of two or more types. Also, certain rock types are usually associated with each type of mountain, and it is often the type of rock, which determines the mountain climbing qualities, and not just the type of mountain.

4. (10 Min) **WEATHERING AND EROSION.** Once mountains have been built, the forces of nature begin a relentless task of tearing them down. As soon as land is raised above the sea and exposed to wind and running water, the forces of weathering and erosion begin to act.

a. Basically, weathering, both mechanical and chemical, breaks the rocks into smaller pieces without moving the pieces very far. Erosion then transports the pieces to another location by gravity, wind, water, or ice.

b. The most important type of weathering in mountainous regions, is called frost wedging. This is the result of moisture in the rocks and crevices freezing and thawing repeatedly. The resulting expansion and contraction wedges off angular flakes and blocks of rock, which fall down the slope and are accumulated as talus and scree. Scree and talus resulting from wedging action are generally poor for climbing due to their instability, although scree may be descended rapidly using certain techniques. (Refer to *MOVEMENT IN MOUNTAINOUS TERRAIN* outline for movement on scree.)

TRANSITION: While the type of mountain and the characteristics of the rock can be important consideration in the mountain movement, you may not be able to use them. In some mountains, only glaciers offer access. In the back of this outline, you can see a drawing of a typical valley glacier.

5. (25 Min) **GLACIERS.** Glaciers are the world's greatest earthmovers. They are rivers of ice flowing under the influence of gravity constantly restructuring the mountains around them. They often provide the best route in otherwise impassable terrain. Glaciers can be very small, or large enough to cover an entire continent, as in Antarctica. They may move from several inches to several hundred feet a year, and when the snow melt rate exceeds the rate of snowfall, a glacier can actually retreat back up the valley. Movement is usually caused by gravity, by basal slippage over the bedrock, or internal flow in the ice.

a. Formation of Glacial Ice. Glaciers can occur in any mountainous terrain where the annual snowfall exceeds snow melt, and the terrain permits deep deposits of snow. Once the snow depths exceed 10 feet or more, the weight of the underlying snow is compacted into firm snow, or neve snow (10 feet of snow makes 1 foot of firm snow). This is re-crystallized into firm ice and then into glacial ice. The region of the glacier where the snow accumulation exceeds the melt year-round is called the accumulation zone. Further down the glacier, where melt exceeds accumulation is an area called the ablation zone. The imaginary line between the two zones is called the firm line.

b. Glacier Classifications

(1) Alpine. Glaciers that form in high mountain bowls by the year round accumulation of snow over a long period of time.

(2) Ice Sheet or Ice Cap. The second and larger type of glacier is the ice sheet or ice cap. The ice cap, a far more significant formation than the alpine glacier, usually covers hundreds of thousands of square miles, though uncommon today, at one time covered as much as 30% of the land. They still blanket Greenland and Antarctica, and smaller ice caps are still present in many highlands.

c. Types of Glaciers

(1) Cirque glacier. Does not advance beyond the bowl in which it is formed.

(2) Hanging glacier. A glacier which is forced out of its origin area and over a cliff or precipice, so that it breaks off at the snout and tumbles down as an ice avalanche.

(3) Valley glacier. A glacier which advances down a valley. Glaciers form U-Shaped valleys, while rivers form V-Shaped valleys.

(4) Piedmont glacier. A glacier which moves out of its valley origins onto an open plain, or into the sea forming a fan-like pattern.

d. Glacial Features. Glaciers are treacherous places, with many features that can impede your travel upon them. The following is a list of them.

(1) Moraines. The most common features of a glacier are moraines. These are piles of rock and other debris, which have either fallen onto the glacier, or been pride loose by the glacier as it moves along. There are four basic types:

(a) Lateral moraine. This is the rock debris along the valley wall.

(b) Medial moraine. This is formed when two glaciers come together and the lateral moraines are forced out into the middle of the glacier.

(c) Terminal moraine. This is the rock debris at the snout of the glacier which is being dredged up and pushed down the valley. Many times a glacial stream will also flow out from under the terminal moraine.

(d) Ground moraine. If the glacier stops and then begins to recede, the rocks and debris under the glacier can be exposed. This is called ground moraine. Generally, the outside of a moraine wall is stable, but the side facing the glacier can be steep and loose and should be avoided. A moraine may' be your only path, but it can be awkward and tiring due to the jumbled rocks. There are also other features that are found on glaciers.

(2) Crevasses

(a) These are the result of irregularities in the bedrock under the glacier or stresses in the ice. They are called transverse, longitudinal, or lateral depending on their orientation to the direction of the glacier's movement. Crevasses can be 200 feet deep, but are normally between 40-100 feet in depth. Crevasses are hindrances in traveling across glaciers because they may have to be bypassed. They can be hazardous when blowing snow conceals them by forming a snow bridge across the top. These bridges must be tested before crossing, and can give way without warning. In summer, the ice is less covered with snow and most crevasses are exposed, but they can still be hazardous at night or during reduced visibility.

(b) Bergschrund. A special crevasse is the bergschrund. This is the separation of the glacial ice at the point where the glacier transitions from the neve snow/ice on the steep mountainsides to the valley floor. Where this crevasse separates the glacier from a rock wall, it is technically not a bergschrund, but a randluft. The bergschrund can be very high and even overhanging and can create a serious obstacle to movement when attempting to move from the valley floor onto the mountainsides.

(3) Seracs. These are ice walls and towers, which have been forced upward due to pressure within the glacier. They are unstable and can fall unexpectedly. This is another hazard and your route planning should give them a wide berth when possible.

(4) Icefalls. These result from the flow of the glacier down a steep ridge. The ridge forces the ice up into a jumbled mass as the glacier flows over the rock. These present formidable obstacles and should be avoided when moving troops.

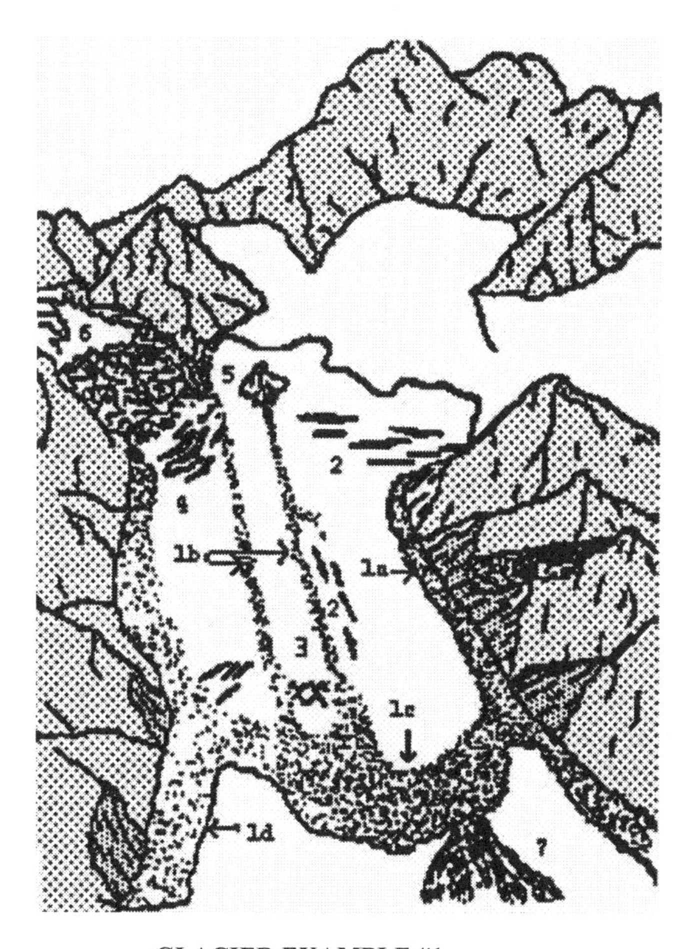

GLACIER EXAMPLE #1

(5) Nunatak. This is bedrock that protrudes up through the glacier creating an obstacle to glacial movement. This can create pressures in the ice and help to bring about some of the hazards that we have mentioned earlier; crevasses, seracs and icefalls.

(6) Rock and ice avalanches. These can occur when seracs and blocks of ice come loose and cascade onto the glacier. These are dangerous when traveling near the valley walls, or under a hanging glacier. Detour around these areas when possible.

(7) Water hazards. This is a mountain or glacier milk. Melt-water from the glacier has found a crevasse and is moving to the ground below the ice. It would be very dangerous to fall into this! The glacial streams above this point must also be treated with respect. They are very cold, can be deep, and create a hazard for a heavily laden Marine. In summer, glacial thaw water can form in troughs, freeze at night, and form glacial swamps. These should also be approached with care.

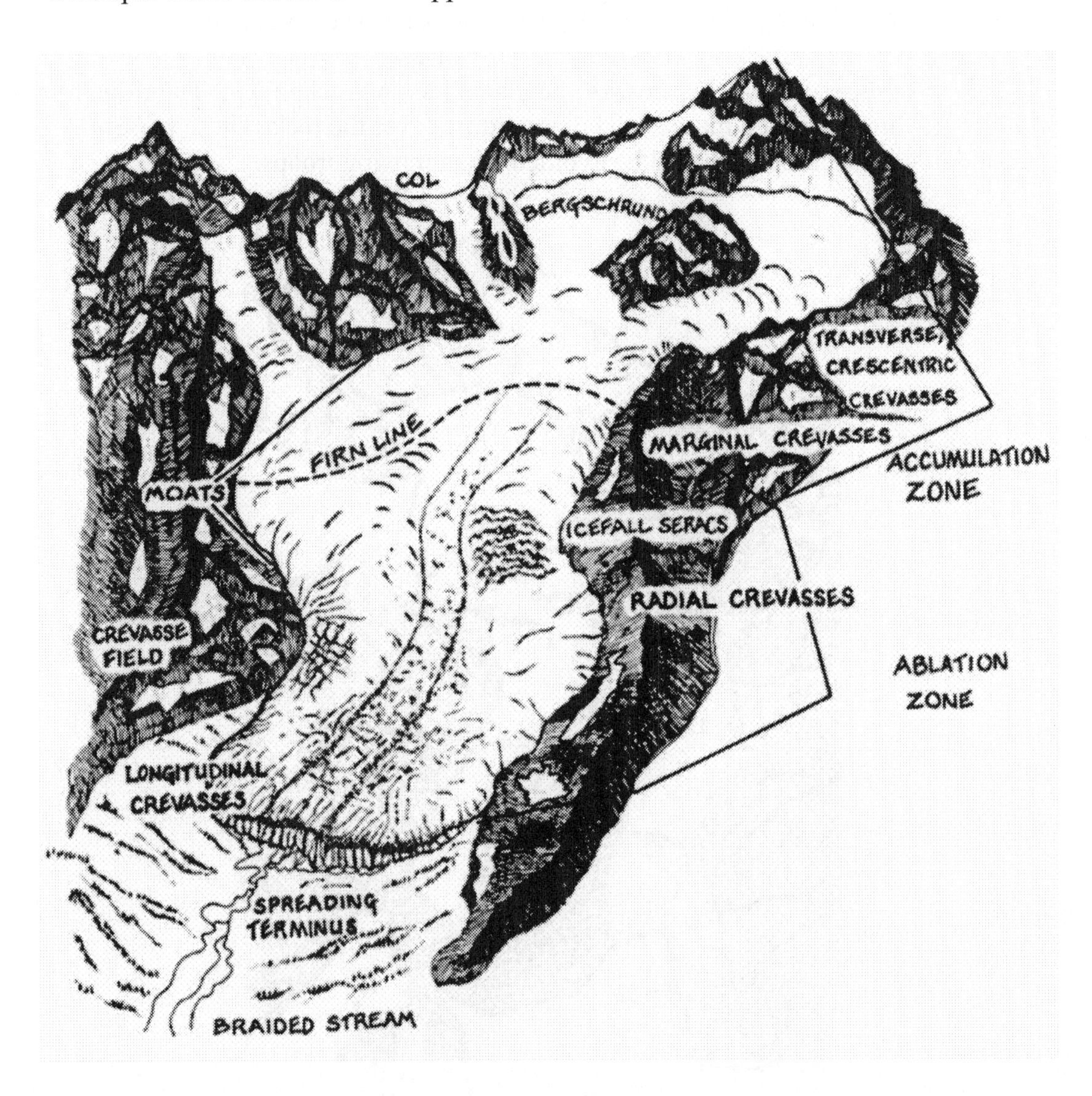

GLACIER EXAMPLE #2

GLACIER EXAMPLE #3

LEGEND TO GLACIER EXAMPLE #3

ICE FEATURES

1. Moat
2. Bergschrund
3. Crescentric crevasses
4. Firn line
5. Nunatak
6. Marginal crevasses
7. Echelon crevasses
8. Terminus or snout
9. Braided outwash stream

MORAINE FEATURES

11. Outwash plain (and ground moraine)
12. Erratic
13. Old lateral moraine
14. Terminal moraine
15. Old terminal moraine
16. Medial moraine
17. Lateral moraine

TRANSITION: We have covered weathering and erosion and glaciers, are there any questions? Obviously, traveling in the mountains and on glaciers can be extremely dangerous. This is why we must employ proper techniques for traveling in these areas.

PRACTICE (CONC)

a. Students will identify the major features of a glacier.

PROVIDE HELP (CONC)

a. The instructors will assist the students when necessary.

OPPORTUNITY FOR QUESTIONS (3 Min)

QUESTIONS FROM THE CLASS

QUESTION TO THE CLASS

Q. What are the three types of rocks?

A. (1) Igneous.
(2) Sedimentary.
(3) Metamorphic.

Q. What are the four types of glaciers?

A. (1) Cirque.
(2) Hanging.
(3) Valley.
(4) Piedmont.

Q. What are the six glacial features that impede movement on a glacier?

A. (1) Moraines.
(2) Crevasses.
(3) Seracs.
(4) Icefalls.
(5) Rock and ice avalanche.
(6) Water hazards.

SUMMARY (2 Min)

a. We have discussed some of the considerations that are used in planning mountain movement, including types of rock, the types of glaciers, and impediments on glaciers.

b. Those of you with IRF's please fill them out at this time and pass them forward to the instructor. We will now take a short break.

UNITED STATES MARINE CORPS
Mountain Warfare Training Center
Bridgeport, California 93517-5001

SML
03/04/02

LESSON PLAN

ICE AXE TECHNIQUES

INTRODUCTION (5 Min)

1. **GAIN ATTENTION**. In order for us to move in the mountains we must be aware of all obstacles and have the knowledge necessary to overcome them. Movement over snow/ice is only one of the many obstacles that can be encountered. In order to efficiently move over snow/ice we must first learn how to use ice axes to ascend snow/ice obstacles.

2. **OVERVIEW**. The purpose of this period of instruction is to introduce the student to the basics of ice axe techniques. This lesson relates to crampon techniques and glacier travel.

INSTRUCTOR NOTE: Have students read learning objectives.

3. **INTRODUCE LEARNING OBJECTIVES**

   a. TERMINAL LEARNING OBJECTIVE

      (1) In glaciated terrain, utilize the ice axe techniques, in accordance with the references

   b. ENABLING LEARNING OBJECTIVE

      (1) In snow/ice covered mountainous terrain, execute ice axe/ice hammer techniques, in accordance with the references.

4. **METHOD/MEDIA**. The material in this lesson will be presented by lecture and demonstration. You will practice what you have learned in upcoming field exercises. Those of you with IRF's will fill them out at the end of this period of instruction.

5. **EVALUATION**. You will be tested by a performance evaluation.

TRANSITION: Are there any questions over the purpose, learning objectives, how the class will be taught, or how you will be evaluated? Let's first discuss how to carry the ice axe/ice hammer when it is not in use.

BODY (55 Min)

1. (5 Min) **STOWAGE OF THE ICE AXE/ICE HAMMER**

   a. The ice axe can be carried on the outside of the pack (some packs have loops provided). The head of the axe should be down, pick inboard with the spike extending up. Covers should be used on the spike and the pick to prevent them from ripping any pack material or clothing.

   b. Ice axes can be placed through a carabiner that is attached to the waist strap of the pack. The adze should be forward with the pick to the rear and the spike facing down. In this position it is readily available for use.

AXE ON OUTSIDE OF PACK

   c. The ice axe can be stowed and retrieved quickly by sliding it diagonally between your back and your pack. The spike is placed down with the pick located between the two shoulder straps. To avoid chances of a runaway axe, remove it before taking off the pack.

   d. The axe/hammer can be carried in hip holsters that are attached to the waist harness.

AXE BEHIND PACK

TRANSITION: Are there any questions over stowing of the ice axe/hammer? When moving across snow it is mandatory that the axe is ready for use to self-arrest, it is also essential when ascending snow/ice cliffs. We will now look at some techniques that can be used for flat through vertical terrain.

2. (5 Min) **GRASPING THE AXE**. How you hold the head of the axe when climbing depends on the climbing situation. There are two basic ways to grasps the axe:

   a. Self-Arrest Grasp. The thumb is placed under the adze and the palm and fingers go over the pick, near the shaft. As you climb, the adze points forward. Sense you can get into the arrest position quickly, this grasp is preferred when the chance of a fall is probable or when working with in experienced climbers.

b. Self-Belay Grasp. The palm sits on top of the adze and the thumb and index finger drop under the pick. As you climb the pick faces forward. This grasp is comfortable because the palm rides firmly on the adze. This grasp is preferred when the chance of a fall is rare.

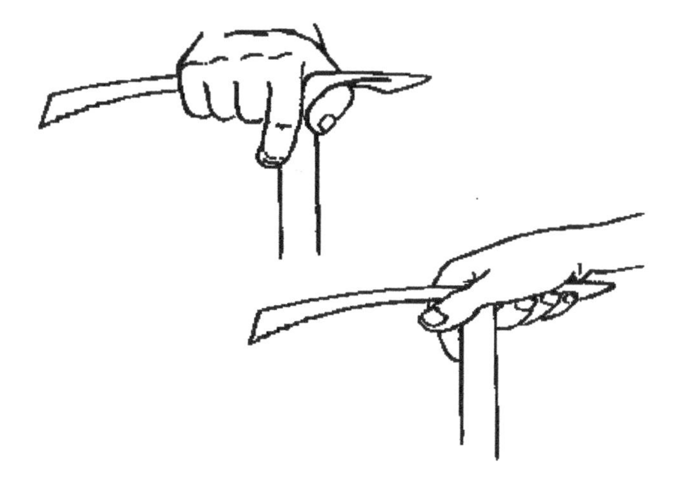

SELF ARREST GRASP AND SELF BELAY GRASP

3. (10 Min) **CLIMBING IN BALANCE**. While ascending slopes, climber must try to stay in balance as much as possible, avoiding any prolonged stance in an unbalanced position. When in the unbalance position, the ice axe acts as a supporting device.

   a. Diagonal Uphill Movement. On a diagonal uphill route, the most balanced position is with the inside (uphill) foot in front of and above the trailing outside (downhill) leg, which is fully extended to make use of the skeleton and minimize muscular effort. In that position, let the trailing leg bear most of your weight. Always grip the ice axe with your uphill hand. The diagonal ascent is performed in the following manner:

      (1) From the in-balance position, place the axe above and ahead into the snow.

      (2) The first step brings the outside (downhill) foot in front of the inside (uphill foot), putting the climber out of balance.

      (3) The second step brings the inside foot up from behind and places it beyond the outside foot, putting the climber back in the balance position.

      (4) Ensure you are back in the balance position before repositioning the ice axe.

      (5) Always keep your weight over your feet and avoid leaning into the slope.

UPHILL DIAGONAL MOVEMENT

b. Changing Direction. The following steps should be performed to remain in balance while changing direction on a diagonal ascent:

(1) Place axe shaft directly uphill of center mass.

(2) Face uphill with feet splayed.

(3) Change axe leash to other wrist, if required.

(4) Turn in the new direction of travel.

CHANGING DIRECTION

4. (25 Min) **USE OF THE ICE AXE**. Using the ice axe while moving over snow or ice covered terrain will provide a third (and sometimes a fourth) point of contact with the surface. Dependent upon whether ascending, descending and/or the angle of the slope, there are various methods in which the axe (or axes) can be used. But regardless of the method used, the ice axe or hammer should be secured to the climber by a leash.

a. Cane Position. Also known as the support position or "Piolet Canne". The cane position is used on slopes of moderate angles from 35 degrees or less.

(1) Ascending. The axe is carried in the uphill hand, with the hand grasping the head, the pick facing to the front or the rear. As you walk along, the axe is used as a cane, and is ideally used to maintain your balance.

CANE POSITION (ASCENDING)

(2) Descending. Holding the axe in either hand, simply face directly downhill, bend the knees slightly, and walk firmly downward. Use the axe as support when needed.

CANE POSITION (DESENDING)

b. Stake Position. Also known as an anchor position. The stake position is used when going directly uphill at a slope angle between 45 and 60 degrees.

(1) Ascending. Before moving upward, the axe is planted firmly into the snow with both hands grasping the head, then take a step upwards, and repeat the movement.

STAKE POSITION (ASCENDING)

(2) Descending. Before moving downward the axe is planted into the snow with both hands grasping the head, then take a step downwards, and repeat the movement. This technique is not usually used for descending.

c. Cross-body Position. Also known as the brace position or "Piolet Ramasse". The cross-body position is used when traversing slope angles of 35 to 50 degrees.

(1) Ascending. The axe is held perpendicular to the angle of the slope, one hand grasping the head, the other holding the shaft. The axe will cross diagonally in front of you, with the pick facing forward. Place the spike into the surface with most of the weight on the shaft. The hand on the head will help stabilize the axe. Move your feet up as in the cane position.

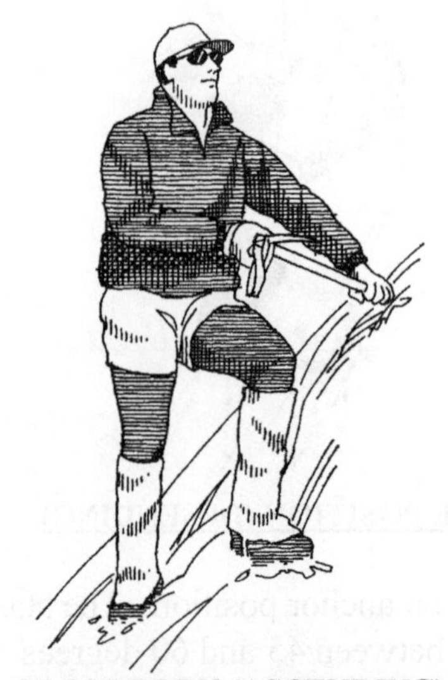

CROSS BODY (ASCENDING)

(2) Descending. Plant the spike of the axe perpendicular to the slope in the cross body position. As descending in the cane position, bend at the knees keeping your body weight over your feet.

CROSS BODY (DESCENDING)

d. Dagger Position. This method is used when ascending 60 degree plus slopes. The head is held like a dagger and thrust into the snow/ice. Remember to grip the head where it meets the surface to avoid an upward leverage. There are two methods of placement using the dagger method, low dagger and high dagger.

(1) Low dagger. Also known as "Piolet Panne". This method is helpful in moving over relatively short but steep sections. Hold the axe by the adze and push the pick into the ice near your waist. The main body weight is thus pressing the pick into the slope taking some of the strain off the legs by using the arms. Move your feet up using your front points.

LOW DAGGER POSITION

(2) High Dagger. Also known as "Piolet Poignard". This method is used for slopes that are too steep to insert the pick at waist level. Hold your hand on the axe head and place the pick into the snow/ice approximately shoulder level. This method also takes the strain off the legs. Move your feet up using your front points.

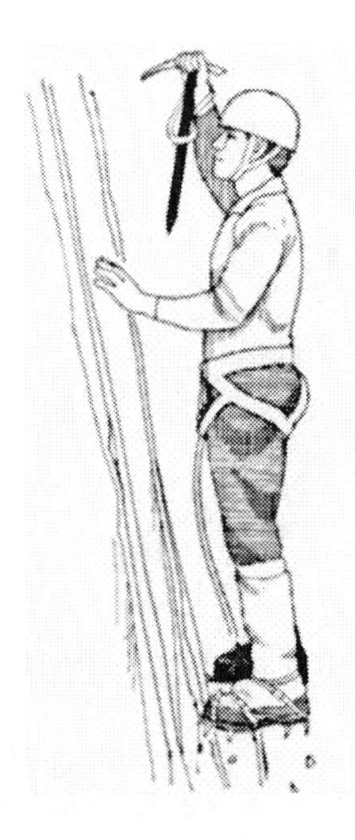

HIGH DAGGER POSITION

(3) Shaft Dagger. This method is used on very steep snow slopes and where the snow is not firm enough to support the pick of the axe. The shaft is held like a dagger in the hand, which is thrust into the snow and used as a hand hold. To avoid leverage, grip the shaft near the surface of the snow.

e. Banisters Position. Also known as "Piolet Rampe". This method is used for descending fairly steep slopes. The axe is planted into the surface down slope and the hand is slid along the shaft while maintaining slight upward pressure. The axe is used as a banister rail for support. Having descended a few steps; stop, extract the axe and repeat the steps.

BANISTERS POSITION

f. Anchor. For harder ice or a steeper slope, you can use this in lieu of the high dagger position. It is used mainly while front pointing but can also be used while flat-footing. Hold the ax shaft near the end and swing the pick in as high as possible without over-reaching. Front point or flat foot upward, holding on higher and higher on the shaft as you progress, adding a self-arrest grasp on the adze with the other hand when you get high enough. Finally, switch hands on the adze, converting to the low-dagger position. When the adze is at waist level, it's time to remove it from the ice and replant it higher.

g. <u>Traction</u>. The steepest and hardest ice calls for the ax to be placed in this position. The ax is held near the spike and planted high; the ice is then climbed by slightly pulling straight down on the ax as you front-point up. When it becomes too difficult to balance on your front points it becomes necessary to use a second tool. This is possible because except for the anchor all ice ax positions require only one hand. Using two tools provides three points of contact (two crampons and one tool or two tools and one crampon) at any given time. The legs should carry most of the weight with the arms helping with both weight and balance. With this two-tool technique you can use the same method for each ax or vary the methods for each. For instance you can climb with both tools in low-dagger position or you can place one tool in high dagger position and the other in traction position.

Fig. 15-26. Front-pointing with ax overhead in traction position

TRANSITION: Are there any questions over the use of the ice axe? If you have none for me, then I have some for you.

PRACTICE (CONC)

a. Students will execute ice axe/hammer techniques.

PROVIDE HELP (CONC)

a. The instructors will assist the students when necessary.

OPPORTUNITY FOR QUESTIONS (3 Min)

1. QUESTIONS FROM THE CLASS

2. QUESTIONS TO THE CLASS

Q. What are the four ways you can carry an ice axe when it is not going to be used?

A. (1) Outside of the pack.
(2) Attached with a carabiner to the waist strap.
(3) Between your pack and your back.
(4) In a hip holster.

Q. What are three of the six ways you can use the ice axe while climbing snow or ice?

A. (1) Cane position.
(2) Stake position.
(3) Cross body position.
(4) Shaft Position.
(5) Banister position.

SUMMARY (2 Min)

a. What we have just covered are the techniques to using an ice axe including ways to carry it both during and not during its use.

b. Those of you with IRF's please fill them out and turn them in to the instructor. We will now take a short break.

UNITED STATES MARINE CORPS
Mountain Warfare Training Center
Bridgeport, California 93517-5001

SML
03/04/02

LESSON PLAN

STEP KICKING AND CUTTING

INTRODUCTION (5 Min)

1. **GAIN ATTENTION.** Crampons are generally used on technical snow and ice routes; however, many mountain routes are of a mixed nature, i.e. rock sections with snow or ice sections that are interspersed. Sometimes on this type of climb crampons are not used, either the snow and ice pitches are short or they are not of a very severe nature. By step kicking we can ascend, descend, or traverse a slope without the aid of crampons. There may also be conditions when the snow is classified as too hard for kicking steps. For this type of condition the climber may use the method of step cutting.

2. **OVERVIEW.** The purpose of this period of instruction is to introduce the student to the basics of glacier movement with regards to step kicking and cutting. This lesson relates to glacier travel.

INSTRUCTOR NOTE: Have students read learning objectives.

3. **INTRODUCE LEARNING OBJECTIVES.**

   a. TERMINAL LEARNING OBJECTIVE

      (1) In glaciated terrain, kick steps in snow, in accordance with the references.

      (2) In glaciated terrain, cut steps in ice, in accordance with the references.

   b. ENABLING LEARNING OBJECTIVES

      (1) In glaciated terrain, execute step kicking, in accordance with the references.

      (2) In glaciated terrain, execute cutting techniques, in accordance with the references.

4. **METHOD/MEDIA.** The material in this lesson will be presented by lecture and demonstration. You will practice what you have learned in upcoming field training exercises. Those of you with IRF's please fill them out at the end of this period of instruction.

5. **EVALUATION.** You will be tested by a performance evaluation.

TRANSITION: Are there any questions over the purpose, learning objectives, how the class will be taught or how you will be evaluated? Now that we know what there is to be accomplished, let's first look at the three methods of step kicking.

BODY (50 min)

1. (10 Min) **STEP KICKING IN SNOW**. Step kicking is a basis of good movement on snow as it develops rhythm, balance and correct body position. It is a fast and effective way of ascending and descending slopes of relatively hard snow. There are three methods we can use to move up and down slopes without crampons.

   a. Diagonal Ascent

      (1) Steps are kicked using the side of the boot. The serrated edge of a lugged-sole boot, i.e., vibram, should be used as saw teeth and the step should be made with a forward motion of the foot, not by pressing down. The sole of the boot should be kept at the horizontal position.

      (2) The ice axe is used as a third leg in the support position on moderate slopes. On easier terrain, the ice axe may be carried in the self-arrest position.

DIAGONAL ASCENT

b. Straight Ascent

(1) The climber must face square to the slope and kick the toe of the boot in at right angles.

(2) If an axe is not carried, handholds may be made by punching a clenched fist into the snow at about shoulder height.

(3) On moderate slopes the axe may be used in the anchor position and on easier terrain the axe should be carried in the support position.

STRAIGHT ASCENT

c. Descent

(1) To descend, face out from the slope and drive a straight leg and heel into the slope, exerting full weight on each foot alternatively. This is also known as the "plunge step".

(2) The ankle is held as rigid as possible to ensure that the heel remains at a sharp angle and drives into the slope.

(3) The ice axe is used in the "banister" position on moderate to steep slopes. On easier slopes, hold the axe in the self-arrest position.

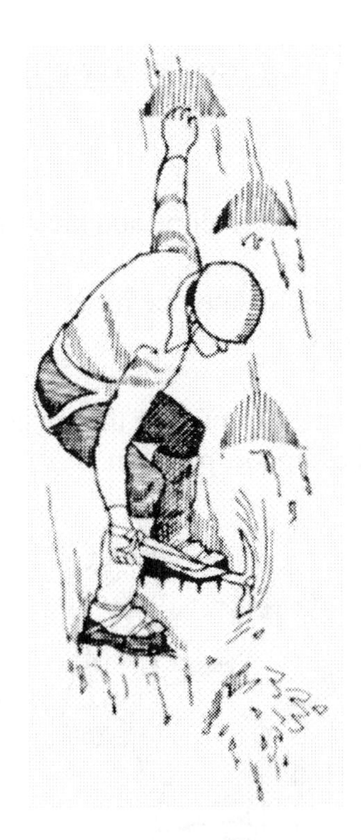

DESCENT

TRANSITION: Now that we have covered step kicking in the snow, are there any questions? We have covered the methods of step kicking, now let's discuss the ways to step cut in snow.

2. (10 Min) **STEP CUTTING IN SNOW.** Normally it is unnecessary to cut steps into a slope if all the members of a patrol are wearing crampons. However, in the event that everyone is not using crampons, you may have to prepare a route for mountain troops by cutting steps. Step cutting is a slow and tiring task in comparison to kicking steps, but with practice, the technique can be carried out in a timely fashion. The essence of good step cutting includes conservation of energy; a good balanced position and a good rhythmic swing of the arm. The weight of the axe will assist you in the work. The type of steps to be cut in the snow are:

   a. Slash Step. The slash step is the most energy economic and fastest way of cutting steps, but it can only be used on snow.

      (1) When cutting on gentle slopes the axe may be held in the hand nearest the slope. On steeper slopes it is more convenient to cut using the outside arm.

      (2) The cutting action should be powerful with the whole arm swinging the axe from a fixed shoulder.

      (3) The step is made with one glancing blow of the axe, which cuts out a 6 inches long "slash" sufficient for the edge of a boot.

      (4) It is important to be in the proper balance when cutting steps.

(5) The slash step is quick and can be used in diagonal ascents and descents of snow covered slopes.

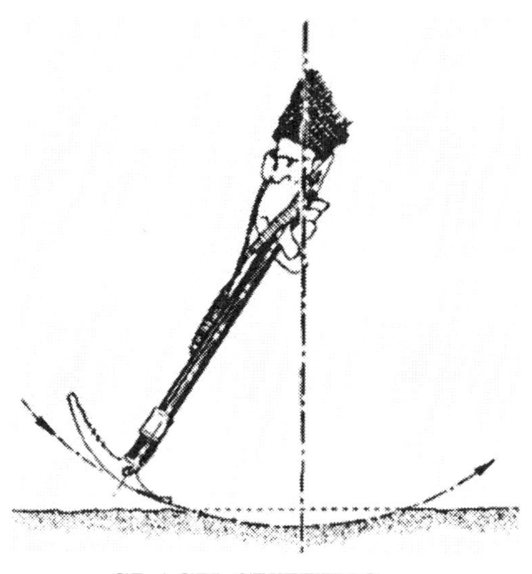

SLASH CUTTING

b. Side Step. The side step is most often used when a more secure cut is needed to enable the whole boot to fit. The cutting position and principles are the same as the slash step but the angle of penetration is steeper. Usually, three blows are sufficient to make a good step.

(1) The step should be sloped in toward the slope angle. A gentle nick is normally enough to start, and the cut section can be flicked away as the axe is withdrawn.

(2) The second and third cuts are made into the existing nick making it larger. There is no reason to waste energy by trying to clear the debris from the steps because they will be packed down by the troops as they cross.

DIAGONAL ASCENT WHILE CUTTTING SIDE STEPS

c. Slab Cut. Under certain conditions, the snow may break apart into slabs and it may be difficult to cut firm steps into the slope. In this type of condition the slab cut should be used.

(1) A horizontal slit is made with the pick of the ice axe and then another slit is cut at a 45 degrees angle to form a "V", pointing to the climber.

(2) The adze is then used to cut out the center of the "V", leaving a triangular step with a flat base.

d. Pigeon Hole. This technique is used for both handholds and footholds when directly ascending a steep snow slope.

(1) The hole is cut and scooped out by using two to three blows from the adze.

(2) It is important that the base is steeply angled so that it allows for a good hold for a gloved hand.

3. (10 Min) **STEP CUTTING IN ICE**. Step cutting in ice is very arduous work, therefore it is important to have a relief frequently to conserve energy. Always look for natural steps in the ice, which can be enlarged by a few blows of the axe.

CUTTING PIGEONHOLE STEPS

a. Side Step. The same method of cutting this step applies as with the snow side step; however, in very hard ice it may be necessary to use the pick to cut a horizontal slit into the ice to start the step.

b. Pigeon Hole. A pigeonhole is cut into the ice in the same way as in snow with the exception that the pick will have to be used. The adze may be used to finish up the hole and to make a more pronounced lip for a good handhold.

c. Ice Nicks/Finger Holds. It is not always necessary to cut a large step in ice. A small nick can be made for a finger hold and can be fashioned with one or two blows of the ice axe. Crampons on thin or brittle ice can be placed on these ice nicks to allow the crampons to work more effectively.

TRANSITION: Now that we have discussed step cutting on ice and snow, are there any questions? Let's talk about the various types of step patterns that can be used.

4. (10 Min) **STEP PATTERNS**. There are three basic patterns that can be cut while on a snow or ice covered slopes. They are the single diagonal line, double diagonal line and the straight up pattern.

   a. Diagonal Line Patterns. The most convenient way of ascending or descending a slope is by a diagonal line because it is easier to cut steps slightly ahead of your position. Both the single and double diagonal line techniques are used in conjunction with the slash step or the side step.

(1) Single Diagonal Line Pattern. The single diagonal line steps are usually cut on gentle slopes where staying in balance is not much of a problem.

(a) Stand in a position of balance with the axe in your downhill hand.

(b) Swing the axe from your shoulder, cutting with the adze and letting the weight of the axe do most of the work.

(c) The swing cut away from your body, starting at the heel-end of the new step and working towards the toe.

(d) Use the adze and pick to finish the step.

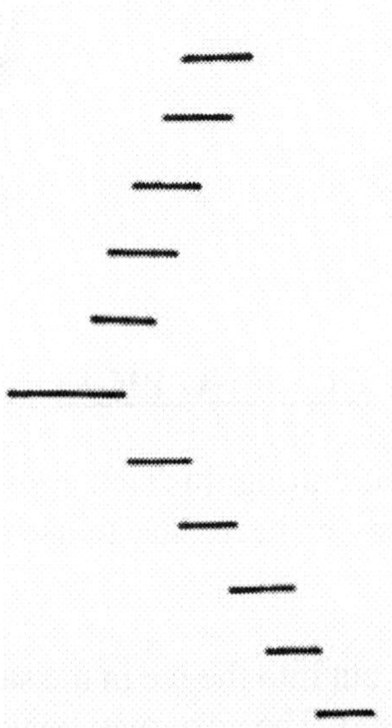

SINGLE DIAGONAL LINE PATTERN

(2) Double Diagonal Line Pattern. The double diagonal line steps are usually cut on moderate slopes where staying in balance is more difficult. This technique does require more time to accomplish but it may prove to be safer to use, especially if the steps are going to be used for the descent.

(a) The same technique for the single pattern is used for the double pattern except that you cut out two foot placements offset of each other.

DOUBLE DIAGONAL LINE PATTERN

b. Straight Up Pattern. The third pattern that can be used when directly ascending steep slopes. This technique is used in conjunction with the pigeon hole step. A set pattern of pigeonhole steps that are spaced at regular intervals will look like the rungs on a ladder. It is important to cut the holes well ahead of your position using this pattern and never climb up to the top steps without already establishing handholds above you for security.

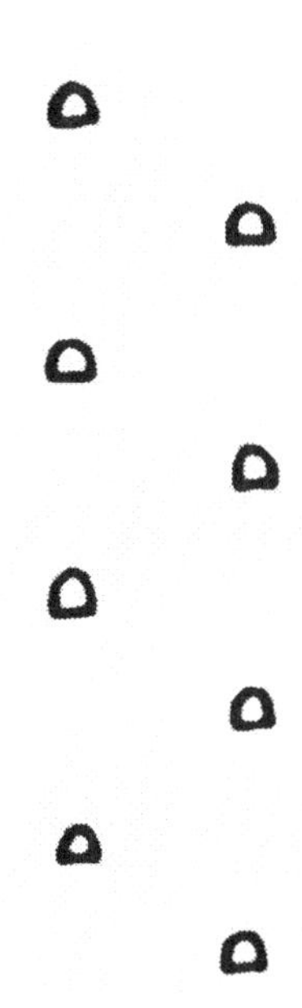

STRAIGHT UP PATTERN

TRANSITION: Are there any questions over step patterns? When ascending a mountain, always keep in mind how you are going to get down and cut steps accordingly. If there are no questions for me, then I have some for you.

PRACTICE (CONC)

a. Students will execute step kicking and cutting techniques.

PROVIDE HELP (CONC)

a. The instructors will assist the students when necessary.

OPPORTUNITY FOR QUESTIONS (3 Min)

1. QUESTIONS FROM THE CLASS

2. QUESTIONS TO THE CLASS

Q. What are the three methods of step kicking?

A. (1) Diagonal ascent
(2) Straight ascent
(3) Descent.

Q. What are two of the four methods of step cutting in snow?

A. (1) Slash step
(2) Sidestep
(3) Slab cut
(4) Pigeon hole

SUMMARY (2 Min)

a. What we have just discussed are the procedures used to kick and cut steps on both snow and ice as well as patterns to be used for both ascending and descending.

b. Those of you with IRF's please fill them in at this time an turn them in to the instructor. We will now take a short break.

UNITED STATES MARINE CORPS
Mountain Warfare Training Center
Bridgeport, California 93517-5001

SML
03/04/02

LESSON PLAN

CRAMPON TECHNIQUES

INTRODUCTION (5 Min)

1. **GAIN ATTENTION.** The decisive winner in the mountains has always been the force that has held the high ground. To do this today, Marine Corps units must be able to move up or over snow and ice obstacles. Using crampons and mastering the techniques of movement on this type of terrain will increase the unit's ability to do so.

2. **OVERVIEW.** The purpose of this period of instruction is to introduce the students to crampon technique, this will teach you how to move across all types of snow or ice covered terrain. This lesson is the primary lesson on snow and ice techniques and relates to all other lessons on snow and ice techniques.

3. **INTRODUCE LEARNING OBJECTIVE**

   a. TERMINAL LEARNING QBJECTIVES. In glaciated terrain, move over a glacier while wearing crampons, in accordance with the reference.

   b. ENABLING LEARNING OBJECTIVES. In glaciated terrain, execute crampon movement, in accordance with the reference.

4. **METHOD/MEDIA**. The material in this lesson will be presented by lecture and demonstration method. You will practice what you have learned in upcoming field training exercise. Those of you with IRF's please fill them out at the end of this period of instruction.

5. **EVALUATION**. You will be tested by a performance evaluation.

TRANSITION: Are there any questions over the purpose, learning objectives, how the class will be taught, or how you will be evaluated? First we will discuss general information regarding the use of crampons.

BODY (50 Min)

1. (5 Min) **GENERAL**. When the snow becomes so hard that you cannot kick steps, you must either cut them or put on crampons. Crampons come with 12 points (two of which are front

points) to provide traction on a snow or ice covered terrain. When walking with crampons on, you use the same basic fundamentals that are used in mountain walking, except that when a leg is advanced it is swung in an outboard motion to prevent the crampons from snagging on each other or your clothes. Crampons should be attached to the boot when on easy terrain. Inexperienced climbers sometimes make the simple and dangerous mistake of getting half way up a route, before attempting to put on crampons on some ridiculously steep slope.

2. (5 Min) **CRAMPON TECHNIQUES**. The basic principle in crampon technique is to utilize the points of the crampons at the correct angles to the snow or ice surface for maximum traction. There are three basic crampon techniques:

   a. The French Technique. This is also known as *pied à plat*, meaning flat footing. This technique is performed by keeping your feet flat against the surface at all times thus keeping all the points of the crampon in contact with the surface.

   b. The German Technique. This is also known as *pied en avant*, meaning front pointing. This technique is performed by using the forward points of the crampon.

   c. The American Technique. This is also known as *pied à trios*, meaning combination of both flat footing and front pointing.

3. (10 Min) **THE FRENCH TECHNIQUE**. The French technique is the easiest and most efficient method of climbing gentle to steep ice/hard snow. Good French technique demands balance, rhythm, and the confident use of crampons and axe. Instead of edging your feet as you would without crampons, allow your ankles to roll out from the slope so that the soles of your boots lie at the same angle as the slope and all 10 points are encouraged to bite. There are three basic foot positions in the French technique: marching, duck walking and flat footing.

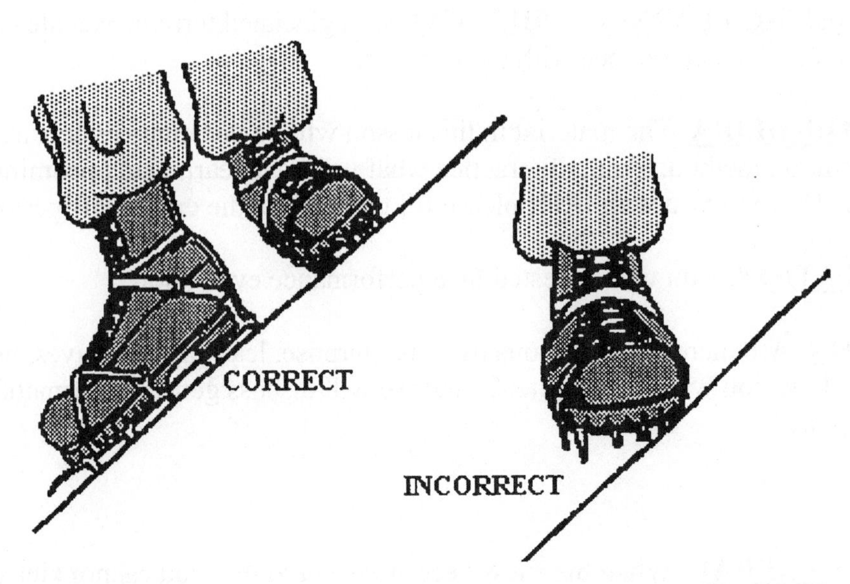

   a. Marching. This is also known as *pied marche*. This technique is performed by walking flat footed in any direction on gentle sloops of 0 to 15 degrees.

b. Duck Walking. This is also known as *pied en canard.* This technique is performed by walking with the feet splayed out and standing back so the body weight is over the crampons. This technique is used on gentle slopes of 15 to 30 degrees.

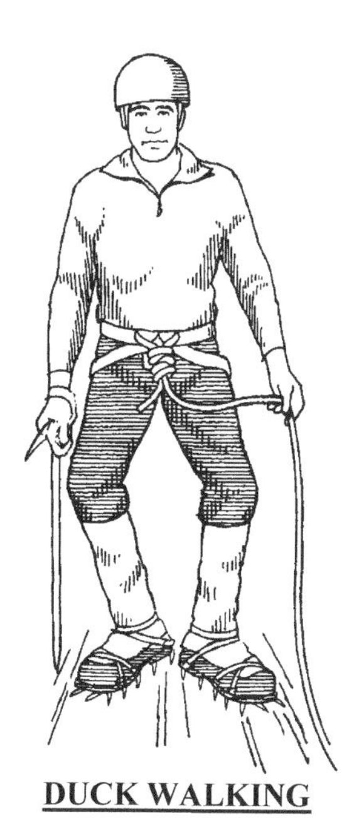

DUCK WALKING

c. Flat Footing. This is also known as *pied à plat.* This technique is performed by keep full sole contact to the surface on moderate to steep slopes of 30 to 65 degrees.

FLAT FOOTING

4. (10 Min) **THE GERMAN TECHNIQUE**. The German technique is much like step kicking straight up a snow slope, but instead of kicking your boot into the snow you kick your front crampon points into the ice and step directly up on them. The German technique is best performed on ice from 45 degrees to vertical. Factors to be considered are:

   a. The boot is placed rather than kicked because a sharp blow may result in a rebound vibration or possibly shatter the ice.

   b. The best placement of the crampon is straight into the ice; you need to avoid splaying your feet, as that tends to rotate the outside front points out of the ice.

   c. Boot heels need to be just above the horizontal. You need to resist the temptation to raise your heels higher, as this will only bring the rear two stabilizing points out of the ice, therefore endangering placement of the front points and tiring your calf muscles.

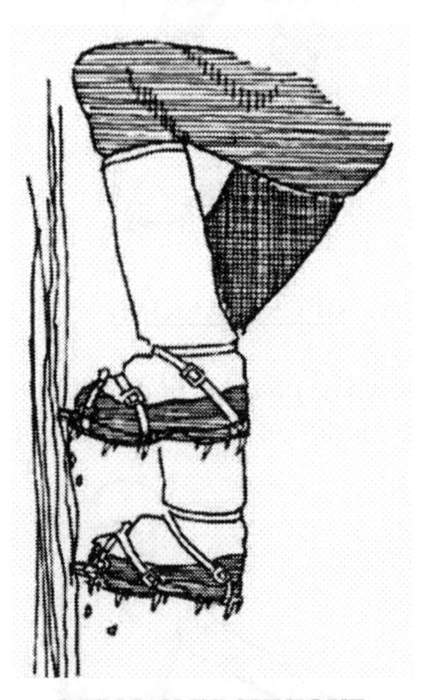

GERMAN TECHNIQUE

   d. Your heels will normally feel lower than what they are, if they do feel low then odds are that they are in fact in the correct position. This is especially important when coming over the top of steep ice onto gentler terrain, where the natural tendency is to raise your heels.

5. (5 Min) **THE AMERICAN TECHNIQUE**. This is a combination of the French and German techniques. One foot is placed flat on the snow/ice using all points, with the other foot only the front points are used. This technique can be used on snow/ice ranging from steep to extremely steep slopes of 55 degree +. It has the advantage of allowing the leg muscles to relax by alternating the techniques, and is also a useful rest position while front pointing.

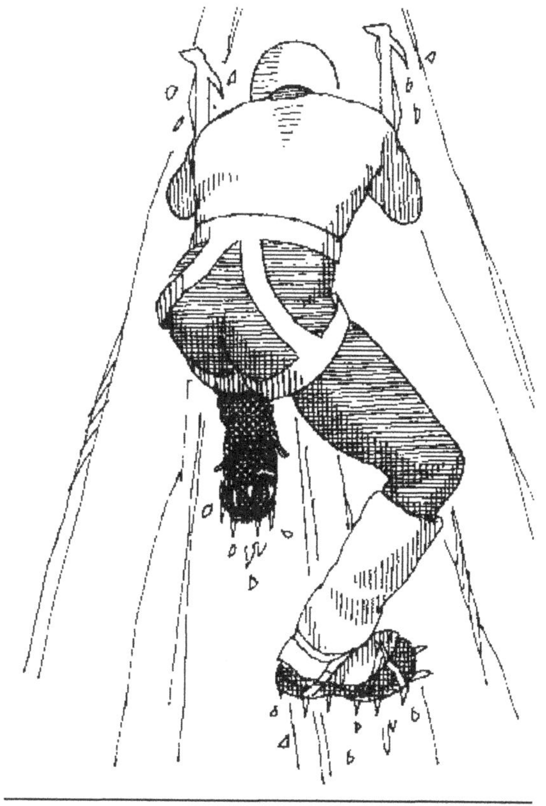

AMERICAN TECHNIQUE

TRANSITION: Now that we have looked at the various methods of ascending different angles of ice and the techniques required, are there any questions? We will now look at descending ice.

6. (5 Min) **THE DESCENT**. As with ascending a slope, the descent technique is determined mainly by the hardness of the surface and the slope angle. In soft snow on a moderate slope the plunge step can be used. On harder snow or as the slope angle increases, simply face outward and walk down.

   a. Plunge Step. The plunge step is a confident, aggressive move. Face outward; step assertively away from the slope and land solidly on your heel with your leg vertical, transferring weight solidly to the new position. Avoid leaning back into the slope, which can result in a glancing blow step, which can result in an unplanned glissade. When plunge stepping, keep the knees bent a bit to maintain control of balance.

   b. Flat Footing. To descend gentle sloping ice/hard snow, simply face directly downhill, bend the knees slightly, and walk downward ensuring that all crampon points are in contact with the surface. As the descent angle increases, bend your knees more and spread them apart, ensuring your body weight is center mast.

TRANSITION: Practice all of these methods on various snow/ice conditions and practice them in an area of safety, while practicing and at all other times, there must be nobody below a climber. These techniques take time and practice to master. Are there any questions? If there are none, then I have some for you.

PRACTICE (CONC)

a. Students will practice what was taught in upcoming field training exercises.

PROVIDE HELP (CONC)

a. The instructors will assist students when necessary.

OPPORTUNITY FOR QUESTIONS (3 Min)

1. QUESTIONS FROM THE CLASS

2. QUESTIONS TO THE CLASS

Q. What are the three techniques of snow/ice climbing when using crampons?

A. (1) The French Technique.
(2) The German Technique.
(3) The American Technique.

SUMMARY (2 Min)

a. This period of instruction covered the three different techniques for using crampons and the ways to employ an ice ax in conjunction with these techniques.

b. Those of you with IRF's please fill them in and turn them into the instructor. We will now take a short break.

UNITED STATES MARINE CORPS
Mountain Warfare Training Center
Bridgeport, California 93517-5001

SML
03/31/01

LESSON PLAN

GLISSADING AND ARREST POSITIONS

INTRODUCTION (5 Min)

1. **GAIN ATTENTION**. There is always two parts to a climb; ascending and descending. Ascending will always be the most time consuming and difficult of the two, while descending normally requires less time and energy especially when utilizing the glissading techniques. But if not cautious, descending from a climb could result in a slip, which may result in a serious injury or even death. Being able to arrest yourself or your rope partner will reduce the chances of serious injuries.

2. **OVERVIEW**. The purpose of this period of instruction is to introduce the student to the techniques used in glissading and arresting in all positions and with different types of equipment on. This lesson relates to glacier travel.

INSTRUCTOR NOTE: Have students read learning objectives.

3. **INTRODUCE LEARNING OBJECTIVES**

   a. TERMINAL LEARNING OBJECTIVES

      (1) In glaciated terrain, execute a glissade, in accordance with the references.

      (2) In glaciated terrain, execute a self-arrest, in accordance with the references.

   b. ENABLING LEARNING OBJECTIVES

      (1) In a mountainous environment, glissade on snow, in accordance with the references.

      (2) In glaciated terrain, execute a self-arrest without crampons on, in accordance with the references.

      (3) In glaciated terrain, execute a self-arrest with crampons on, in accordance with the references.

4. **METHOD / MEDIA**. The material in this lesson will be presented by lecture and demonstration method. You will practice what you have learned in upcoming field exercises. Those of you with IRF's please fill them out at the conclusion of this period of instruction.

5. **EVALUATION**. You will be tested by a performance evaluation.

TRANSITION: Are there any questions over the purpose, learning objectives, how the class will be taught, or how you will be evaluated? Let's begin this lesson with glissading.

BODY (50 Min)

1. (5 Min) **GLISSADING IN GENERAL**. The term comes from the French vocabulary meaning to slide. Glissading is a method used to quickly descend a slope with little use of energy. It should only be used on slopes where you can keep your speed under control. When preparing to glissade, the following considerations for equipment should be applied:

   a. Uniform. An outer water-resistant layer, such as Gore-Tex, should be worn and zipped up to prevent saturation of clothing. Material made of nylon also provides a good sliding surface. Ensure that the gaiters are worn on the outside of the trousers to prevent snow from entering.

   b. Axe. The axe is used as a rudder to help steer. Placing pressure on the axe's spike helps to control the climber's speed of descent. If the climber loses control of his speed, the pick of the axe will be used as in the self-arrest position.

   c. Harness. If a harness is worn, all gear on the rear loops should be removed to prevent them from becoming detached or interfering in the glissade.

   d. Pack. If a pack is worn, it should be inspected to ensure all equipment is properly attached and that all sharp objects are buffed or stowed inside.

   e. Crampons. Crampons will not be worn when conducting glissading drills. The points may get stuck into the snow and cause the glissader to trip and fall out of control.

TRANSITION: Are there any questions about glissading in general? If not, let us talk about the methods of glissading.

2. (10 Min) **METHODS OF GLISSADING**. There are three principal methods of glissading. The ideal snow for glissading is a firmly packed slope with a soft top layer of snow, such as a slope that has been warmed by the early afternoon sun. It should be reasonably uniform and free of protruding obstacles with a reasonable run out zone. Steep, icy slopes should be avoided due to the lack of control they allow. Before you glissade you must be able to observe the run out zone.

a. Standing Glissade. The standing glissade offers the earliest look at hazards of the route and is the most maneuverable. It also saves your clothes from wetness and abrasion. The standing technique is very similar to downhill skiing.

(1) Center yourself over your feet with bent knees and outspread arms for balance. The ice axe is carried in either hand so that it can be quickly brought into the self-arrest position.

(2) The feet can be placed together or spread as needed for stability. One foot can be advanced for improved stability.

(3) To increase speed, bring your feet closer together, toes tilted downward while leaning further down slope.

(4) Always maintain your speed of descent. The following are methods to slow down or stop:

(a) Stand all the way up and dig your heals into the snow.

(b) Turn your feet sideways using edging as with downhill skiing.

(c) Crouch down and drag the axe spike into the snow.

STANDING GLISSADE

b. Squatting Glissade. This method is much like the standing position except the climber holds the axe in the self-arrest position to one side of the body. The axe will act as a third point of contact to provide stability.

(1) Center yourself over your feet in a semi crouched position with both knees bent.

(2) The feet can be placed together or spread as needed for stability.

(3) To increase speed lean forward down the slope.

(4) To decrease in speed, drag the spike of the axe into the snow.

SQUATTING GLISSADE

c. Sitting Glissade. This method is the easiest method to learn and works on very soft snow where you might bog down using the other methods.

(1) The glissader will sit on the snow surface with his legs flat on the snow and his heels lifted off the slope.

(2) The ice axe is grasped with one hand on the head with the pick outboard and the other hand is on the shaft. The hand that is on the shaft should be locked into your side at the hip.

(3) To begin the glissade, push yourself down the slope either by throwing your weight forward or by pushing off with the ice axe.

SITTING GLISSADE

(4) As you descend the slope, use the spike of the ice axe as a rudder to control your direction.

(5) Downward pressure applied near the axe's spike will control the speed of your descent.

TRANSITION: Are there any questions over glissading? If not, we will now discuss the arrest of a fallen rope partner.

3. (10 Min) **ARRESTING A FALLEN ROPE PARTNER**. The arrest technique serves as a method to brace yourself solidly in the snow if you have to hold the fall of a roped partner. The most important factor when performing an arrest is to act quickly. When performing an arrest, the following body positions must be applied:

   a. Hands. One hand will grasp the head of the axe with the thumb under the adze and the fingers over the pick. The other hand will be placed on the shaft just above the spike.

      (1) Press the pick into the snow just above the shoulder so that the adze is near the angle formed by the neck and shoulder.

      (2) The shaft will cross the chest diagonally with the spike of the axe near the hip. Pulling upward on the shaft will sink the pick firmly into the snow.

   b. Chest and Shoulders. These parts of the body should be pressed firmly against the axe's shaft.

   c. Back. The spine should be arched slightly away from the surface. This will assist in sinking the pick into the snow.

   d. Knees. The knees should be pressed against the snow to help stabilize the body position.

   e. Feet. Dig your feet into the snow approximately shoulder width apart to provide added support.

TRANSITION: Are there any questions over arresting a fallen climbing partner? Let us now discuss self-arresting.

4. (15 Min) **SELF-ARREST POSITIONS**. Self-arrest is the life saving technique of using the ice axe to stop your own uncontrolled slide down a slope. As with arresting a fallen rope partner, quick reaction by setting the pick into the snow will prevent you from accelerating to an unstoppable speed. Your body position during the slide may be in one of four positions: head uphill or head downhill and, in either case, face down or on your back.

   a. Feet First, Face Down. This is the desired position for self-arrest.

      (1) Position the axe diagonally beneath the chest with the spike of the axe near the hip.

      (2) Arch the back while pulling upward on the shaft. This will place weight toward the shoulder by the axe head.

      (3) Press the knees into the slope to help slow the fall. This will also assist in stabilizing the body.

(4) If not wearing crampons, keep the legs stiff and spread apart with the toes digging in. If wearing crampons, keep the legs bent at the knees with your feet in the air. As your momentum slows down, dig your feet into the snow aggressively.

b. <u>Feet First, On Back</u>. Recovery from this slide is decided on which side the head of the axe is located.

(1) Roll toward the head of the axe and insert the pick into the snow aggressively. If the head of the axe is on the left hand side, roll to the left. If the head of the axe is on the right hand side, roll to the right.

(2) Position your body over the axe with the shaft diagonally beneath the chest with the spike of the axe near the hip.

(3) Arch the back while pulling upward on the shaft.

(4) Press the knees into the slope to help slow the fall.

(5) The feet placement are the same as discussed earlier.

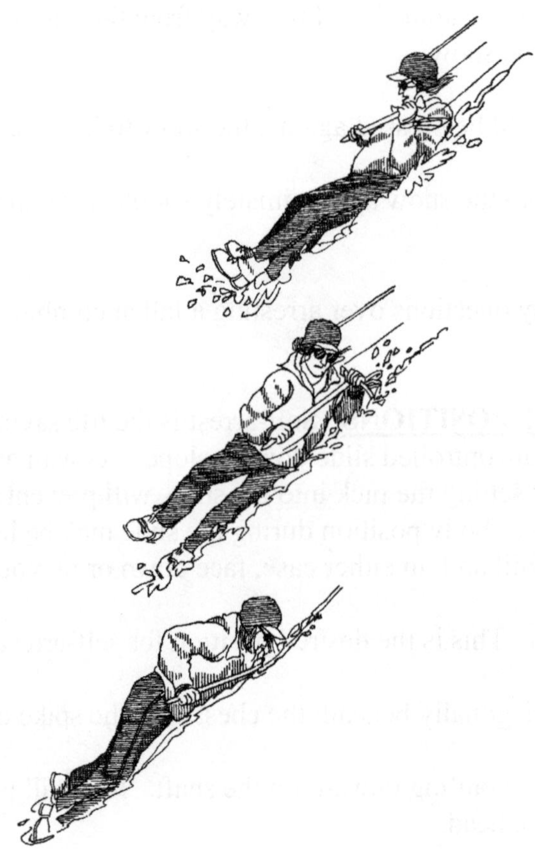

<u>FEET FIRST, ON BACK SELF ARREST</u>

c. Head First, Face Down. This is a more difficult slide to recover from because the feet need to be swung downhill.

(1) Arch the back upward while reaching downhill and off to the axe head side.

(2) Aggressively insert the pick of the axe into the snow to serve as a pivot to swing the body around.

(3) Swing the legs around so that they are pointed downhill.

(4) Position your body over the axe with the shaft diagonally beneath the chest with the spike of the axe near the hip.

(5) Arch the back while pulling upward on the shaft.

(6) Press the knees into the slope to help slow the fall.

(7) The feet placement are the same as discussed earlier.

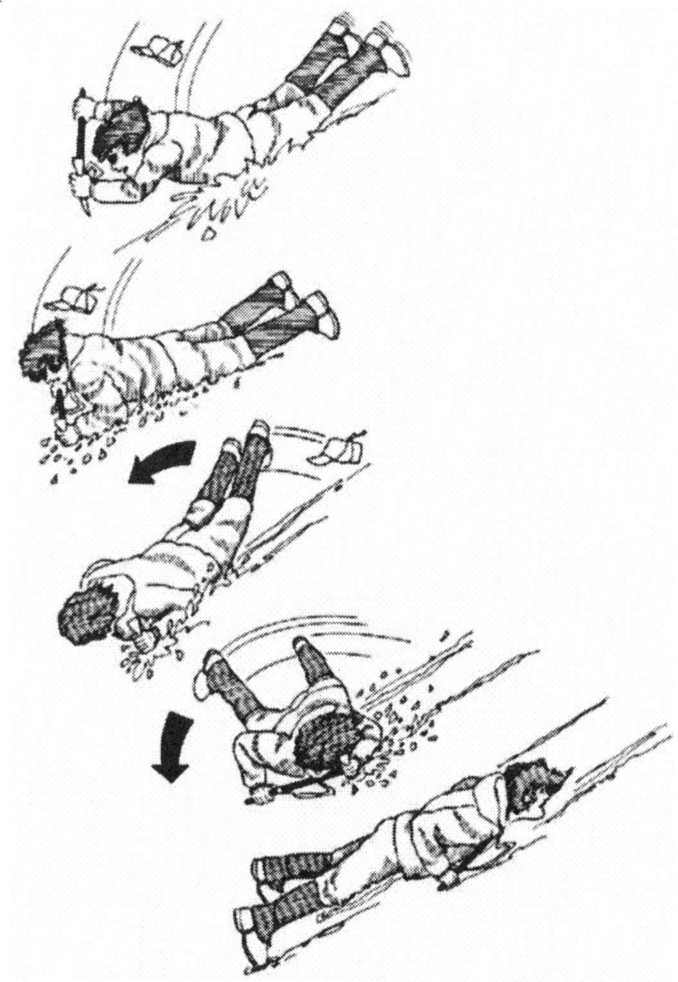

HEAD FIRST, FACE DOWN SELF ARREST

d. <u>Head First, On Back</u>. Not being able to see possible obstacles in the slide makes this a dangerous situation.

(1) Hold the axe across the torso and slide the pick into the snow. This will serve as a pivot point. A sitting-up motion will assist you in this maneuver.

(2) Twist and roll your chest toward the axe head while swinging your legs around so that they are pointed downhill.

(3) Position your body over the axe with the shaft diagonally beneath the chest with the spike of the axe near the hip.

(4) Arch the back while pulling upward on the shaft.

(5) Press the knees into the slope to help slow the fall.

(6) The feet placement are the same as discussed earlier.

HEAD FIRST, ON BACK SELF ARREST

TRANSITION: Now that we have discussed how to self-arrest, are there any questions? Self-arrest is one of the most important techniques you can learn while working in a glaciated terrain. If there are no questions for me, then I have some for you.

PRACTICE (CONC)

a. Students will practice glissading techniques and the different methods of arresting.

PROVIDE HELP (CONC)

a. The instructors will assist the students when necessary.

OPPORTUNITY FOR QUESTIONS (3 Min)

1. QUESTIONS FROM THE CLASS

2. QUESTIONS TO THE CLASS

Q. What are the two types of arrests?

A. (1) Arrest of a fallen rope partner
(2) Self-arrest

Q. What are the three methods of glissading?

A. (1) Standing
(2) Squatting
(3) Sitting

Q. Should crampons be worn while glissading?

A. No.

SUMMARY (2 Min)

a. This lesson has covered the procedures for glissading and for arresting.

b. Those of you with IRF's please fill them out at this time and turn them in to the instructor. We will now take a short break.

UNITED STATES MARINE CORPS
Mountain Warfare Training Center
Bridgeport, California 93517-5001

SML
03/04/02

LESSON PLAN

GLACIER TRAVEL

INTRODUCTION

1. **GAIN ATTENTION**. Glacier travel is an important aspect of Alpine mountaineering. The most serious part of an Alpine climb is often the approach up a glacier or the descent at the end of the day. There is nothing one can do to completely eliminate dangers in the mountain no matter how much experience or knowledge one may possess. However, an understanding of types of hazards likely to be encountered when traveling in the mountains makes it possible to minimize the dangers and improve the chances of success.

2. **OVERVIEW**. The purpose of this period of instruction is to introduce the student to glacier travel, including how to rope up and how to perform crevasse rescues. This lesson relates to all other lessons on snow and ice work.

3. **INTRODUCE LEARNING OBJECTIVES**

   a. TERMINAL LEARNING OBJECTIVE: On glaciated terrain, move over the glacier, in accordance with the references.

   b. ENABLING LEARNING OBJECTIVES:

      (1) In glaciated terrain, rope up for glacier travel, in accordance with the references.

      (2) In glaciated terrain, pre-rig for glacier travel, in accordance with the references.

      (3) Without the aid of references, perform crevasse rescue, in accordance with the references.

      (4) Without the aid of references, in glaciated terrain, perform a self-rescue from a crevasse using the Texas kick method, in accordance with the references.

4. **METHOD/MEDIA**. The material in this lesson will be presented by lecture and demonstration method. You will practice what you have learned in the upcoming field training exercises. Those of you with IRA's please fill them out at the end of this period of instruction.

5. **EVALUATION**. You will be tested later by a performance evaluation.

TRANSITION: Are there any questions over the purpose, learning objectives, how the class will be taught, or how you will be evaluated? Lets move on to why this information is important.

BODY (60 Min)

1. (5 Min) **UNDERSTANDING THE MOUNTAIN.** The first and most important fact a climber must realize is that mountains undergo constant and relentless change. The process of mountain building and decay is wholly dynamic and very violent. The climber is himself is in the midst of this change. At issue is the question of control; that is, how much control the climber can exert over his situation. An individual has little or no control over the weather or snow conditions that influence avalanche activity. He cannot know which rocks may loosen through the effects of frost wedging and hurtle down upon him. On the other hand, a climber aware of the forces at play around him may be able to establish a pattern of events that helps him predict when the danger is most serious and when travel may be conducted in relative safety. Only experience and study can produce the judgment required to intelligently minimize the risks. The following are hazard considerations:

   a. Objective Hazards. The Mountain environment:

      (1) The natural processes that exist whether humans are involved or not.

      (2) Darkness, storms, lightning, cold, precipitation, high altitude, avalanches, rock falls, cornices, ice fall, crevasses, wind, fog and whiteouts are all examples of objective hazards.

   b. Subjective Hazards. The climber:

      (1) The human factor shares the blame for many accidents.

      (2) Knowledge. A climber with inadequate knowledge can mean danger to a climbing party.

      (3) Skill. The climber's skill level should match the difficulty of the climb.

      (4) Judgment. Good judgment is the quality of using knowledge gained from study and experience to make sound decisions.

TRANSITION: Are there any questions over the hazards? Now let's look at some equipment needed for glacier travel.

2. (5 MIN) **EQUIPMENT NEEDED FOR GLACIER TRAVEL**. Like climbing, glacier travel

can be easy or complicated and difficult to negotiate. However, there is no system of categorizing the difficulty of glacier travel. In the majority of the cases a fall into a crevasse is only up to the chest level and extraction is a simple matter. In some instances though, the fallen climber may lose contact with his companions on the surface. The following are essential equipment necessary to rescue a fallen climber:

a. Helmet. For protection of the head area.

b. Harness. Ensure that the harness can be adjusted to fit over several layers of clothing.

c. Crampons. To gain appropriate traction and purchase upon the glacier.

d. Ice Axe. For possible arresting of self or climbing partner(s).

e. Snow / Ice Protection. To establish a necessary anchor during a crevasse rescue, if required.

f. Rope. Since crevasse falls are usually not free falls, a 9mm diameter rope of at least 150 feet should be used.

g. Type II cordage. Two separate lengths of cordage will be utilized:

(1) A three foot piece of cordage constructed into an endless loop using a double fisher man knot. This is known as a body prusik.

(2) An eighteen foot piece of cordage constructed into a Texas Kick in the following manner:

(a) Locate the middle of the cordage and tie a figure eight loop with the loop's diameter approximately eight inches. The two strands of cordage created from an overhand knot are known as "Legs".

(b) Tie a figure of eight loop at the end of each leg.

<u>TRANSITION</u>: Now that we have talked about the equipment needed to travel thorough glaciated terrain, are there any questions? We will now discuss roping up techniques.

3. (10 Min) **<u>ROPING-UP TECHNIQUE</u>**. In glacier travel, a minimum of three climbers on the rope will be the standard. Three climbers traveling together will assure that two climbers should be able to hold a fall and anchor the rope. No more than five climbers should be on one rope's length, although a larger party provides more person-power holding ability, it also puts them closer together. The basis in safety lies in keeping the rope slack free between the climbers. Each climber on a rope should be spaced evenly apart, however this will depend on the terrain and the distance between crevasses. The method of tying the climbers into the rope is known as the Kiwi Coil method. The following is an example of three on a rope:

a. With the end of the rope, the first and third climber will tie into the rope in the same method as for two party climbing.

b. The climbers will then place a locking carabiner into their harness's hard point.

c. The second climber will locate the middle of the rope and tie a three foot over hand knot. With the bight he has now created, tie a figure of eight loop. He will clip the loop into his harness's hard point using a locking carabiner. He is now secured to the rope.

d. The first and second climbers will begin to wrap coils from over the shoulder to under the arm ensuring that the coils reach just to the bottom of their rib cages.

NOTE: If wearing a pack, the first coil will be attached to the pack's hard point with a carabiner. The hard point on the pack is constructed with the use of a web runner woven around the pack straps.

e. Again depending upon the situation, they will continue to wrap the coils until they are within 24 to 30 feet away from the second climber.

f. After the desired distance has been reached, the climbers will reach under the coils and pull through a bight of the live rope running to the middle climber. The length of this bight is important, as it should just reach the locking carabiner on the hard point.

g. Using the bight, tie an overhand knot around the coils ensuring to include the rope from the harness.

h. The climbers will clip this bight into the locking carabiner on the harness's hard point and lock it down.

i. The climber should now test pull the live end of the rope to ensure it does not cinch around the climber's shoulder.

NOTE: If the rope cinches around the climber's neck/shoulder area, the overhand needs to be reconstructed.

j. All climbers are now secured into the rope.

NOTE: If more than three climbers are on the rope, all climbers will be evenly spaced ensuring that the first and last climber's have at least 15 feet of coil wrap around them for construction of the anchor.

KIWI COIL

TRANSITION: Are there any questions over roping up techniques? If not, let us discuss pre-rigging for glacier travel.

4. (5 Min) **PRE-RIGGING FOR GLACIER TRAVEL**. In the event of a climber falling through a crevasse, speed is essential to rescue him. Pre-rigging gives the climbers this advantage. The first and last climber will:

   a. From the retrace figure of eight on his harness, attach the Texas Kick cordage to the live rope by tying a middle of the rope prusik using the loop of the figure of eight. Place the legs of the Texas Kick in a convenient pocket.

   b. Directly in front of the Texas Kick, tie a middle of the rope prusik using the body prusik cordage. Attach this cordage to the harness's hard point using another locking carabiner.

TRANSITION: Pre-rigging is an important ingredient for speed of a rescue, are there any questions? If not, let's begin talking about movement techniques.

5. (10 Min) **MOVEMENT TECHNIQUES**. Route finding on a glacier involves finding a path around or over crevasses safeguarding against possible falls at all times. The crossing is seldom accomplished without its detours. The following are techniques used to safely cross a field of crevasses:

   a. End Run. Crossing over a crevasse is seldom the preferred choice. The safest and most dependable technique is to go around it as in the end run.

      (1) The leader will make a wide swing around the corner while probing carefully for any hidden crevasses.

      (2) Be aware of adjacent crevasses, they may be just an extension of the crevasse your maneuvering around.

   b. Echelon Formation. If the route demands travel that is parallel to the crevasses, the echelon formation is preferred technique to maneuver safely. Do not use this technique where hidden crevasses are likely.

(1) The leader will choose a lane parallel to the crevasses and travel directly down it.

(2) The rest of the rope team will take up lanes also parallel to the crevasses and travel behind and offset from the leader.

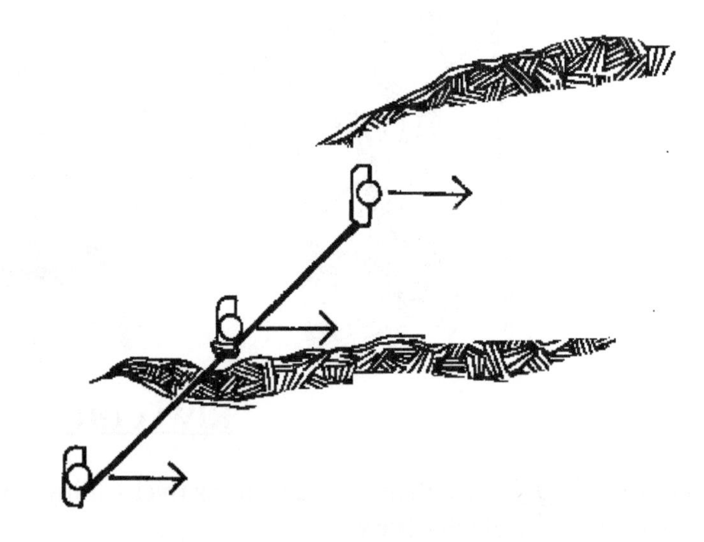

ECHELON FORMATION

c. Snow Bridges. If an end run is impractical, the next technique is to cross a snow bridge. The strength of a snow bridge will vary depending on the temperature. Do not assume a bridge crossed early in the morning is safe to cross during the afternoon.

(1) The leader will visually study the bridge for all angles before putting any faith in it.

(2) If there is doubt, the leader can probe the bridge with his walking axe while the rest of the rope team stay braced on a taut rope to help safeguard against a possible breakthrough.

(3) After the leader crosses the bridge, the rest of the rope team will follow exactly in the leader's footsteps.

d. Jumping. Most jumps across crevasses are short, simple leaps. If planning a larger jump, ensure that all other alternatives have been ruled out.

(1) The leader will determine the true edge of the crevasse by performing the probing method as with snow bridges.

(2) The leader may want to trample down the snow near the crevasse's edge to create a better jumping platform.

(3) The leader should make final equipment preparations, i.e. adjust gear, clothing, check prusik cords and harness.

(4) The leader will determine the amount of rope required to clear the crevasse and back stack it near the next climber in the rope team.

(5) All the other climbers will assume the arrest position.

(6) The leader will jump the crevasse with his axe in the self-arrest position.

(7) Upon clearing the crevasse, the leader will assume the arrest position slightly offset so that the rest of the rope team doesn't land on him.

(8) The remaining members of the rope team will adjust their position to allow enough rope for the next climber to jump.

(9) Repeat this process until all members of the rope successfully complete the jump.

JUMPING A CREVASSE

e. Crossing a Bergshrund.

(1) Moving over a Bergschrund. This may require a belay stance to ensure safety of the climbers. The lead climber will climb the route as in a party climb.

(2) Moving down a Bergshrund. A rope may have to be deployed as a rappel lane. Establish the anchor as taught in *ESTABLISHING ANCHORS ON SNOW AND ICE.*

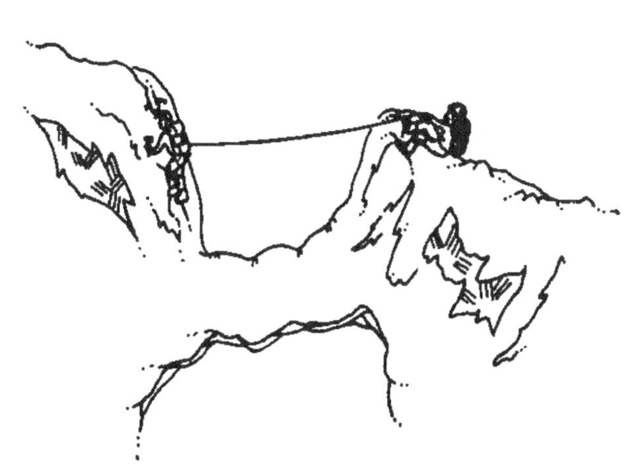

CROSSING BERGSCHRUND

TRANSITION: Now that we have discussed movement techniques, are there any questions? If not, let us talk about crevasse rescuing.

6. (15 Min) **CREVASSE RESCUE**. Even when a rope team takes all the necessary precaution when traveling through a crevassed glacier field, the chance of a fall through a crevasse still

INITIAL REACTION FOR STOPPING A CREVASSE FALL

exists. The following methods for rescue are designed for a three man rope team assuming that the lead climber is the victim. Keep in mind that these methods can also be used for any member of the rope team who has fallen through or for even larger climbing parties, with some adjustment to the organization. Regardless of the situation, each member of the rope team must react quickly to assure of a safe rescue.

a. Anchoring a Semi-Submerged Climber. The most common fall into a crevasse is when the climber punches through the snow layer up to the waist or chest. If this happens, the rest of the rope team will immediately assume the arrest position. There are two methods of rescuing the fallen climber:

(1) The fallen climber will perform self rescue by pull himself out using his axe.

(2) The other members of the rope team will provide assistance to him by establishing a sound anchor and belay the fallen climber to safety.

b. Anchoring a Totally Submerged Climber. This happens when a climber disappears beneath the snow. Again, the rest of the rope team will assume the arrest position to stop and hold the fallen climber. Once the fallen climber has been stabilized, the following actions will be taken:

(1) The #2 (middle) man on the rope team takes the entire load onto his arrest position. This is accomplished by the #3 (last) man easing off his arrest position slowly assuring not to shock load the #2 man's arrest position.

(2) The #3 man will move down toward the #2 man's position while simultaneously trying to establish communication with the fallen climber to determine his condition.

(3) Once at the #2 position, he will place an anchor (as taught in *ESTABLISHING ANCHORS ON SNOW AND ICE*) offset and forward of the #2.

(4) The #3 will now attach a knotted runner with a single knot in the middle to the anchor.

(5) The #3 will attach the #2's short prusik to the end of the knotted runner.

(6) The #3 will now back up the anchor by placing a second anchor no closer than 18 inches from the initial anchor. He can equalize the system by either:

(a) Place the second anchor behind and in line with the first anchor. Clip a nonlocking carabiner into both anchors. Connect the anchors together using a long runner or piece of cordage. If the first anchor is a T – trench or fluke an extra runner must be added to it in order to reach the rear anchor. Attach the runner to the first anchor carabiner, then attach the runner to the second anchor carabiner using a round turn and two half hitches to tension and secure.

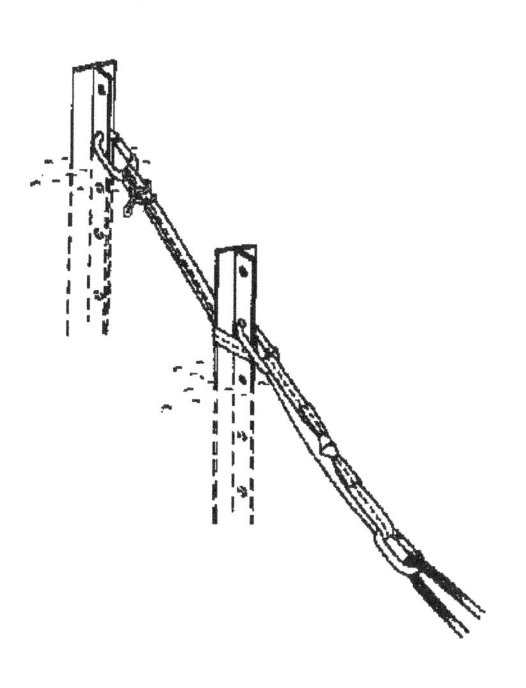

EQUALIZING INLINE ANCHORS

(b) Place the second anchor on line with the primary anchor. Attach a carabiner to each anchor. Clip a long runner or piece of cordage into both anchors. Equalize in the direction of pull and tie an overhand or figure of eight knot. Clip the #2 man's short prusik into it.

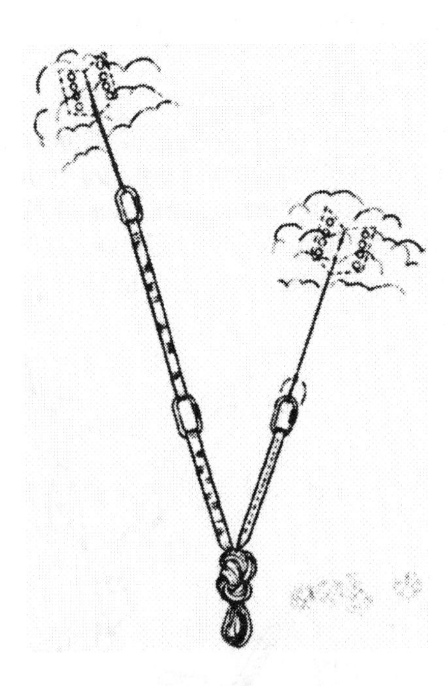

EQUALIZING SIDE BY SIDE ANCHORS

(7) The #3 will place a locking carabiner onto the end of the knotted runner. Then grabbing the rope above where the #2 is tied in, attach it to the locking carabiner with a Munter hitch.

(8) The #2 will slowly ease off the loaded rope, ensuring not to shock load the anchors. He will then anchor himself by his axe or clip into one of the anchors by some means.

(9) The #2 will unclip from the rope and untie the figure of eight loop and the overhand knot. Meanwhile, the #3 will take in the slack. When the system is taut, the Munter hitch is tied off with a round turn and two half hitches.

NOTE: The Munter hitch is used in case that once the fallen climber is checked on, it might be preferred to just lower him for ease of another passage out of the crevasse, i.e. a shallow depth crevasse, possible ledge, etc…

(10) Next, the #2 and #3 switch jobs. The #2 sits at the anchor and the #3 attaches his short prusik cord to the rope below the anchor and toward the crevasse with a six finger prusik knot and connect it to the hard point of his harness with a locking carabiner. This will serve as his safety prusik.

(11) The #2 can body belay the #3. The #3 will now move toward the crevasse with his axe

and/or pack. The #3 should probe his path enroute to the edge of the crevasse to ensure there are no other hidden dangers.

(12) Once at the edge of the crevasse, the #3 climber will check on the condition of the fallen climber. If the fallen climber is unconscious, the #3 will ensure that he is not entangled in the rope, which may be strangling him, submerged in water and/or in immediate need of first aid. If so, a rappel line into the crevasse may be needed. This rappel line will be a separate rope.

(13) If immediate first aid isn't required, the #3 will prepare the lip of the crevasse by padding it to minimize further entrenchment of the rope. This can be accomplished by sliding the axe shaft, pack or other piece of equipment under he rope. Be careful to not cut the rope on sharp edges and to anchor the equipment to avoid it falling onto the victim and into the crevasse.

c. Team Rescue. If the fallen climber is unconscious or is unable to rescue himself and immediate application of first aid isn't necessary then the rope team must construct a mechanical advantage system or use brute force to haul him up.

(1) Z-Pulley System (3:1 Mechanical Advantage).

(a) Once the #3 has padded the lip of the crevasse, he will move back to the anchor points and unclip from the safety prusik. The #3 is still attached to the end of the rope that is attached to the anchors. However, he should be careful. Any fall will be long because of the slack. Use the ice axe for safety and self belay.

(b) The #2 will take the #3 off of the body belay. The rope from the Munter hitch will be clipped into the safety prusik carabiner.

(c) The #3 will then cycle the prusik knot as far forward as the rope will allow.

(d) The #2 will untie the Munter hitch ensuring that the rope now runs freely through the carabiner. Attempt to do this without completely unclipping the rope from the carabiner.

(e) Both climbers will now attempt to pull the victim out of the crevasse with the #2 man attending to the safety prusik as well.

(f) If necessary, recycle the system as needed, ensuring that the safety brake is set before doing so.

NOTE: A pulleys placed on the carabiners that the rope runs over can reduce friction and ease the hoisting the victim.

(2) Z-Z Pulley System (9:1 Mechanical Advantage). This system is constructed in the same manner as taught in *ONE ROPE BRIDGE* except for the Munter hitch

attachment.

(3) Muscle Method. This method requires nothing more than brute force to pull the fallen climber to safety. The decision to use this method will depend on the situation i.e. fallen climber is conscious and able to assist in his extraction, the victim is in immediate need of medical care, another team in the area to assist, etc…

(a) If the anchor is already in place, the #2 and #3 climber can attach themselves to the live rope with a prusik knot clipped into their harness's hard point.

(b) Both climbers will face into the opposite direction of pull and "walk" the victim out of the crevasse.

NOTE: This method can also be used before setting up an anchor but the victim should be conscious to ensure that you are not causing increased harm to him. Also be aware of the potential hazard of the rope digging into the lip of the crevasse.

d. Self-Rescue. The fallen climber performs this method. He will ascend out of the crevasse by using the cordage already pre-rigged from his harness to the rope. This method is also referred to as the Texas Kick Method.

(a) The fallen climber will untie the overhand knot tied around the coils.

(b) He will now remove the coils from around his neck and drop them into the crevasse ensuring not to get the rope entangled.

(c) If he is wearing a pack and it becomes cumbersome, he will ensure that it is attached to the rope and also drop it down the crevasse.

(d) As soon as he is able he should tie in short. Do this by grabbing the rope well below short prusik and tying a figure of eight knot. Clip that into your hard point of your harness. If the prusiks fail this knot will keep you from going to the end of the rope or to the bottom of the crevasse.

(e) The fallen climber will slide his body prusik up until it is taut.

(f) He will now place the Texas Kick on the rope below the body prusik if he did not already do so. The climber may have to do this before dropping the coils and pack in order to release tension on the knot holding the coils.

(g) The fallen climber will place his feet into the leg loops of the Texas Kick.

(h) Hanging from the body prusik, he will now push the Texas Kick up the rope until it meets the body prusik.

(i) The fallen climber will stand up onto the Texas Kick, grabbing the rope for stability if needed.

(j) Again he will push the body prusik up until it is taut and repeat the same with the Texas Kick until he reaches the top of the crevasse. Every 15 feet or so he should tie in short. Do this by grabbing the rope below the Texas Kick and tying a figure of eight knot. Clip that into your hard point of your harness.

(k) Once he reaches the lip, he may have to use his axe in the high dagger or traction method (as taught in *ICE AXE TECHNIQUES*) to gain a perch to get out of the crevasse.

NOTE: If the #2 climber happens to be the one to fall into a crevasse, he may be suspended between the two walls of the crevasse due to the rope being taut by the first and last members of the rope team. Normally, he will be lowered toward the wall in the direction of travel. For this to be accomplished, the last man needs to carefully move toward the victim. Constant communication must take place between the climbers to ensure that this maneuver is done safely.

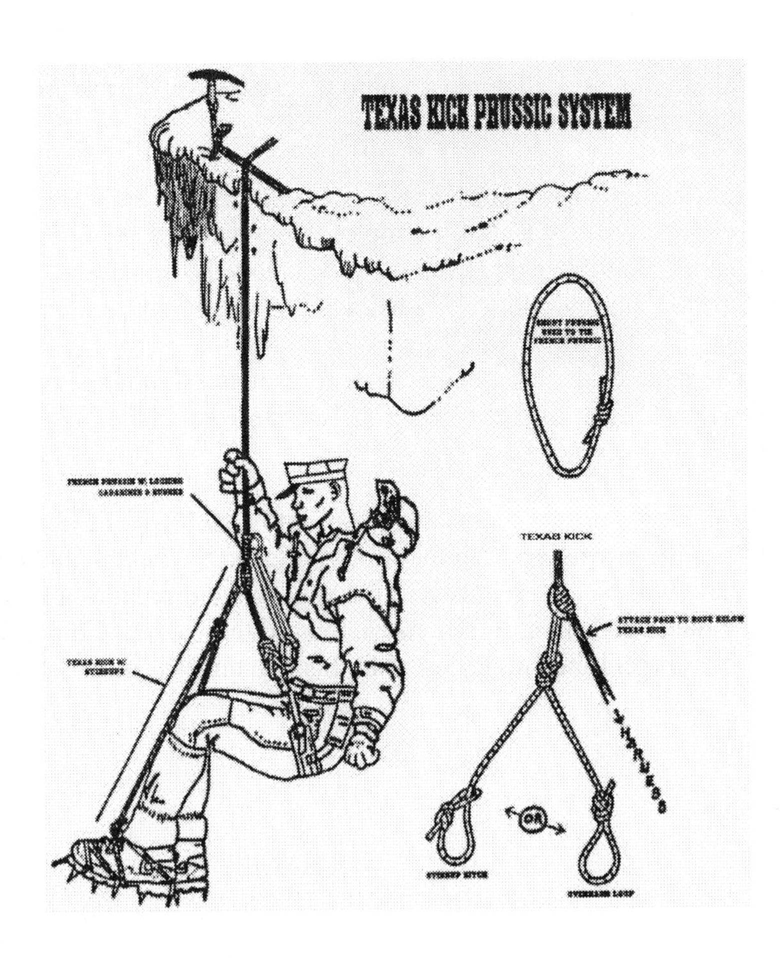

TEXAS KICK METHOD

TRANSITION: Now that we have talked about the crevasse rescuing, are there any questions? Now let's talk about basic safety rules.

7. (5 Min) **BASIC SAFETY RULES**. There are four basic safety rules for glacier travel. These are as follows:

   a. Always rope up with one or more climbers.

   b. The climbers must have the knowledge to be able to anchor the rope after their partner falls into a crevasse.

   c. The fallen climber must have the knowledge, and the means available to get himself out of a crevasse.

   d. The climbers must know how to extricate an injured or unconscious victim.

TRANSITION: Are there any questions over rescue techniques? Movement across glaciers is a dangerous undertaking, but it need not be deadly. With just a few precautions and sound training we can greatly reduce the chances of being another statistic. If you have no questions for me, then I have some for you.

PRACTICE (CONC)

a. Students will travel across a glacier.

PROVIDE HELP (CONC)

a. The instructors will assist students when necessary.

OPPORTUNITY FOR QUESTIONS (3 Min)

QUESTIONS FROM THE CLASS

QUESTIONS TO THE CLASS

Q. What are the two types of hazards encountered on a glacier?

A. Objective and Subjective.

Q. What are the five basic safety rules for glacier travel?

A. (1) Always rope up with one or more climbers.
(2) The climbers must be able to anchor the rope after the fall.
(3) The falling climber must be prepared to get himself out.
(4) The belayer must know how to extricate an unconscious victim.

Q. Prior to a rescue should you buff the edge of the crevasse?

A. Yes, to keep the rope from cutting into the lip of the crevasse causing unnecessary friction.

SUMMARY (2 Min)

a. During this period of instruction we've covered the types of hazards encountered on a glacier, the basic safety rules, method of roping up, actual movement on a glacier, and procedures for rescue of a fall victim.

b. Those of you with IRF's please fill them out at this time and pass them to the instructor, we will now take a short break.

UNITED STATES MARINE CORPS
Mountain Warfare Training Center
Bridgeport, California 93517-5001

SML
2/21/02

LESSON PLAN

SPECIFIC SNOW AND ICE EQUIPMENT

INTRODUCTION

(5 Min)

1. **GAIN ATTENTION.** The decisive winner in combat in the mountains has been the force that occupies the high ground. In order to occupy the high ground you must of course get there, no matter what the conditions. As a mountain leader it is essential for you to know how to get over any type of terrain. In this class we will discuss the equipment necessary to move over snow/ice.

2. **OVERVIEW.** The purpose of this period of instruction is to introduce the students to those pieces of equipment that are specifically designed to be used in a snow or ice covered environment. This lesson relates to glacier travel.

INSTRUCTOR NOTE: Have students read learning objectives.

3. **INTRODUCE LEARNING OBJECTIVES**

    a. TERMINAL LEARNING OBJECTIVE

    (1) In glaciated terrain, maintain snow equipment, in accordance with the references.

    (2) In glaciated terrain, maintain ice equipment, in accordance with the references.

    b. ENABLING LEARNING OBJECTIVES

    (1) Without the aid of references and given a diagram of an ice ax, label the parts of the ice ax, in accordance with the references.

    (2) Without the aid of references, execute proper care of snow equipment, in accordance with the references.

    (3) Without the aid of references, execute proper care of ice equipment, in accordance with the references.

4. **METHOD/MEDIA.** The material in this lesson will be presented by lecture and demonstration method. You will practice what you have learned in upcoming field training exercises. Those of you with IRF's please fill them out at the conclusion of this period of instruction.

5. **EVALUATION.** You will be tested by a written examination.

TRANSITION: Are there any questions over the purpose, learning objectives, how the class will be taught, or how you will be evaluated? The first item we will discuss is the ice ax.

BODY (50 Min)

1. (10 Min) **ICE AXES**. There are different types of axes for different types of climbing. The longer walking axe is the preferred tool when alpine climbing or for moving across glaciated terrain. The shorter technical axe, which has a "hooked in" shaft, is ideal for vertical ice. Regardless of the type of axe the nomenclature will remain the same:

   a. Nomenclature

      (1) Pick. The pick is curved so that it follows the natural swing of the arm when in use on steep ice. The teeth on the underside prevent it from slipping out of the ice.

      (2) Adze. This is the blade which forms a right angle to the shaft, it is used in crusty snow and for cutting steps.

      (3) Head. This is the top part of the axe that holds the adze, and pick to the shaft. Some axes have an eyelet where the leash may be attached.

      (4) Shaft. This is the handle of the ax; length varies from 40 to 90 cm. The shaft is oval in shape, which enables the climber to achieve a solid grip, and aids in directing the ax when placing it in the ice. The shaft should be rough to ensure that a good grip is maintained, if necessary you can tape the shaft with surgical tape. It is constructed from either wood, metal or fiber glass.

      (5) Ferrule. This is the point just above the spike, on some axes it is a separate band of metal.

      (6) Spike. This is the point of the axe located at the base.

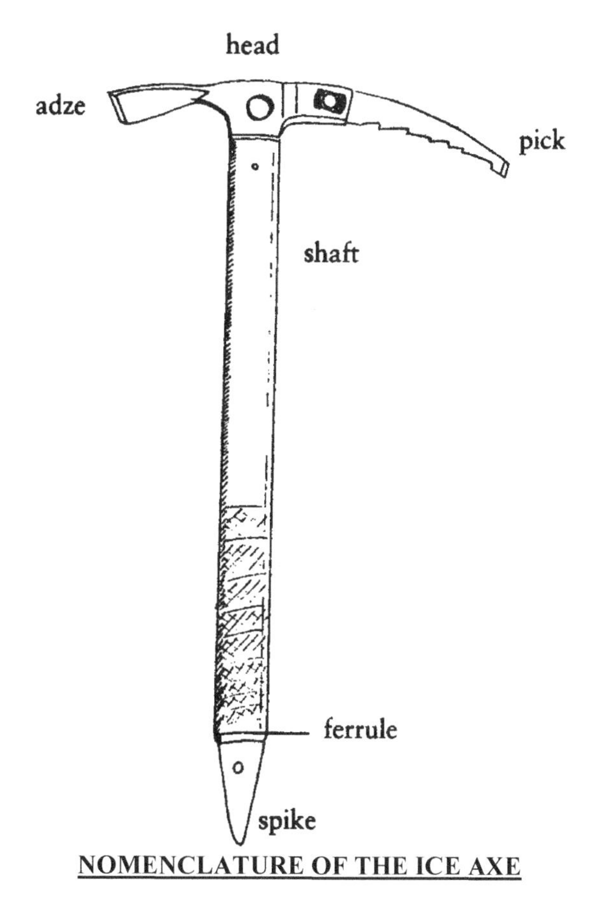

NOMENCLATURE OF THE ICE AXE

b. Ice Axe Leashes. The purpose of ice axe leashes is to act as a dummy cord or to provide support to a climber who is negotiating high angle ice.

(1) A long sling (also known as an umbilical cord) can be attached from the head of the axe to a climber's harness to act as a dummy cord. This is designed for easier/faster exchange of the axe between hands while traversing.

(2) A wrist loop can be tied to the eyelet of the axe using one inch nylon tubular webbing. With a wrist loop the climber does not have to grip the shaft as tightly, thus reducing the chance of cutting off circulation which may result in frostbite. To provide a better stability, the wrist loop can be tied to the shaft or even wrapped around it.

(3) A sliding wrist loop attaches to the shaft of the axe with a metal ring and stopper screw. The metal ring allows you to move it up/down the length of the shaft thus enabling you to have access to quick adjustment if necessary. The stopper screw in the shaft prevents the metal ring from becoming detached from the axe.

(4) There are holsters available for the ice axe, however these should only be used by experienced climbers.

c. Care and Maintenance of the Ice Axe

(1) Check the shaft for splits or cracks and apply linseed oil to wooden shafts.

(2) Ensure the shaft is rough so that a good grip can be maintained.

(3) Inspect the leash and its knots for cuts or abrasions.

(4) Ensure that the pick and adze are sharp.

NOTE: Do not use electrical grinders to sharpen the pick or the adze as they may remove the temper of the metal and greatly weaken the tool. The pick should be sharpened from the sides toward the tip with a file held at 20 degrees. The adze should be sharpened from the underside only. The spike should never be sharpened.

(5) Rubber protectors should be placed on the pick and the spike when not in use to prevent unnecessary damage.

d. Ice Hammers. These tools are the same as an ice axe except that a hammer replaces the adze. They are used to insert snow pickets and ice pitons. They again differ in length depending on their intended use.

TRANSITION: Now that we have discussed the ice ax, are there any questions? Let's now discuss crampons.

2. (10 Min) **CRAMPONS**. Crampons are designed as an attachment to the boots to provide traction while traveling on crusted snow and/or ice. Each crampon has 12 points with two projecting front points for climbing very steep snow or ice. Crampons are adjustable and should always be fitted to a full shank boot. There are two types of crampons: flexible and rigid.

a. Flexible Crampons. These crampons have two sections (toe section and heel section), which are attached by an adjustable bar that can be extended or retracted to accommodate a boot size. The adjustment bar may have to be sawed to a shorter length to adjust for a extremely small boot size.

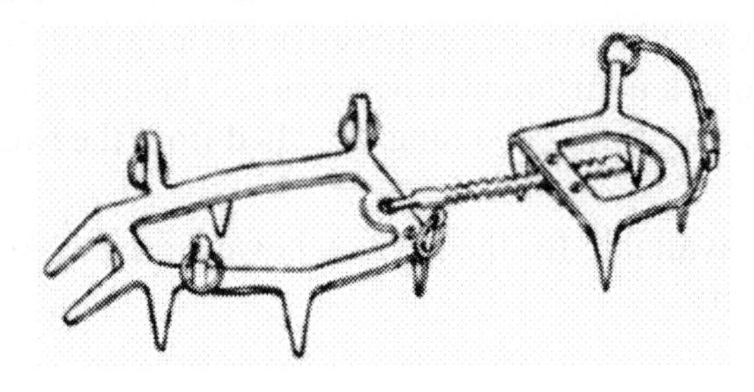

FLEXIBLE CRAMPON

b. Adjustment of Flexible Crampons

(1) Place boot on the crampon and with a screwdriver, loosen the flexible adjusting bar.

(2) Slide the heel piece from the toe piece until heel of boot is even

(3) Then place on so that the piece on the crampon and the toe of the boot covers the two front points leaving 1/2 - 3/4 inches of the front points exposed.

(4) Now take an Allen wrench and make your width adjustments by taking out the screws on your toe piece and heel piece.

(5) Slide in or out until the sole of your boot is snug.

(6) Line up the holes for the screws and tighten (boot should stay in crampon without leather straps).

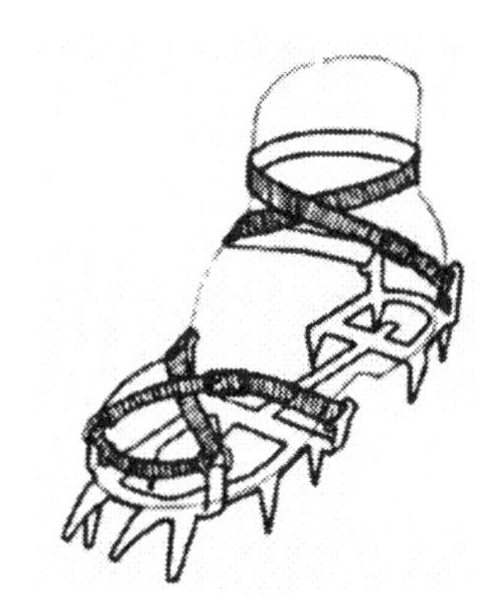

FLEXIBLE CRAMPON ON BOOT

c. Rigid Crampons. These crampons either come in two or four sections dependent upon the manufacture. Unlike flexible crampons, rigid crampons have a left and a right crampon, one for each foot. Adjustments are made by removing the screws to extend or retract the sections.

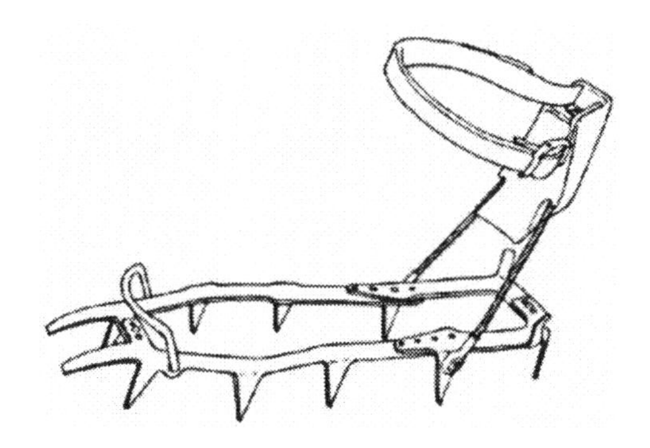

RIGID CRAMPON

d. Adjustment of Rigid Crampons

(1) Rigid crampons are shaped just like boots. You have a left and a right crampon. All you will use is an Allen wrench.

(2) Just like the flexible crampon you loosen screws; adjust for length first, then width. The adjustment procedures are the same as for the flexible crampons.

e. Fitting of Crampons

(1) Select and use a full shank boot.

(2) Ensure that the heel strap is long enough to go around the ankle once.

(3) The toe strap must be secured so that it does not slip over the toe of the boot.

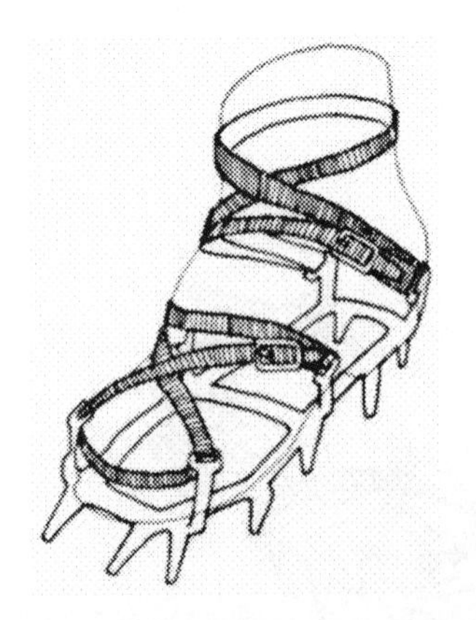

RIGID CRAMPON ON BOOT

f. Care and Maintenance of the Crampons

(1) Keep the front points sharp. Use on ice can dull and wear down the points. Careful hand filing will sharpen the crampons and reduce the possibility of a slip or fall. Lightly file down the top of the front points. File down on the other points from the rear only. **Don't use power tools to sharpen your crampons as this may cause the metal to weaken**.

(2) The adjusting nuts should be checked frequently and re-tightened as necessary.

(3) Check all rivets, straps and buckles and replace or repair as necessary.

(4) A piece of plastic can be taped to the underside of the crampon to help prevent snow buildup.

(5) To avoid unnecessary damage to your crampons cover the points or place inside your pack.

(6) Always carry a crampon repair kit in case of emergency. The repair kit should have the following items:

(a) Allen wrench

(b) Screwdriver

(c) Heavy Gauge Wire

(d) Pliers

(e) File

(f) Spare straps, nuts and bolts

TRANSITION: Now that we've discussed crampons and how to adjust them, are there any questions? If not, let's discuss snow protection.

3. (10 Min) **SNOW PROTECTION**. There are two types of artificial snow protection: flukes and pickets.

a. Snow Flukes: Dead-Man and Dead-Boy. These are aluminum plates with a pointed base, reinforced top and a metal cable attached to its middle. The snow flukes come in various sizes, and their holding power generally increases with size. The large one is known as the Dead-Man and the small one is known as the Dead-Boy. They also come in different configurations, some have angled flanges, and some have holes to make them lighter.

(1) Characteristics.

(a) Dead Man: 11"x 7 ½" with 24 inches of wire cable.

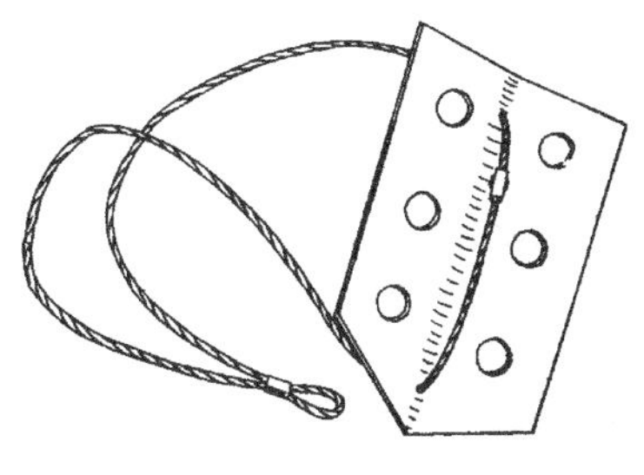

DEAD MAN

(b) Dead Boy: 8" x 5" with 24 inches of wire cable.

(2) Serviceability Checks of Snow Flukes.

(a) The body of the Dead-Man/Dead-Boy should be checked for fractures and obvious deformities.

(b) The attached cable should be checked for any cuts or frays.

NOTE: On the Dead-Boy, the brass fitting on the cable should be inspected and the cables should equalize when stretched.

b. Snow Pickets. These are aluminum in construction and come in various style and sizes.

(1) Characteristics

(a) Snow Stakes: "V" or "T" shaped, 24 – 36 inches in length with a hole at the top and pointed at the bottom.

SNOW STAKE

(b) Coyotes: "T" shaped, 18 - 36 inches in length with various holes throughout its construction, same angle on both ends.

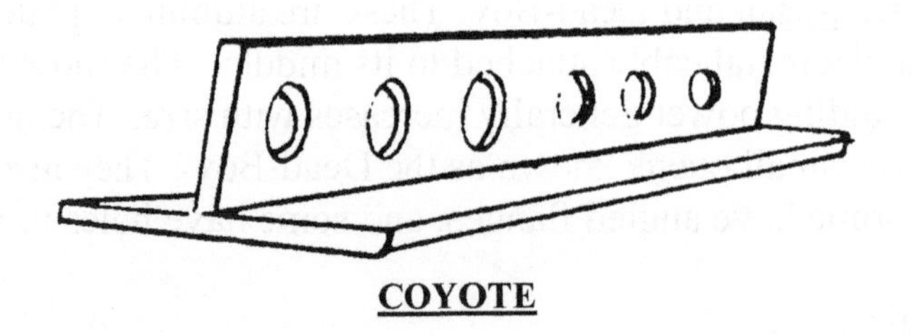

COYOTE

(2) Serviceability Checks of Snow Pickets.

(a) Check for fractures and obvious deformities.

(b) The tops of the stakes may become smashed or splintered due to hammering. Tape the tops to prevent the stakes from snagging on gear.

4. (10 Min) **ICE PROTECTION**. There are two types of artificial ice protection: ice pitons and ice screws.

a. Ice Pitons. These are constructed from steel or a metal alloyed and come in various sizes. All ice pitons are Hammered-In, but dependent upon which type, they are removed by either Prying–Out or Screwing-Out.

(1) Characteristics

(a) Spectre: Hammered-In / Pry-out, hook-shaped with teeth.

(b) Snarg: Hammered-In / Screw-out, tubular in shape with a flanged top and an open slot on its side.

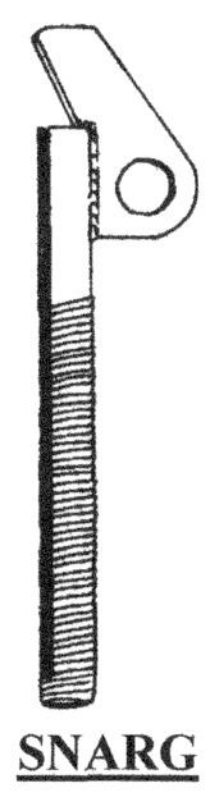

SNARG

(2) Serviceability Checks of Ice Pitons.

(a) Check the shaft for fractures and visible signs of obvious deformities.

(b) Inspect the threads to ensure that they are not stripped.

(c) Keep the points of the pitons sharp.

b. Ice Screws. These are constructed from a metal alloyed or titanium, tubular in shape and come in various lengths from 4 inches to 7 inches. Most types include a

(1) Characteristics.

(a) Teeth on the bottom for a quicker start.

(b) Built-in ratchet for easier placement.

(c) Hollow design to minimize fracturing of the ice.

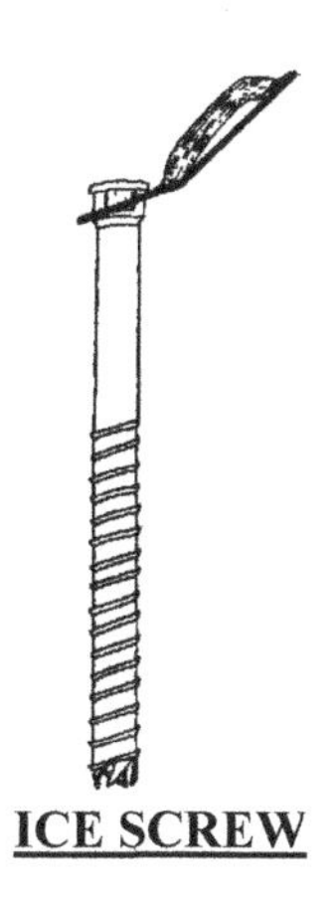

ICE SCREW

(2) Serviceability Checks of Ice Screws

(a) Check the shaft for fractures and visible signs of obvious deformities.

(b) Inspect the threads to ensure that they are not stripped.

(c) Keep the teeth of the screw protected and sharp.

(d) Inspect the ratchet to ensure that it functions properly.

TRANSITION: We have covered the protection used in snow and ice climbing. Due to the treacherousness of glacier travel, it is highly important that you know and care for your equipment. Are there any questions? If there are no questions for me, then I have some for you.

PRACTICE (CONC)

a. Students will me and maintain their specific snow/ice equipment when applicable during the course.

PROVIDE HELP (CONC)

a. The instructors will assist the students when necessary.

OPPORTUNITY FOR QUESTIONS (3 Min)

1. QUESTIONS FROM THE CLASS

2. QUESTIONS TO THE CLASS

Q. Should you use an electrical grinder to sharpen an ice axe?

A. No, because it can weaken the metal.

Q. What are the two different types of crampons?

A. Flexible and rigid.

Q. What are the two basic types of ice pitons?

A. (1) Hammer-in, Pry-out
(2) Hammer-in, Screw out

SUMMARY (2 Min)

a. During this class we talked about the different types of equipment that is used in snow/ice climbing. Including ice axes, protection devices, and mechanical ascenders. We also discussed the usage and care of these devices.

b. Those of you with IRF's please fill them out and turn them in to the instructor. We will now take a short break.

UNITED STATES MARINE CORPS
Mountain Warfare Training Center
Bridgeport, California 93517-5001

SML
03/31/01

LESSON PLAN

ESTABLISHING ANCHORS AND BELAYS ON SNOW AND ICE

INTRODUCTION (5 Min)

1. **GAIN ATTENTION.** When a team is climbing on snow or ice and the slope angle increases, the team needs the ability to place anchors or establish belays in order to protect each other. Keep in mind that belays placed in the snow and ice are not always secure, this is due mainly to the lack of stability in the snow pack and ice. In the event of a fall, a climber could possibly drag his partner to his death if an anchor failed. It is important to have the knowledge of placing good solid anchors when on snow and ice. It is also important to choose the appropriate belay for the environment.

2. **OVERVIEW.** The purpose of this period of instruction is to introduce the student to the establishment of anchors on snow and ice and executing a belay from these anchors. This lesson relates to glacier travel.

INSTRUCTOR NOTE: Have students read learning objectives.

3. **INTRODUCE LEARNING OBJECTIVES**

   a. TERMINAL LEARNING OBJECTIVE. In glaciated terrain, establish anchors and belays on snow and ice, in accordance with the references.

   b. ENABLING LEARNING OBJECTIVES

      (1) In glaciated terrain, establish anchors in snow, in accordance with the references.

      (2) In glaciated terrain, establish anchors in ice, in accordance with the references.

      (3) In glaciated terrain, execute a belay on snow, in accordance with the references.

      (4) In glaciated terrain, execute a belay on ice, in accordance with the references.

4. **METHOD/IMEDIA**. The material in this lesson will be presented by lecture and demonstration. You will practice what you have learned in upcoming field evaluations. Those of you with IRF's please fill them out at the conclusion of this period of instruction.

5. **EVALUATION.** You will be tested by a performance evaluation.

TRANSITION: Are there any questions over the purpose, learning objectives, how the class will be taught, or how you will be evaluated ? First let's discuss the establishment of snow anchors.

BODY (60 min)

1. (10 Min) **ESTABLISHMENT OF SNOW ANCHORS**. Snow anchors will vary in strength depending on snow conditions and placement. Their strength will also continue to change throughout the day due to sun exposure and snow metamorphism. A sitting ledge can be constructed below each of these anchors for the purpose of belaying.

   a. Emplacement of Deadman / Deadboy. In theory, these flukes serve as a dynamic anchor, burrowing deeper into the snowpack when subjected to a load.

      (1) With the adze of an axe, dig a one-foot long trench perpendicular to the intended load and about a foot deep. As you dig, undercut the trench toward the load.

      (2) With the pick of the axe, dig a slot from the middle of the trench running toward the load. This slot, known as the T-slot, is for the fluke's wired cable and should be as deep as the original trench. This completes a T-shaped site also known as a T-trench.

      (3) The fluke is now pushed or hammered in as far as possible ensuring that the cable runs down the T- slot and is fully stretched with no slack.

      (4) To seat the fluke properly, pull the cable in it's intended direction of pull. This will allow the fluke to burrow even deeper into the snowpack. If the fluke rises upward then the T-slot needs to be dug deeper.

      (5) Once the fluke has been seated, cover it and the T-slot with snow.

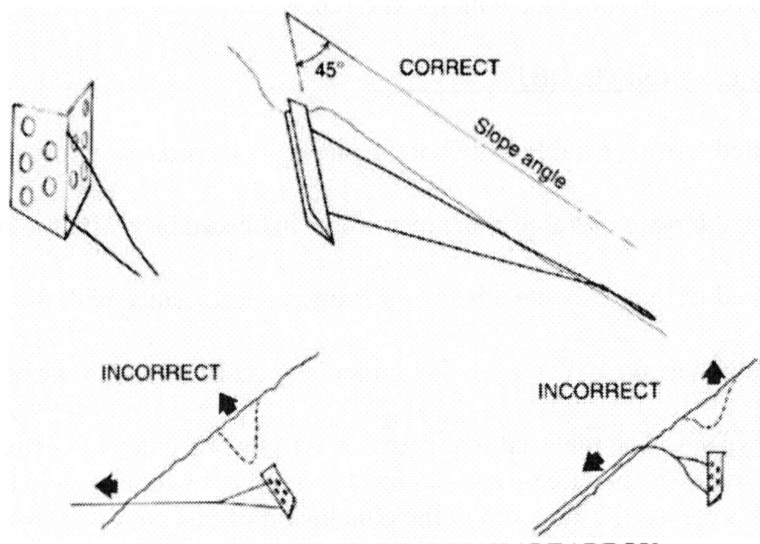

EMPLACEMENT OF DEADMAN / DEADBOY

b. Emplacement of the Snow Stake / Coyote. The other well-known snow anchor is the picket. The snow picket is a relatively easy piece of gear to place in snow as either perpendicular or as a dead man.

(1) Perpendicular Placement.

(a) Place the snow picket angled back into the slope at approximately 45°. Hammer the snow stake in as far as possible. Attach a carabiner or runner to the picket near the snow surface to prevent possible leveraging.

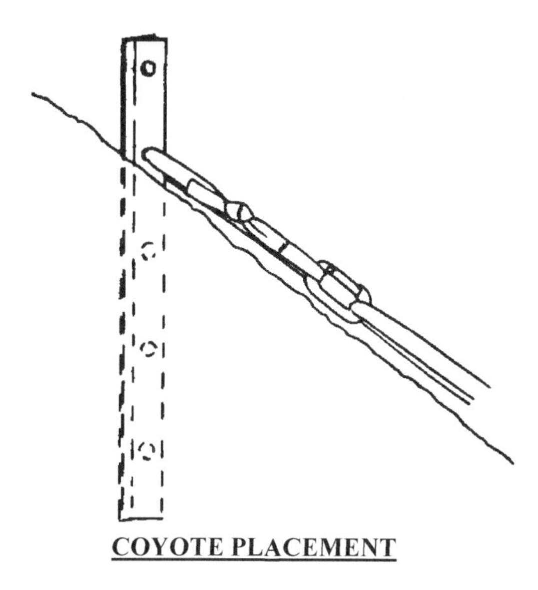

COYOTE PLACEMENT

(b) Another perpendicular method is to cut or dig a large step into the slope and then hammer the snow stake into the step at a right angle to the step. The side with the holes facing in the direction of the fall line. Now either attach a carabiner or girth hitch a web runner directly to one of the holes on the snow stake nearest the surface of the snow.

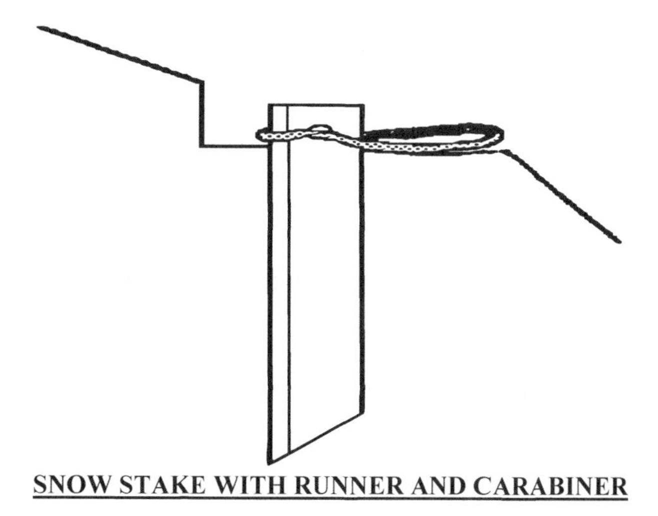

SNOW STAKE WITH RUNNER AND CARABINER

(2) Dead Man Placement.

(a) Dig a T-trench perpendicular to the intended load and about a foot deep and as long as the picket. As you dig, undercut the trench toward the load.

(b) Girth-hitch a runner around the center of the picket then stamp it into the trench ensuring that the runner runs down the T- slot and is fully stretched with no slack.

(c) Finally bury and stamp down everything except the tail of the runner, which will be used as the anchor point. You can strengthen this anchor by plunging an axe or two immediately in front of the picket.

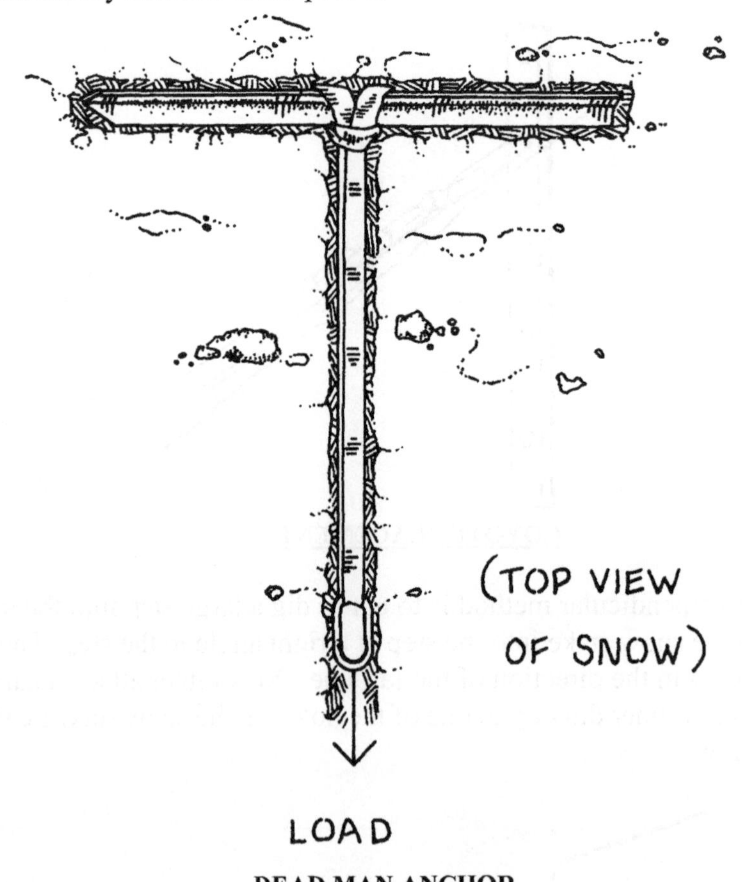

DEAD MAN ANCHOR

c. Emplacement of the Ice Axe. An ice axe could be used as a dead man anchor.

(1) Dig a T-trench perpendicular to the intended load and about a foot deep and as long as the ice axe. As you dig, undercut the trench toward the load.

(2) Girth-hitch a runner around the center of the ice axe then place it into the trench with the pick facing downward. Ensure that the runner runs down the T- slot and is fully stretched with no slack.

(3) Finally, bury and stamp down everything except the tail of the runner, which will be used as the anchor point. You can strengthen this anchor by plunging an axe or two immediately in front of the buried axe.

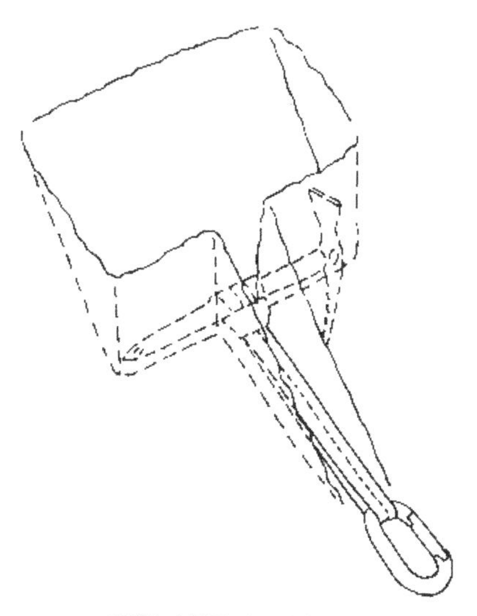

ICE AXE ANCHOR

d. Snow Bollard. A solid bollard distributes the load around a fairly even curve without the anchor rope running over any high spots.

(1) Construction of a Snow Bollard. A mound of snow is the basis of this type of anchor. These teardrop shaped anchors offer sturdy and reliable protection but are time consuming to construct.

(a) Look for a relatively high spot where the load will naturally pull downward, then scrape away any superficial snow that won't hold its shape.

(b) Create the mound by chopping the teardrop outline, gradually concentrating more on the three sides that will bear most of the weight.

(c) The minimum dimensions of a snow bollard are 4-10 feet wide, 12-15 feet long and 18 inches deep.

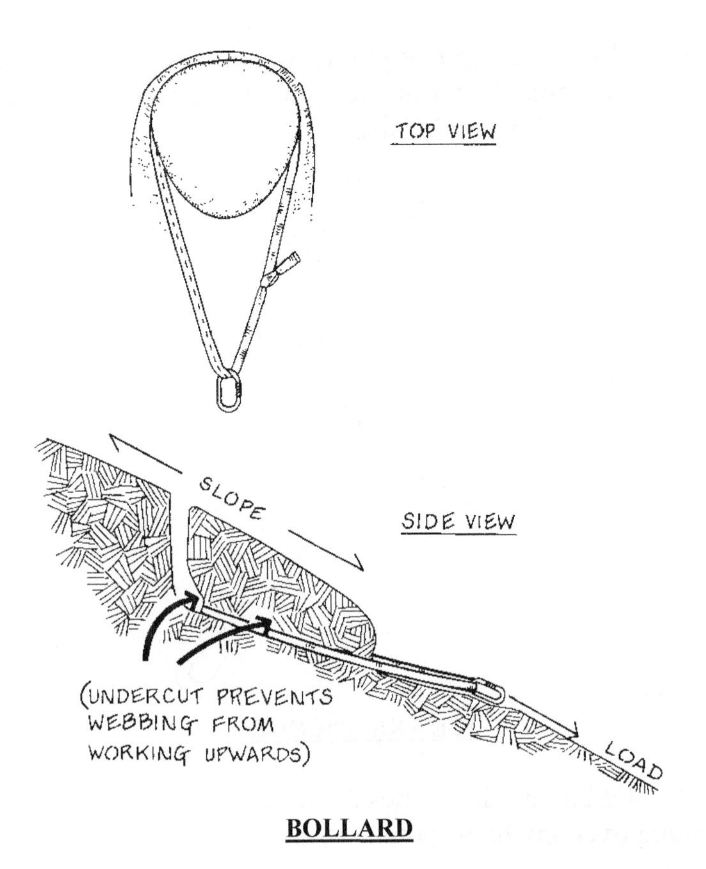

BOLLARD

(d) As the groove gets deeper, small blows with the axe's pick can define the undercut lip.

(e) Place the rope around the bollard to see if any protrusions might lift off and eliminate them.

(f) The rope is placed around the bollard only once and secured with a suitable anchor knot.

(g) An axe placed spike first in the back of the bollard can provide additional support.

<u>TRANSITION</u>: Now that we have talk about establishment of snow anchors, are there any questions? Placing anchors on snow can be adapted to icy areas, but utilizes different equipment. Let's discuss how to place anchors on ice.

2. (25 Min) **<u>EMPLACEMENT OF ICE SCREWS/PITONS</u>**. A favorable location for an ice-screw placement is the same as for an axe. A good choice is a natural depression where fracture lines caused by the screw are less likely to reach the surface.

   a. <u>Emplacement of Ice Screws</u>

      (1) If there is any unstable surface ice, it should be cut away so that the screw can be inserted into solid ice.

(2) Punch out a small starting hole with the spike of an axe to create a good grip for the starting threads or teeth.

(3) Ice screws are inserted at an angle of 90 degrees to the slope and then tilted back 10-20 degrees away from the load direction.

(4) Press the screw firmly and twist it into the ice at the same time.

(5) If you are unable to place the ice screw through the ice all the way then a hero's loop can be tied to the screw to prevent leverage.

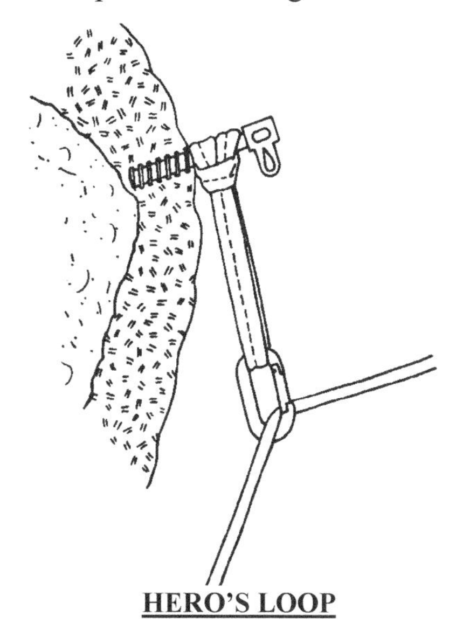

HERO'S LOOP

(6) If you are unable to screw the length of the screw through the ice then use the pick or spike of an axe to assist you.

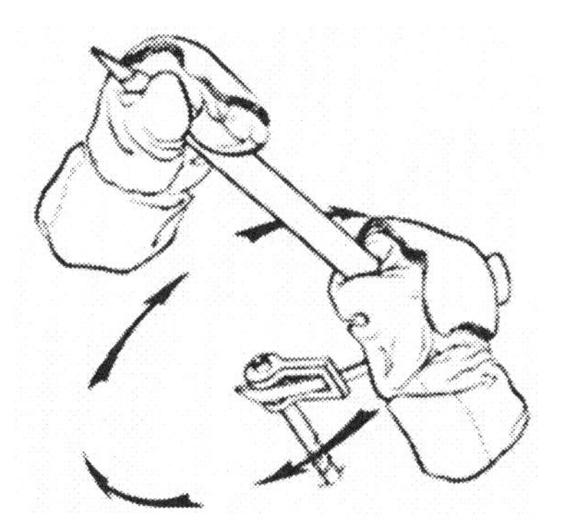

SCREWING IN WITH AN ICE TOOL

(7) When inserting the screw and you notice the ice "starring" or cracking, the ice screw should be removed and placed in another area. These signs would indicate that the ice is unstable and unsafe for an anchor point.

(8) To slow the process of melt-out due to soft summer ice or direct sunlight, pack snow or ice around the head of the screw.

b. Emplacement of Ice Pitons.

(1) When using a Snarg, begin the hole gently with light taps to prevent fracturing of the ice.

(2) Before hammering the Snarg into the ice, ensure that the piton is tilted back 10-20 degrees away from the load direction.

(3) When inserting the screw and you notice the ice "starring" or cracking, the ice screw should be removed and placed in another area. These signs would indicate that the ice is unstable and unsafe for an anchor point.

(4) To remove, just unscrew counter clock wise. You can use the pick of the ice axe to assist in the removal.

3. **ESTABLISHMENT OF ICE ANCHORS**. Ice anchors will vary in strength depending on ice conditions and strength. Their strength will also continue to change throughout the day due to the ice melting. A sitting ledge can be constructed below each of these anchors for the purpose of belaying.

a. Equalized Anchor System. This anchor system is similar to artificial anchors on rock, but instead it is being placed on ice.

(1) Place two ice screws offset approximately 18 inches apart. Ensure that the ice screws are placed parallel in relation to the direction of the belay.

(a) Attach carabiners to each of the screws.

(b) Attach a web runner through each carabiner and equalize the system. After it has been equalized, tie an overhand loop into the runner and attach a carabiner into the overhand loop.

(c) The anchor is now prepared to perform either a direct or indirect belay.

NOTE: The rope itself can be tied into each carabiner with a clove-hitch if no web runner is available.

b. Natural Anchors. Natural ice anchors such as ice pillars/ice pinnacles can provide a very secure anchor point for a belay position. If there are no natural ice archers available, ice can be cut and shaped in order to provide you with a solid anchor point.

c. Ice bollard. It is very similar to the snow bollard, the only exception being that it is cut out of ice instead of snow.

(1) The bollard should be cut out of a solid base of ice that is approximately 18 inches wide and 24 inches long. At a minimum the trench should be 6 inches deep. The backside of

the bollard must be undercut to accommodate the rope, this will help keep the rope in place.

(2) If there is any "starring" or opaqueness in the center of the bollard, remove the belay and place it in another area. These signs would indicate that the ice is unstable and therefore unsatisfactory for a safe anchor point.

NOTE: You may also back up the bollard by placing a piece of ice protection to the rear of the bollard with a runner and a carabiner attached to the rope.

TRANSITION: Are there any questions over placing protection on ice? Now we will discuss belays on snow.

4. (15 Min) **BELAYING ON SNOW**. Snow belays can be constructed from established snow anchors or even as simple as the use of an ice axe. No matter what belaying technique is employed, every snow belay should be as dynamic as possible to help limit the force on the anchor.

a. Quick Belays. These belays are used when the consequences of a fall would not be of great fall factor.

(1) The Boot Axe Belay. This is a fast and easy method to provide protection as a rope team moves up together. It is conducted in the following manner:

(a) Stamp a firm platform in the snow.

(b) Jam the axe shaft as deeply into the rear of the platform with the axe head perpendicular to the fall line.

(c) Stand below the axe at a right angle to the fall line and facing the side on which the climber's route lies.

(d) Plant your uphill boot into the snow against the downhill side of the shaft, bracing it against a downward pull.

(e) Plant the downhill boot in a firmly compacted step far enough below the other boot so that the downhill leg is straight providing a stiff brace.

(f) Flip the rope around the axe. The final configuration will have the rope running from the direction of the potential load, across the toe of the uphill boot, around the uphill side of the axe then back across the boot above the instep.

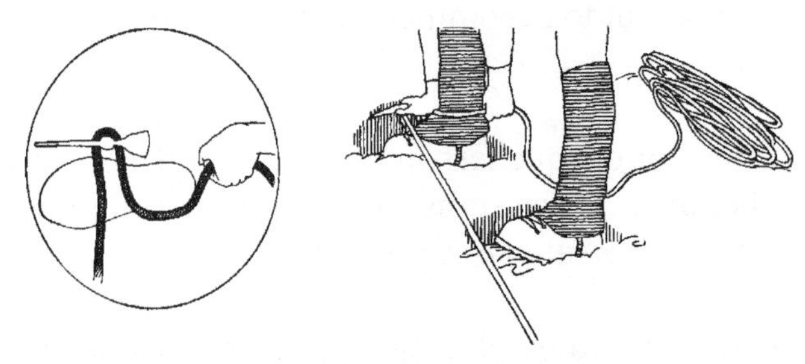

BOOT AXE BELAY

(g) Hold the rope with the downhill hand, applying extra friction by bringing the rope uphill behind the heel, forming an S-bend. The braking hand must never leave the rope.

(h) Use the uphill hand for two jobs; to grasp the head of the axe to further brace the shaft and then, as the belayed climber moves upward, to take in rope.

(2) The Carabiner ice axe belay. This system provides the same level of security as a boot-axe belay, with easier rope handling. It is conducted in the following manner:

(a) Stamp a firm platform in the snow.

(b) Jam the axe shaft as deeply into the rear of the platform with the axe head perpendicular to the fall line.

(c) Attach a short sling to the axe shaft at the surface snow line and clip a carabiner through it.

(d) Stand below the axe at a right angle to the fall line and facing the side on which the climber's route lies.

(e) Brace the axe with the uphill boot, standing on top of the axe head but leaving the sling exposed.

(f) The rope runs from the potential direction of pull, up through the carabiner, then around the back of your shoulder and into your uphill hand.

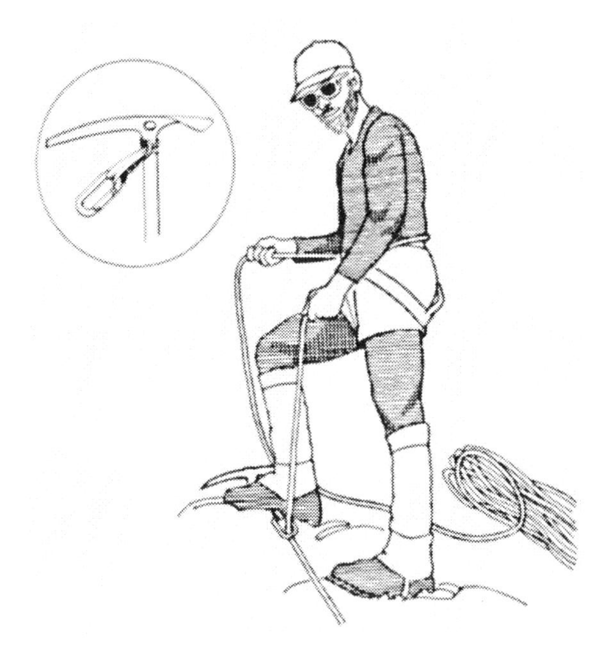

CARABINER ICE AXE BELAY

NOTE: Instead of placing the rope around the body, if wearing a harness the rope can go straight to the hard point onto a stitch plate.

b. Anchored Belays. These belays require formal anchors.

(1) The Sitting Hip Belay. This system is very dynamic and secure on hard or deep, heavy, wet snow.

(a) Stamp or chop out a belay seat in the snow plus a platform to brace each boot against.

(b) As in rock climbing, from your retrace figure-8 on the harness, tie into the anchor point.

SITTING HIP BELAY

(c) Pull in all the slack from the number two.

(d) Place the number 2 on a belay devise.

(e) You are now ready to bring up the number 2 climber.

NOTE: The belayer may want to insulate himself from the snow by using a isomat or pack.

(2) Standing Hip Belay. This is easier to set up than a sitting hip belay because only slots for the boots will need to be dug. However, it is far less secure because the belayer may topple over under the force of a fall.

(a) Stamp or chop out a belay ledge to stand on.

(b) From your retrace figure-8 on the harness, tie into the anchor point.

(c) Pull in all the slack from the number two.

(d) Place the number 2 on a belay devise.

(e) Stand sideways and facing the same side as the climber's route, bring up the number 2.

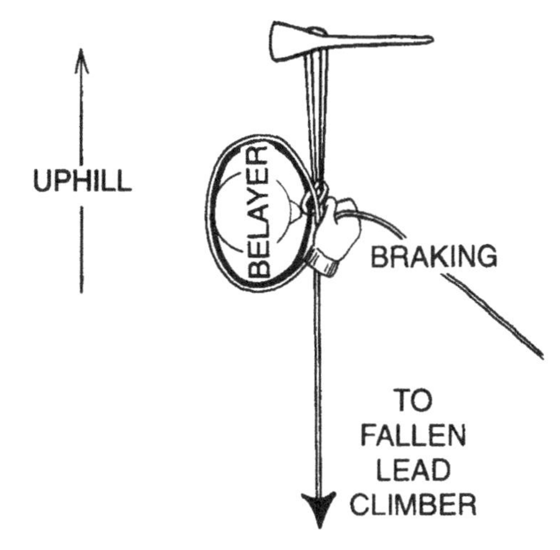

STANDING HIP BELAY

NOTE: Ice belay anchors are constructed similar to rock climbing belay anchors.

TRANSITION: Now that we have discussed the types of snow belays, are there any questions? It is very important that movement on glaciers is conducted safely, as a slip can send a Marine hurtling thousands of feet down a slope to his death. If you have no questions for me, then I have some for you.

PRACTICE (CONC)

a. The students will place protection on ice and execute a belay on ice.

PROVIDE HELP (CONC)

a. The instructors will assist the students when necessary.

OPPORTUNITY FOR QUESTIONS (3 Min)

1. QUESTIONS FROM THE CLASS

2. QUESTIONS TO THE CLASS

Q. At what angle are ice screws inserted into the ice face?

A. Ice screws are inserted at an angle of 90 degrees to the slope and then tilted back 10-20 degrees away from the load direction.

SUMMARY (2 Min)

a. This period of instruction has covered how to place anchors on snow and ice and how to belay off of these anchors.

b. Those of you with IRF's please fill them out at this time and turn them in to the instructor We will now take a short break.

UNITED STATES MARINE CORPS
Mountain Warfare Training Center
Bridgeport, California 93517-5001

MSVX
WMLC
WMO
09/31/00

LESSON PLAN

REQUIREMENTS FOR SURVIVAL

INTRODUCTION (5 Min)

1. **GAIN ATTENTION**. Although you are receiving this class as a course requirement from a formal school at MCMWTC, it can apply to you and/or any of your loved ones in many situations. If you, or they, ever backpack, cross country ski, fly in an aircraft, or simply drive or ride in a vehicle that passes through remote or isolated areas, then this class pertains to you.

2. **PURPOSE**. The purpose of this period of instruction is to introduce the student to the requirements for survival by discussing those aspects of survival that are absolutely essential for survival. This period of instruction relates to Mountain Safety.

3. **INTRODUCE LEARNING OBJECTIVES**

   a. TERMINAL LEARNING OBJECTIVE. In a summer mountainous environment, apply the requirements for survival, in accordance with the references.

   b. ENABLING LEARNING OBJECTIVES

      (1) Without the aid of references and given the acronym "SURVIVAL", describe in writing the acronym "SURVIVAL", in accordance with the references.

      (2) Without the aid of references, list in writing the survival stressors, in accordance with the references.

      (3) Without the aid of references, list in writing the priorities of work in a survival situation, in accordance with the references.

4. **METHOD / MEDIA**. The material in this class will be presented by lecture. You will practice what you have learned in upcoming field training exercises. Those of you with IRF's please fill them out at the end of this lesson and return them to the instructor.

5. **EVALUATION**. You will be tested later in this course by written and performance evaluations.

TRANSITION: The term "SURVIVE" brings to mind a very rough and demanding situation. In such a situation, Marines must develop a certain "mind-set" to survive.

BODY (50 Min)

1. (10 Min) **REQUIREMENTS FOR SURVIVAL**

   a. This mental "mind-set" is important in many ways. We usually call it the "will to survive" although you might call it "attitude" just as well. This basically means that, if you do not have the right attitude, you may still not survive.

   b. A guideline that can assist you is the acronym "SURVIVAL".

   (1) Size up

   (a) Size up the situation.

   1. Conceal yourself from the enemy.

   2. Use your senses to hear, smell, and see to determine and consider what is developing on the battlefield before you make your survival plan.

   (b) Size up your surroundings.

   1. Determine the rhythm or pattern of the area.

   2. Note animal and bird noises and their movement.

   3. Note enemy traffic and civilian movement.

   (c) Size up your physical condition.

   1. Check your wounds and give yourself first aid.

   2. Take care to prevent further bodily harm.

   3. Evaluate condition of self and unit prior to developing survival plan.

(d) Size up your equipment.

1. Consider how available equipment may affect survival senses; tailor accordingly.

(2) **U**ndue haste makes waste.

(a) Plan your moves so that you can move out quickly without endangering yourself if the enemy is near.

(3) **R**emember where you are.

(a) If you have a map, spot your location and relate it to the surrounding terrain.

(b) Pay close attention to where you are and where you are going. Constantly orient yourself.

(c) Try to determine, at a minimum, how your location relates to the following:

1. The location of enemy units and controlled areas.

2. The location of friendly units and controlled areas.

3. The location of local water sources.

4. Areas that will provide good cover and concealment.

(4) **V**anquish fear and panic.

(a) Realistic and challenging training builds self-confidence and confidence for a unit's leadership.

(b) The feeling of fear and panic will be present. The survivor must control these feelings.

(5) **I**mprovise and Improve.

(a) Use tools designed for one purpose for other applications.

(b) Use objects around you for different needs. (i.e. use a rock for a hammer)

(6) **V**alue living.

(a) Place a high value on living.

(b) Refuse to give into the problem and obstacles that face you.

(c) Draw strength from individuals that rise to the occasion.

(7) **A**ct like the natives.

(a) Observe the people in the area to determine their daily eating, sleeping, and drinking routines.

(b) Observe animal life in the area to help you find sources of food and water.

NOTE: Remember that animal reactions can reveal your presence to the enemy. Animals cannot serve as an absolute guide to what you can eat and drink.

(8) **L**ive by your wits, but for now, learn basic skills.

(a) Practice basic survival skills during all training programs and exercises.

TRANSITION: Now that we have covered the acronym "SURVIVAL", let's move on to stress.

2. (5 Min) **STRESS**. Stress has many positive benefits. Stress provides us with challenges: it gives us chances to learn about our values and strengths. Too much stress leads to distress. While many of these signs may not be self-identified, it remains critical that all survivors remain attentive to each other's signs of distress. Listed are a few common signs of distress found when faced with too much stress

a. Difficulty in making decisions (do not confuse this sign for a symptom of hypothermia).

b. Angry outbursts.

c. Forgetfulness.

d. Low energy level.

e. Constant worrying.

f. Propensity for mistakes.

g. Thoughts about death or suicide.

h. Trouble getting along with others.

i. Withdrawing from others.

j. Hiding from responsibilities.

k. Carelessness.

TRANSITION: Now that we understand stress, let's discuss survival stressors.

3. (10 Min) **SURVIVAL STRESSORS**. Any event can lead to stress. Often, stressful events occur simultaneously. These events are not stress, but they produce it and are called "stressors". In response to a stressor, the body prepares either to "fight or flee". Stressors add up. Anticipating stressors and developing strategies to cope with them are the two ingredients in the effective management of stress. It is essential that the survivor be aware of the types of stressors they will encounter.

   a. Injury, Illness, or Death. Injury, illness, and death are real possibilities a survivor has to face. Perhaps nothing is more stressful than being alone in an unfamiliar environment where you could die from hostile action, an accident, or from eating something lethal. Illness and injury can also add to stress by limiting your ability to maneuver, get food and drink, find shelter, and defend yourself.

   b. Uncertainty and Lack of Control. Some people have trouble operating in settings where everything is not clear-cut. The only guarantee in a survival situation is that nothing is guaranteed. This uncertainty and lack of control also add to the stress of being ill, injured, or killed.

   c. Environment. A survivor will have to contend with the stressors of weather, terrain, and the variety of creatures inhabiting an area. Heat, cold, rain, winds, snow, mountains, insects, and animals are just a few of the challenges awaiting the Marine working to survive. Depending on how a survivor handles the stress of environment, his surroundings can be either a source of food and protection or a cause of extreme discomfort leading to injury, illness, or death.

   d. Hunger and Thirst. Without food and water a person will weaken and eventually die. Getting and preserving food and water takes on increasing importance as the length of time in a survival setting increases. With the increased likelihood of diarrhea, replenishing electrolytes becomes critical. For a Marine used to having his provisions issued, foraging can be a significant source of stress.

   e. Fatigue. It is essential that survivors employ all available means to preserve mental and physical strength. While food, water, and other energy builders may be in short supply, maximizing sleep to avoid deprivation is a very controllable factor. Training data collected at MCCDC demonstrates that individuals who lack sleep for 24 hours or more suffer a 25% decrease in effective performance. Further, sleep deprivation directly correlates with increased fear. Effective survival is highly unlikely when fatigue builds to a level where staying awake becomes a stressful evolution.

   f. Isolation. Although Marines complain about higher headquarters, we become used to the information and guidance it provides, especially during times of confusion. Being in contact with others provides a greater sense of security and a feeling someone is available to help if problems occur. A significant stressor in survival situations is that often a person or team has to rely solely on its own resources.

TRANSITION: Now that we understand stressors, let's discuss natural reactions.

4. (5 Min) **NATURAL REACTIONS**. Man has been able to survive many shifts in his environment throughout the centuries. His ability to adapt physically and mentally to a changing world kept him alive. The average person will have some psychological reactions in a survival situation. These are some of the major internal reactions you might experience with the survival stressors.

   a. Fear. Fear is our emotional response to dangerous circumstances that we believe have the potential to cause death, injury, or illness. Fear can have a positive function if it encourages us to be cautious in situations where recklessness could result in injury. Unfortunately, fear can also immobilize a person. It can cause us to become so frightened that we fail to perform activities essential for survival.

   b. Anxiety. Anxiety can be an uneasy, apprehensive feeling we get when faced with dangerous situations. When used in a healthy way, anxiety urges us to act to end, or at least master, the dangers that threaten our existence. A survivor reduces his anxiety by performing those tasks that will ensure his coming through the ordeal alive. Anxiety can overwhelm a Marine to the point where he becomes easily confused and has difficulty thinking.

   c. Anger and Frustration. Frustration arises when a person is continually thwarted in his attempts to reach a goal. The goal of survival is to stay alive until you can reach help or until help can reach you. To achieve this goal, Marines must complete some tasks with minimal resources. One outgrowth of frustration is anger. Getting lost, damaged or forgotten equipment, the weather, inhospitable terrain, enemy patrols, and physical limitations are just a few sources of frustration and anger. Frustration and anger encourage impulsive reactions, irrational behavior, poorly thought-out decisions, and, in some instances, an "I quit" attitude. If the Marine does not properly focus his angry feelings, he can waste much energy in activities that do little to further either his chances of survival or the chances of those around him.

   d. Depression. Depression is closely linked with frustration and anger when faced with the privations of survival. A destructive cycle between anger and frustration continues until the person becomes worn down-physically, emotionally, and mentally. When a person reaches this point, he starts to give up, and his focus shifts from "What can I do" to "There is nothing I can do." If you allow yourself to sink into a depressed state, then it can sap all your energy and, more important, your will to survive.

   e. Loneliness and Boredom. Man is a social animal and enjoys the company of others. Loneliness and boredom can be another source of depression. Marines must find ways to keep their minds productively occupied.

f. Guilt. The circumstances leading to your survival setting are sometimes dramatic and tragic. It may be the result of an accident or military mission where there was a loss of life. Perhaps you were the only, or one of a few, survivors. While naturally relieved to be alive, you simultaneously may be mourning the deaths of others who were less fortunate. Do not let guilt feelings prevent you from living.

TRANSITION: Now that we have covered stress, let's move on to the priorities of work in a survival situation.

5. (5 Min) **PRIORITIES OF WORK IN A SURVIVAL SITUATION.** Each survival situation will have unique aspects that alter the order in which tasks need to be accomplished. A general guideline is to think in blocks of time.

   a. First 24 hours. The first 24 hours are critical in a survival situation. You must make an initial estimate of the situation. Enemy, weather, terrain, time of day, and available resources will determine which tasks need to be accomplished first. They should be the following:

      (1) Shelter

      (2) Fire

      (3) Water

      (4) Signaling.

   b. Second 24 hours. After the first 24 hours have passed, you will now know you can survive. This time period needs to be spent on expanding your knowledge of the area. By completing the following tasks, you will be able to gain valuable knowledge.

      (1) Tools and weapons. By traveling a short distance from your shelter to locate the necessary resources, you will notice edible food sources and game trails.

      (2) Traps and snares. Moving further away from your shelter to employ traps and snares, you will be able to locate your shelter area from various vantage points. This will enable you to identify likely avenues of approach into your shelter area.

      (3) Path guards. Knowing the likely avenues of approaches, you can effectively place noise and casualty producing path guards to ensure the security of your shelter area.

   c. Remainder of your survival situation. This time is spent on continuously improving your survival situation until your rescue.

TRANSITION: Now that we understand the concept behind individual survival, let's talk about group survival.

6. (5 Min) **GROUP SURVIVAL.** In group survival, the group's survival depends largely on its ability to organize activity. An emergency situation does not bring people together for a common goal; rather, the more difficult and disordered the situation, the greater are the disorganized group's problems.

   a. Groups Morale. High morale must come from internal cohesiveness and not merely through external pressure. The moods and attitudes can become wildly contagious. Conscious, well-planned organization and leadership on the basis of delegated or shared responsibility often can prevent panic. High group morale has many advantages.

      (1) Individual feels strengthened and protected since he realizes that his survival depends on others whom he trusts.

      (2) The group can meet failure with greater persistency.

      (3) The group can formulate goals to help each other face the future.

   b. Factors that Influence Group Survival. There are numerous factors that will influence whether a group can successfully survive.

      (1) Organization of Manpower. Organized action is important to keep all members of the group informed; this way the members of the group will know what to do and when to do it, both under ordinary circumstances and in emergencies.

      (2) Selective Use of Personnel. In well-organized groups, the person often does the job that most closely fits his personal qualifications.

      (3) Acceptance of Suggestion and Criticisms. The senior man must accept responsibility for the final decision, but must be able to take suggestion and criticisms from others.

      (4) Consideration of Time. On-the-spot decisions that must be acted upon immediately usually determines survival success.

      (5) Check Equipment. Failure to check equipment can result in failure to survive.

      (6) Survival Knowledge and Skills. Confidence in one's ability to survive is increased by acquiring survival knowledge and skills.

TRANSITION: Keeping a proper attitude is very important in surviving. Without the will, there is no way.

PRACTICE (CONC)

   a. The students will implement what has been taught concurrently throughout the course.

PROVIDE HELP (CONC)

a. The instructors will assist the students when necessary.

OPPORTUNITY FOR QUESTIONS (3 Min)

1. QUESTIONS FROM THE CLASS

2. QUESTIONS TO THE CLASS

Q. What does the "S" in the acronym SURVIVAL stand for?

A. Size up the situations.

Q. What priorities of work are accomplished in the first 24 hours of a survival situation?

A. (1) Shelter
(2) Fire
(3) Water
(4) Signaling

SUMMARY (2 Min)

a. We have discussed many items very briefly in this class. Many of these items will be covered in greater detail in classes you will receive later in the course of instruction you are undergoing.

b. Those of you with IRF's please fill them out at this time and turn them in. We will now take a short break.

UNITED STATES MARINE CORPS
Mountain Warfare Training Center
Bridgeport, California 93517-5001

MSVX
WML
WMO
03/31/01

LESSON PLAN

WATER PROCUREMENT

INTRODUCTION (5 Min)

1. **GAIN ATTENTION**. More than three-quarters of the human body is composed of liquids. Heat, cold, stress, and exertion can cause a loss or expenditure of body fluids that must be replaced, if you are to function effectively in a survival situation.

2. **PURPOSE**. The purpose of this period of instruction is to introduce the student to the basics of water procurement. This will be accomplished by discussing locating, gathering, and disinfecting water. This lesson relates to all other lessons on survival here at MWTC.

3. **INTRODUCE LEARNING OBJECTIVES**

   a. TERMINAL LEARNING OBJECTIVE. In a summer mountainous environment and given water procurement materials, obtain potable water, in accordance with the references.

   b. ENABLING LEARNING OBJECTIVES

      (1) Without the aid of references, list in writing the types of incidental water, in accordance with the references.

      (2) Without the aid of references, list in writing the hazardous fluids to avoid substituting for potable water, in accordance with the references.

      (3) Without the aid of references, list in writing the methods for disinfecting water, in accordance with the references.

(4) Without the aid of references and given a military bottle of water purification tablets, state in writing its self-life, in accordance with the references.

(5) Without the aid of references, and given the water temperature and chemical concentration, state in writing the contact time, in accordance with the references.

(6) Without the aid of references, construct a solar still, in accordance with the references.

4. **METHOD / MEDIA**. The material in this lesson will be presented by lecture and demonstration. You will practice what you have learned during upcoming field exercises. Those of you with IRF's please fill them out at the end of this period of instruction.

5. **EVALUATION**. You will be tested later in the course by written and performance evaluations.

TRANSITION: Now that we know what is expected of us, let's talk about what determines water intake.

BODY (45 Min)

1. (5 Min) **WATER INTAKE**

a. Thirst is not a strong enough sensation to determine how much water you need.

b. The best plan is to drink, utilizing the OVER DRINK method. Drink plenty of water anytime it is available and particularly when eating.

c. Dehydration is a major threat. A loss of only 5 % of your body fluids causes thirst, irritability, nausea, and weakness; a 10% loss causes dizziness, headache, inability to walk, and a tingling sensation in limbs; a 15% loss causes dim vision, painful urination, swollen tongue, deafness, and a feeling of numbness in the skin; also a loss of more than 15% body fluids could result in death.

d. Your water requirements will be increased if:

(1) You have a fever.

(2) You are experiencing fear or anxiety.

(3) You evaporate more body fluid than necessary. (i.e., not using the proper shelter to your advantage)

(4) You have improper clothing.

(5) You ration water.

(6) You overwork.

TRANSITION: Now that we have a general idea of what determines water intake, let's move on to techniques of locating water.

2. (5 Min) **INCIDENTAL WATER**

a. During movement, you may have to ration water until you reach a reliable water source. Incidental water may sometimes provide opportunities to acquire water. Although not a reliable or replenished source, it may serve to stretch your water supply or keep you going in an emergency. The following are sources for incidental water:

(1) Dew. In areas with moderate to heavy dew, dew can be collected by tying rags or tuffs of fine grass around your ankles. While walking through dewy grass before sunrise, the rags or grass will saturate and can be rung out into a container. The rags or grass can be replaced and the process is repeated.

(2) Rainfall. Rainwater collected directly in clean container or in plants that contain no harmful toxins is generally safe to drink without disinfecting. The survivor should always be prepared to collect rainfall at a moments notice. An inverted poncho works well to collect rainfall.

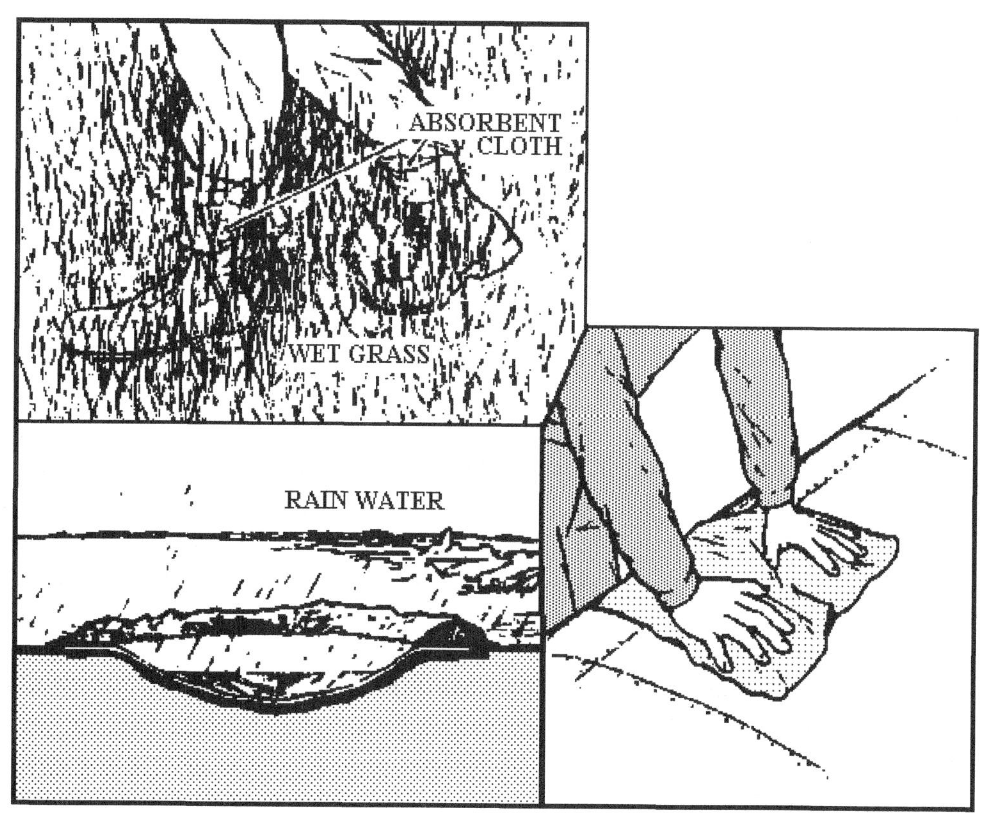

TRANSITION: Now that we know where to obtain water, let's discuss some hazards of substituting other fluids for potable water.

3. (5 Min) **HAZARDOUS FLUIDS**

a. Survivors have occasionally attempted to augment their water supply with other fluids, such as alcoholic beverages, urine, blood, or seawater. While it is true that each of these fluids has a high water content, the impurities they contain may require the body to expend more fluid to purify them. Some hazardous fluids are:

(1) Sea water. Sea water in more than minimal quantities is actually toxic. The concentration of sodium and magnesium salts is so high that fluid must be drawn from the body to eliminate the salts and eventually the kidneys cease to function.

(2) Alcohol. Alcohol dehydrates the body and clouds judgment. Super-cooled liquid, if ingested, can cause immediate frostbite of the throat, and potential death.

(3) Blood. Blood, besides being salty, is a food. Drinking it will require the body to expend additional fluid to digest it.

(4) Urine. Drinking urine is not only foolish, but also dangerous. Urine is nothing more than the body's waste. Drinking it only places this waste back into the body, which requires more fluid to process it again.

TRANSITION: Now that we know what fluids to avoid, let's talk about water quality.

4. (10 Min) **WATER QUALITY**. Water contains minerals, toxins, and pathogens. Some of these, consumed in large enough quantities may be harmful to human health. Pathogens are our primary concern. Pathogens are divided into Virus, Cysts, Bacteria, and Parasites. Certain pathogens are more resistant to chemicals and small enough to move through microscopic holes in equipment (i.e., T-shirt, parachute). Certain pathogens also have the ability to survive in extremely cold water temperatures. Pathogens generally do not live in snow and ice. Water quality is divided into three levels of safety with disinfection as the most desired level, then purified, followed by potable.

a. Disinfection. Water disinfection removes or destroys harmful microorganisms. Giardia cysts are an ever-present danger in clear appearing mountain water throughout the world. By drinking non-potable water you may contract diseases or swallow organisms that could harm you. Examples of such diseases or organisms are: Dysentery, Cholera, Typhoid, Flukes, and Leeches.

b. Remember, impure water, no matter how overpowering the thirst, is one of the worst hazards in a survival situation.

c. The first step in disinfecting is to select a treatment method. The two methods we will discuss are as follows:

(1) Heat. The Manual of Naval Preventive Medicine (P-5010) states that you must bring the water to a rolling boil before it is considered safe for human consumption. This is the most preferred method.

(a) Bringing water to the boiling point will kill 99.9% of all Giardia cysts. The Giardia cyst dies at 60°C and Cryptosporidium dies at 65°C. Water will boil at 14,000' at 86°C and at 10,000' at 90°C. With this in mind you should note that altitude does not make a difference unless you are extremely high.

(2) Chemicals. There are numerous types of chemicals that can disinfect water. Below are a few of the most common. In a survival situation, you will use whatever you have available.

(a) Iodine Tablets.

(b) Chlorine Bleach.

(c) Iodine Solution.

(d) Betadine Solution.

(e) Military water purification tablets. These tablets are standard issue for all DOD agencies. These tablets have a shelf-life of four years from the date of manufacture, unless opened. Once the seal is broken, they have a shelf-life of one year, not to exceed the initial expiration date of four years.

49703
Month / Year / Batch Number

(3) Water Disinfection Techniques and Halogen Doses

Iodination Techniques Added to 1 liter or quart of water	Amount for 4 ppm	Amount for 8 ppm
Iodine tablets Tetraglcine hydroperiodide EDWGT Potable Aqua Globaline	½ tablet	1 tablet
2% iodine solution (tincture)	0.2 ml 5 gtts	0.4 ml 10 gtts
10% povidone-iodine solution*	0.35 ml 8 gtts	0.70 ml 16 gtts

Chlorination Techniques	Amount for 5 ppm	Amount for 10 ppm
Household bleach 5%	ml	ml

Sodium hypochlorie	2 gtts	4 gtts
AquaClear Sodium dichloroisocyanurate		1 tablet
AquaCure, AquaPure, Chlor-floc Chlorine plus flocculating agent		8 ppm 1 tablet

*Providone-iodine solutions release free iodine in levels adequate for disinfection, but scant data is available.

Measure with dropper (1 drop=0.05 ml) or tuberculin syringe

Ppm-part per million gtts-drops ml-milliliter

Concentration of Halogen	**Contact time in minutes at various water temperatures**		
	5°C / 40°F	15°C / 60°F	30°C / 85°F
2 ppm	240	180	60
4 ppm	180	50	45
8 ppm	60	30	15

NOTE: These contact times have been extended from the usual recommendations to account for recent data that prolonged contact time is needed in very cold water to kill *Giardia* cysts.

NOTE: Chemicals may not destroy Cryptosporidium.

d. Purification. Water purification is the removal of organic and inorganic chemicals and particulate matter, including radioactive particles. While purification can eliminate offensive color, taste, and odor, it may not remove or kill microorganisms.

(1) Filtration. Filtration purifying is a process by which commercial manufacturers build water filters. The water filter is a three tier system. The first layer, or grass layer, removes the larger impurities. The second layer, or sand layer, removes the smaller impurities. The final layer, or charcoal layer (not the ash but charcoal from a fire), bonds and holds the toxins. All layers are placed on some type of straining device and the charcoal layer should be at least 5-6 inches thick. Layers should be changed frequently and straining material should be boiled. Remember, this is not a disinfecting method, cysts can possibly move through this system.

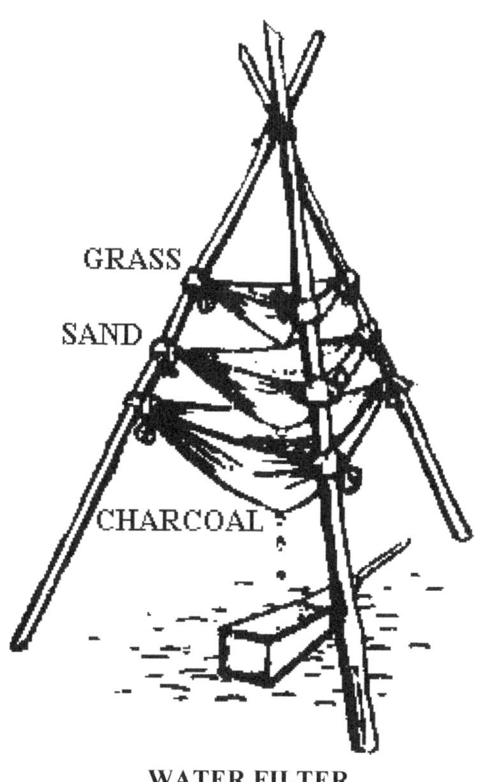

WATER FILTER

(2) Commercial Water Filters. Commercial water filters are generally available in most retail stores and may be with you. Understanding what the filter can do is the first step in safeguarding against future illnesses.

(a) A filter that has a .3 micron opening or larger will not stop Cryptosporidium.

(b) A filter system that does not release a chemical (i.e., iodine) may not kill all pathogens.

(c) A filter that has been overused may be clogged. Usage may result in excessive pumping pressure that can move harmful pathogens through the opening.

e. Potable. Potable indicates only that a water source, on average over a period of time, contains a "minimal microbial hazard," so the statistical likelihood of illness is acceptable.

(1) Sedimentation. Sedimentation is the separation of suspended particles large enough to settle rapidly by gravity. The time required depends on the size of the particle. Generally, 1 hour is adequate if the water is allowed to sit without agitation. After sediment has formed on the bottom of the container, the clear water is decanted or filtered from the top. Microorganisms, especially cysts, eventually settle, but this takes longer and the organisms are easily disturbed during pouring or filtering. Sedimentation should not be considered a means of disinfection and should be used only as a last resort or in an extreme tactical situation.

TRANSITION: Since locating water may be difficult, let's look at two stills that may supplement your existing water supply.

5. (10 Min) **SOLAR STILLS**. Solar stills are designed to supplement water reserves. Contrary to belief, they will not provide enough water to meet the daily requirement for water.

   a. Above-Ground Solar Still. This device allows the survivor to make water from vegetation. To make the aboveground solar still, locate a sunny slope on which to place the still, a clear plastic bag, green leafy vegetation, and a small rock.

      (1) Construction

         (a) Fill the bag with air by turning the opening into the breeze or by "scooping" air into the bag.

         (b) Fill the bag half to three-quarters full of green leafy vegetation. Be sure to remove all hard sticks or sharp spines that might puncture the bag.

CAUTION

Do not use poisonous vegetation. It will provide poisonous liquid.

         (c) Place a small rock or similar item in the bag.

         (d) Close the bag and tie the mouth securely as close to the end of the bag as possible to keep the maximum amount of air space. If you have a small piece of tubing, small straw, or hollow reed, insert one end in the mouth of the bag before tying it securely. Tie off or plug the tubing so that air will not escape. This tubing will allow you to drain out condensed water without untying the bag.

         (e) Place the bag, mouth downhill, on a slope in full sunlight. Position the mouth of the bag slightly higher than the low point in the bag.

         (f) Settle the bag in place so that the rock works itself into the low point in the bag.

         (g) To get the condensed water from the still, loosen the tie and tip the bag so that the collected water will drain out. Retie the mouth and reposition the still to allow further condensation.

         (h) Change vegetation in the bag after extracting most of the water from it.

         (i) Using 1 gallon zip-loc bag instead of trash bags is a more efficient means of construction.

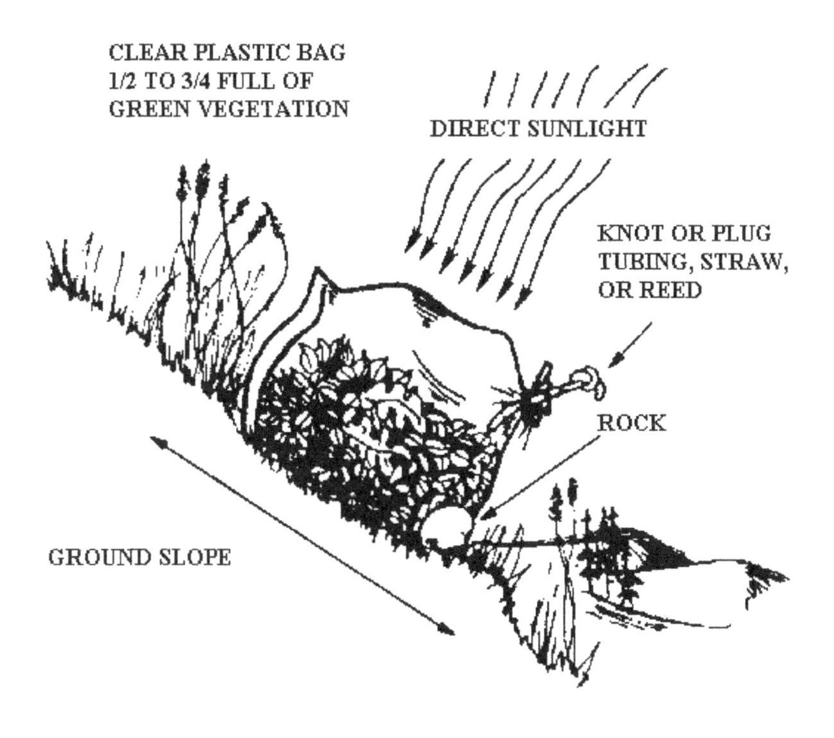

ABOVE GROUND SOLAR STILL

b. Below-Ground Solar Still. Materials consist of a digging stick, clear plastic sheet, container, rock, and a drinking tube.

(1) Construction

(a) Select a site where you believe the soil will contain moisture (such as a dry streambed or a low spot where rainwater has collected). The soil should be easy to dig, and will be exposed to sunlight.

(b) Dig a bowl-shaped hole about 1 meter across and 24 inches deep.

(c) Dig a sump in the center of the hole. The sump depth and perimeter will depend on the size of the container you have to place in it. The bottom of the sump should allow the container to stand upright.

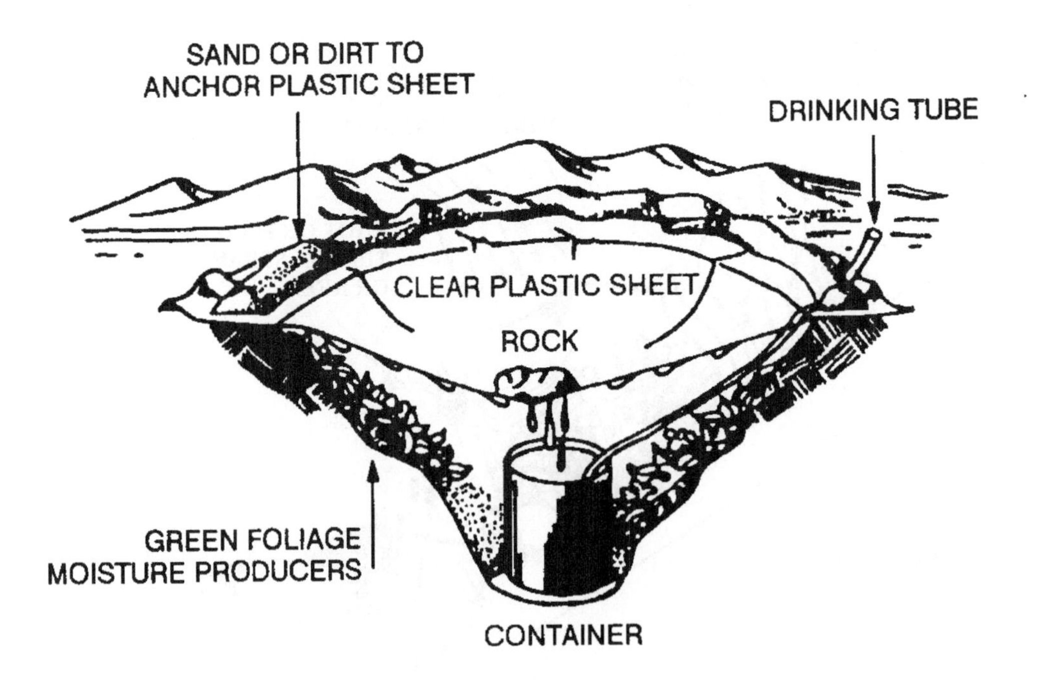

BELOW GROUND SOLAR STILL

(d) Anchor the tubing to the container's bottom by forming a loose overhand knot in the tubing. Extend the unanchored end of the tubing up, over, and beyond the lip of the hole.

(e) Place the plastic sheet over the hole, covering its edges with soil to hold in place. Place a rock in the center of the plastic sheet.

(f) Lower the plastic sheet into the hole until it is about 18 inches below ground level. Make sure the cone's apex is directly over the container. Ensure the plastic does not touch the sides of the hole because the earth will absorb the moisture.

(g) Put more soil on the edges of the plastic to hold it securely and prevent the loss of moisture.

(h) Plug the tube when not in use so that moisture will not evaporate.

(i) Plants can be placed in the hole as a moisture source. If so, dig out additional soil from the sides.

(j) If polluted water is the only moisture source, dig a small trough outside the hole about 10 inches from the still's lip. Dig the trough about 10 inches deep and 3 inches wide. Pour the polluted water in the trough. Ensure you do not spill any polluted water around the rim of the hole where the plastic touches the soil. The trough holds the polluted water and the soil filters it as the still draws it. This process works well when the only water source is salt water.

NOTE: Three stills will be needed to meet the individual daily water intake needs.

PRACTICE (CONC)

a. The students will practice these skills in a survival situation.

PROVIDE HELP (CONC)

a. The instructors will provide help as necessary.

OPPORTUNITY FOR QUESTIONS (3 Min)

1. QUESTIONS FROM THE CLASS

2. QUESTIONS TO THE CLASS

Q. What factors increase your need for water?

A. (1) You have a fever.
(2) You are in fear of something.
(3) You evaporate more body fluid than necessary by not using available shelter to your advantage.
(4) You overwork.

Q. What are the four hazardous fluids to replace water with?

A. (1) Sea water.
(2) Alcohol.
(3) Blood.
(4) Urine.

SUMMARY (2 Min)

a. During this period of instruction we have covered what determines water intake, what hazardous fluids to avoid substituting for potable water, how to procure, and how to purify water.

b. Those of you with IRF's please fill them out at this time and turn them in to the instructor. We will now take a short break.

UNITED STATES MARINE CORPS
Mountain Warfare Training Center
Bridgeport, California 93517-5001

MSVX
WML
WMO
03/31/01

LESSON PLAN

SURVIVAL NAVIGATION

INTRODUCTION (5 Min)

1. **GAIN ATTENTION**. In almost every MOS there is a chance that a Marine may become separated from his unit. To survive, he will need some type of method to find his way back. Survival navigation could very well be the method he needs.

2. **PURPOSE**. The purpose of this period of instruction is to introduce the student to the principles and techniques of survival navigation. This lesson relates to Requirements for Survival.

3. **INTRODUCE LEARNING OBJECTIVES**

   a. TERMINAL LEARNING OBJECTIVE. In a summer mountainous environment, navigate in a survival situation, in accordance with the references.

   b. ENABLING LEARNING OBJECTIVES

      (1) Without the aid of references, list in writing the considerations for travel, in accordance with the references.

      (2) Without the aid of references, describe in writing the seasonal relationship of the sun and its movement during the equinox and solstice, in accordance with the references.

      (3) Without the aid of references, and given a circular navigational chart and operating latitude, determine the bearing of the sun at sunrise and sunset, in accordance with the references.

      (4) Without the aid of references, construct a pocket navigator, in accordance with the references.

(5) Without the aid of references, describe in writing the two methods for locating the North Star, in accordance with the references.

4. **METHOD / MEDIA**: The material in this lesson will be presented by lecture and demonstration. You will practice what you have learned during upcoming field training exercises. Those of you with IRF's please fill them out at the end of this period of instruction and turn them in to the instructor.

5. **EVALUATION**: You will be tested later in the course by written and performance evaluations.

TRANSITION: The decision to stay or travel needs to be well thought out. Let's discuss some considerations to making that decision.

BODY (50 Min)

1. (5 Min) **CONSIDERATIONS FOR STAYING OR TRAVELLING**

a. Stay with the aircraft or vehicle if possible. More than likely somebody knows where it was going. It is also a ready-made shelter.

b. Leave only when:

(1) Certain of present location; have known destination and the ability to get there.

(2) Water, food, shelter, and/or help can be reached.

(3) Convinced that rescue is not coming.

c. If the decision is to travel, the following must also be considered:

(1) Which direction to travel and why.

(2) What plan is to be followed.

(3) What equipment should be taken.

(4) How to mark the trail.

(5) Predicted weather.

d. If the tactical situation permits leave the following information at the departure point:

(1) Departure time.

(2) Destination.

(3) Route of travel/direction.

(4) Personal condition.

(5) Available supplies.

TRANSITION: Once the decision to move is made, the first step is to orient yourself to your surroundings. Mother Nature has provided us a reliable method of using the sun and stars to help.

2. (15 Min) **DAYTIME SURVIVAL NAVIGATION**

a. Sun Movement. It is generally taken for granted that the sun rises in the east and sets in the west. This rule of thumb, however, is quite misleading. In fact, depending on an observer's latitude and the season, the sun could rise and set up to 50 degrees off of true east and west.

b. The following diagram and terms are essential to understanding how the sun and stars can help to determine direction:

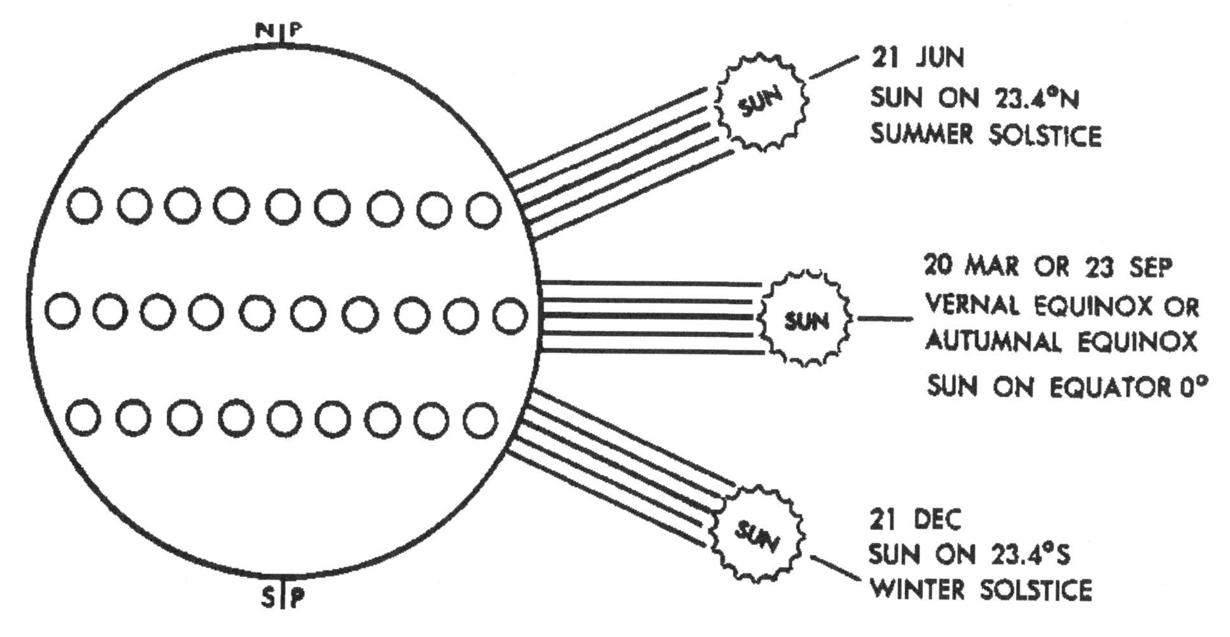

POSITION OF THE SUN AT EQUINOX AND SOLSTICE

(1) Summer/Winter Solstice: (21 June/21 December) Two times during the year when the sun has no apparent northward or southward motion.

(2) Vernal/Autumnal Equinox: (20 March/23 September) Two times during the year when the sun crosses the celestial equator and the length of day and night are approximately equal.

c. Sun's Movement. As reflected in the diagram above, the sun continuously moves in a cycle from solstice to equinox; throughout each day, however, the sun travels a uniform arc in the sky from sunrise to sunset. Exactly half-way along its daily journey, the sun will be directly south of an observer (or north if the observer is in the Southern Hemisphere). This rule may not apply to observers in the tropics (between 23.5 degrees north and south latitude) or in the polar regions (60 degrees latitude). It is at this point that shadows will appear their shortest. The time at which this occurs is referred to as "local apparent noon."

d. Local Apparent Noon. Whenever using any type of shadow casting device to determine direction, "local apparent noon" (or the sun's highest point during the day) must be known. Local apparent noon can be determined by the following methods.

(1) Knowing sunrise and sunset from mission orders, i.e., sunrise 0630 and sunset 1930. Take the total amount of daylight (13 hours), divide by 2 (6 hours 30 minutes), and add to sunrise (0630 plus 6 hours 30 minutes). Based on this example, local apparent noon would be 1300.

(2) Using the string method. The string method is used to find two equidistant marks before and after estimated local apparent noon. The center point between these two marks represents local apparent noon.

e. Sun's Bearing. With an understanding of the sun's daily movement, as well as its seasonal paths, a technique is derived that will determine the true bearing of the sun at sunrise and sunset. With the aid of a circular navigational chart, we can accurately navigate based on the sun's true bearing:

(1) Determine the sun's maximum amplitude at your operating latitude using the top portion of the chart.

(2) Scale the center baseline of the chart where 0 appears as the middle number; write in the maximum amplitude at the extreme north and south ends of the baseline.

(3) Continue to scale the baseline; you should divide the baseline into 6 to 10 tick marks that represent equal divisions of the maximum amplitude.

(4) From today's date along the circumference, draw a straight line down until it intersects the baseline.

(5) The number this line intersects is today's solar amplitude. If the number is left of 0, it is a "north" amplitude; if the number is right of 0, it is a "south" amplitude. Use the formula at the bottom of the chart to determine the sun's bearing at sunrise or sunset.

LATITUDE (N OR S)	5	10	15	20	25	30	35	40	45	50	55	60
MAXIMUM AMPLITUDE	24°	24°	24°	25°	26°	27°	29°	31°	34°	38°	44°	53°

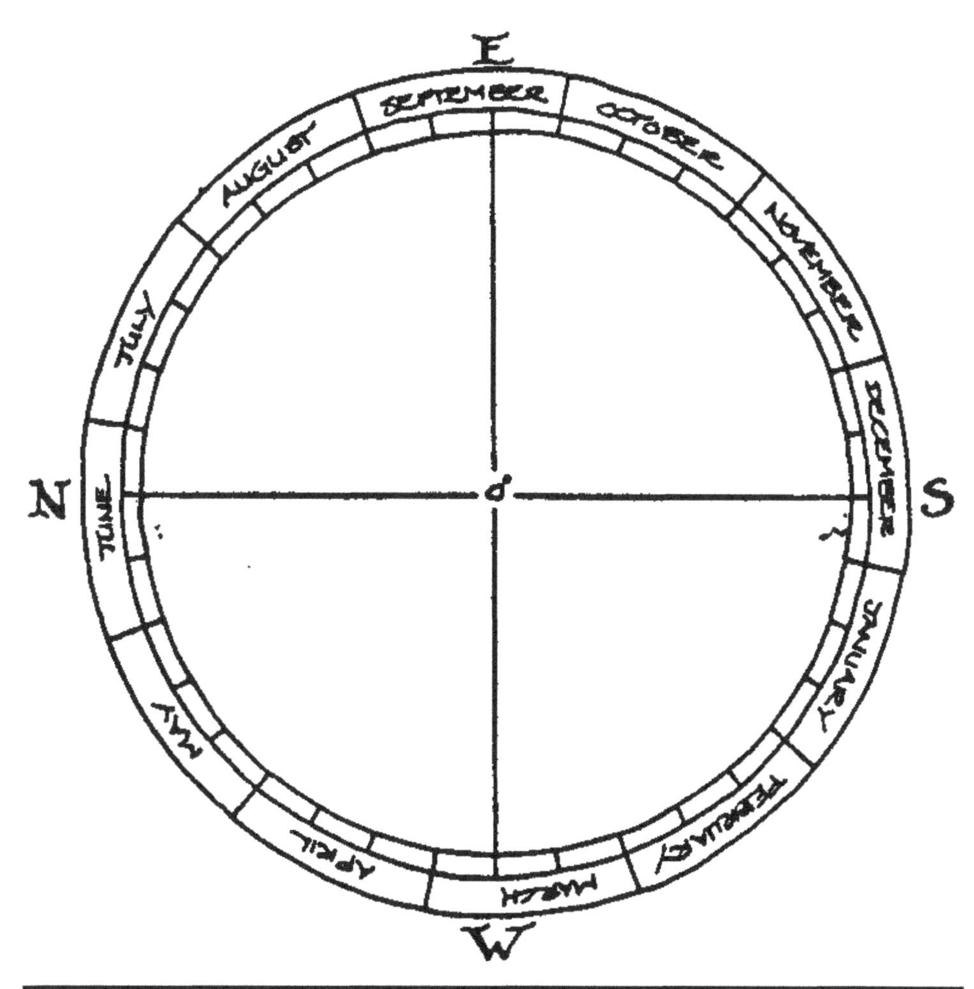

NORTH AMPLITUDES	SOUTH AMPLITUDES
SUNRISE BEARING = 90° − AMPLITUDE	SUNRISE BEARING = 90° + AMPLITUDE
SUNSET BEARING = 270° + AMPLITUDE	SUNSET BEARING = 270° − AMPLITUDE

CIRCULAR NAVIGATIONAL CHART

f. <u>Shadow Stick Construction</u>. This technique will achieve a cardinal direction within 10 degrees of accuracy if done within two hours of local apparent noon. Once again, this technique may be impractical near the polar regions as shadows tend to be very long; similarly, in the tropics shadows are generally very small.

(1) Get a straight, 3-6 foot stick free of branches and pointed at the ends and 3-5 small markers: i.e., sticks, rocks, or nails.

(2) Place the stick upright in the ground and mark the shadow tip with a marker.

(3) Wait 10-15 minutes and mark shadow tip again with a marker.

(4) Repeat this until all of the markers are used.

<u>SHADOW STICK METHOD</u>

(5) The markers will form a West—East line.

(6) Put your left foot on the first marker and your right foot on the last marker, you will then be facing north.

TRANSITION: Another method of navigating can be worked out before hand and will make for a quick and easy orientation.

3. (5 Min) **POCKET NAVIGATOR**. The only material required is a small piece of paper or other flat-surface material upon which to draw the trace of shadow tips and a 1 to 2 inch pin, nail, twig, wooden matchstick, or other such device to serve as a shadow-casting rod.

   a. Set this tiny rod upright on your flat piece of material so that the sun will cause it to cast a shadow. Mark the position where the base of the rod sits so it can be returned to the same spot for later readings. Secure the material so that it will not move and mark the position of the material with string, pebbles, or twigs, so that if you have to move the paper you can put it back exactly as it was. Now, mark the tip of the rod's shadow.

   b. As the sun moves, the shadow-tip moves. Make repeated shadow-tip markings every 15 minutes. As you make the marks of the shadow tip, ensure that you write down the times of the points.

   c. At the end of the day, connect the shadow-tip markings. The result will normally be a curved line. The closer to the vernal or autumnal equinoxes (March 21 and September 23) the less pronounced the curvature will be. If it is not convenient or the tactical situation does not permit to take a full day's shadow-tip markings, your observation can be continued on subsequent days by orienting the pocket navigator on the ground so that the shadow-tip is aligned with a previously plotted point.

   d. The markings made at the sun's highest point during the day, or solar noon, is the north—south line. The direction of north should be indicated with an arrow on the navigator as soon as it is determined. This north-south line is drawn from the base of the rod to the mark made at solar noon. This line is the shortest line that can be drawn from the base of the pin to the shadow-tip curve.

   e. To use your pocket navigator, hold it so that with the shadow-tip is aligned with a plotted point at the specified point. i.e.; if it is now 0900 the shadow-tip must be aligned with that point. This will ensure that your pocket navigator is level. The drawn arrow is now oriented to true north, from which you can orient yourself to any desired direction of travel.

   f. The pocket navigator will work all day and will not be out of date for approximately one week.

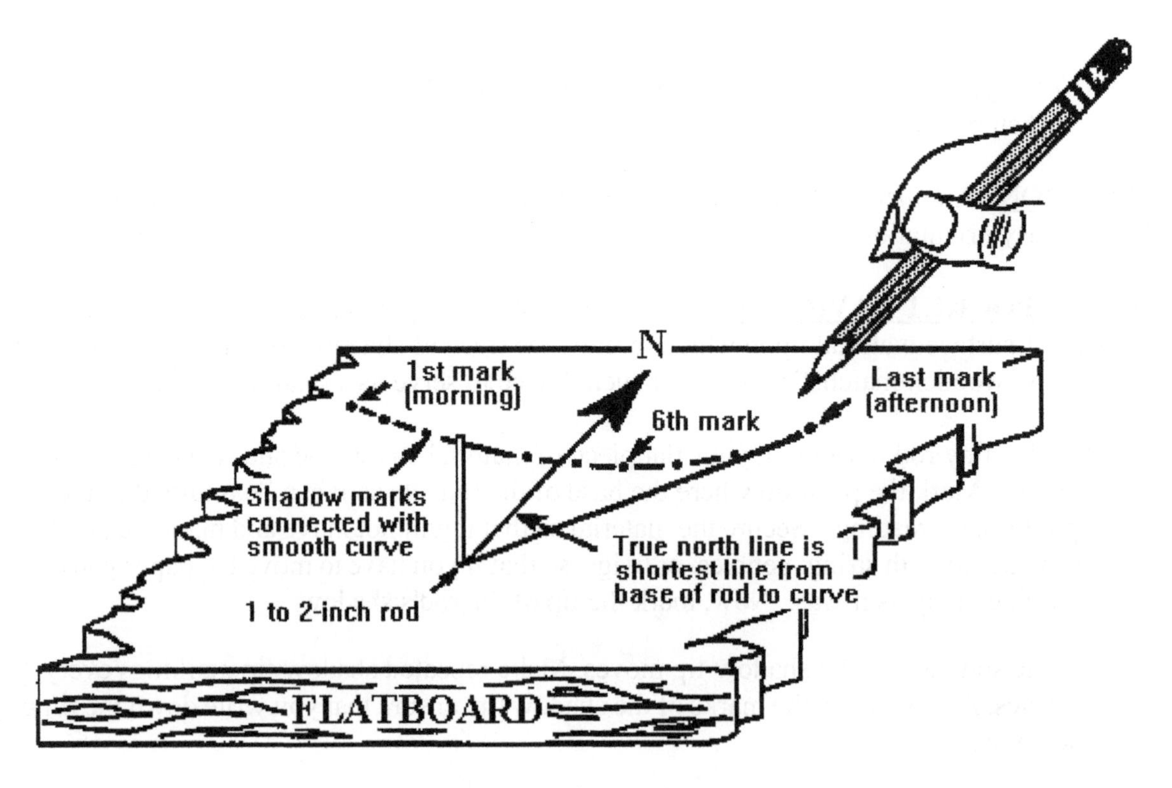

POCKET NAVIGATOR

TRANSITION: Now that we have covered some of the methods used for navigating during the daytime, let's talk about navigating at night.

4. (5 Min) **NIGHTTIME SURVIVAL NAVIGATION**

   a. Mark North. To aid you in navigating at night, it is beneficial to watch where the sun goes down. If you're going to start moving after dark mark the northerly direction.

   b. Locating the North Star. There are two methods used in locating the North Star.

      (1) Using the Big Dipper (*Ursa Major*). The best indictors are the two "dippers ". The North Star is the last star in the handle of the little dipper, which is not the easiest constellation to find. However, the Big Dipper is one of the most prominent constellations in the Northern Hemisphere. The two lowest stars of the Big Dipper's cup act as pointers to the North Star. If you line up these two stars, they make a straight line that runs directly to the North Star. The distance to the North Star along this line is 5 times that between the two pointer stars.

      (2) Using Cassiopeia (*Big M or W*). Draw a line straight out from the center star, approximately half the distance to the Big Dipper. The North Star will be located there.

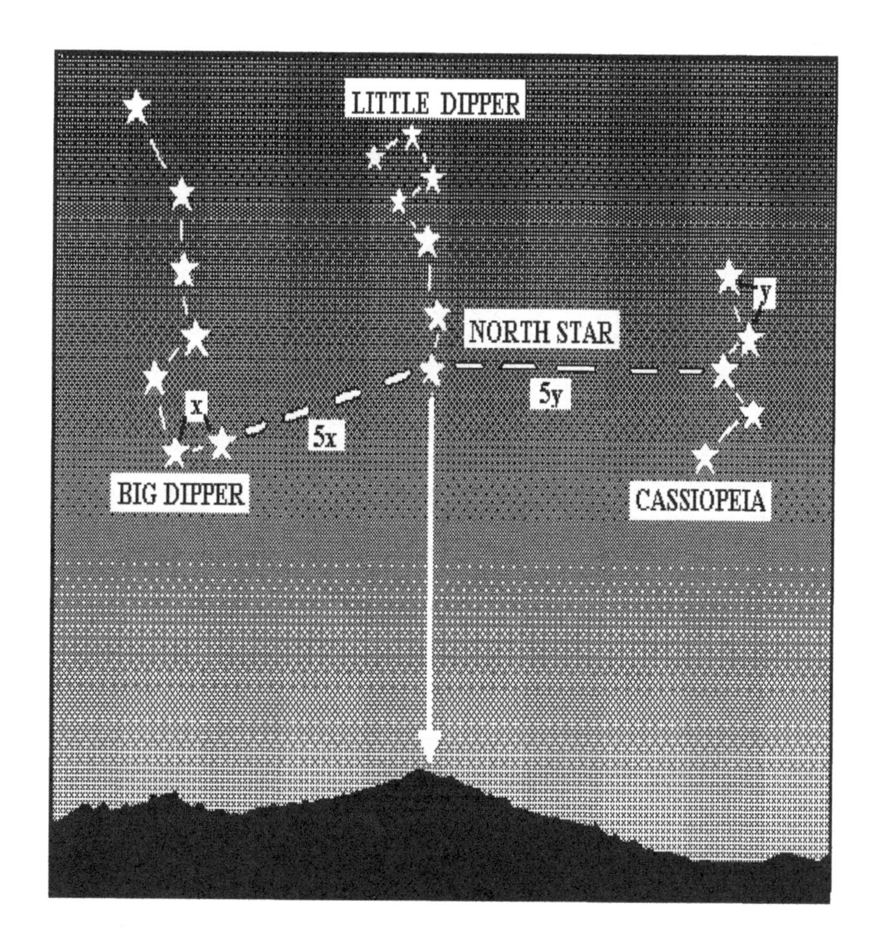

LOCATING THE NORTH STAR

NOTE: Because the Big Dipper and Cassiopeia rotate around the North Star, they will not always appear in the same position in the sky. In the higher latitudes, the North Star is less effective for the purpose of orienting because it appears higher in the sky. At the center of the Arctic circle, it would be directly overhead, and all directions lead South.

c. Southern Cross. In the Southern Hemisphere, Polaris is not visible. There, the Southern Cross is the most distinctive constellation. An imaginary line through the long axis of the Southern Cross, or True Cross, points towards a dark spot devoid of stars approximately three degrees offset from the South Pole. The True Cross should not be confused with the larger cross nearby know as the False Cross, which is less bright, more widely spaced, and has five stars. The True Cross can be confirmed by two closely spaced, very bright stars that trail behind the cross piece. These two stars are often easier to pick out than the cross itself. Look for them. Two of the stars in the True Cross are among the brightest stars in the heavens; they are the stars on the southern and eastern arms. The stars on the northern and western arms are not as conspicuous, but are bright.

NOTE: The imaginary point depicted in the picture is the dark spot devoid of stars.

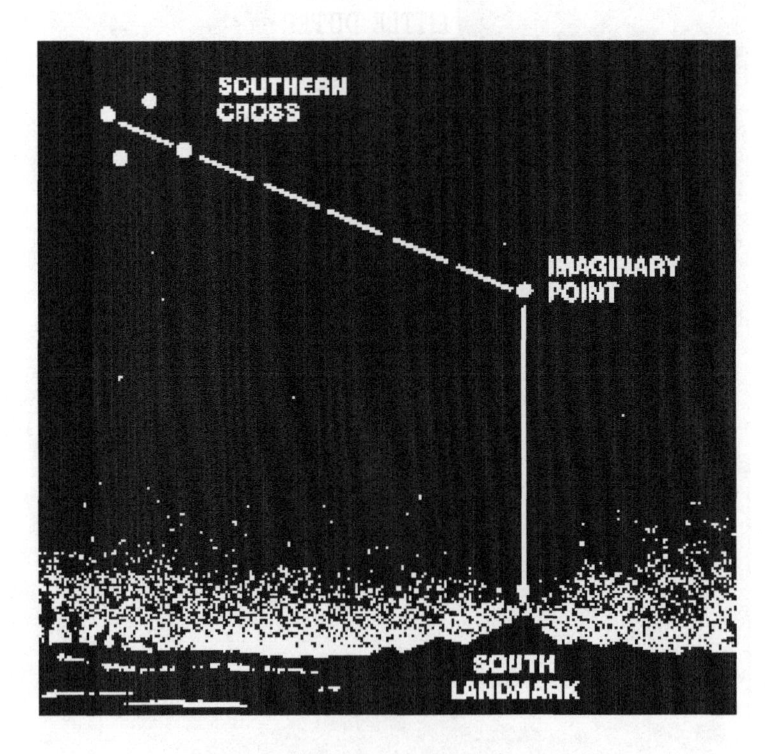

<u>**SOUTHERN CROSS**</u>

d. <u>Moon Navigator</u>. Like the sun, the moon rises in the east and sets in the west. Use the same method of the shadow stick as you did during the day.

<u>TRANSITION</u>: Some day you could find yourself without the aid of celestial bodies which will require alternative navigational tools.

5. (5 Min) <u>**IMPROVISED COMPASSES**</u>. There are three improvised techniques to construct a compass.

a. <u>Synthetic technique</u>. The required items are a piece of synthetic material, (i.e., parachute cloth), and a small piece of iron or steel that is long, thin, and light. Aluminum or yellow metals won't work (only things that rust will do). A pin or needle is perfect, but a straightened paper clip, piece of steel baling wire, or barbed wire could also work.

(1) Stroke the needle repeatedly in one direction against the synthetic material. Ensure that you lift the material a few inches up into the air at the end of each stroke, returning to the beginning of the needle before descending for another stroke in the same direction. Do this approximately 30 strokes. This will magnetize the needle.

(2) Float the metal on still water using balled up paper, wood chip, or leaf. Gather some water in a non-magnetic container or a scooped out recess in the ground, such as a puddle. Do not use a "tin can" which is made of steel. (An aluminum can would be fine.) Place the float on the water, then the metal on it. It will slowly turn to orient itself.

b. Magnet technique. You will achieve the same results by using a magnet. Follow the same steps as you did with the synthetic material. The magnets you are most likely to have available to you are those in a speaker or headphones of a radio.

c. Magnetization through a battery. A power source of 2 volts or more from a battery can be used with a short length of insulated wire to magnetize metal. Coil the wire around a needle. If the wire is non-insulated, wrap the needle with paper or cardboard. Attach the ends to the battery terminals for 5 minutes.

d. Associated problems with improvised compasses. The following are common problems with all improvised compasses.

(1) Soft steel tends to lose its magnetism fairly quickly, so you will have to demagnetize your needle occasionally, though you should not have to do this more than two or three times a day.

(2) Test your compass by disturbing it after it settles. Do this several times. If it returns to the same alignment, you're OK. It will be lined up north and south, though you will have to determine by other means which end is north. Use the sun, stars, or any other natural signs in the area.

(3) Remember, this will give magnetic north. In extreme northern latitudes, the declination angle can be extreme.

6. (5 Min) **SURVIVAL NAVIGATION TECHNIQUES**

a. Navigator

(1) Employ a navigation method.

(2) Find the cardinal direction.

(3) Pick a steering mark in the desired direction of travel.

b. Maintain a Log. The possibility may arise when you will not have a map of the area. A log will decrease the chance of walking in circles.

(1) Construction.

(a) Use any material available to you i.e., paper, clothing, MRE box, etc.

(b) Draw a field sketch annotating North, prominent terrain features, and friendly/enemy position.

(2) Maintenance

(a) Annotate distance traveled, elevation gained and lost, and cardinal directions.

(b) Maintain and update field sketch as movement progresses.

(c) Ensure readability of your field sketch. (i.e.; don't clutter the sketch so much that it can't be read.)

c. <u>During Movement Constantly Refer To</u>

(1) Log

(2) Steering marks

d. <u>Actions If You Become Lost</u>

(1) Immediate action

(a) Orient your sketch. This will probably make your mistake obvious.

(2) Corrective action

(a) Backtrack using steering marks until you have determined the location of your error.

(b) Re-orient your sketch.

(c) Select direction of travel and continue to march.

<u>TRANSITION</u>: Knowing where you are is just as important as knowing where you want to be, one leads naturally to another. Are there any questions?

PRACTICE (CONC)

a. Students will practice survival navigation.

PROVIDE HELP (CONC)

a. The instructors will assist the students when necessary.

OPPORTUNITY FOR QUESTIONS (3 Min)

1. QUESTIONS FROM THE CLASS

2. QUESTIONS TO THE CLASS

Q. What are the two daytime survival navigation techniques?

A. (1) The shadow stick method.
(2) The pocket navigator method.

Q. What are the two methods in locating the North Star?

A. (1) Big Dipper (Ursa Major)
(2) Cassiopeia (Big M or W)

Q. What is the immediate action if you become lost using navigating techniques?

A. Orient the sketch.

SUMMARY (2 Min)

a. We have covered survival navigation to include day and night techniques, methods for stationary and moving situations, and some general techniques and tips to prevent you from going astray.

b. Those of you with IRF's please fill them out at this time and turn them in to the instructor We will now take a short break.

UNITED STATES MARINE CORPS
Mountain Warfare Training Center
Bridgeport, California 93517-5001

SMO
03/31/01

LESSON PLAN

EXPEDIENT SHELTERS AND FIRES

INTRODUCTION (5 Min)

1. **GAIN ATTENTION**. Imagine yourself in this situation. While being inserted by helicopter, it suddenly loses power and crashes. Being the only survivor of the crash all of your equipment was destroyed except what you have on your body. Due to the weather your chances of being found within hours are slim. A cold front is now moving in on your position which may bring snow. What will be the first survival consideration? The first consideration in this survival situation is going to be shelter and fire. The human body can survive an amazing length of time without food or water. Without protection from the elements, particularly in a harsh environment, you will only survive for a short time. You will die from exposure to the elements long before you would ever die from lack of food or water. It is essential that every Marine has sufficient knowledge to construct some type of shelter and fire.

2. **PURPOSE**. The purpose of this period of instruction is to introduce the student to building expedient shelters and fires. This will be accomplished by breaking the class into two parts dealing with shelters and fires separately. We will discuss good shelter, hazards, methods for preparing and starting fires. This lesson relates to Requirements for Survival and Survival Medicine.

3. **INTRODUCE LEARNING OBJECTIVES**

   a. TERMINAL LEARNING OBJECTIVE.

      (1) In a summer mountainous environment, construct expedient shelters, in accordance with the references.

      (2) In a summer mountainous environment, construct fires, in accordance with the references.

   b. ENABLING LEARNING OBJECTIVES

      (1) Without the aid of references, list in writing the characteristics of a safe expedient shelter, in accordance with the references.

(2) Without the aid of references, list in writing the hazards to avoid when using natural shelters, in accordance with the references.

(3) Without the aid of references, list in writing the types of man-made expedient shelters, in accordance with the references.

(4) Without the aid of references, list in writing the tactical fire lay, in accordance with the references.

(5) Without the aid of references, start a fire using a primitive method, in accordance with the references.

4. **METHOD/MEDIA**. The material in this lesson will be presented by lecture and demonstration. You will practice what you have learned during upcoming field training exercises. Those of you with IRF's please fill them out at the end of this period of instruction.

5. **EVALUATION**. You will be tested later in the course by written and performance evaluations on this period of instruction.

TRANSITION: There are several criteria to be considered when building a survival shelter.

BODY (60 Min)

1. (5 Min) **BASIC CHARACTERISTICS FOR SHELTER** Any type of shelter, whether it is a permanent building, tentage, or an expedient shelter must meet six basic characteristics to be safe and effective. The characteristics are:

   a. Protection From the Elements. The shelter must provide protection from rain, snow, wind, sun, etc.

   b. Heat Retention. It must have some type of insulation to retain heat; thus preventing the waste of fuel.

   c. Ventilation. Ventilation must be constructed, especially if burning fuel for heat. This prevents the accumulation of carbon monoxide. Ventilation is also needed for carbon dioxide given off when breathing.

   d. Drying Facility. A drying facility must be constructed to dry wet clothes.

   e. Free from Natural Hazards. Shelters should not be built in areas of avalanche hazards, under rock fall or "standing dead" trees have the potential to fall on your shelter.

   f. Stable. Shelters must be constructed to withstand the pressures exerted by severe weather.

TRANSITION: There are two categories of expedient shelters, natural and man-made. The type you choose will be dictated by the terrain, weather, availability of materials and tools, time available and tactical situation. We will first discuss natural shelters.

2. (5 Min) **NATURAL SHELTERS**. Natural shelters are usually the preferred types because they take less time and materials construct. The following may be made into natural shelters with some modification.

   a. Caves or Rock Overhangs. Can be modified by laying walls of rocks, logs or branches across the open sides.

   b. Hollow Logs. Can be cleaned or dug out, then enhanced with ponchos, tarps or parachutes hung across the openings.

   c. Hazards of Natural Shelters.

      (1) Animals. Natural shelters may already be inhabited (i.e. bears, coyotes, lions, rats, snakes, etc.). Other concerns from animals may be disease from scat or decaying carcasses.

      (2) Lack of Ventilation. Natural shelters may not have adequate ventilation. Fires may be built inside for heating or cooking but may be uncomfortable or even dangerous because of the smoke build up.

      (3) Gas Pockets. Many caves in a mountainous region may have natural gas pockets in them.

      (4) Instability. Natural shelters may appear stable, but in reality may be a trap waiting to collapse.

TRANSITION: Because of the gear Marines take to the field with them and their natural knowledge of "field living", man-made shelters are more likely to be used.

3. (5 Min) **MAN-MADE SHELTERS**. Many configurations of man-made shelters may be used. Over-looked man-made structures found in urban or rural environments may also provide shelter (i.e. houses, sheds, or barns). Limited by imagination and materials available, the following man-made shelters can be used in any situation.

   a. Poncho Shelter.

   b. Sapling Shelter.

   c. Lean-To.

   d. Double Lean-To.

e. A-frame Shelter.

f. Fallen Tree Bivouac.

TRANSITION: We will now learn how to construct each of the man-made shelters.

4. (15 Min) **CONSTRUCTION OF MAN-MADE SHELTERS**. To maximize the shelter's effectiveness, Marines should take into consideration the following prior to construction.

a. Considerations.

(1) Group size.

(2) Low silhouette and reduced living area dimensions for improved heat conservation

(3) Avoid exposed hill tops, valley floors, moist ground, and avalanche paths.

(4) Create a thermal shelter by applying snow, if available, to roof and sides of shelter.

(5) Location of site to fire wood, water, and signaling, if necessary.

(6) How much time and effort needed to build the shelter.

(7) Can the shelter adequately protect you from the elements (sun , wind, rain, and snow). Plan on worst case scenario.

(8) Are the tools available to build it. If not, can you make improvised tools?

(9) Type and amount of materials available to build it.

(10) When in a tactical environment, you must consider the following:

(a) Provide concealment from enemy observation.

(b) Maintain camouflaged escape routes.

(c) Use the acronym BLISS as a guide.

1. B – Blend in with the surroundings.

2. L – Low silhouette.

3. I – Irregular shape.

4. S – Small.

5. S – Secluded located.

b. Poncho Shelter. This is one of the easiest shelters to construct. Materials needed for construction are cord and any water-repellent material (i.e. poncho, parachute, tarp). It should be one of the first types of shelter considered if planning a short stay in any one place.

(1) Find the center of the water-repellent material by folding it in half along its long axis

(2) Suspend the center points of the two ends using cordage.

(3) Stake the four corners down, with sticks or rocks.

c. Sapling Shelter. This type of shelter is constructed in an area where an abundance of saplings are growing. It is an excellent evasion shelter.

(1) Find or clear an area so that you have two parallel rows of saplings at least 4' long and approximately 1 ½' to 2' apart.

(2) Bend the saplings together and tie them to form several hoops which will form the framework of the shelter.

(3) Cover the hoop with a water-repellent covering.

(4) The shelter then may be insulated with leaves, brush, snow, or boughs.

(5) Close one end permanently. Hang material over the other end to form a door.

d. Lean-To. A lean-to is built in heavily forested areas. It requires a limited amount of cordage to construct. The lean-to is an effective shelter but does not offer a great degree of protection from the elements.

(1) Select a site with two trees (4-12" in diameter), spaced far enough apart that a man can lay down between them. Two sturdy poles can be substituted by inserting them into the ground the proper distance apart.

(2) Cut a pole to support the roof. It should be at least 3-4" in diameter and long enough to extend 4-6" past both trees. Tie the pole horizontally between the two trees, approximately 1 meter off the deck.

(3) Cut several long poles to be used as stringers. They are placed along the horizontal support bar approximately every 1 ½' and laid on the ground. All stringers may be tied to or laid on the horizontal support bar. A short wall or rocks or logs may be constructed on the ground to lift the stringers off the ground, creating additional height and living room dimensions.

(4) Cut several saplings and weave them horizontally between the stringers. Cover the roof with water-repellent and insulating material.

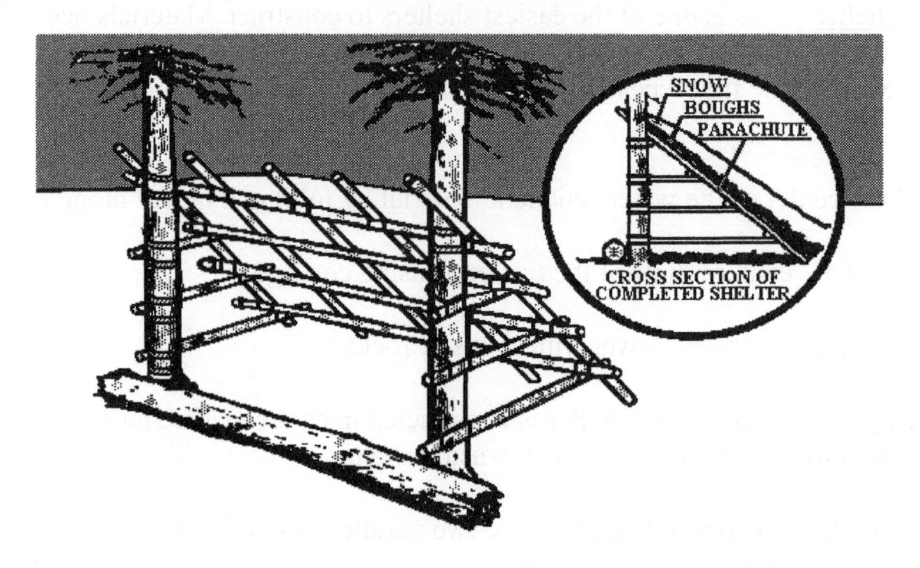

LEAN-TO

e. Double Lean-To. The double lean-to shelter is constructed for 2-5 individuals. It is constructed by making two lean-to's and placing them together.

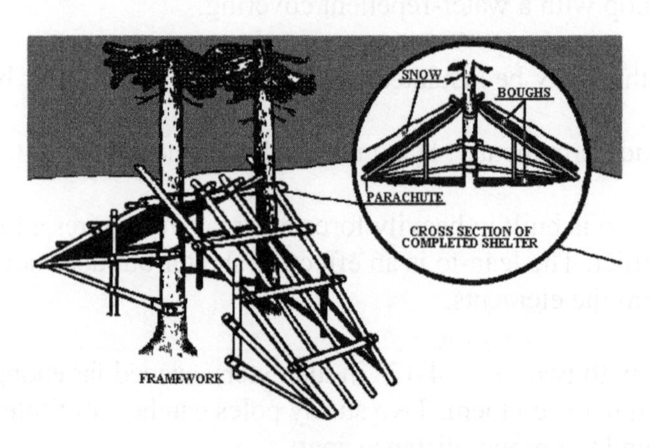

DOUBLE LEAN-TO

f. A-Frame Shelter. An A-Frame shelter is constructed for 1-3 individuals. After the frame work is constructed, bough/tentage is interwoven onto the frame and snow, if available, is packed onto the outside for insulation.

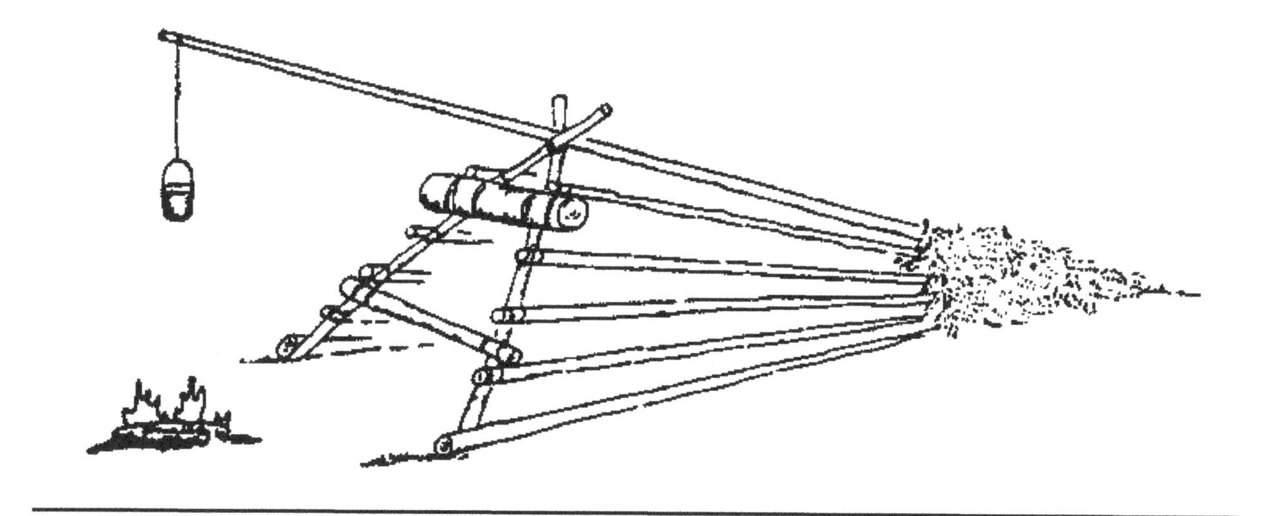

A-FRAME

g. Fallen Tree Bivouac. The fallen tree bivouac is an excellent shelter because most of the work has already been done.

(1) Ensure the tree is stable prior to constructing.

(2) Branches on the underside are cut away making a hollow underneath.

(3) Place additional insulating material to the top and sides of the tree.

(4) A small fire is built outside of the shelter.

FALLEN TREE BIVOUAC

5. (5 Min) **REFLECTOR WALLS.** Heating a shelter requires a slow fire that produces lots of steady heat over a long period of time. A reflector wall should be constructed for all open ended shelters. A reflector wall is constructed with a flat rock or a stack of green logs propped behind the fire. A surprising amount of heat will bounce back from the fire into the shelter.

TRANSITION: Now that we've discussed shelters, let's look at the methods we can use to warm ourselves.

6. (15 Min) **FIRES**. Fires fall into two main categories: those built for cooking and those built for warmth and signaling. The basic steps are the same for both: preparing the fire lay, gathering fuel, building the fire, and properly extinguishing the fire.

   a. Preparing the fire lay. There are two types of fire lays: fire pit and Dakota hole. Fire pits are probably the most common.

      (1) Create a windbreak to confine the heat and prevent the wind from scattering sparks. Place rocks or logs used in constructing the fire lay parallel to the wind. The prevailing downwind end should be narrower to create a chimney effect.

      (2) Avoid using wet rocks. Heat acting on the dampness in sandstone, shale, and stones from streams may cause them to explode.

      (3) Dakota Hole. The Dakota Hole is a tactical fire lay. Although no fire is 100% tactical, this fire lay will accomplish certain things:

         (a) Reduces the signature of the fire by placing it below ground.

         (b) Provides more of a concentrated heat source to boil and cook, thus preserving fuel and lessening the amount of burning time.

         (c) By creating a large air draft, the fire will burn with less smoke than the fire pit.

         (d) It is easier to light in high winds.

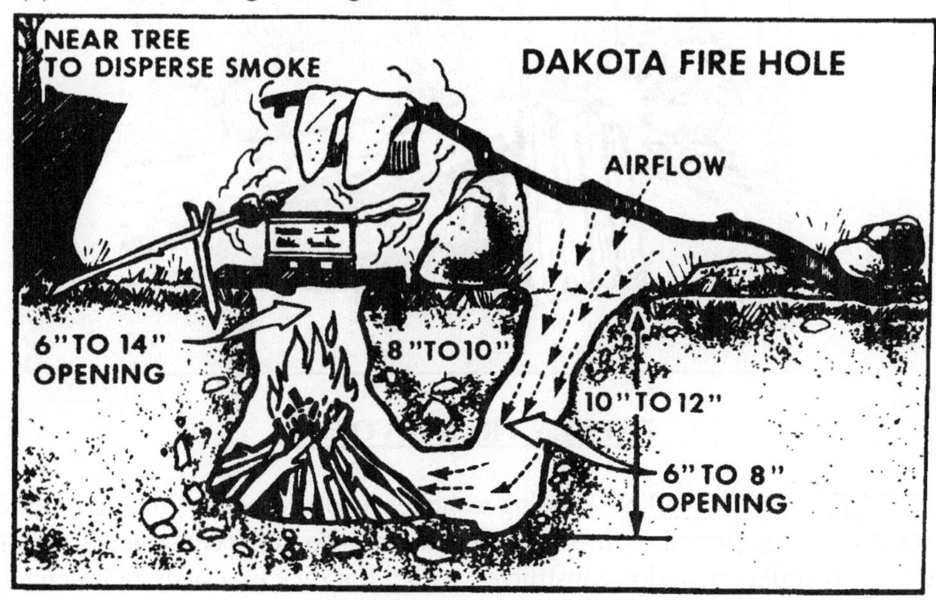

DAKOTA HOLE

b. Gather Fuel. Many Marines take shortcuts when gathering firewood. Taking a few extra minutes can mean the difference between ease and frustration when building a fire.

(1) Tinder. Tinder is the initial fuel. It should be fine and dry. Gather a double handful of tinder for the fire to be built and an extra double handful to be stored in a dry place for the following morning. Dew can moisten tinder enough to make lighting the fire difficult. Some examples are:

(a) Shredded cedar/juniper bark, pine needles.

(b) Dry grass.

(c) Slivers shaved from a dry stick.

(d) Hornet's nest.

(e) Natural fibers from equipment supplemented with pine pitch (i.e., cotton battle dressing).

(f) Cotton balls and petroleum jelly or Char-cloth.

NOTE: Sticks used for tinder should be dry and not larger than the diameter of a toothpick.

(2) Kindling. This is the material that is ignited by the tinder that will burn long enough to ignite the fuel.

(a) Small sticks/twigs pencil-thick up to the thickness of the thumb. Ensure that they are dry.

(b) Due to a typically large resin content, evergreen limbs often make the best kindling. They burn hot and fast, but typically do not last long.

(3) Fuel Wood. Fuel Wood is used to keep the blaze going long enough to fulfill its purpose. Ideally, it should burn slow enough to conserve the wood pile, make plenty of heat, and leave an ample supply of long-lasting coals.

(a) Firewood broken from the dead limbs of standing trees or windfalls held off the ground will have absorbed less moisture and therefore should burn easily.

(b) Refrain from cutting down live, green trees.

(c) Softwoods (evergreens and conifers) will burn hot and fast with lots of smoke and spark, leaving little in the way of coals. Hardwoods (broad leaf trees) will burn slower with less smoke and leave a good bed of coals.

(d) Learn the woods indigenous to the area. Birch, dogwood, and maple are excellent fuels. Osage orange, ironwood, and manzanita, though difficult to break up, make terrific coals. Aspen and cottonwood burn clean but leave little coals.

(e) Stack your wood supply close enough to be handy, but far enough from the flames to be safe. Protect your supply from additional precipitation.

(f) If you happen to go down in an aircraft that has not burned, a mixture of gas and oil may be used. Use caution when igniting this mixture.

c. Building the Fire. The type of fire built will be dependent upon its intended use; either cooking or heating and signaling.

(1) Cooking Fires. The following listed fires are best used for cooking:

(a) Teepee Fire. The teepee fire is used to produce a concentrated heat source, primarily for cooking. Once a good supply of coals can be seen, collapse the teepee and push embers into a compact bed.

TEEPEE FIRE

(2) Heating and Signaling Fires.

(a) Pyramid Fire. Pyramid fires are used to produce large amounts of light and heat. They will dry out wet wood or clothing.

PYRAMID FIRE

d. Starting Fires. Starting a fire is done by a source of ignition and falls into two categories; modern igniters and primitive methods.

(1) Modern Methods. Modern igniters use modern devices we normally think of to start a fire. Reliance upon these methods may result in failure during a survival situation. These items may fail when required to serve their purpose.

(a) Matches and Lighters. Ensure you waterproof these items.

(b) Convex Lens. Binocular, camera, telescopic sights, or magnifying lens are used on bright, sunny days to ignite tender.

(c) Flint and Steel. Sometimes known as metal matches or "Mag Block". Scrape your knife or carbon steel against the flint to produce a spark onto the tinder. Some types of flint & steel designs will have a block of magnesium attached to the device which can be shaved onto the tinder prior to igniting. Other designs may have magnesium mixed into the flint to produce a higher quality of spark.

(2) Primitive Methods. Primitive fire methods are those developed by early man. There are numerous techniques that fall into this category. The only method that will be taught at MCMWTC is the Bow & Drill.

(3) Bow & Drill. The technique of starting a fire with a bow & drill is a true field expedient fire starting method which requires a piece of cord and knife from your survival kit to construct. The components of the bow & drill are bow, drill, socket, fire board, ember patch, and birds nest.

(a) Bow. The bow is a resilient, green stick about 3/4 of an inch in diameter and 30-36 inches in length. The bow string can be any type of cord, however, 550 cord works best. Tie the string from one end of the bow to the other, without any slack.

(b) Drill. The drill should be a straight, seasoned hardwood stick about 1/2 to 3/4 of an inch in diameter and 8 to 12 inches in length. The top end is tapered to a blunt point to reduce friction generated in the socket. The bottom end is slightly rounded to fit snugly into the depression on the fire board.

(c) Socket. The socket is an easily grasped stone or piece of hardwood or bone with a slight depression on one side. Use it to hold the drill in place and to apply downward pressure.

(d) Fire board. The fire board is a seasoned softwood board which should ideally be 3/4 of an inch thick, 2-4 inches wide, and 8-10 inches long. Cut a depression 3/4 of an inch from the edge on one side of the fire board. Cut a U-shape notch from the edge of the fire board into the depression. This notch is designed to collect and form an ember which will be used to ignite the tinder.

(e) Ember Patch. The ember patch is made from any type of suitable material (i.e., leather, aluminum foil, bark). It is used to catch and transfer the ember from the fire board to the birds nest. Ideally, it should be 4 inches by 4 inches in size.

(f) Birds Nest. The birds nest is a double handful of tinder which will be made into the shape of a nest. Tinder must be dry and finely shredded material (i.e., outer bark from juniper/cedar/sage brush or inner bark from cottonwood/aspen or dry grass/moss). Lay your tinder out in two equal rows about 4 inches wide and 8-12 inches long. Loosely roll the first row into a ball and knead the tinder to further break down the fibers. Place this ball perpendicular onto the second row of tinder and wrap. Knead the tinder until all fibers of the ball are interwoven. Insert the drill half way into the ball to form a partial cylinder. This is where the ember will be placed.

(4) Producing a fire using the bow & drill.

(a) Place the ember patch under the U-shaped notch.

(b) Assume the kneeling position, with the left foot on the fire board near the depression.

(c) Load the bow with the drill. Ensure the drill is between the wood of the bow and bow string. Place the drill into the depression on the fire board. Place the socket on the tapered end of the drill.

(d) Use the left hand to hold the socket while applying downward pressure.

(e) Use the right hand to grasp the bow. With a smooth sawing motion, move the bow back and forth to twirl the drill.

(f) Once you have established a smooth motion, smoke will appear. Once smoke appears, apply more downward pressure and saw the bow faster.

(g) When a thick layer of smoke has accumulated around the depression, stop all movement. Remove the bow, drill, and socket from the fire board, without moving the fire board. Carefully remove your left foot off the fire board.

(h) Gently tap the fire board to ensure all of the ember has fallen out of the U-shaped notch and is lying on the ember patch. Remove the fire board.

(i) Slowly fan the black powder to solidify it into a glowing ember. Grasping the ember patch, carefully drop the ember into the cylinder of the birds nest.

(j) Grasp the birds nest with the cylinder facing towards you and parallel to the ground. Gently blow air into the cylinder. As smoke from the nest becomes thicker, continue to blow air into the cylinder until fire appears.

(5) <u>Trouble Shooting the Bow & Drill</u>

(a) Drill will not stay in depression- Apply more downward pressure and/or increase width/depth of depression.

(b) Drill will not twirl- Lessen the amount of downward pressure and/or tighten bow string.

(c) Socket smoking- Lessen the amount of downward pressure. Wood too soft when compared to hardness of drill. Add some lubrication: animal fat, oil, or grease.

(d) No smoke- Drill and fire board are the same wood. Wood may not be seasoned. Check drill to ensure that it is straight. Keep left hand locked against left shin while sawing.

(e) Smoke but no ember- U-shaped notch not cut into center of the depression.

(f) Bow string runs up and down drill- Use a locked right arm when sawing. Check drill to ensure that it is straight. Ensure bow string runs over the top of the left boot.

(g) Birds nest will not ignite- Tinder not dry. Nest woven too tight. Tinder not kneaded enough. Blowing too hard (ember will fracture).

e. <u>Extinguishing the Fire</u>. The fire must be properly extinguished. This is accomplished by using the drown, stir, and feel method.

(1) <u>Drown</u> the fire by pouring at water in the fire lay.

(2) <u>Stir</u> the ember bed to ensure that the fire is completely out.

(3) Check the bed of your fire by feeling for any hot spots.

(4) If any hot spots are found, start the process all over again.

TRANSITION: Fires are very important to our well-being. They keep us warm, cook our foods, signal for help, give us a sense of accomplishment, force us to work, and really warm your spirits when you are in a bad situation.

PRACTICE (10.5 hrs)

a. Students will construct survival shelters and build expedient fires.

PROVIDE HELP (CONC)

a. The instructors will assist the students when necessary.

OPPORTUNITY FOR QUESTIONS (3 Min)

1. QUESTIONS FROM THE CLASS

2. QUESTIONS TO THE CLASS

Q. What are the six characteristics for a safe shelter?

A. (1) Heat retention.
(2) Protection from the elements.
(3) Ventilation.
(4) Drying facility.
(5) Free from hazards.
(6) Stable.

Q. What are the types of man made shelters?

A. (1) Poncho Shelter.
(2) Sapling Shelter.
(3) Lean-To.
(4) Double Lean-To.
(5) A-Frame.
(6) Fallen Tree Bivouac.

SUMMARY (2 Min)

a. During this period of instruction we have discussed the six basic characteristics for a safe expedient shelter, natural shelters, man-made shelters, and fires to include such things as site selection and fire starting methods.

b. Those of you with IRF's please fill them out at this time and turn them in to the instructor. We will take a short break.

UNITED STATES MARINE CORPS
Mountain Warfare Training Center
Bridgeport, California 93517-5001

SMO
03/31/00

LESSON PLAN

SURVIVAL DIET

INTRODUCTION (5 Min)

1. **GAIN ATTENTION**. In almost every military occupational specialty, there exists the possibility that Marines may find themselves in a survival situation. With the information received in the period of instruction, your knowledge of gathering food will be greatly increased.

2. **OVERVIEW**. The purpose of this period of instruction is to familiarize the student to plants and insects that may be foraged for survival uses. This lesson relates to all other survival classes.

INSTRUCTOR NOTE: Have students read learning objectives.
SMO CUE: TC 1

3. **INTRODUCE LEARNING OBJECTIVES**

   a. TERMINAL LEARNING OBJECTIVE.

      (1) In a summer mountainous environment, subsist on plant resources, in accordance with the reference.

      (2) In a summer mountainous environment, subsist on insects resources, in accordance with the reference.

      (3) In a summer mountainous environment, prepare game, in accordance with the reference.

   b. ENABLING LEARNING OBJECTIVES

      (1) With the aid of the references, list orally the plants to be avoided, in accordance with the reference.

(2) With the aid of references, describe an edible plant, in accordance with the reference.

(3) With the aid of references, describe the plant testing procedure, in accordance with the reference.

(4) With the aid of references, list orally insects to be avoided, in accordance with the reference.

(5) With the aid of references, describe orally an edible insect, in accordance with the reference.

(6) With the aid of references, list the two types of cooking, in accordance with the reference.

4. **METHOD/MEDIA**. The material in this lesson will be presented by lecture and demonstration. You will practice what you have learned in upcoming field training exercises. Those of you with IRF's please fill them out at the conclusion of this period of instruction.

5. **EVALUATION**. You will be tested by an oral exam later in this course.

TRANSITION: Are there any questions over the purpose, learning objectives, how the class will be taught or how you will be evaluated? First let us cover some general information.

BODY (100 Min)

SMO CUE: TC 2

1. (5 Min) **GENERAL CONSIDERATIONS**. There are few places without some type of edible vegetation. Plants contain vitamins, minerals, protein, carbohydrates, and dietary fiber. Some plants also contain fats. The following are general considerations:

a. Do not assume that because birds or animals have eaten a plant it is edible by humans.

b. Poor plant recognition skills will seriously limit your ability to survive.

c. Plant dormancy and snowfall make foraging plants during the winter months difficult.

d. Plants generally poison by:

(1) Ingestion. When a person eats a part of a poisonous plant.

(2) Contact. When a person makes contact with a poisonous plant that cause any type of skin irritation or dermatitis.

(3) Absorption. When a person absorbs the poison through the skin, which can interrupt the bodies functions.

(4) Inhalation. Poisoning can occur through the inhalation of smoke that contains poisonous properties.

e. Plant properties can change throughout the growing season. Plants can be eatable during certain periods while poisonous in others.

TRANSITION: Are there any questions over the general considerations? Now, let's look at some plants that should be avoided.

SMO CUE: TC 3

2. (5 Min) **PLANT RECOGNITION**. Plant recognition can take years of study to learn. The following is used as a general guideline:

a. Types of plants that should be avoided are:

(1) All mushrooms.

(2) Plants with milky sap.

(3) White and yellow berries should be avoided as they are almost poisonous.

(4) Plants with shiny leaves should be considered poisonous.

(5) Plants that are irritants to the skin should not be eaten.

(6) All beans and peas should be avoided.

SMO CUE: TC 4

b. Edible Plants. There are several edible plants found at the Marine Corps Mountain Warfare Training Center. All are easy to recognize and many of the same plants are located throughout the United States and abroad.

(1) Wild onions – Complete plant

(2) Dandelion – Roots and leaves

(3) Cattail – Root, stalk and stem

(4) Bull Thistle – The flower

(5) Juniper Tree – Berries and cambium layer.

(6) All Conifers – Cambium layer, needles, and nuts within the cones.

(7) Currants – Berries

SMO CUE: TC 5

c. Preparing an edible plant. All edible plants should be boiled twice with a change of water to remove any bitterness, acids, or bacteria the plant may contain.

TRANSITION: Now that we have discussed plants that should be avoided and several that we can eat safely, are there any questions? Let's discuss the plant testing procedure.

SMO CUE: TC 6

3. (10 Min) **PLANT TESTING PROCEDURE**. Always adopt the following procedure when tying unknown plants as food.

a. Wait 8 hours without eating.

b. Select a plant that grows in sufficient quantity in the local area. Separate the part of the plant that you wish to test: root, stem, leaf, or flower. Certain parts of plants are poisonous while the other parts may be edible.

c. Rub a portion of the plant you have selected on your inner forearm. Wait 15 minutes and look for any swelling, rash, or irritation.

d. Boil the plant or plant part in changes of water. The toxic properties of many plants are water soluble or destroyed by heat. Cooking and discarding two changes of water can lessen the amount of poisonous material or remove it completely. These boiling periods should last at least 5 minutes each.

e. Place 1 teaspoon of the prepared plant food in the mouth and chew for 5 minutes, but do not swallow. If unpleasant effects occur (burning, bitter, or nauseating taste), remove the plant from the mouth at once and discard it as a food source. If no unpleasant effects occur, swallow the plant material and wait 8 hours.

f. If after 8 hours no unpleasant effects have occurred (nausea, cramps, diarrhea), eat two table spoonfuls and wait 8 hours.

g. If no unpleasant effects have occurred at the end of this 8 hour period, the plant may be considered edible.

h. Completely document and sketch the plant in a log book to refer to for future use. This will aid in future procurement of this plant. If plant properties have changed, you will have to repeat the plant testing procedure.

NOTE: Have charcoal ready to eat. If any side effects occur, attempt to induce vomiting and immediately eat charcoal. Charcoal will absorb toxins and reduce the chance of death.

TRANSITION: Now that we have discussed the plant testing procedure, are there any questions? Let's talk about two plants than can be used for medicinal purposes.

SMO CUE: TC 7

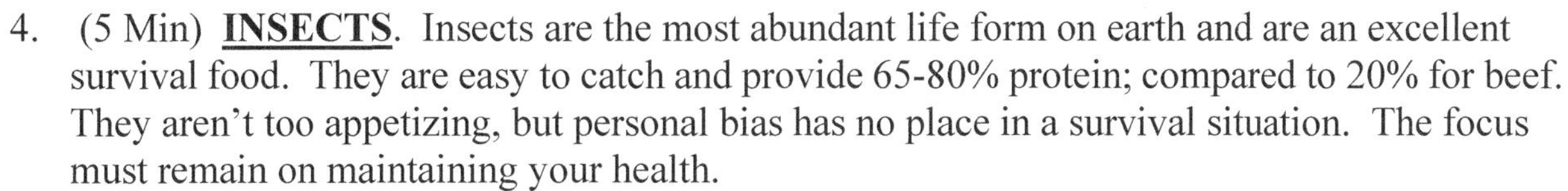

4. (5 Min) **INSECTS**. Insects are the most abundant life form on earth and are an excellent survival food. They are easy to catch and provide 65-80% protein; compared to 20% for beef. They aren't too appetizing, but personal bias has no place in a survival situation. The focus must remain on maintaining your health.

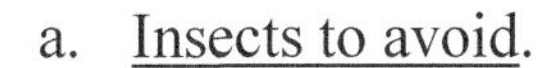

a. Insects to avoid.

(1) All adults that sting or bite.

(2) Hairy or brightly colored insects.

(3) All caterpillars.

(4) Insects that have a pungent odor.

(5) All spiders.

(6) Disease carriers like ticks, flies, or mosquitoes.

SMO CUE: TC 8

b. Edible insects.

(1) Insect larvae.

(2) Grasshoppers.

(3) Beetles.

(4) Grubs.

(5) Ants.

(6) Termites.

(7) Worms.

c. Foraging for Insects. One must be careful not to expend more energy harvesting food than can be replaced. For example, catching insects such as grasshoppers can become exasperating and tiring.

(1) At night grasshoppers climb tall plants and cling to the stalks near the top. They can be picked from the plants in the early morning while they are chilled and dormant.

(2) Dig for worms in damp humus soil, under rock/logs or look for them on the ground after it has rained.

(3) Carpenter ants are found in dead trees and stumps and can be gathered by hand.

(4) Most other insects can be found in rotten logs, under rocks, and in open grassy areas.

SMO CUE: TC 9

d. Preparing Edible Insects.

(1) Insects with a hard outer shell have parasites. Remove the wings and barbed legs before cooking.

(2) Drop worms into potable water for at least a half hour. They will naturally purge themselves. You can either cook or eat them raw.

(3) Most other insects can be eaten raw. Cooking insects will improve their taste. If the thought of eating insects is unbearable, grind them into a paste and mix with other foods.

TRANSITION: Now that we have discussed insects for food, are there any questions? Next we will briefly discuss traps and snares.

INSTRUCTORS NOTE: Inform the students that this next section is just to give them an idea of traps and snares, and that this subject is too large to teach in this class.

SMO CUE: TC 10

5. (5 Min) **GENERAL TIPS AND TECHNIQUES FOR TRAPPING**

a. General Tips. Knowing a few general hints and tips will make the trapping of animals much easier and considerably more effective. The eight general tips for trapping are:

(1) Know your game. Knowing the habits of the animal you want to trap will help lure it into your trap. Such things as when and where they move, feed, and water will help you determine where the trap can be most effectively placed.

(2) Keep things simple. You don't have time in a survival situation to construct elaborate traps and they do not necessarily do a better job.

(3) Set traps in the right place. Animals will travel and stop in certain locations. That is where to best employ traps.

(4) Cover up your scent. Animals will avoid an area where smells are strange to them. Smoke from your fire is the best cover to use.

(5) Use the right type of trap. Some traps work better than others do (i.e. deadfall works better on squirrels than a spring pole).

(6) Use the right size trap. Adjust the size of your traps to the size of the animal.

(7) Check traps. Check your traps twice daily; morning and evening. Checking your trap less than twice a day can allow your game to escape, rot, or be taken by other predators.

(8) Bait your traps. Bait of any type will add to your chances of success.

SMO CUE: TC 11

b. General Technique. A general technique is the method in which the trap is intended to kill or hold the animal.

(1) Strangulation. This method strangles the animal, such as a snare.

(2) Mangling. This method crushes the animal, such as a dead fall.

(3) Entanglement. This method entangles the animal, such as a net.

(4) Live. This method holds the animal, such as a box trap.

SMO CUE: TC 12

6. (5 Min) **KILLING GAME**. There are a few different ways to kill your game.

a. Nose Tap and Heart Stomp. This method is best used for game that is alive and caught in a trap. Using a club, hit the animal on the nose. This will knock it unconscious. Lay the club across the animal's neck. Placing one foot on the club to keep the animal down, use the heel of your other boot to give the animal several sharp blows to the chest area (behind the front legs). This causes the heart to swell up and the animal bleeds internally. This method will not work well on animals with thick skulls (i.e. raccoons).

b. Bleeding. This method is used by wither slitting the throat or puncturing the animal's chest cavity with a spear.

c. Bludgeoning. This method is used by hitting the animal with a club numerous times until it no longer moves.

d. Strangling. This method normally occurs if snares are properly employed.

TRANSITION: Now that we have discussed how to kill the game, are there any questions? Let's cover how to dress your game.

SMO CUE: TC 13

7. (15 Min) **PREPARING GAME**

a. Dressing. Once the animal is dead, dressing should occur immediately. This allows the chest cavity to cool, thus slowing the decay and bacteria rate.

(1) Game

(a) Using a well sharpened pocket knife, cut around anus. Be careful not to puncture intestines or kidneys.

(b) Cut the hide straight up the belly towards the chest cavity. The hide is what you are cutting through, not the layer of skin that holds the organs.

(c) With two fingers inserted under the skin just above the anus, place the blade of the pocket knife in between fingers and again cut up to the chest cavity. This prevents you from puncturing the urinary tract, gall bladder and intestines.

(d) Reach in and pull out the heart, lungs, and liver; keeping them separate from the guts. These organs are good to eat. Check the liver for white spots. If white spots appear on the liver, the animal may have tularemia. A sausage like swelling is the best indication of the disease. Immediately disinfect your hands. Continue to dress the animal, provided you are using a barrier between yourself and the animal. The animal is still edible provided the meat is properly cooked.

SMO CUE: TC 14

(2) Fish

(a) With a pocket knife, scrape the scales off the fish, going back and forth from tail to head.

(b) With your knife, cut the fish open starting at the anus and work towards the gills.

(c) With your finger or thumb, push all guts out and wash thoroughly. Look throughout the intestines to find out what the fish has been eating. It may aid you in procuring more fish.

(3) Birds

(a) Pluck feathers while body is warm. Hot water will help loosen the feathers.

(b) Make incision from vent to tail, insert hand and draw out guts. Retain heart, liver, kidneys, and gizzards. Cut off head and feet.

SMO CUE: TC 15

(4) Reptiles/Amphibians

(a) Cut off head well down behind poison sacs.

(b) Cut open skin from anus to neck. Pull out internal organs and discard.

NOTE: Box turtles, brightly colored frogs, frogs with "X" mark on their backs, and toads should be avoided.

SMO CUE: TC 16

b. Skinning. Depending on the situation, skinning will occur 1-3 days after the animal has been bled. Blood is a valuable food, rich in vitamins and minerals, including salt, that are essential to survive. The hide serves as a protective layer for the meat during the bleeding process. Bleeding is essential for preserving meat.

(1) Large Game

(a) You will find the Achilles Tendon just above the feet, cut a hole between the bone and tendon. This is done so you can thread a rope, string, etc., through the hole in order to hang the animal upside down from a tree branch or a make shift rack.

(b) Cut completely around the hind legs, just below where the animal is suspended. Now cut from the tail up on the inside portion of the hind legs to the original cut on the hind legs.

(c) Pull hide straight down towards the head, pop the front legs through the hide, then pull hide over the skull cutting when necessary.

(2) Small Game

(a) Small game can be skinned like large game or it can be cased. Casing a hide, means to pull the entire skin off the carcass from rear forward, with cuts made only around the feet of the animal and from the back legs to the tail. This method allows the skin to be made into mittens, bags, and other holding materials.

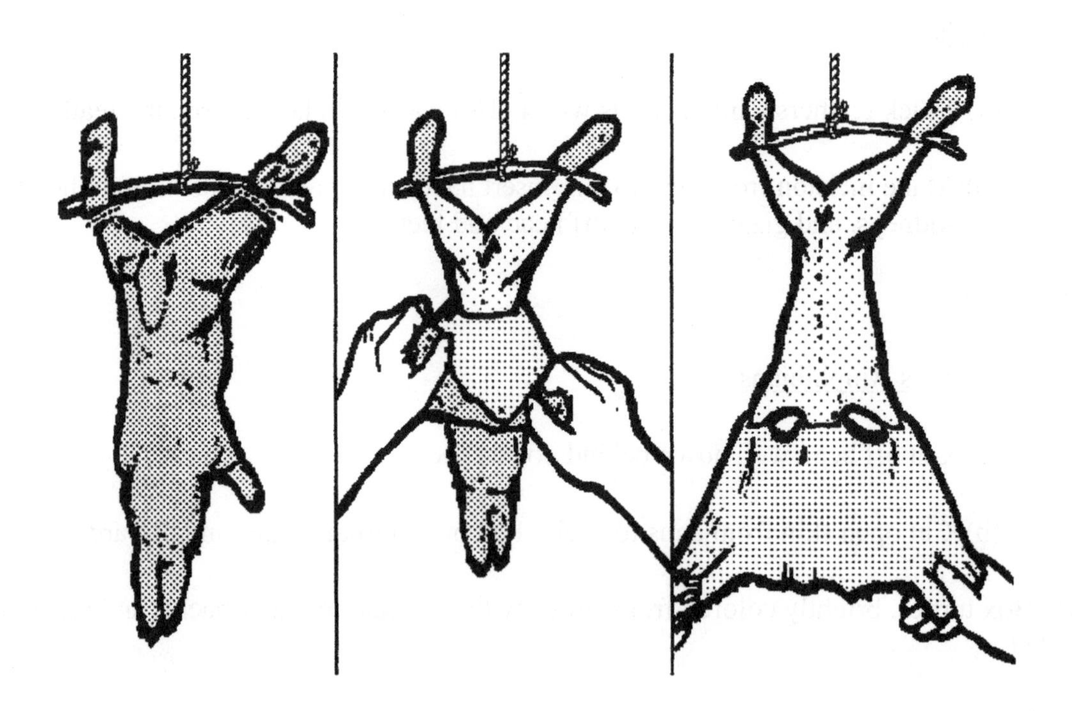

SMO CUE: TC 17

(3) Fish

(a) The skin of fish is usually left on.

(4) Birds

(a) The skin of birds should be left on. There is a heavy layer of fatty tissue between skin and meat which is essential to survival.

(5) Reptiles/Amphibians

(a) The skin of reptiles is left on.

TRANSITION: Now that we have skinned the animal, now we must cut up the meat.

c. Butchering. Butchering entails the cutting of larger animals into manageable chunks or quartering as shown. Smaller animals are generally best left whole.

NOTE: Animals that were killed by the use of poisons should have a 2" cubic size square of meat removed at the point of insertion.

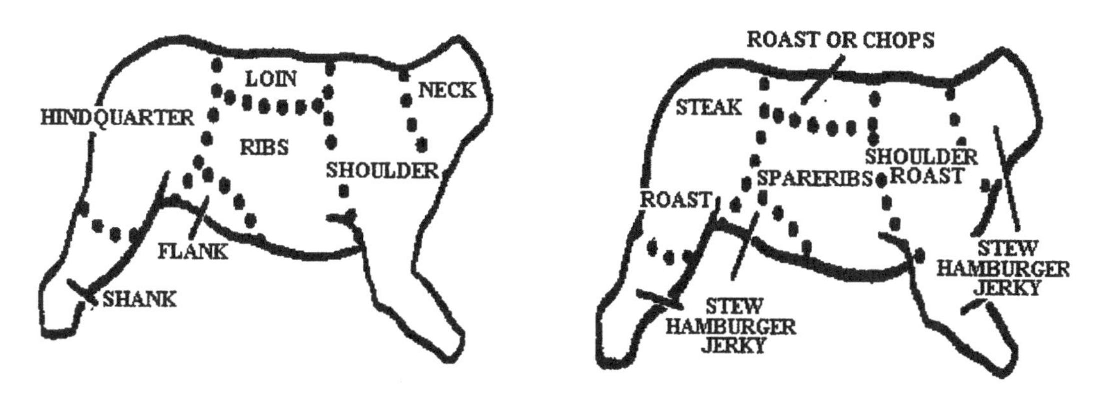

BUTCHERING

d. Washing. The final step in preparing game is to wash the meat.

TRANSITION: Now that we are ready to eat, let's talk about cooking.

SMO CUE: TC 18

8. (5 Min) **COOKING OF MEATS**

a. Cooking. Cooking of meats will kill bacteria/parasites and is only to be used for immediate consumption. There are two methods for cooking.

(1) Boiling. This is the best method of cooking meats. It's easy, requires less fuel, and if you drink the broth, you get full food value.

(a) In order to assure all meat is properly done, cut the meat of equal size.

(b) Place meat into a metal container and either suspend it over a fire or place it on coals.

(c) Bring broth to a rapid boil to ensure all meat is completely done.

(d) This method is best for fish and reptiles. It ensures all bacteria/parasites have been killed.

(2) Roasting. This method is easy and produces the tastiest results, though it causes the most waste.

(a) Cut the meat into very thin strips. Thicker strips of meat requires more time to cook, resulting in higher fuel usage.

(b) Turn meat occasionally to ensure it is thoroughly cooked.

TRANSITION: Now that we have discussed cooking, are there any questions? Let's discuss what to do with the hide.

SMO CUE: TC 19

9. (5 Min) **TANNING HIDES**. The tanning process is a seven step procedure.

   a. Stretching. A fresh green or soaked hide must be stretched immediately. Stretching is accomplished by either making a frame or using the ground.

      (1) Frame stretching involves lacing the hide to a frame with cordage and pulling it tight.

      (2) Ground stretching involves staking the hide tight to the ground.

      (3) Remember, the tighter the hide is stretched, the softer it will be at the end of the process.

   b. Fleshing. Fleshing is the actual removal of meat, tissue, and fat form the hide.

      (1) Soak or wet the hide.

      (2) Using a blunt knife, sharp stone or bone tool scrapper, scrape and hack off all the flesh. Be careful not to make holes in the more tender parts of the belly.

      (3) Scrub and wash bloody areas as you scrape.

      (4) Restretch the hide and let it dry.

   c. Dehairing. Dehairing is the removal of the hair. This step can be eliminated if the hair is to be left on.

      (1) Using a sharp stone scrapper, scrape off the hair. Two things come off in the this process: the hair and the epidermis or first layer of the skin.

   d. Braining. The agent that breaks down the glue or glycerin and loosens the fibers of the skin is found in the brains of the animal.

      (1) Soak the hide on the stretcher.

      (2) Extract the brain from the animal.

      (3) Place the brain on a cupped rock in a fire and cook like "scrambled eggs".

      (4) Once the brains are warm and thoroughly mixed, remove from the hot rock and place on the hide. Firmly rub into the complete hide with your hand, only rubbing into

dehaired sides.

(5) Restretch and let stand for at least six hours.

e. Graining.

(1) Sponge on water to further dampen hide.

(2) Using a blunt end of a pole, apply pressure over every inch, scrapping and stretching the fibers until most of the water is gone from the skin. Allow it to dry.

(3) Cut the skin from the frame around the perimeter, leaving only the lacing holes and hair that could not be removed.

f. Rubbing. The next critical step is the high friction rubbing needed to create a little heat and finish the drying, stretching, and breaking of the grain. Either method can be used.

(1) Use a half inch rope attached between two trees. Grasp the skin at different points all around its perimeter and pull, pull, pull.

(2) If rope is not available, rub the skin by sitting on the ground and hooking the skin over your feet.

g. Smoking. When a white tanned skin gets wet, it will often dry stiff and need rework to restore the softness. Smoking the hide will prevent this and also gives it a buckskin brown color and pleasant wood smoke aroma.

(1) Sew up the hide into a tube with the tail and head ends open.

(2) Attach head to a cross pole and stake the base down over a small but deep fire pit.

(3) Add wet or green wood chips to the fire. Sage or willow are good woods. The object is to get the chips to smoke, not burn.

(4) It only takes a few minutes to smoke, but be careful to prevent flame from ruining the hide.

h. Animal Hide Uses. Animal hide uses are limited only be imagination. Listed below are a few ideas:

(1) Deer, elk, moose, antelope, horse, cow, and goat hides with hair removed: used for clothing, footgear, straps, snowshoes, bowstrings, snare lines, pack sacks.

NOTE: Hallow-haired animals such as deer, elk, and antelope do not tan well with hair left on and shed badly.

(2) Horse, cow, sheep, goat, bear, cougar, bobcat, wolf, marten, coyote, weasel, mink, beaver, otter, raccoon, and skunk hides with hair left on: used for robes, clothing, hats, cold weather footgear.

(3) Badger, wolverine, and porcupine hides with hair left on or removed: used for moccasin soles, hats, cold weather footgear. Badger is noted as a tough hide for moccasin soles.

(4) Rabbit, muskrat, mice, and other small rodent hides with hair left on: used for bags, hats, mittens, fur liners for footgear.

TRANSITION: Now that we have discussed hides, are there any questions? Let's discuss what to do with the excess meat.

SMO CUE: TC 20

10. (5 Min) **PRESERVING MEATS**. To have a supply of food in excess of daily consumption is the object of the survivor. Steps must be taken to avoid spoilage and theft by small animals.

a. Freezing. In winter this is no great problem.

(1) Before freezing, cut the meat into pieces of a size that can be used one at a time.

(2) Keep it frozen until ready to use. Remember, meat will spoil if thawed and refrozen.

b. Cooling. In the summer, small quantities can be kept for a day or so.

(1) Place meat in a metal or wooden container with a lid. The container should be ventilated.

(2) Set it in water or bury it in damp earth, preferably in a shaded location.

(3) Do not throw moldy meat away; cut or scrape off the mold and cook as usual.

c. Jerky. Jerky allows the meat to last a couple of weeks while reducing the weight of meat by dehydrating it. Jerky is made from the meat only.

(1) Cut meat into thin strips about ¼ inch thick. Remove all thick portions of fat.

(2) Place meat by a fire to lightly smoke it. You are attempting to develop a thin crust layer on the meat. This serves to deter the bugs and insects. Remember to use hard woods and not conifer type wood. You want to smoke it, not cook the meat.

(3) Once the meat has a crust layer, remove the meat and place strips on a hanger or rock in the sun for approximately 24 hours. Once dry, break down fibers by slightly pulling

apart the meat and allow it to dry another 24 hours.

(4) When it becomes hard and brittle, it is taken down and stored in bags. It is then used in stews, soups, or roasted lightly on coals and eaten.

(5) Small animals, fish, and birds are dried whole. After they are skinned, the back is cracked between the legs, a stick is inserted to hold the body cavity open. The animal is lightly smoked and laid out in the sun to dry. When thoroughly dried, they are pounded until the bones are crushed. Another day in the sun will dry the marrow and ensure preservation.

d. Pemmican.

(1) Dry berries and pound into a paste.

(2) Dried jerky is added to the paste.

(3) Melted suet (the hard fatty tissues around the kidneys) is mixed with the berries and jerky.

(4) Roll the mixture into small balls and place in the cleaned intestines of a large animal.

(5) The intestine sack is ties shut, sealed with suet and stored in plastic or leather bags.

TRANSITION: Now that we have talked about preserving the meat, are there any questions? Let's cover the specific uses of all the parts of the animal.

SMO CUE: TC 21

11. (5 Min) **SPECIFIC PARTS**

a. Other than the actual meat on game there ate other parts of it that can be eaten. They are the:

(1) Brain

(2) Eyes

(3) Tongue

(4) Liver

(5) Heart

(6) Lungs

(7) Kidneys

(8) Gizzards

TRANSITION: It is very important that you eat sufficiently to keep up your strength. Hungry survivors are dead survivors, so use all edible parts of what you catch. Are there any questions? If you have none for me, then I have some for you.

PRACTICE (CONC)

a. The students will conduct a vertical assault.

PROVIDE HELP (CONC)

a. The instructors will assist the students if necessary.

OPPORTUNITY FOR QUESTIONS (3 Min)

1. QUESTIONS FROM THE CLASS

2. QUESTIONS TO THE CLASS

Q. What are the four of the six types of insects to be avoided?

A. (1) All adults that sting or bite.
(2) Hairy or brightly colored insects.
(3) All caterpillars.
(4) Insects that have a pungent odor.
(5) All spiders.
(6) Disease carriers like ticks, flies, or mosquitoes

Q. What are four of the six types of plants to be avoided?

A. (1) All mushrooms.
(2) Plants with milky sap.
(3) White and yellow berries should be avoided as they are almost poisonous.
(4) Plants with shiny leaves should be considered poisonous.
(5) Plants that are irritants to the skin should not be eaten.
(6) All beans and peas should be avoided.

SUMMARY (2 Min)

a. During this period of instruction, we have talked about edible plants and insects, plants and insects that should be avoided, the plant testing procedure, and the two plants that can be used to poison our survival weapons.

b. Those of you with IRF's please fill them out at this time and turn them in to the instructor. We will now take a short break.

UNITED STATES MARINE CORPS
Mountain Warfare Training Center
Bridgeport, California 93517-5001

MSVX
WML
WMO
03/31/01

LESSON PLAN

SIGNALING AND RECOVERY

INTRODUCTION (5 Min)

1. **GAIN ATTENTION**. In a survival situation, your basic concern is to establish communications with other friendly units. In a survival/evasion situation, your basic problem is to establish communications with only the right people. Communication is generally interpreted as "giving and receiving information". The signals you use as a survivor or evader must make it easier for the rescue crew to find you. The types of signals you and the rescue crew will use depends on the ground situation. Selecting the correct signaling method will assist in the rescue.

2. **PURPOSE**. The purpose of this period of instruction is to introduce the student to signaling, specifically those devices, methods, and specific means of communicating with rescuers that are used in a survival situation. This lesson relates to Survival Fires.

3. **INTRODUCE LEARNING OBJECTIVES**

   a. TERMINAL LEARNING OBJECTIVE. In a summer mountainous environment, conduct recovery, in accordance with the references.

   b. ENABLING LEARNING OBJECTIVES

      (1) Without the aid of references, describe in writing the audio international distress Signal, in accordance with the references.

      (2) Without the aid of references, describe in writing the visual international distress signal, in accordance with the references.

      (3) Without the aid of references, construct an improvised visual signaling device, in accordance with the references.

(4) Without the aid of reference, utilize a hoist recovery device, in accordance with the references.

4. **METHOD / MEDIA**. The material in this lesson will be presented by lecture and demonstration. You will practice what you have learned during upcoming field training exercises. Those of you with IRF's please fill them out at the end of this period of instruction.

5. **EVALUATION**. You will be evaluated on this material by written and practical examinations in future field evolutions.

TRANSITION: Now let's look at some signaling devices that you may have at your disposal.

BODY (50 Min)

1. (10 Min) **SIGNALING DEVICES**. The equipment listed below are items that may be on your body or items inside an aircraft. Generally, these items are used as signaling devices while on the move. They must be accessible for use at any moment's notice. Additionally, in a summer mountainous environment, Marines may experience areas that are snow covered and must be familiar with the effects that snow will have on specific signaling devices.

   a. Pyrotechnics. Pyrotechnics include star clusters and smoke grenades. When using smoke grenades in snow pack, a platform must be built. Without a platform, the smoke grenade will sink into the snow pack and the snow will absorb all smoke. A rocket parachute flare or a hand flare have been sighted as far away as 35 miles, with an average of 10 miles. Pyrotechnic flares are effective at night, but during daylight their detectability ranges are reduced by 90 percent.

   b. M-186 Pen Flare. The M-186 Pen Flare is a signaling device carried in the vest of all crew chiefs and pilots. Remember to cock the gun prior to screwing in the flare.

   c. Strobe Light. A strobe light is generally carried in the flight vests of all crew chiefs and pilots. It can be used at night for signaling. Care must be taken because a pilot using goggles may not be able to distinguish a flashing strobe from hostile fire.

   d. Flashlight. By using flashlights, a Morse code message can be sent. An SOS distress call consists of sending three dots, three dashes, and three dots. Keep repeating this signal.

   e. Whistle. The whistle is used in conjunction with the audio international distress signal. It is used to communicate with forces on the ground.

   f. AN/PRC-90 & AN/PRC-112. The AN/ PRC 90 survival radio is a part of the aviator's survival vest. The AN/PRC-112 will eventually replace the AN/PRC-90 . Both radios can transmit either tone (beacon) or voice. Frequency for both is 282.8 for voice, and 243.0 for beacon. Both of these frequencies are on the UHF Band.

g. Day/Night Flare. The day/night flare is a good peacetime survival signal. The flare is for night signaling while the smoke is for day. The older version flare is identified by a red cap with three nubbins while the new generation has three rings around the body for identification during darkness. The flare burns for approximately 20 second while the smoke burns for approximately 60 seconds.

NOTE: Once one end is used up, douse in water to cool and save the other end for future use.

h. Signal Mirror. A mirror or any shiny object can be used as a signaling device. It can be used as many times as needed. Mirror signals have been detected as far away as 45 miles and from as high as 16,000', although the average detection distance is 5 miles. It can be concentrated in one area, making it secure from enemy observation. A mirror is the best signaling device for a survivor, however, it is only as effective as its user. Learn how to use one now, before you find yourself in a survival situation.

(1) Military signal mirrors have instructions on the back showing how to use it. It should be kept covered to prevent accidental flashing that may be seen by the enemy.

(2) Any shiny metallic object can be substituted for a signal mirror.

(3) Haze, ground fog, or a mirage may make it hard for a pilot to spot signals from a flashing object. So, if possible, get to the highest point in your area when flashing. If you can't determine the aircraft's location, flash your signal in the direction of the aircraft noise.

AIMING THE SIGNAL MIRROR

TRANSITION: There are basically two ways to communicate, audio and visual.

2. (5 Min) **METHODS OF COMMUNICATION**

   a. Audio. Signaling by means of sound may be good, but it does have some disadvantages.

      (1) It has limited range unless you use a device that will significantly project the sound.

      (2) It may be hard to pinpoint one's location due to echoes or wind.

      (3) International Distress Signal. The survivor will make six blasts in one minute, returned by three blasts in one minute by the rescuer.

   b. Visual. Visual signals are generally better than audio signals. They will pinpoint your location and can been seen at a greater distances under good weather conditions.

      (1) The visual international distress symbol is recognized by a series of three evenly spaced improvised signaling devices.

TRANSITION: Now, let's cover improvised signaling devices.

3. (10 Min) **IMPROVISED SIGNALING DEVICES**. Improvised signaling devices are generally static in nature. They must be placed in a position to be seen by rescuers. They are made from any resources available, whether natural or man-made.

   a. Smoke Generator. The smoke generator is an excellent improvised visual signaling device. It gives the survivor the flexibility to signal in either day or night conditions. This type of signal has been sighted as far away as 12 miles, with an average distance of 8 miles. Smoke signals are most effective in calm wind conditions or open terrain, but effectiveness is reduced with wind speeds above 10 knots. Build them as soon as time and the situation permits, and protect them until needed.

      (1) Construct your fire in a natural clearing or along the edge of streams (or make a clearing). Signal fires under dense foliage will not be seen from the air.

      (2) Find two logs, 6 - 10 inches in diameter, and approximately five feet long. Place the two logs parallel to each other with 3 - 4 feet spacing.

      (3) Gather enough sticks, approximately two inches in diameter and four feet long, to lay across the first two logs. This serves as a platform for the fire.

      (4) Gather enough completely-dry branches to build a pyramid fire. The pyramid fire should be 4 feet by 4 feet by 2 feet high.

      (5) Place your tinder under the platform.

      (6) Gather enough pine bough to lay on top of the pyramid fire. This serves to protect the fire and the tinder.

(7) To light, remove the pine bough and ignite the tinder. If available, construct a torch to speed up the lighting process, especially for multiple fires.

SMOKE GENERATOR

(8) To create a smoke effect during the day light hours, place the pine bough on the ignited fire.

(9) Placing a smoke grenade or colored flare under the platform will change the color of the smoke generated. Remember, you want the fire to draw in the colored smoke which will create a smoke color that contrasts with the back ground will increase the chances of success.

TRANSITION: Next we'll discuss the arrangement or alteration of natural materials.

b. Arrangement or alteration of natural materials. Such things as twigs or branches, can be tramped into letters or symbols in the snow and filled in with contrasting materials. To attract more attention, ground signals should be arranged in big geometric patterns.

(1) International symbols. The following symbols are internationally known.

Number	Message	Code symbol
1	REQUIRE ASSISTANCE	V
2	REQUIRE MEDICAL ASSISTANCE	X
3	NO OR NEGATIVE	N
4	YES OR AFFIRMATIVE	Y
5	PROCEED IN THIS DIRECTION	↑

INTERNATIONAL SYMBOLS

(a) Shadows. If no other means are available, you may have to construct mounds that will use the sun to cast shadows. These mounds should be constructed in one of the International Distress Patterns. To be effective, these shadow signals must be oriented to the sun to produce the best shadows. In areas close to the equator, a North—South line gives a shadow anytime except noon. Areas further north or south of the equator require the use of East—West line or some point of the compass in between to give the best result.

(b) Size. The letters should be large as possible for a pilot or crew to spot. Use the diagram below to incorporate the size to ratio for all letter symbols.

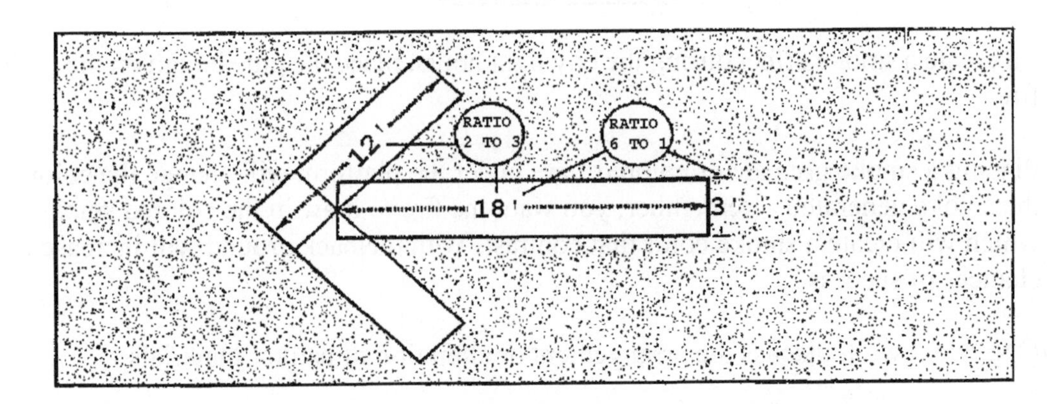

SIZE AND RATIO

(c) Contrast. When constructing letter symbols, contrast the letter from the surrounding vegetation and terrain. Ideally, bring material from another location to build the letter. This could be clothing, air panels, space blanket, etc.

1. On snow, pile bough or use sea dye from a LPP (Life preserver, personal). Fluorescent sea dye markers have been sighted as far away as 10 miles, although the average detection distance is 3 miles.

TRANSITION: Now that we know how to signal an aircraft, let us find out how an aircraft will signal us.

4. (5 Min) **AIR TO GROUND COMMUNICATIONS**. Air to ground communications can be accomplished by standard aircraft acknowledgments.

   a. Aircraft will indicate that ground signals have been seen and understood by:

      (1) Rocking wings from side to side. This can be done during the day or in bright moonlight.

   b. Aircraft will indicate that ground signals have been seen but not understood by:

      (1) Making a complete, clockwise circle during the day or in bright moonlight.

TRANSITION: Now that we have an understanding of signaling devices, remember, you are still in a survival situation until recovered.

5. (10 Min) **RECOVERY**. Marines trapped behind enemy lines in future conflicts may not experience quick recovery. Marines may have to move to a place that minimizes risk to the recovery force. No matter what signaling device a Marine uses, he must take responsibility for minimizing the recovery force's safety.

   a. Placement Considerations. Improvised signaling devices, in a hostile situation, should not be placed near the following areas due to the possibility of compromise:

      (1) Obstacles and barriers

      (2) Roads and trails.

      (3) Inhabited areas.

      (4) Waterways and bridges.

      (5) Natural lines of drift.

      (6) Man-made structures.

      (7) All civilian and military personnel.

   b. Tactical Consideration. The following tactical considerations should be adhered to prior to employing an improvised signaling device.

(1) Use the signals in a manner that will not jeopardize the safety of the recovery force or you.

(2) Locate a position which affords observation of the signaling device and facilitates concealed avenues of escape (if detected by enemy forces). Position should be located relatively close to extract site in order to minimize "time spent on ground" by the recovery force.

(3) Maintain continuous security through visual scanning and listening while signaling devices are employed. If weapon systems are available, signaling devices should be covered by fire and/or observation.

(4) If enemy movement is detected in the area, attempt to recover the signaling device, if possible.

(5) Employ improvised signaling devices only during the prescribed times, if briefed in the mission order.

c. Recovery Devices. In mountainous terrain, a helicopter landing may be impossible due to ground slope, snow pack, or vegetation. The survivor must be familiar with recovery devices that may be aboard the aircraft.

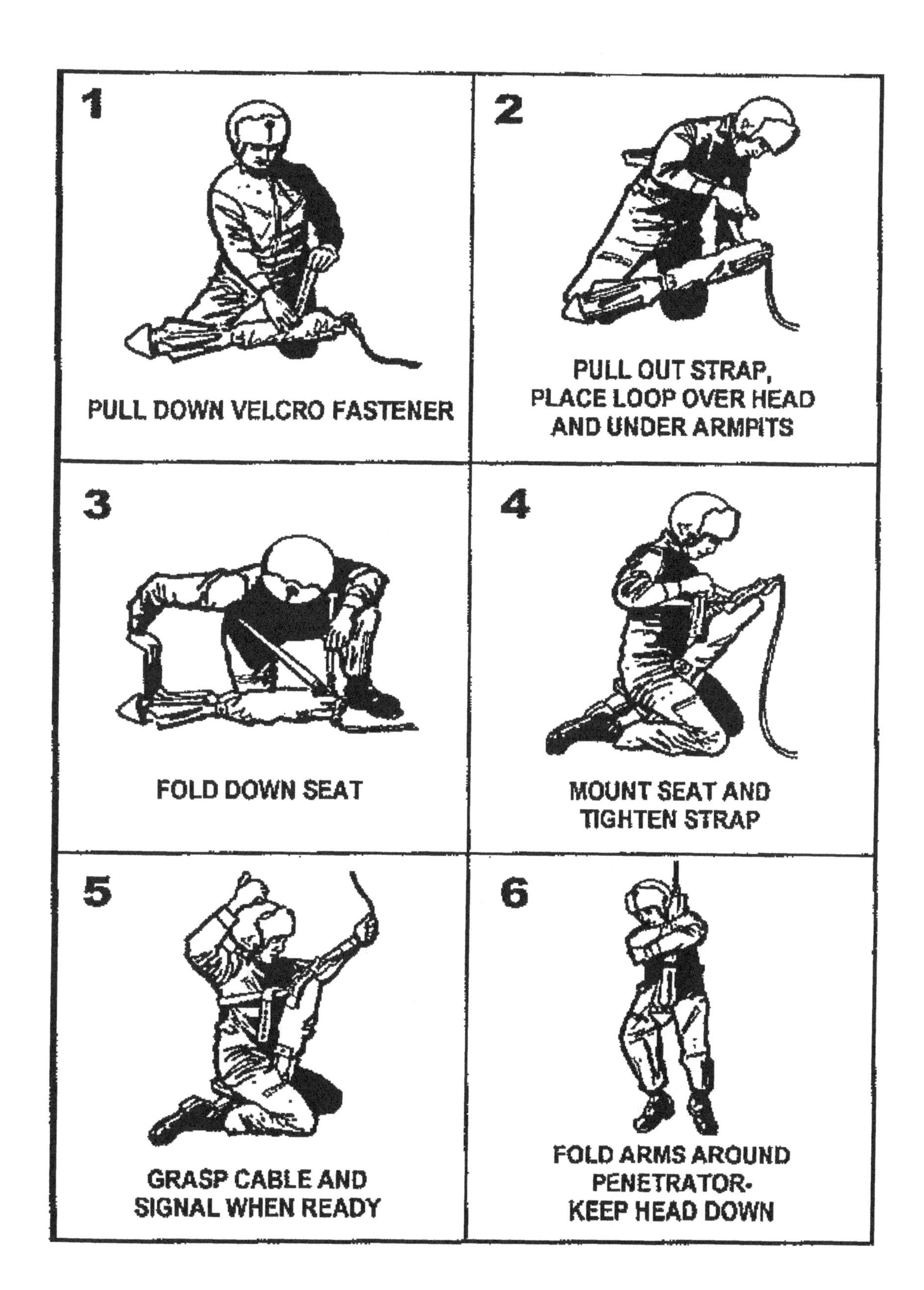
1
PULL DOWN VELCRO FASTENER
2
PULL OUT STRAP,
PLACE LOOP OVER HEAD
AND UNDER ARMPITS
3
FOLD DOWN SEAT
4
MOUNT SEAT AND
TIGHTEN STRAP
5
GRASP CABLE AND
SIGNAL WHEN READY
6
FOLD ARMS AROUND
PENETRATOR-
KEEP HEAD DOWN

<u>JUNGLE PENETRATOR</u>

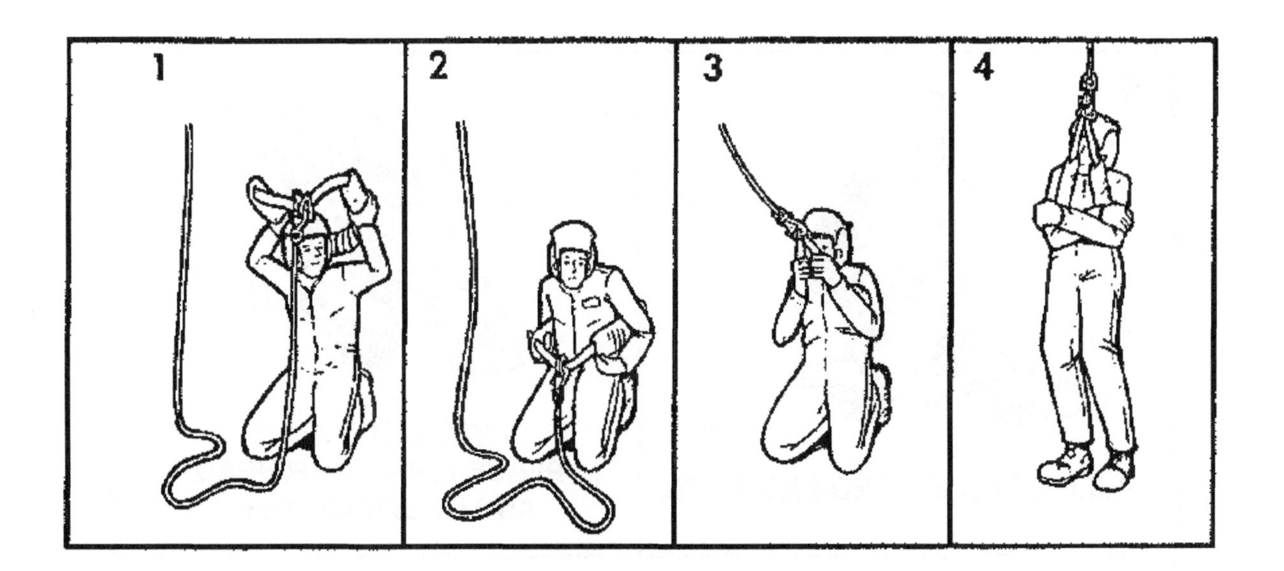

SLING HOIST

d. Recovery by other than aircraft. Recovery by means other than aircraft may occur. Unit SOP's should include signaling and link-up with forces at the following locations:

(1) Border Crossings. The evader who crosses into a neutral country is subject to detention by that country for the duration of the war.

(2) FEBA/FLOT.

(a) Static. Recovery along a static FEBA is always difficult. Under these conditions, enemy and friendly forces can be expected to be densely arrayed and well camouflaged, with good fields of fire. Attempts to penetrate the FEBA should be avoided.

(b) Advancing. Individuals isolated in front of advancing friendly units should immediately take cover and wait for the friendly units to overrun their position.

(c) Retreating. Individuals between opposing forces should immediately take cover and wait for enemy units to pass over their position. After most enemy units have moved on, evaders should try to link up with other isolated friendly elements and return to friendly forces.

(d) Link-up with friendly patrols. Unit authentication numbers and/or locally developed codes may assist the evader to safely make contact in or around the FEBA and when approached by friendly forces.

TRANSITION: If you are lost and incapable of moving, signaling may be the only way that you are going to get saved. Are there any questions?

PRACTICE (CONC)

a. Students will practice what was taught in upcoming field evolutions.

PROVIDE HELP (CONC)

a. The instructors will assist the students when necessary.

OPPORTUNITY FOR QUESTIONS (3 Min)

1. QUESTIONS FROM THE CLASS

2. QUESTIONS TO THE CLASS

Q. Describe the International Distress Signal using audio?

A. The survivor makes six blasts in one minute, returned by three blasts in one minute by the rescuer.

SUMMARY (2 Min)

a. During this period of instruction we have discussed survival signaling to include devices, techniques, procedures, and recovery used to signal rescuers in a survival situation.

b. Those of you with IRF's please fill them out at this time and turn them in to the instructor. We will now take a short break.

UNITED STATES MARINE CORPS
Mountain Warfare Training Center
Bridgeport, California 93517-5001

WML
WMO
02/21/02

LESSON PLAN

COMMUNICATIONS CONSIDERATIONS IN A COLD WEATHER/MOUNTAINOUS ENVIRONMENT

INTRODUCTION

(5 Min)

1. **GAIN ATTENTION**. Communicating in a cold weather, mountainous environment presents unique challenges that must be identified and overcome. Weather, geography, and altitude are all factors that can effect equipment, personnel, and communications organization. The key to success in combat is reliable, secure, rapid, and flexible communications. Commanders must understand cold's effects on their communications systems and their personnel, and know the procedures to counteract these effects.

2. **PURPOSE**. The purpose of this period of instruction is to familiarize communicators with the problem they will encounter in a cold weather environment, techniques that can use to combat these problems and recommendations for command group communication systems for foot mobile and mechanized winter mountain operations. This class relates to Wave Propagation, Antenna Theory, and Field Expedient Antennas.

3. **INTRODUCE LEARNING OBJECTIVES**

   a. TERMINAL LEARNING OBJECTIVE. In a cold weather/mountainous environment, plan cold weather/mountain operations, in accordance with the references.

   b. ENABLING LEARNING OBJECTIVE. In a cold weather/mountainous environment, demonstrate preventive measures for operating communication equipment in a cold weather/mountainous environment, in accordance with the references.

4. **METHOD / MEDIA**. The material in this lesson will be presented by lecture and demonstration. You will practice what was learned during upcoming field training exercises. Those of you with IRF's please fill them out at the conclusion of this period of instruction.

5. **EVALUATION**. You will be evaluated by performance evaluation during the upcoming field exercises.

TRANSITION: Let's first take a look at some considerations for planning.

BODY (75 Min)

1. (5 Min) **PLANNING**. The primary communication planning factors that must be considered by battalion and company staff personnel during cold weather communication planning are:

    a. Communication equipment.

    b. Communication maintenance and supplies.

    c. Safety.

    d. The equipment load communications personnel must carry in addition to their required personal equipment.

    e. Communication plans.

    f. Additional personnel and equipment needed to man retransmission sites and to conduct mountain-picketing operations.

    g. Command group communication system configurations.

TRANSITION: The first consideration in using communications we will discuss, is the radio.

2. (10 Min) **RADIO**. Radios are the most common means of communicating. They are subject to many problems in the cold. The two major problems are reduced battery power and increased equipment failure.

    a. Battery power

        (1) Dry cell batteries lose efficiency and produce less power if not protected from cold weather.

            (a) Spare batteries. Store spare batteries inside heated shelters. They should be stored at a temperature above 10°F and gently warmed before use. Carry one spare set of batteries in a parka or trouser pocket between your body and your outside layer of protective clothing.

(b) Snow. Never place the battery in the snow or unprotected against the shell of a sled. If snow covers the pressure release cover on the radio battery box, ice may form over it, which may restrict it from air exchange.

(c) Rotate batteries every 4 hours with a spare and label them with the amount of time used. Keep logs entries when batteries are changed.

(d) Always use cold weather batteries.

(2) Lithium batteries. Lithium batteries are superior to magnesium batteries in the cold. They are lighter, last longer, and perform best when kept cool (but not cold or freezing temperatures). These are hazardous material because of an explosive chemical reaction that may occur when not ventilated or when they are in contact with water or fire.

(a) Optimum storage temperatures are 35°F or slightly colder.

(b) These batteries should be serialized and will be accounted for during and after each exercise.

(c) Radio operators will keep the plastic bag that lithium batteries are issued in and repack the battery for turn-in when battery is depleted. This protects the battery from moisture.

(3) Cold weather batteries

Dry Cell	**Expected Life**	**Transmit/Receive Life in Hours**	**C/W Battery**
BA 4386	36-48 hrs	1/9	BA 5598
BA 30	24-48 hrs		BA 3030
BA 1588	18-36 hrs	1/9	BA 5588
No corresponding dry cell battery	72-96 hrs	1/9	BA 5590

b. Material Failures. When temperatures are below 10°F, radio equipment becomes brittle and is very susceptible to breakage.

(1) AT-271A(10 foot whip)

(a) Difficulty during movement in thick vegetation.

(b) Antenna may break if radio operator falls.

(c) Company/platoon radio operators should carry a spare AT-271A.

(2) RC-292. Coaxial cable, connectors, and antenna elements must have a thin coat of silicone lubricant spread on components.

(3) AN/GRA-39. The radio remote cables, and connectors must have a thin coat of silicone lubricant.

(4) H-250 or H-189. The handset cable, and connector, must have a thin coat of silicone lubricant.

(a) Press-to-talk button is subject to sticking from freezing. Company radio operators should carry a spare handset.

(b) Microphone. Moisture cover for handsets should be used to prevent moisture from freezing in microphone, or a plastic bag can be wrapped around the handset.

(5) Ice and snow. Keep radios, remote sets, telephones and cryptographic equipment off of the ice and snow.

(a) Radio remote antenna stations should use an ECW tent at the antenna station to keep the radio equipment warm.

(b) Fabricate insulated cold weather bags for radios, and equipment, if tent cannot be used.

(6) Condensation. Do not bring radio equipment from 0°F into warm tents above temperatures of 40°F, because the equipment will sweat, causing moisture to short the radio circuitry.

(a) Remove frost from the equipment, before bringing it into the tent.

(b) Operators must gradually warm equipment and batteries.

(7) Nighttime. Do not turn radio equipment off at night, if on line, and if needed for operation in the morning, unless equipment is in temperatures above 10 to -20°F.

(8) Operator maintenance. Because the polar regions are subject to disturbances, which affect radio reception, it is important to get the very best performance from radio sets. Operators must be intimately familiar with their equipment and should keep their radio equipment clean, dry, and where possible, warm.

(a) Always keep plugs and jacks clean.

(b) Antenna connections must be tight.

(c) Keep insulators dry and clean.

(d) Always remove snow and ice.

(e) Power connections must be tight.

(f) Motors and fans should turn freely.

(g) Knobs and control should operate easily.

(h) Keep batteries fresh, warm, and spares on hand.

(i) Install breath shields on all handsets.

(j) Coat cables and wires with silicone.

TRANSITION: The next topic we will discuss, will be wire communications. Wire is one of the most preferable types of communication in the cold, because there are less problems with actual communication; although, installing and maintaining the network can create several problems.

3. (5 Min) **WIRE (TELEPHONE)**

a. Battery Power

(1) BA 30 dry cell battery. It is used in field telephones, and is not a reliable battery in temperatures below 30°F.

(2) BA 3030. This is the cold weather battery replacement.

b. Material Failures. Like radio equipment when temperatures are below 10°F, telephone equipment materials become brittle, and are very susceptible to breakage.

(1) Field telephone handset cables. Cables must have a thin coat of silicone lubricant.

(2) TA 312 field phone. This provides the best wire communications; however, a microphone moisture cover must be installed in the telephone.

(3) TA 1 sound powered telephones have a carbon element microphone which freezes, and needs to be kept warm and dry to operate.

c. Planning Considerations

(1) Wire line route maps should be drawn by the wire chief at the battalion headquarters alpha command, so wiremen have a recommended wire laying path to each company, and attachment.

(2) Vehicles. Laying wire may take time unless a over-the-snow capable vehicle is available.

(3) Helicopter and a wire dispenser case CY 1064A. Which holds five rolls of 1/2 mile communication wire. Allowing for a slack factor, the helicopter can lay about 1 1/2 miles of wire.

(4) Crossing roads. Either bury the wire 6-12 inches down, or overhead 18 feet above the road.

(5) Wire may be laid by over-the-snow mobile troops on snow shoes or skis using the DR-8: 15 lbs., MX306A: 25 lbs., or RL-159 and RL-27B (idiot stick): 70 lbs.

(6) FM 9-3. Gives guidance on laying wire for infantry units in relation to the time a unit will be stationary, and the distance between units.

(7) Plans. There are several plans that an infantry battalion can use t lay, and retrieve wire. One plan is as follows: The infantry companies move forward into position, when a company takes a position, the alpha command secures a position, and dispatches an over the snow vehicle to the companies position. During this time, the bravo command runs wire to the alpha command and other attachments. When the alpha command prepares the companies to move forward, the bravo command sends vehicles, and wiremen to receive the wire line route map, so wire retrieving can be started by the bravo command.

(8) Unit retrieving wire. The unit retrieving the wire must carry empty RL-159 and DR-5 reels to retrieve the wire for reuse.

TRANSITION: The most secure method of communication will be the messenger. I the cold, the messenger will have several problems to contend with, so before it is decided to employ a messenger, these things must be considered.

4. (5 Min) **MESSENGER**. This is the most secure means; however, it is limited by the terrain, weather, etc. Basically every man is a messenger, especially commanders that attend regular meetings. The use of messengers should be preplanned and they must NEVER travel alone.

   a. Problems Encountered. The following considerations in using messengers in a cold weather environment.

      (1) The enemy.

      (2) Personal survival. Messengers must have the proper equipment. An estimated time of arrival and return must be set and contingency plans made.

      (3) Transportation over the snow. Use an over-the-snow capable vehicle, if possible. Ensure the messenger is properly trained in the use of skis/snowshoes, familiar with the terrain and arctic navigation.

      (4) Wild animals.

TRANSITION: The most basic type of field communications is visual communications. We will begin by discussing the problems inherent to this type of communication in a cold weather environment.

5. (5 Min) **VISUAL COMMUNICATION**. Visual communications are an accepted method in most situations, but in cold weather environment, it can be rendered ineffective by blowing snow, such as whiteouts. Visual signals should be prearranged, and in your operations order.

   a. Air Panel Markers. Air panel markers contain one set of white and black markers, with 13 markers per set. Each marker is 2 feet wide, and 3 feet long. The black marker will show up well against a snow background.

   b. Fluorescent Panel Markers. One set contains 60 red and yellow, 18 inches wide, by 26 inches long markers. These markers are excellent for ground to air signals, by tactical air control party teams.

   c. Semaphore Flags. Red and white are used on land, and red and yellow are used on water.

   d. Pyrotechnics. Red and green colors can be most easily seen against a snow-covered background. A red signal is the International Signal for Distress or emergency.

TRANSITION: Let's now discuss audio communications, and considerations for employment in a cold weather environment.

6. (5 Min) **AUDIO**. Sound signals are satisfactory only for short distances, and their effectiveness is greatly reduced by battle noise. There are three types that are used in a cold weather environment. Other sound signals can be devised through the use of your imagination. They must be kept brief and simple to prevent misunderstanding.

   a. Whistles

   b. Sirens

   c. UIQ 10 Loudspeaker. Can project sound over a large area. A deception plan using tape recordings of mechanized vehicles, can be projected by a two-man team in front of the enemy location.

TRANSITION: To assure our equipment carries out its purpose, we must ensure proper maintenance techniques are performed.

7. (5 Min) **MAINTENANCE**. Performance of preventive maintenance is essential to ensure the proper operation of communication equipment.

   a. Limited Technical Inspection (LTI). All communications equipment to be used must have a second echelon LTI performed on it before going to the field.

b. Daily Preventive Maintenance (PM). By operator. Personnel operating communication equipment in the field must perform daily PM.

c. Communication Contact Team. Should attach to the log train to provide maintenance support for the infantry companies and attachments.

d. Maintenance Personnel. The communication contact team, and headquarters groups must have maintenance personnel attached, and sufficient amount of pre-expended bin items and supplies, such as handsets, coaxial cable, connectors, whip and base antennas, silicone lubricant, plastic bags, duct tape, dry cloth, erasers, pencils, special cold weather electrical tape, and batteries.

TRANSITION: Safety while installing and operating communications equipment is most important. If safety is not practiced, we create the possibility of injury to ourselves and others.

8. (5 Min) **SAFETY**. The four safety precautions in operating communication equipment in cold weather are:

a. Do not touch metal parts on communications equipment with bare hands if temperatures are below 0°F to 10°F.

b. Construct antennas to be windproof.

c. Ensure that the HF equipment is grounded properly.

d. If the MRC vehicle is operating constantly, check the exhaust system to ensure proper ventilation.

TRANSITION: Great equipment is no good unless it can be transported to where it is needed.

9. (5 Min) **EQUIPMENT LOAD**

a. Foot Mobility. During foot mobile operations, the battalion/company communication personnel may carry the following communication equipment in addition to the required cold weather pack:

QUANTITY	EQUIPMENT	WEIGHT
One	PRC 77 with spare battery	23 lbs.
One	KY 57 with spare battery	6 lbs.
One	KY 38 with spare battery	27 lbs.
*One	RC 292 antenna	45 lbs.
One	DR 8 with handle 1/4 mile reel of wire	18 lbs.
One	MX 306 1/2 mile wire	25 lbs.
One	RL 159 with handle 1 mile reel of wire	73 lbs.
One	TA 312 field telephone	10 lbs.
One	TAI field telephone	3 lbs.

One	PRC 104 with spare batteries	25 lbs.
One	PRC 75 with spare battery	10 lbs.
One	PRC 68 with spare battery	4 lbs.
One	PRC 18 with spare battery	27 lbs.

* The weight of the RC 292 antenna can be reduces to 20 pounds by not carrying the 12 metal mast sections. Parachute cord 70 feet in length must be carried so that the RC 292 can be tree-topped.

b. Methods of reducing the communication equipment load on the individual Marine are:

(1) Log trains. Utilize the log trains to provide re-supply of batteries, wire, preventive maintenance material, and maintenance support for the interchanging of equipment that is inoperable.

(2) Spread load communication equipment and cold weather equipment between the Marines of each command group tent team.

c. MRC 138 vehicles must carry their pioneer gear: a pick, sledge hammer, two grounding stakes (4-6 feet in length) and salt.

TRANSITION: Let us now talk about the actual battalion communication plan.

10. (5 Min) **CONFIGURATIONS**. Staff personnel must know the radio circuits that are used for each operation. Depending on the task organization, a regiment or battalion operation at MCMWTC may use the following radio circuits:

RLT RADIO CIRCUITS/BATTALION RADIO CIRCUITS
RLT Link A/TAC VHF Battalion TAC 1 VHF
RLT Link B/Command HF Battalion TAC 2 VHF
RLT Link D/Intel VHF 81 mm mortar VHF
RLT FSC VHF Artillery COF VHF
Artillery Command/Fire Direction VHF TACP Local VHF
TAR/HR 1 I-IF Company TACs VHF
TAD/HD 1 UHF Platoon TACs VHF
LZ Control VHF LZ Control VHF
CSSD Request VHF/HF
MCMWTC SAR/MEDEVAC VHF

NOTE: The Battalion Tactical 2 radio net is by doctrine a VHF circuit. In a mountainous environment, battalions should establish the Battalion Tactical 2 net as an HF circuit. If a high frequency Tactical 2 net is used, additional PRC 104 radios for foot mobile communications must be requested from higher headquarters. The battalion communication platoon table of equipment is authorized only five PRC 104 radios and three MRC 138 radio vehicles which is not sufficient to provide additional high frequency communications for the Battalion Tactical 2 net.

TRANSITION: Let's now talk about a mobile communications system, concentrating specifically on that vehicle which we will most likely use in combat.

11. (10 Min) **MECHANIZED COMMUNICATIONS SYSTEMS**

a. Mechanized vehicles may provide additional equipment not identified in the infantry regiment T/E. All vehicles must be winterized. They also provide heated areas and as additional mobility capability.

b. MRC vehicles. When MRC vehicles remain off in temperatures below 0°F for over 4-5 hours, operators must allow 10-15 minutes at a constant idle for the vehicle to warm. Once the MRC vehicle is operating, allow 5-10 minutes for the High Frequency (HF), Very High Frequency (VHF), or Ultra High Frequency (UHF) mobile radios to warm up. If your radio does not key out, the problem may be that the radio set is not warm enough. You should start the MRC vehicles every 1-2 hours for 5-10 minutes to prevent freezing of vehicle, and radio components.

c. AAV C-7. The C-7 will provide a capability to move in the unfrozen waters of fjords and streams. If snow is considerable, the AAV will be roadbound. However, these vehicles are relatively mobile and displace rapidly.

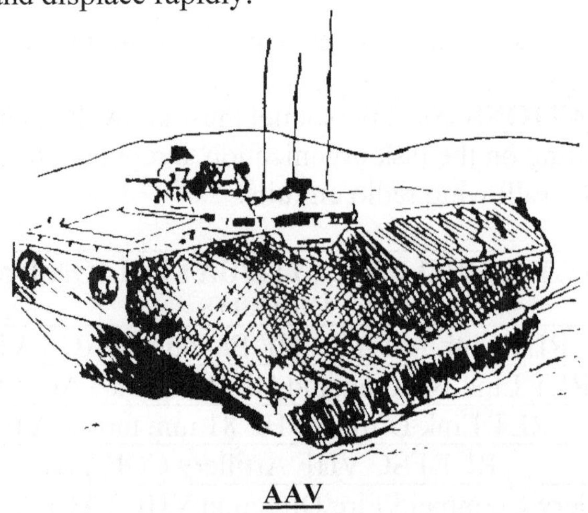

AAV

d. LAV. The LAV, equipped with chains, is fast and mobile both on and off the roads in moderate snow depths. Self-recovery capabilities of LAVs make them ideal for quick displacement. In deep snow and on ice roads, they will be roadbound.

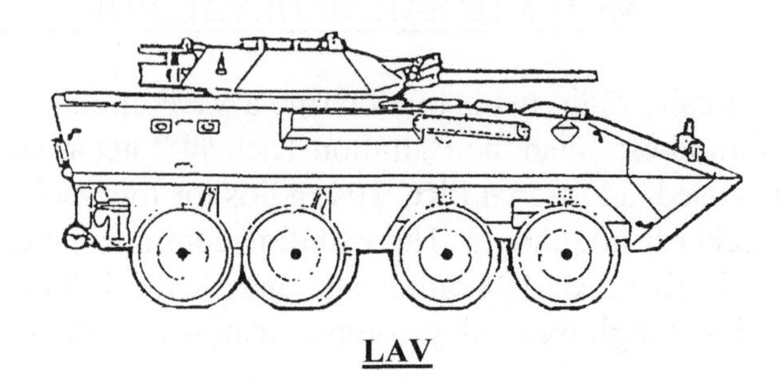

LAV

e. BV-202 / BV-206 C2 Variant. These vehicles provide exceptional COC/CP displacement capabilities and their off-road capability is unsurpassed. The BV-202s, which are dedicated to the USMC, are pre-positioned in Norway. They come with drivers/signalmen who can be used to augment USMC needs, especially in maintaining communications with allied units. Radios used in the BV-206 C2 variant will need to be provided from the unit's T/E. Mechanized infantry operations in snow-covered mountainous terrain should be supported by BV-206 over-the-snow vehicles. Using the BV-206 vehicle, Combat Operations Center (COC) for the forward, alpha, bravo, and log trains can be configured to best meet the characteristics of the BV-206 as shown.

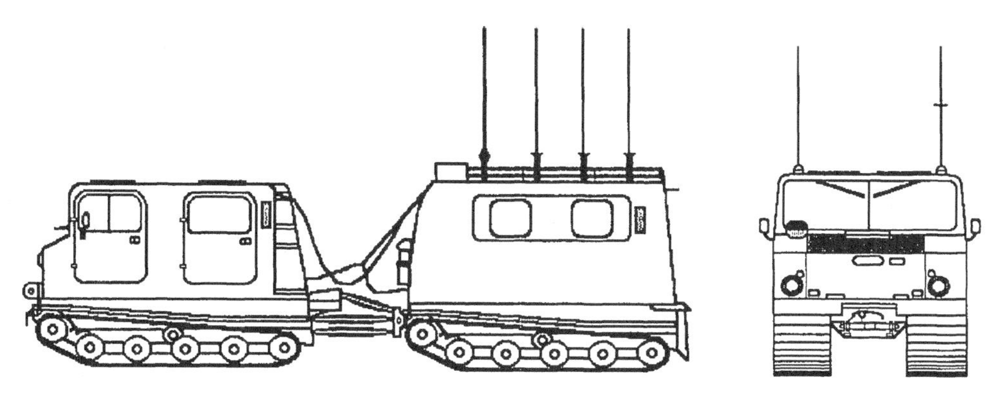

BV-206 SIDE AND FRONT VIEWS

(1) The BV-206 troop carrier vehicle is not configured with communications equipment. The antennas constructed on the rear cab of the BV-206 roof are recommended for the battalion alpha and bravo command group communication system configurations.

(2) Front Cab Radio Configuration

(a) The two sets behind the commander and driver were removed and two RT-254 radio sets with KY-57 cryptographic mounts were installed. Two sections of 2x4's were bolted to the seat frame to provide a base for the mounts. The RT-524 mounts were wired to the BV-206 vehicle battery to power the radios.

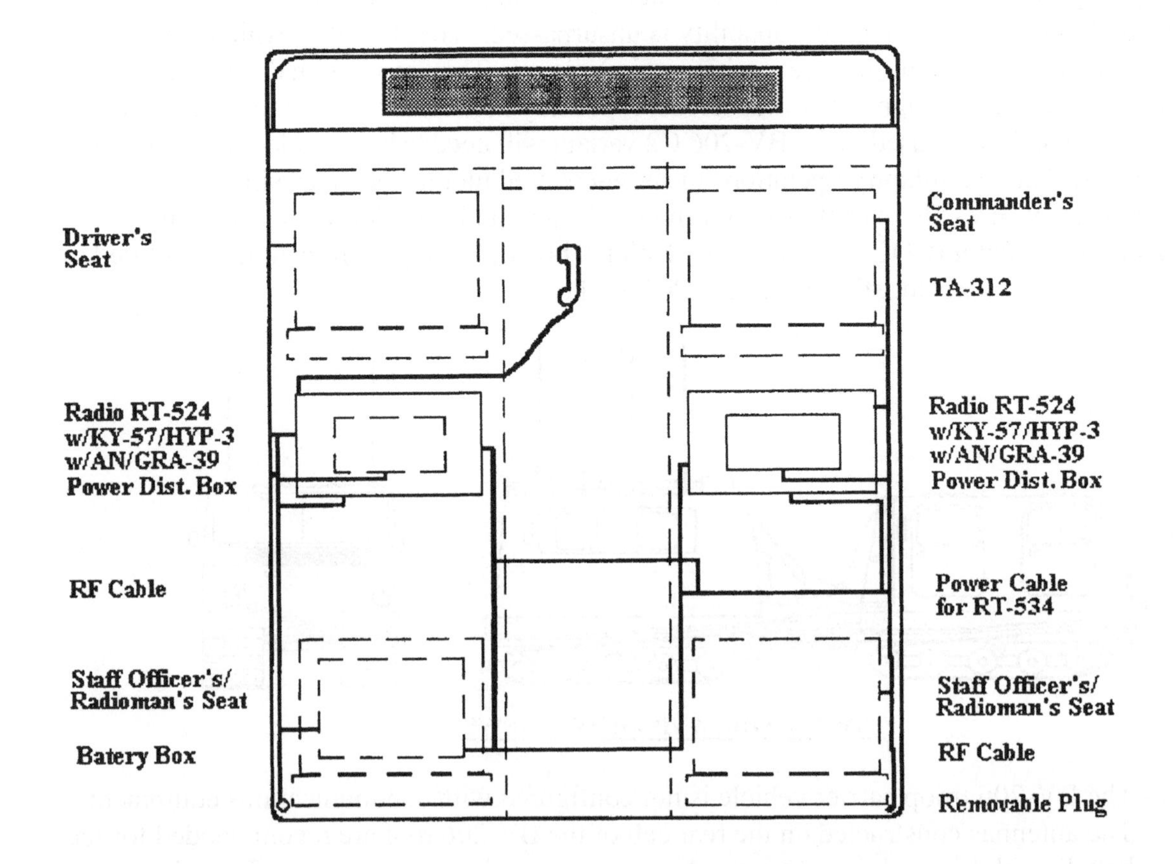

(b) The antenna system for the two radio sets consists of two MX-6707 matching units mounted on the rear cab roof. 15 feet of RF cable was used to connect each radio set to the MX-6707. The RF cable was run through the two removable plugs in the rear corners of the front cab t the MX-6707 matching units.

(c) One TA-312 telephone line was installed from the front cab to the rear cab serving as the intercom system.

(3) Rear Cab Radio Configuration

(a) The wooden seat on the side opposite the door entrance was removed and a 2x4 inch, ten (10) foot wooden mount for eight (8) radio/remote sets was constructed and bolted into the rear vehicle frame. A 2x4 inch, ten (10) foot wooden seat was constructed and bolted into the vehicle frame above the wooden mounts for the eight (8) radio/remote sets. The radio and cryptographic sets were secured in the wooden mounts by bungie cords. This permitted easy access to set frequencies, change batteries and troubleshoot the equipment.

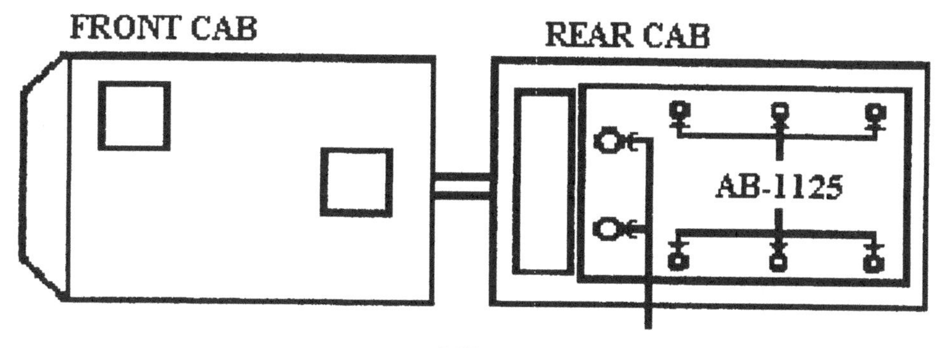

MX-6707
REAR CAB ANTENNA CONFIGURATION

(b) The antenna system for the six (6) radio sets was constructed using six (6) AB-1125 antenna stakes mounted on the rear cab roof. Fifteen (15) foot sections of RF cable were used to connect each radio to the AB-1125 stakes. The RF cable was run under the wooden mounts for the radio sets through the two (2) removable plugs in the corners of the rear cab. The cables should be marked on both ends to facilitate troubleshooting. The six (6) AB-1125 (TRC-166 base antennas) should be mounted to the rack on the rear cab within metal U-bolts. Additionally, six (6) AT-271 ten (10) foot whip antennas are needed to attach to the AB-1125 base antennas to provide radio communications.

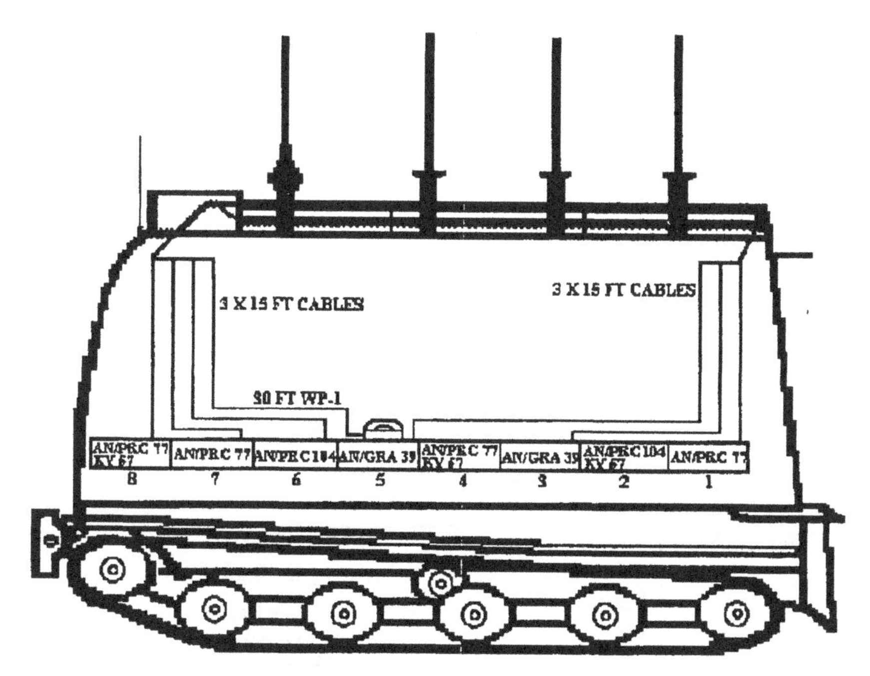

RADIO INSTALLATION IN REAR CAB

(c) One TA-312 telephone line was installed from the front cab to the rear cab which served as the intercom system.

(4) Radio Circuits Configuration

(a) Radio circuit requirements depend on a battalion or regiment's organization and mission. A reinforced battalion would ideally use four BV-206 vehicles during mechanized operations to support its communications plan.

1. Forward/Jump COC.

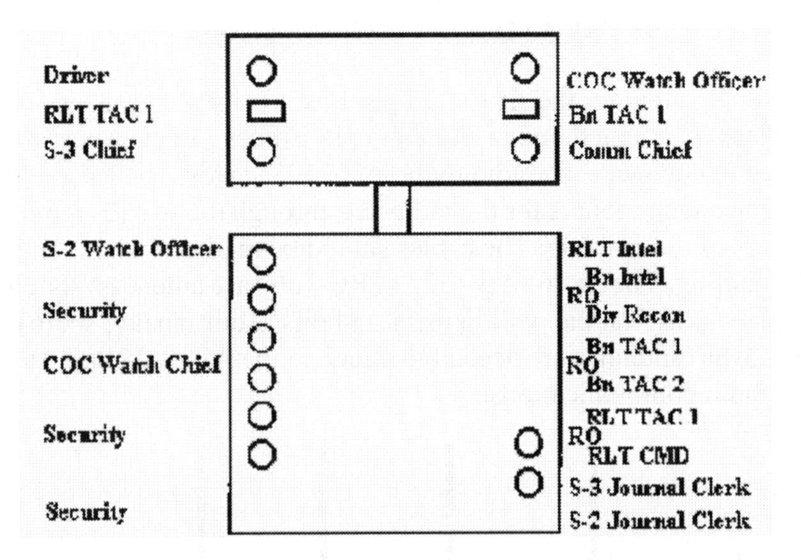

2. Alpha/Tactical COC.

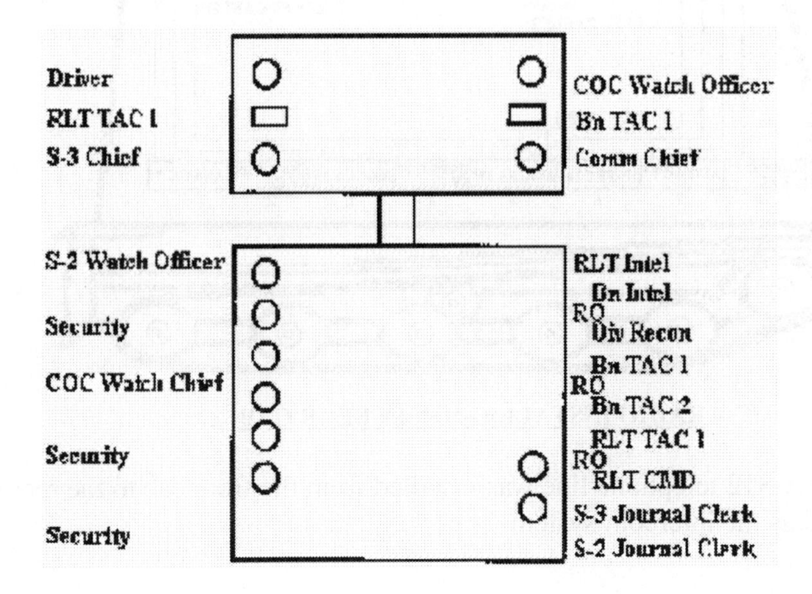

3. Bravo/Main COC.

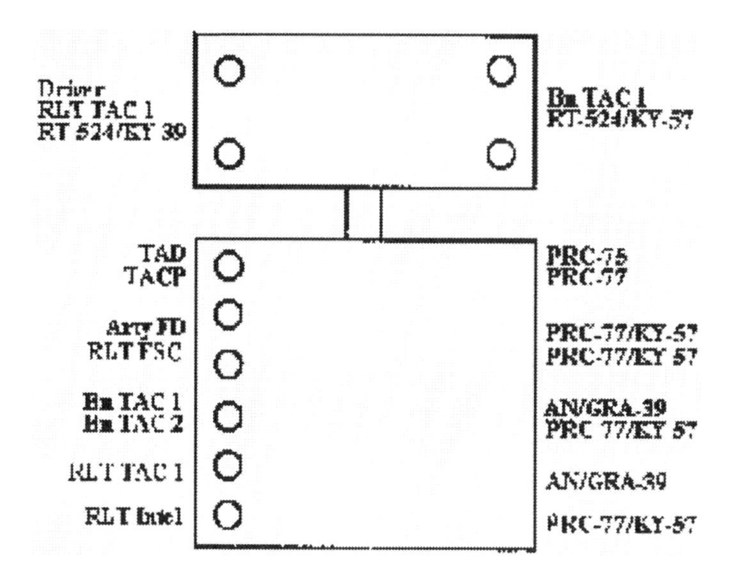

4. Log Trains.

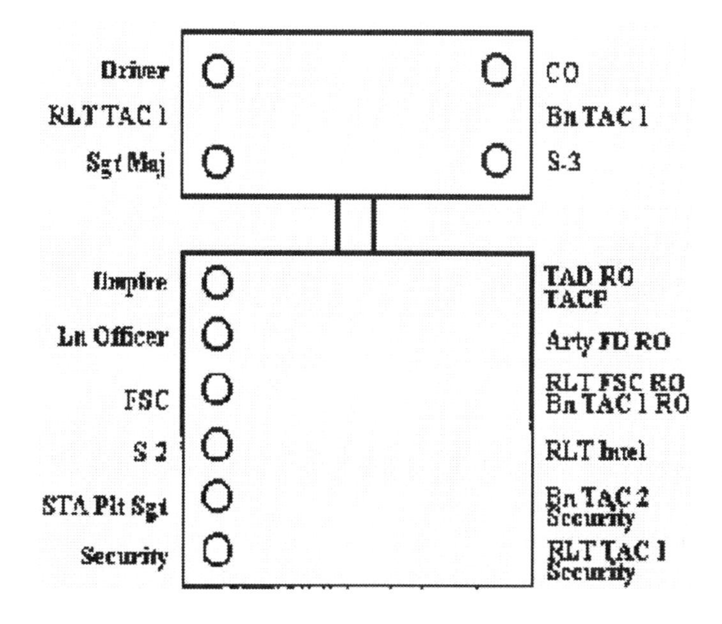

(b) If four BV-206 vehicles are not available, the battalion modifies its communications plan. Depending on the task organization of the battalion and the number of BV-206 vehicles available, the battalion radio circuits can be configured to best meet their operational doctrine.

(5) Chase BV-206

(a) The chase BV-206 vehicle will carry additional communication technicians, radio operators, wiremen, and security personnel. The chase vehicle will support the COC BV-206 vehicle with:

1. Tentage.

2. Provide security for the COC configuration.

3. Additional communication equipment and maintenance items.

4. Radio operators to relieve the actuals on radio circuits.

5. Wiremen to run and retrieve telephone lines to company positions.

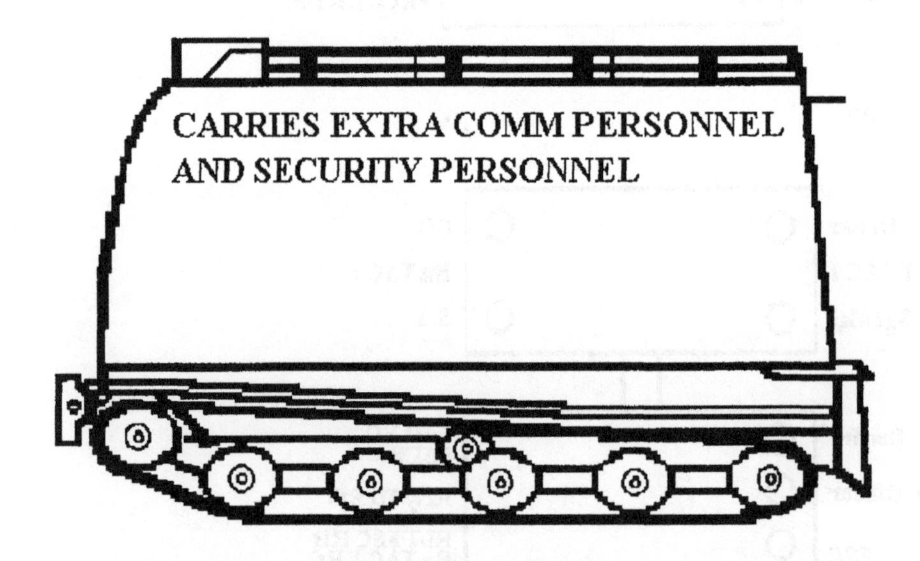

(b) The chase vehicle is recommended if an additional BV-206 is available for use.

(6) Following are diagrams of recommended battalion COC using a 10-man tent. Again, it is emphasized to configure the communications systems to best meet the battalion's requirements.

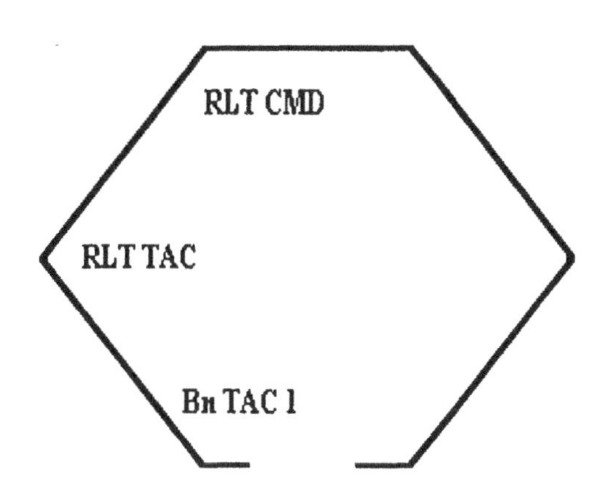

TENT CONFIGURATION FOR FORWARD/JUMP COC

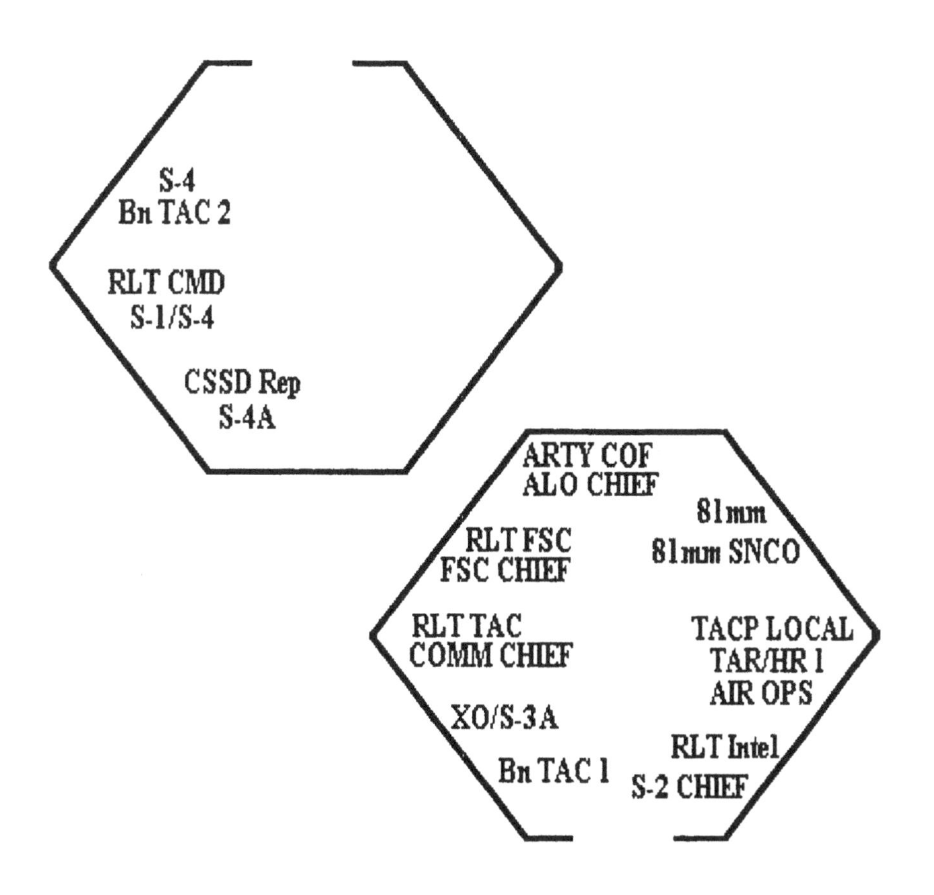

TENT CONFIGURATION FOR MAIN/BRAVO COMMAND

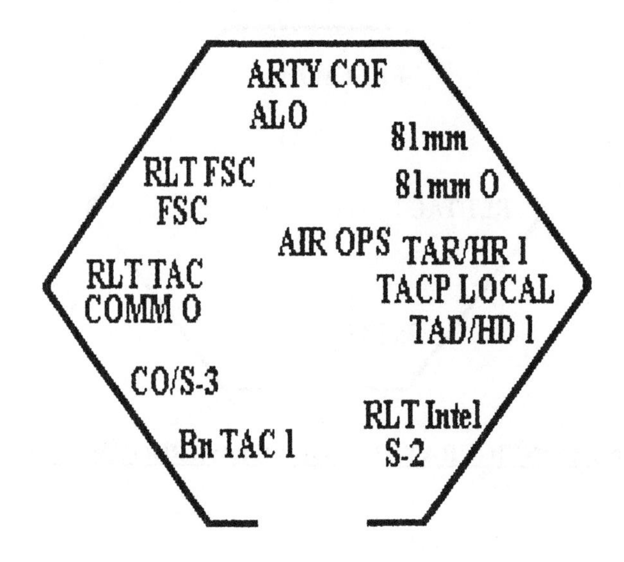

TENT CONFIGURATION FOR TACTICAL/ALPHA COMMAND

TRANSITION: At this time are there any questions?

PRACTICE (CONC)

a. Students will practice what was taught in upcoming training evolutions.

PROVIDE HELP (CONC)

a. The instructors will assist the students when necessary.

OPPORTUNITY FOR QUESTIONS (3 Min)

1. QUESTIONS .FROM THE CLASS

2. QUESTIONS TO THE CLASS

Q. What are two considerations in using radio communications?

A. (1) Battery power.
(2) Material failures.

Q. What are two considerations in using wire communications?

A. (1) Battery power.
(2) Material failures.

Q. What are the problems that a messenger may encounter in a cold weather environment?

A. (1) Personal survival.
(2) Transportation over the snow.
(3) Wild animals.
(4) The enemy.

SUMMARY (2 Min)

a. During this period of instruction we have covered the problems encountered when using radio, wire, messenger, visual, audio, maintenance, and safety procedures in a cold weather/mountainous environment. We have also discussed the techniques that should be used to combat these problems.

b. Those of you with IRF's please fill them out at this time and turn them into the instructor. We will now take a short break.

UNITED STATES MARINE CORPS
Mountain Warfare Training Center
Bridgeport, California 93517-5001

WML
WMO
03/31/01

LESSON PLAN

WAVE PROPAGATION, ANTENNA THEORY, AND FIELD EXPEDIENT ANTENNAS

INTRODUCTION (5 Min)

1. **GAIN ATTENTION**. As the eyes of the battalion, you will find yourself far in front of your unit, often well outside of voice or visual communications. You become increasingly reliant on your radios to relay what you see to higher ups. No communications means no information going back.

2. **OVERVIEW**. The purpose of this period of instruction is to familiarize the student in the principles and techniques used to construct and employ field expedient antennas. This lesson relates to patrolling in the mountains.

3. **METHOD/MEDIA**. The material in this lesson will be presented by lecture. You will practice what you have learned during upcoming field training exercises. Those of you with IRF's please fill them out at the conclusion of this period of instruction.

4. **EVALUATION**. This lesson is a review and has no evaluation.

BODY (90 Min)

1. (10 Min) **HF AND VHF PROPAGATION FUNDAMENTALS**

   a. HF comm (2-30 MHz) is accomplished by either ground wave (out to 30 km with an AN/PRC-104) or sky-wave propagation (out 1000's of kms).

      (1) Ground Wave Propagation. Involves transmission of a radio signal along or near the earth's surface, and is divided into three parts:

(a) Direct Wave travels along line-of-sight (LOS) from one antenna to another. Max LOS distance for comm depends on height of antenna above ground. Any terrain obstructions block direct wave. For a 10' whip antenna, max LOS distance is approx 8 Km.

(b) Reflected Wave travels through the atmosphere but reflects off the earth as it goes from the transmitting to the receiving antenna. Together, the direct and reflected waves are called the Space Wave.

(c) Surface Wave travels along the earth's surface and is very dependant on the terrain between the two antennas. With a flat, good conducting surface, such as sea water, long ground-wave distances are possible. Shorter distances can be expected from sandy, frozen, heavily vegetated, or mountainous terrain.

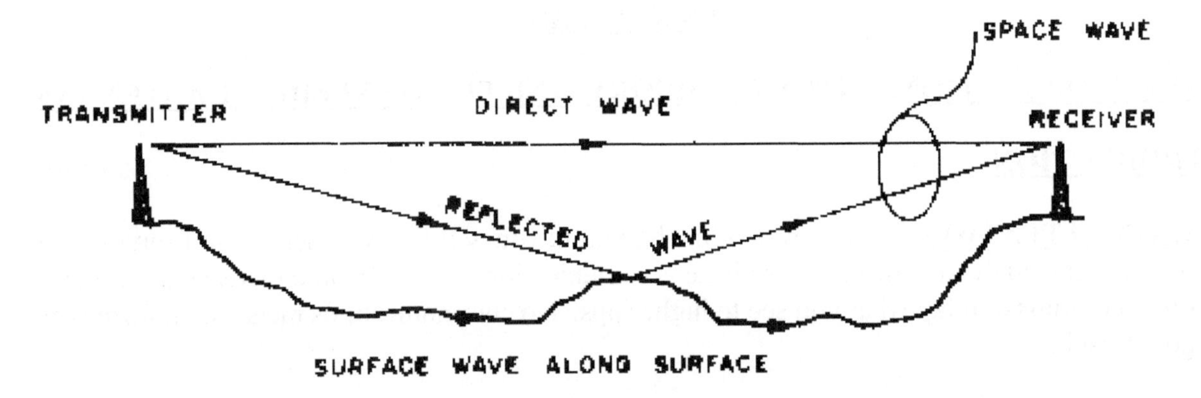

(2) Sky Wave Propagation. Made possible through the bending or refraction of radio signals by the ionosphere (an electrically charged region of the atmosphere that extends from 60 Km to 1000 Km). This area is divided into mainly four regions:

(a) D Region (75-90 Km) only exists during daylight hours, and absorbs energy from radio signals and frequencies to low to pass through it.

(b) E Region (90-130 Km) is present 24 hours a day but is much weaker at night. It is the first region with enough charge to bend radio signals. At times, mostly in summer, it becomes highly charged and either helps or blocks HF comm. This is called Sporadic E Region.

(c) F1 and F2 Regions (200-250 Km and 250-500 Km) are the most important for HF sky-wave communications. Success depends on the frequency of the signal, degree of ionization of the atmosphere, and the angle at which the signal strikes the ionosphere (take-off angle).

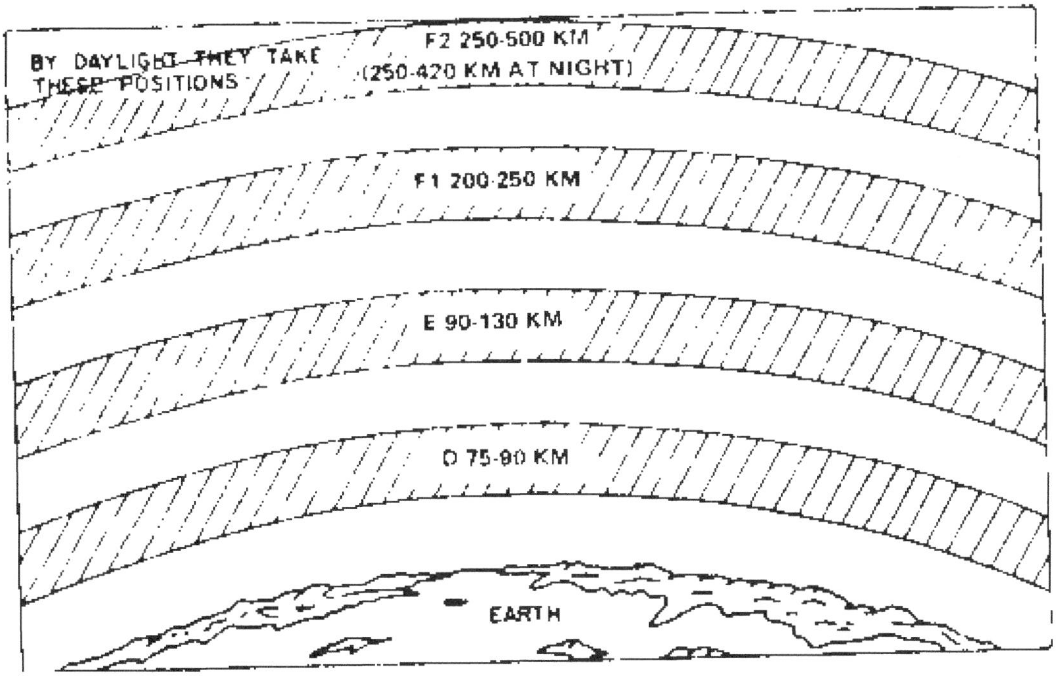

BY DAYLIGHT THEY TAKE THESE POSITIONS

F2 250-500 KM
(250-420 KM AT NIGHT)

F1 200-250 KM

E 90-130 KM

D 75-90 KM

EARTH

(3) HF Sky-wave Considerations

(a) At a vertical angle (90 degrees), the highest frequency that can be bent back to earth for a given time of day is called the critical angle or critical frequency.

(b) For any given transmission distance, the highest frequency that will be bent back to the intended receiver is called the maximum usable frequency (MUF).

(c) For any given transmission distance, the lowest frequency that will be bent back to the intended receiver is called the lowest usable frequency (LUF).

(d) Because the MUF varies rapidly during short periods of time, rising during the day as the ionosphere "heats up" and falling at night, and due to variations during different seasons and sun-spot activity, the frequency of optimum transmission (FOT) is used to account for fluctuations. The (FOT) is 85% of the MUF at any given time.

(e) A skip zone, or region of no usable signal, can exist between maximum range of the ground wave signal and the minimum distance that the sky-wave signal will return to earth. The skip zone can be reduced or eliminated by lowering the operating frequency and/or using antennas with more gain at higher take-off angles.

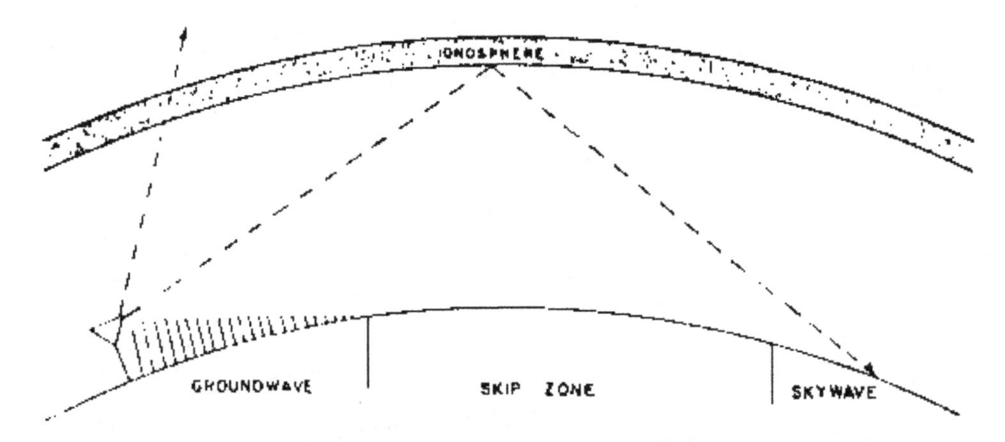

(f) The take off angle of an antenna is the angle above the horizon that an antenna radiates the largest amount of energy. HF sky wave antennas are designed for specific take off angles, which can determine whether a circuit will be successful or not. High (near vertical) take off angles are needed for short range sky-wave communication (0-300 Km).

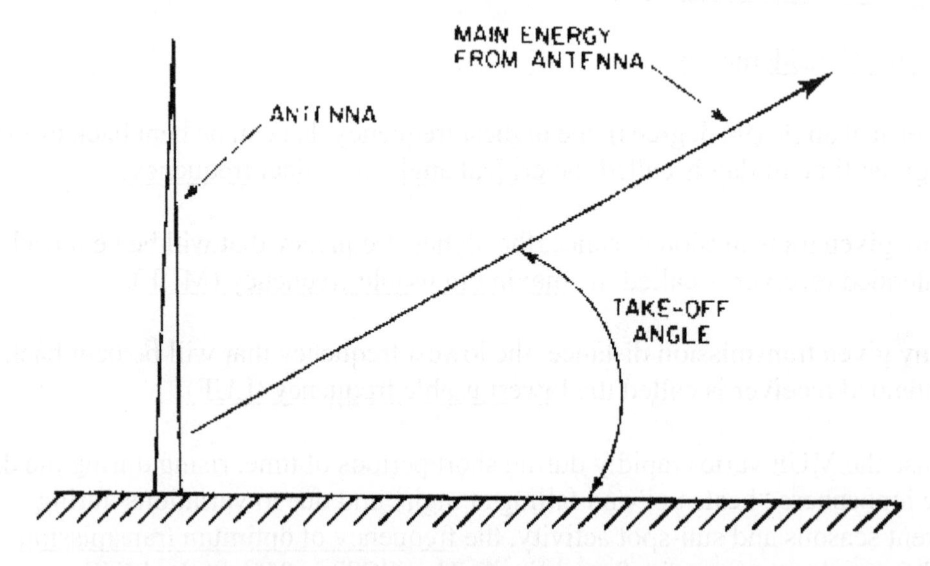

(g) Radio signals arriving at the receiver from multiple transmission paths, will arrive at different times causing echoes, "motor boating", and sort term fading. This multi-path interference can be reduced by using operating frequencies near the MUF.

(h) Multiple frequencies are usually required to maintain sky wave comm. At a minimum, primary and alternate day frequencies (5-10 MHz) and night frequencies (2.5-5 MHz) will be needed. Ensure 10 KHz separation between frequencies.

(i) The sun's energy is the major contributor to ionization of the atmosphere. Increased solar disturbance, such as solar flares (sun spots) increases UV radiation and ion density and raises optimum frequencies. Sunspot activity is cyclic, repeating every 11.1 years. Sunspot numbers (SSN) range from lows of 5 to highs of 130. During maximum sunspot conditions, magnetic storms and sudden ionospheric disturbances (SID) can drastically increase signal absorption by the D and E layers.

(j) During the summer, the suns higher zenith angle will increase critical frequencies and their duration as compared to the winter. Summer night critical frequencies are also higher and are subject to greater variation than in the winter.

(4) Band Segments. The following is a breakdown and a general description of the behavior of each segment. There are exceptions but the descriptions hold true in most cases.

(a) 2 thru 5 MHz: This band is useful during daylight hours for intermediate and short-range sky-wave communications and good for long range (several thousand miles) during the night the static level is highest during the summer.

(b) 5 thru 10 MHz: This band is similar to the 2 thru 5 MHz band except long-range communication is possible during daylight hours under good conditions (high sunspot activity). The signals follow the darkness past best and during winter it is possible to communicate with stations on the other side of the world. The winter months are better than in the summer months because of the high summer static level, especially in equatorial parts of the world.

(c) 10 thru 15 MHz: This band is best choice for reliable intermediate and long-range communications during all propagation conditions. During very low levels of sunspot activity, the band is not useful at night but improves greatly at dawn and holds up well until dusk.

(d) 15 thru 25 MHz: This band is highly variable and dependant on sunspot activity. When conditions are good, it is useful during day and early night; but if sunspot conditions are poor, it may not be useable at all. This band is suitable for short range (less than 10 miles) surface wave communication using a whip antenna.

(e) 25 thru 30 MHz: This band is useful as very-short range communication band (less than 10 miles) and is excellent for long-range (several thousand miles) communication during good propagation conditions. It is generally unusable for intermediate range communication.

(f) As a general rule for good frequencies to transmit on, if the suns up then the freqs up; if the suns down, then the freqs down.

b. VHF Comm (30 to 88 MHz) propagates LOS. VHF-LOS is influenced by four separate components that determine the receiving signal: The direct ray, reflected ray, refracted ray, and diffracted ray.

(1) Direct Ray. Travels straight line distance from transmitting to receiving antenna. The

higher the antennas are above the earth, the longer the effective range.

(2) Reflected Ray. Skips off the earth's surface as it travels from one antenna to another. The path of a reflected signal is longer than a direct signal, and the two may arrive out of phase, degrading comm or creating a "null". Moving antennas either closer or farther apart or changing the height of one should improve the signal.

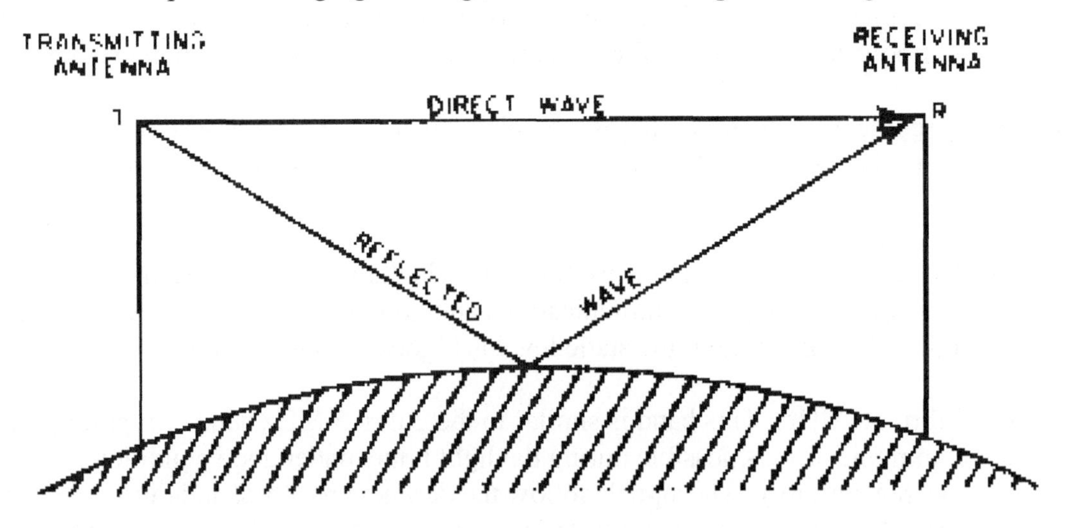

(3) Refracted Ray. Signal bent slightly by lower atmosphere back to earth, allowing it to travel further than a direct ray (much like the trajectory of a rifle bullet).

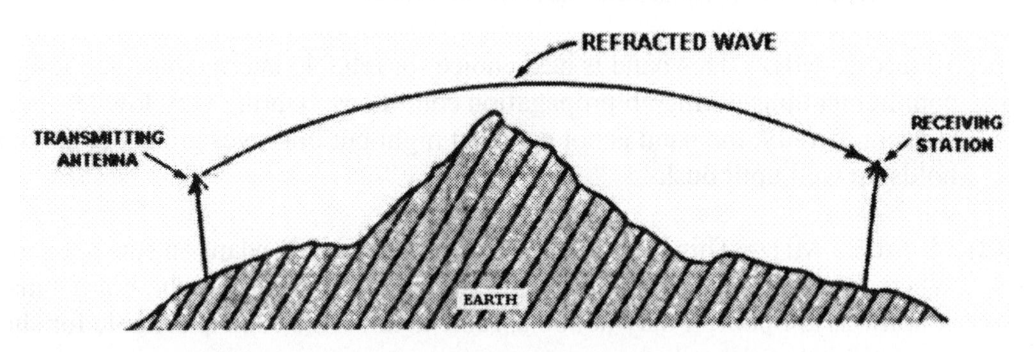

(4) Diffracted Ray. Signal scatters around obstacles, reaching the shadow region behind them. Lower frequencies diffract better then higher ones.

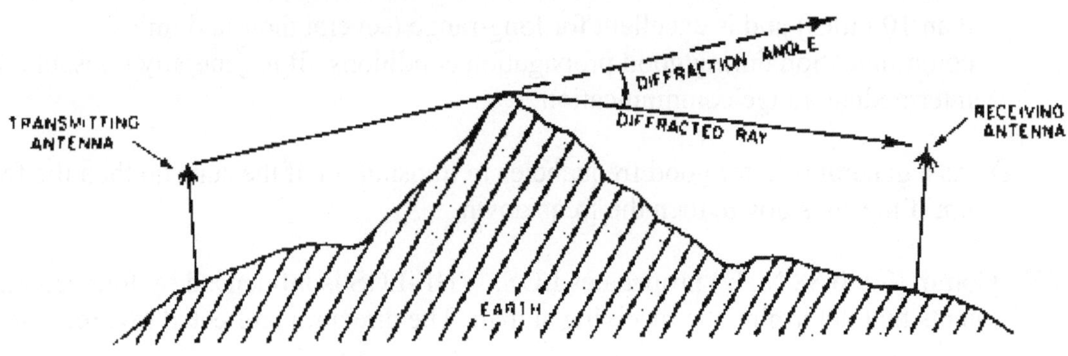

2. (5 Min) **OPTIMIZING COMMUNICATION**

a. Site selection

(1) Locate the antenna as high as tactically possible for ground wave HF and VHF comm. When possible put intervening terrain between your position and the enemy's. Avoid topographical crests.

(2) When using HF near vertical incidence sky wave (NVIS) comm, select a valley or draw to shield the transmission from hostile intercept and to protect the circuit from ground and low incidence sky wave interference.

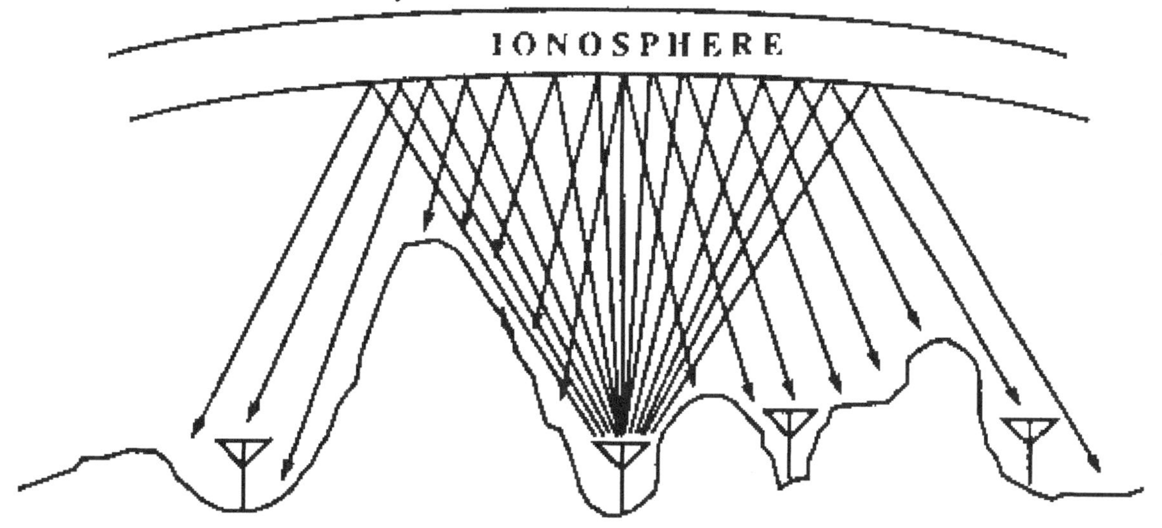

(3) Select a position away from metal objects, or large structures having metallic content, radar sites, and electrical power lines.

(4) Select a position as free of obstructions as tactically possible, with good ground conductivity, such as a wet forest clearing. Trees within approx 30 feet of even a NVIS antenna will have some effect on its signal pattern and functioning. A slight change in antenna location can make substantial difference.

(5) Directional antennas work better when placed on the far side of a clearing from the receiving station. Omni-directional antennas work best when placed in the center of a clearing.

(6) In dense forest, select a location that allows ground wave transmission over the canopy with a vertically polarized antenna. Get the antenna as high as possible. If this is not practical or longer transmission distance is required, use a horizontally polarized antenna with a high take off angle.

(7) Whenever possible, position yourself to allow use of a direction antenna, to increase signal strength in the intended direction, and away from enemy.

(8) Properly aligned directional antennas and antennas with the same polarization (vertical or horizontal) talk best. To ensure the most radiation is being directed at the receiving station, have them slowly turn their antenna until the signal is strongest. Next, repeat

the procedure, orienting your antenna while distant station transmits.

(9) By placing high ground between the antenna and the enemy, not only is the enemy's observation blocked, but radiation from the antenna is blocked, reducing the enemy's intercept capability.

(10) A clearing in the forest can be used to improve propagation if the antenna can be placed so that the clearing is between the antenna and the distant station (for a directional antenna). For an omni-directional antenna, placement should be in the center of the clearing if possible.

3. (10 Min) **ANTENNA FUNDAMENTALS**

a. Polarization is the relationship of the radio energy radiated by an antenna to the earth. The most common polarizations are horizontal and vertical, but others exist such as circular and elliptical.

(1) Vertical Antennas, such as a 10' whip, radiate a vertically polarized signal, and are the best choice in the following situations:

(a) For ground wave propagation (terrain dependent). The longest ranges possible when beaming over salt water.

(b) When omni-directional transmissions or reception is required.

(c) Where long range sky-wave, or low altitude atmospheric refraction is required, particularly when a horizontal antenna cannot be raised to sufficient height.

(d) For better reception, when a vertically polarized antenna is being used by the transmitting station.

(e) When sky-wave propagation is not possible due to magnetic storms or sudden ionospheric disturbances, or made difficult by extremely low ionization levels. If these conditions occur outside ground-wave range, a high HF frequency (25-29.9999 MHz) can be used with a low take-off angle, in order to bend the radio signal back to earth using the lower atmosphere.

(2) Horizontal Antennas, such as a Dipole, radiate a horizontally signal, and are the best choice in the following situations:

(a) On short range sky-wave circuits (0-300 miles) due to lower noise and higher gain at near vertical radiation (take-off) angles.

(b) When directional antennas are required or tactically appropriate.

(c) Where ground conductivity is low, such as in Okinawa, due to more favorable reflection characteristics of horizontal polarization.

(d) Near cities, where man-made noise propagation by surface waves may be a problem.

(e) Except for in special cases, the refracted sky-wave from horizontally polarized transmitting antenna will be circularly polarized, allowing even vertically polarized antennas to receive it.

b. Wavelength and Frequency. There is a direct relationship between antenna length and transmitter frequency wavelength. This fundamental is important to the construction of field expedient antennas, custom-made to provide superior performance over factory antennas under specific tactical and environmental conditions, and over a given transmitting distance.

(1) Determining Antenna Length. Radio signals travel at the speed of light in free space. The wavelength of a radio signal is the distance it takes to travel one complete cycle and is equal to the speed of light divided by the frequency.

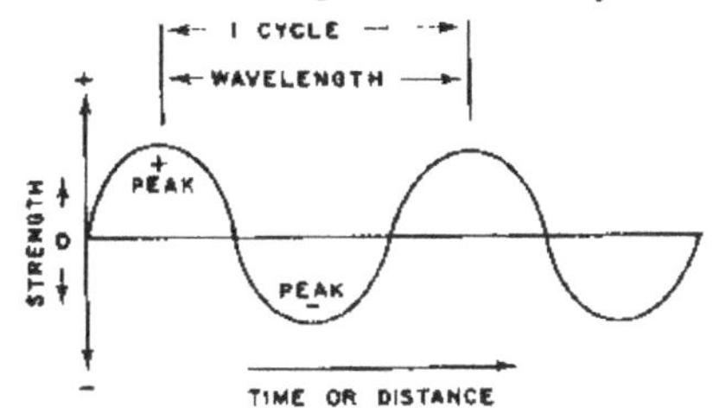

(a) In free space, radio signals travel faster than when traveling through an antenna wire because of less resistance. The following free space wavelength formula can be used to determine how high to erect a horizontally polarized antenna off the ground, or how close a vertically polarized antenna can be to ground obstructions for optimum transmission:

$$1 \text{ WAVELENGTH (free space)} = \frac{984 \text{ (ft/sec)}}{\text{FREQUENCY (MHz)}}$$

$$1/2 \text{ WAVELENGTH (free space)} = \frac{492 \text{ (ft/sec)}}{\text{FREQUENCY (MHz)}}$$

$$1/4 \text{ WAVELENGTH (free space)} = \frac{246 \text{ (ft/sec)}}{\text{FREQUENCY (MHz)}}$$

(b) 12-20 gauge Copper Wire, such as the speaker wire or firing wire, should be used in antenna construction. The following formula matches the radiating length to the frequency:

$$1 \text{ WAVELENGTH (copper)} = \frac{936 \text{ (ft/sec)}}{\text{FREQUENCY (MHz)}}$$

$$1/2 \text{ WAVELENGTH (copper)} = \frac{468 \text{ (ft/sec)}}{\text{FREQUENCY (MHz)}}$$

$$1/4 \text{ WAVELENGTH (copper)} = \frac{234 \text{ (ft/sec)}}{\text{FREQUENCY (MHz)}}$$

(c) WD1/TT Slash Wire can also be used when pure copper wire is not available but, due to greater resistance, the following formula must be used for determining antenna length:

$$1 \text{ WAVELENGTH (slash)} = \frac{902 \text{ (ft/sec)}}{\text{FREQUENCY (MHz)}}$$

$$1/2 \text{ WAVELENGTH (slash)} = \frac{451 \text{ (ft/sec)}}{\text{FREQUENCY (MHz)}}$$

$$1/4 \text{ WAVELENGTH (slash)} = \frac{226 \text{ (ft/sec)}}{\text{FREQUENCY (MHz)}}$$

(d) The tactical situation often dictates use of smaller antennas, built lower to the ground, even though they will be less efficient. The formulas above for 1/2 and 1/4 wavelength antennas show those smaller antennas. The end product is the length of wire in feet. Be exact as possible leaving a little extra for tie offs for insulators, cobra heads, etc.

(e) On the PRC-119 the frequency is already in MHz format (Example: 39.95). This is the number that you divide by in the above formulas. On the PRC-104 the frequency is in KHz. To make a KHz frequency into a MHz you move the decimal point three places to the left. (Example: 18,525.0 KHz frequency would turn into a 18.525 MHz frequency and this is the number you use in the above formulas.)

(2) Frequency and Take-off Angle Selection. Usable frequencies for tactical radio propagation, particularly HF NVIS transmissions (0-300 Km), may be restricted to a narrow band, outside which communication is not possible. Additionally, take-off angles between 80-90 degrees will normally be required to complete these HF circuits, especially at night and in restrictive terrain.

(a) The take-off angle of an antenna is the angle above the horizon than an antenna radiates the largest amount of energy. In VHF communications, antennas are designed so that the energy is radiated parallel to the earth (do not confuse take-off angle and polarization). In HF communications, the take off angle of an antenna can determine whether a circuit is successful or not. HF sky-wave antennas are designed for specific take-off angles depending on the circuit distance. High take-off angles are used for short-range communications and low take-off angles are used for long-range communications.

(b) Computer generated HF propagation studies should be requested as a routine part of patrol planning. Principal factors affecting the range of usable frequencies are:

1. The latitude and distance between stations. The farther the distance, the lower the ideal take-off angle will be. The closer to the equator and the more hours of daylight, the higher band of usable frequencies will be.

2. The time of day/year. The higher the sun is in the sky and the higher its zenith, the higher the optimum frequency.

3. SSN and ionospheric conditions. The higher the sun spot number, the higher usable frequencies. If ionization gets great, however, all frequencies may be absorbed.

4. Station locations with respect to the ground as discussed earlier. Antenna elevations, intervening terrain and vegetation and ground conductivity will all impact frequency performance.

5. Weather conditions. Atmospheric noise from rain and electrical storms degrade the lower end of the HF spectrum. Higher frequencies within the usable spectrum will provide better communication in inclement weather.

(c) Take-off angles needed to complete a circuit are a function of the operating frequency vs. current ionospheric conditions, and the distance between stations. Factors affecting take-off angles are the type of antenna selected, antenna height, and the position of the radiating elements in relation to the ground and nearby obstructions.

(d) On the next page is an example propagation study from which you request your frequencies. The FOT should be the one of the frequencies that you request. But you can also tell is your frequency is above or below the usable frequencies for that area, time of year, etc.

FREQUENCY PREDICTION TABLE

UNCLASSIFIED

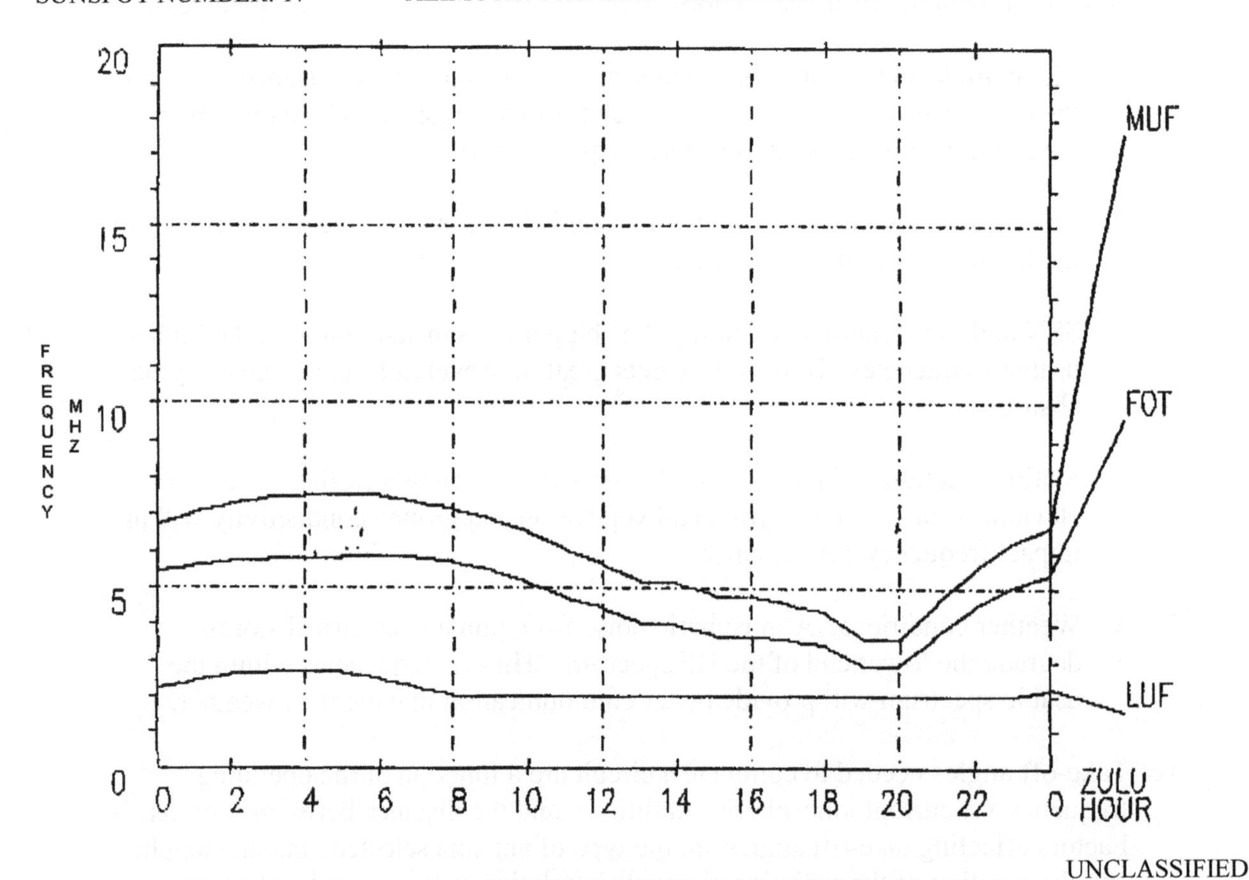

UNCLASSIFIED

HFPLAN SOLAR AND IONOSPHERIC PARAMETERS

DATE: 25APR95
MAXIMUM OBSERVED FREQUENCY: .0
FREQUENCY OF TRANSMISSION: 10.0
SUNSPOT NUMBER: 15.
10.7 COM FLUX: 74.9
X-RAY FLUX: .000001
FLARE START TIME:
FLARE PEAK TIME:
PEAK FLUX OF FLARE:
TRANSMITTER ANTENNA HEIGHT: 10F
RECEIVER ANTENNA HEIGHT: 10F
INTERFERER ANTENNA HEIGHT: 10F
ANTENNA POLARIZATION: V

ATMOSPHERIC NOISE MODEL INCLUDED: YES
BACKGROUND NOISE LEVEL: RURAL
TRANSMISSION MODE: 100HA1A
TERRAIN TYPE: FORESTATION
TERRAIN COVER: JUNGLE

c. Antenna Resonance. Antennas can be classified as either resonant or non-resonant depending on their design.

(1) In a resonant antenna, almost all of the radio signal fed to the antenna is radiated. A resonant antenna will effectively radiate a signal usually within only 2% of its design frequency. Outside this bandwidth, large losses in signal power occur. Signal energy is "turned back" from the antenna, causing standing waves in the feed-line. The Standing Wave Ratio (SWR) is used to determine if an antenna resonates a particular frequency. An SWR of 1:1 is perfect, but 2:1 acceptable. As most field expedient antennas are resonant, separate antennas should be customized for each frequency used. A resonant antenna is cut for a specific freq.

(2) Non-resonant antennas will effectively radiate a broad band of frequencies with lower efficiency. The 10' whip, RC-292, AS-2259 antennas are examples of non-resonant antennas. The principal advantage that these antennas provide is convenience, although only the 10' whip is routinely carried on patrols. A non-resonant antenna is cut for many frequencies.

d. Ground Effects can greatly enhance or degrade signal output because the earth below an antenna acts like a large reflector. The relationship between the direct wave and the ground reflected wave is additive. When they are 180 degrees out of phase, it creates a null. Particularly for vertically polarized antennas, if the electrical characteristics below the antenna are poor, there will be large losses to the ground, resulting in poor radiation characteristics. Ground screens, radials, or planes can be used to artificially produce good ground effects.

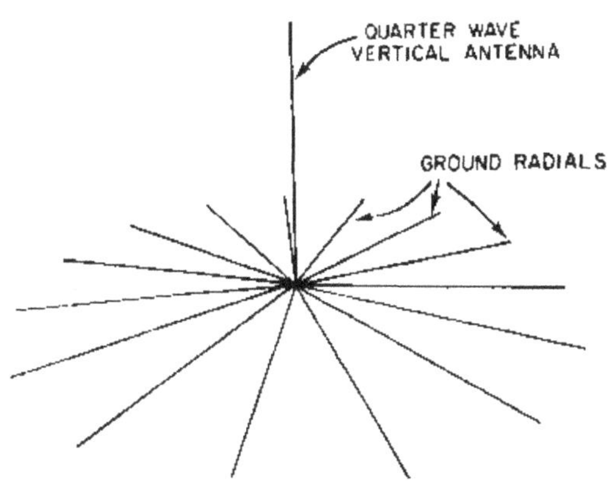

e. Feed-lines, Baluns, and Chokes

(1) A feed-line is used to feed power from the radio to the antenna when the radiating elements of the antenna is not connected directed to the antenna base. Most tactical antennas are fed with a coaxial cable, which is a reasonable compromise between efficiency, convenience, and durability. When coax cable is not available, twist the section of wire you do not wish to radiate upon itself to negate signal transmission. A feed-line is fed into a center-fed connector (cobra-head).

(2) A balun is used to prevent unwanted current flow on the coax cable, which can cause the radio to be "hot" and shock the operator. A balun is also used to change the impedance of coax cable to match an antenna which allows radio energy to pass into the antenna. If a balun is not available, ten 6" coils should be made and taped together in the end of the feed-line towards the cobra-head. This is called a choke.

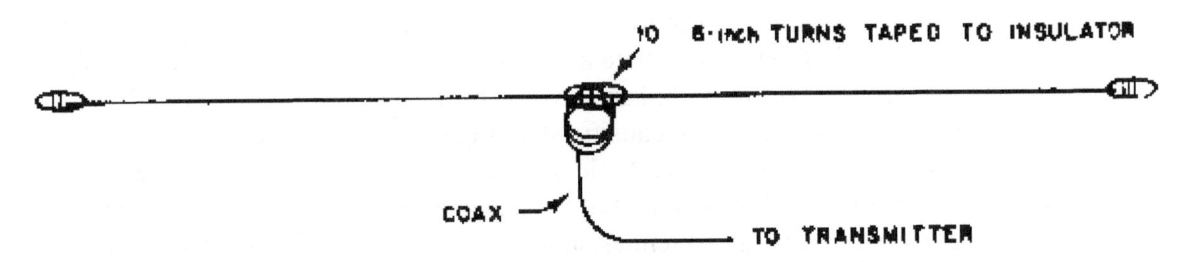

f. Multi-skip Phenomenon. There will be instances when the sky-wave is reflected between the ionosphere and the earth's surface several times. This multi-skip phenomenon occurs frequently making global communications possible in the HF band. In addition to multi-skip, there also exists a multi-path propagation characteristic. If this occurs, two waves that took different paths reach the receiver simultaneously. Depending on their relative phases, the signal strength is enhanced or reduced. If the two waves are of equal amplitude and phase shifted 180 degrees, total cancellation results and the received signal fades out. Operating near the FOT minimizes multi-path degradation.

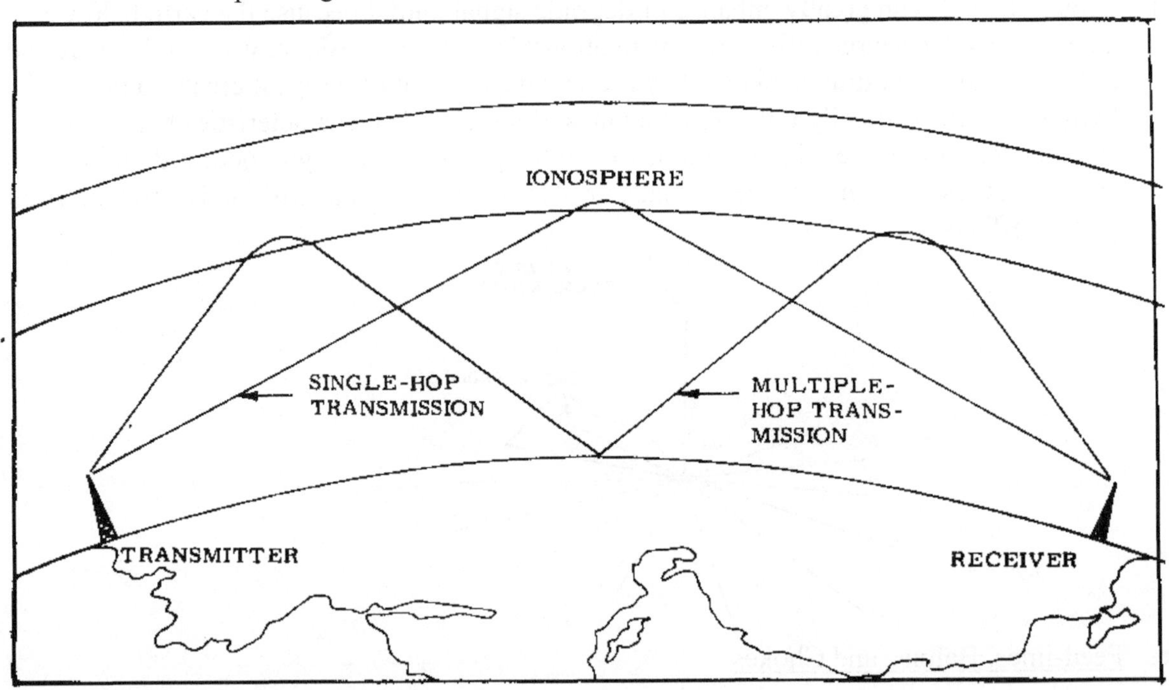

4. (10 Min) **DIRECTIONAL ANTENNAS**

a. Antennas can be made to radiate in all or specific directions. This is accomplished using no or a certain number of resistors. Some antennas are naturally omni-directional or bi-directional without resistors.

(1) Omni-directional. Radiating energy equally well in all directions, the omni-directional antenna is used when it is necessary to communicate in several separated directions at once. It will also receive in all directions.

(2) Bi-directional. Antennas produce a stronger signal in two favored directions while

reducing the signal in other directions. Tactical bi-directional antennas are usually sloping wires, and dipoles. It creates nulls in the areas not receiving energy. These antennas have to positioned correctly (azimuth) in order for them to work.

(3) Directional. Much like a bi-directional antenna except one of its transmission lobes are cut off. Many bi-directional antennas are made directional by the addition of a resistor that sucks up the second lobe.

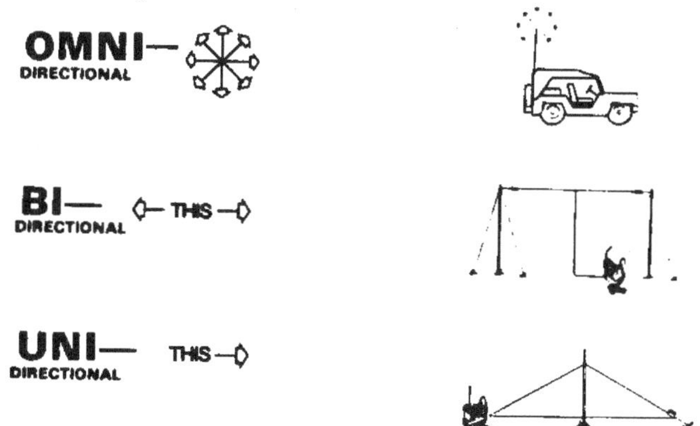

b. Antenna Gain. Gain is the term used to describe how well an antenna radiates power. It is necessary to know what the gain of an antenna is being compared to before two antennas can be compared. In some cases, an antenna is said to have gain compared to an isotropic antenna and the gain is expressed in dBi.

c. Patterns. Antenna patterns graphically show the radiation for a specific antenna.

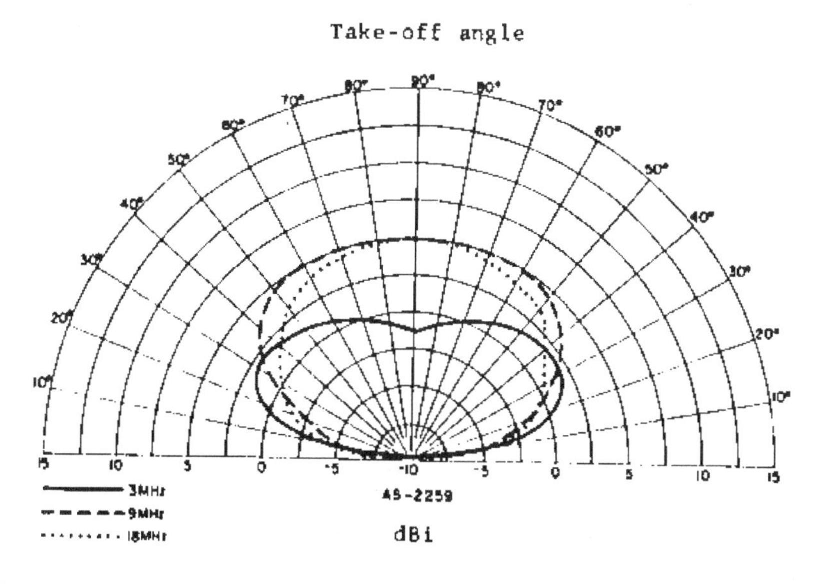

d. Insulators prevent the energy being transmitted from the RT of the radio from escaping out of the end of the wire that you are using as an antenna. If the radio's energy goes out of the end of the wire, your transmission goes nowhere. Tie the end of the wire of the expedient antenna to a insulator and your transmission energy is safe.

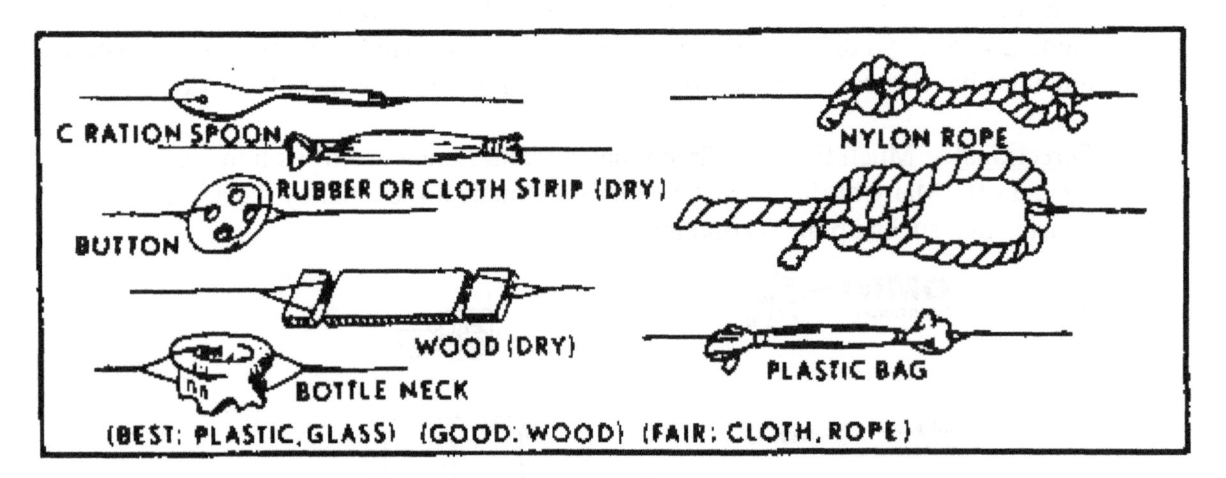

e. Resistors make antennas directional and are ideal for expedient antennas. They draw energy from the RT towards it and then shoots it off in the transmission. Ideally you want a resistor with a wattage of 1/2 of the max of the radio you are using it on and 600-800 ohm. You can add resistors together to add them together or subtract from each. Each color band on fabricated resistors represent a certain ohm factor.

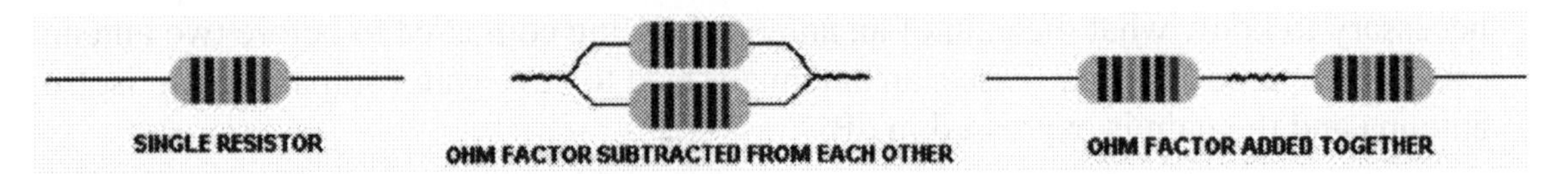

f. Grounding. A good electrical ground is needed so that antennas and their radios functions properly. Ground using a copper stake with a wire attached to it and drive it into the ground. The wire secures to the grounding post on the radio. It is ideal if the ground around the grounding stake is wet and salty. You can make a solution of Epsom salt/table salt and water and pour it onto the ground around your grounding stake. If that is not available, urine can be used. Multiple grounding rods like the "star ground" can be used also.

5. (25 Min) **ANTENNA SELECTION**

a. Antenna Planning

(1) Mission Analysis (METT-TS-L). The characteristics of the mission as it relates to communication must be thoroughly analyzed in order to plan for and construct the proper antennas. This includes the circuit direction and distance to all stations (location of all friendly forces that can relay information or provide support), all usable operating frequencies, the effects of terrain, weather, and the enemy, the type and duration of the patrol, and the materials available.

(2) Antenna characteristics. Based on this analysis, the following antenna characteristics are used to select the best antennas for the mission: ground-wave vs. sky-wave, horizontal vs. vertical polarization, narrow vs. wide frequency band, short range vs. long range, omni-directional vs. direction, mobile vs. fixed, feed-line vs. no feed-line, short vs. long setup time required, external supports vs. height required.

b. HF Antennas

(1) The Vertical Whip, a component of all radio sets, quick and easy to use, is probably the worst antenna when sky-wave communication is required. Non-resonant, omni-directional, and vertically polarized, it suffers from radiation losses and interference even on ground-wave circuits. A field expedient ground radial and/or reflector can greatly improve 10' whip performance, however. You can also make an resonant insulated wire and put it up vertically. This is known as the Vertical Wire.

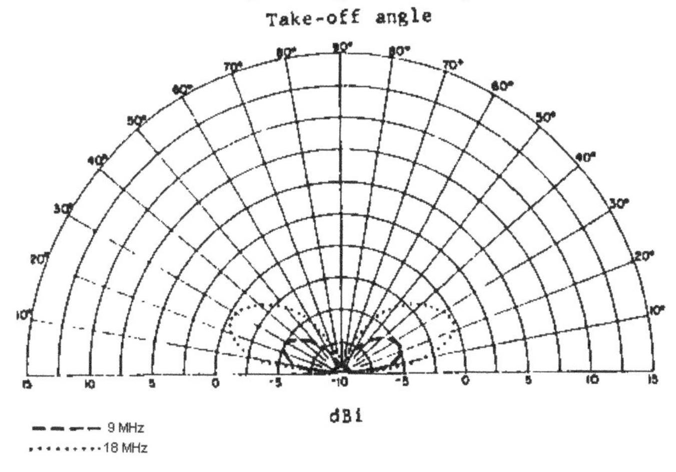

Ten-foot vertical whip vertical antenna pattern.

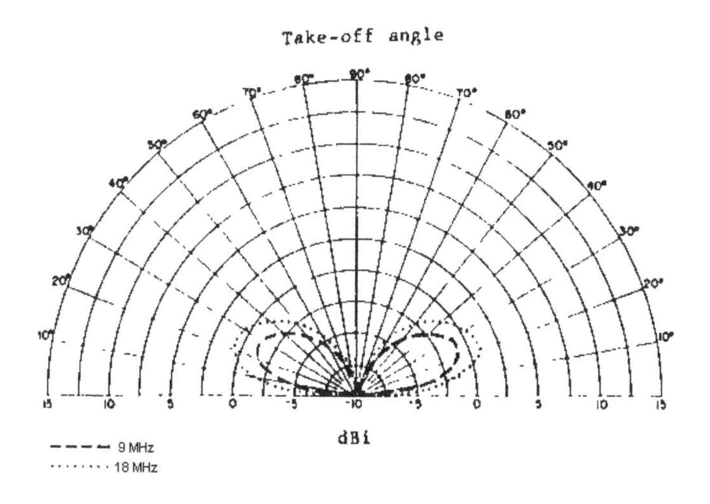

Fifteen-foot vertical whip vertical antenna pattern.

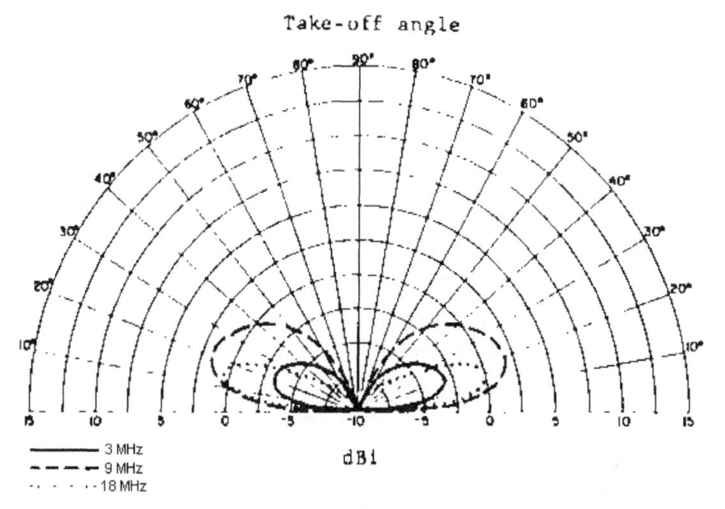

Thirty-two-foot vertical whip vertical antenna pattern.

(a) A <u>ground radial</u> can be easily constructed with twenty 4' to 5' lengths of slash wire with 6" of insulation removed from one end. Bundle the exposed ends together and attach them to a grounding rod. Spread the wires in a circle like spokes in a wheel, then stand or place the radio in the center and communicate.

(b) A <u>reflector</u> can be placed approx 1/4 wavelength (free space) behind a vertical whip, to increase the energy radiated in the opposite direction. The reflector is simply a vertical wire, cut 1/4 wavelength longer than the antenna, and insulated off the ground. Adjust the position of the reflector, first while listening then while transmitting, until the strongest signals are exchanged. Note: a <u>director</u>, which draws energy towards it, can also be made, by placing a wire which is 1/4 wavelength shorter than the antenna 1/4 wavelength away in the direction of the distant station.

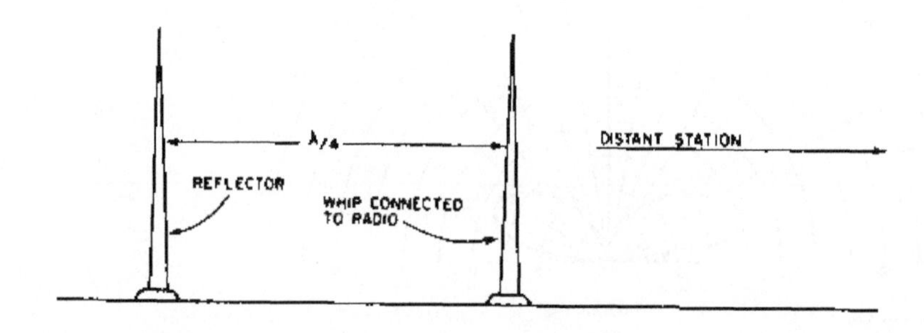

(2) The <u>Dipole</u> is a highly efficient, horizontally polarized, bi-directional, resonant antenna used for NVIS and medium length signal paths out to 1200 miles. Although the ideal antenna for a SARC in situations where recon teams are operating in the same general direction, it is, however, difficult for teams to erect tactically in the field. Maximum vertical gain for this antenna is achieved when it is raised at 1/2 wavelength off the ground. In tactical situations however, this will be sometimes impossible. In a 1/2 wavelength dipole, each radiating side is 1/4 wavelength long. Each leg is half of what the total wavelength is.

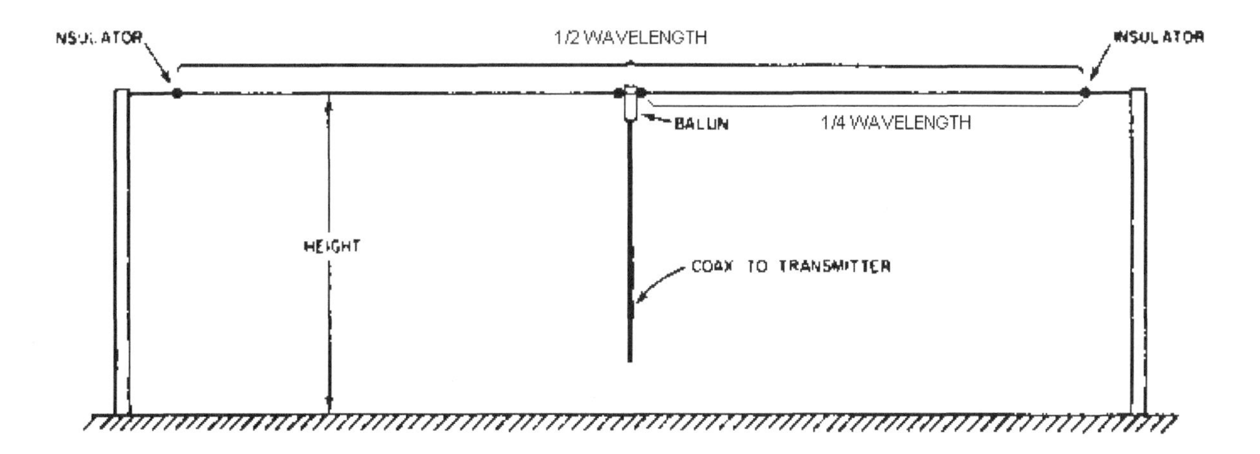

It has a “broadside” pattern normally associated with a dipole. The illustration shows this pattern in polar plot format A. The radiation off the ends of the wire is shown in B If the antenna were lowered to only 1/4 wavelength above ground, the pattern in C results and it’s radiation off the ends is shown in D.

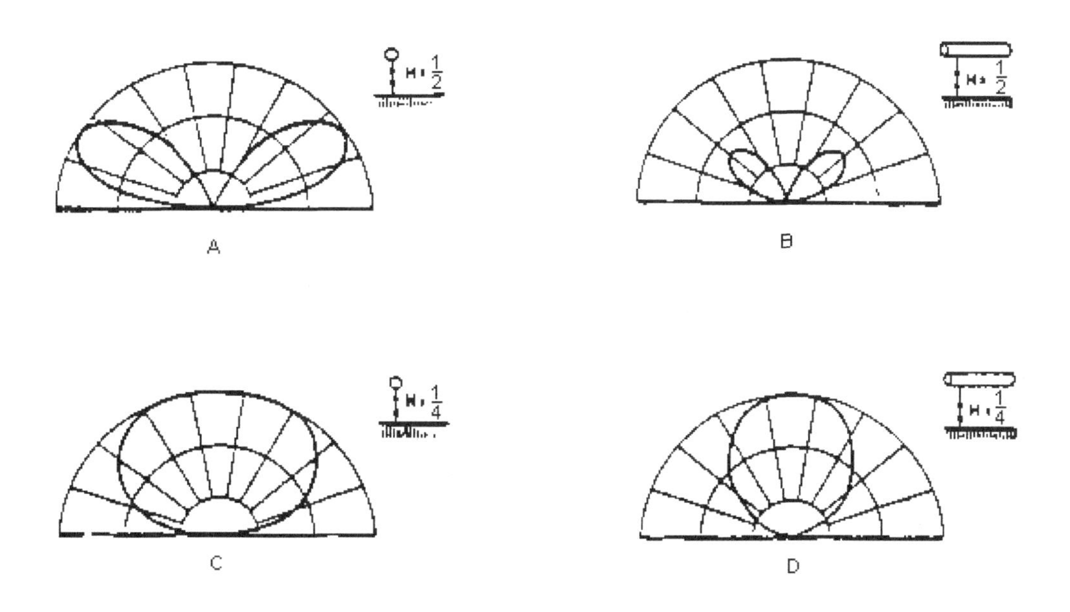

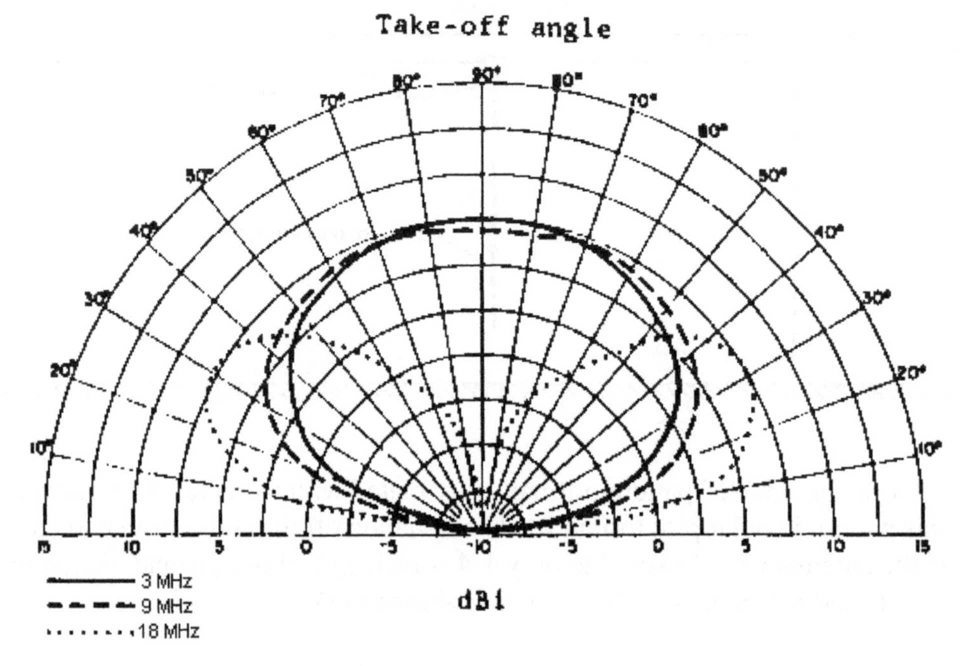

Half-wave dipole antenna vertical pattern, height 8 meters.

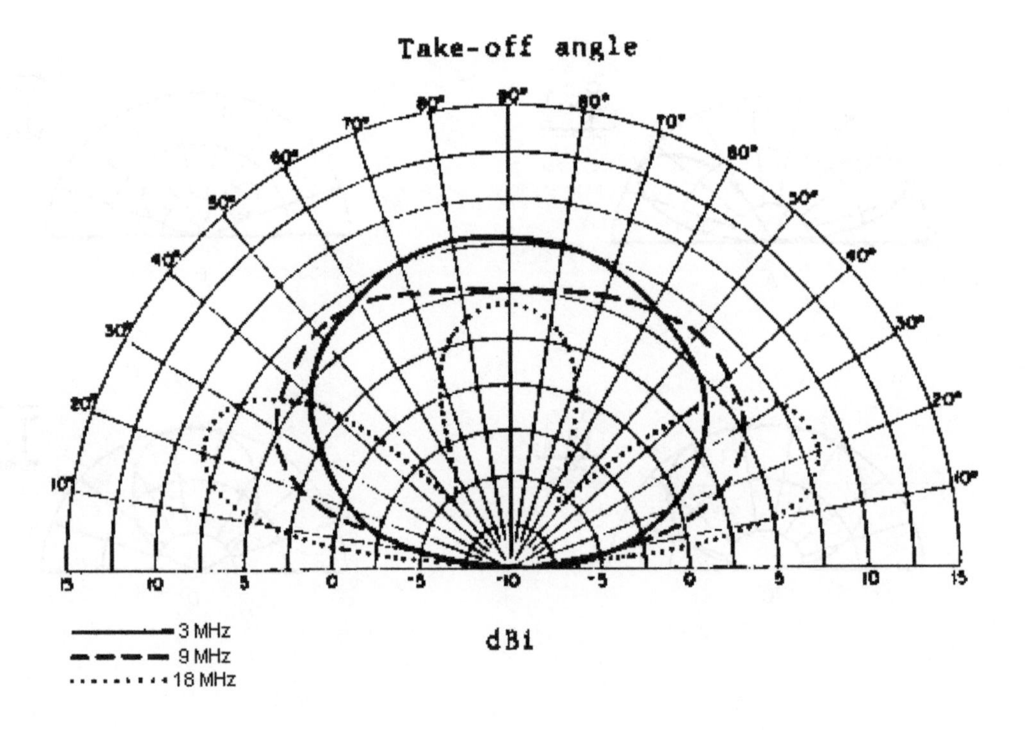

Half-wave dipole antenna vertical pattern, height 10 meters

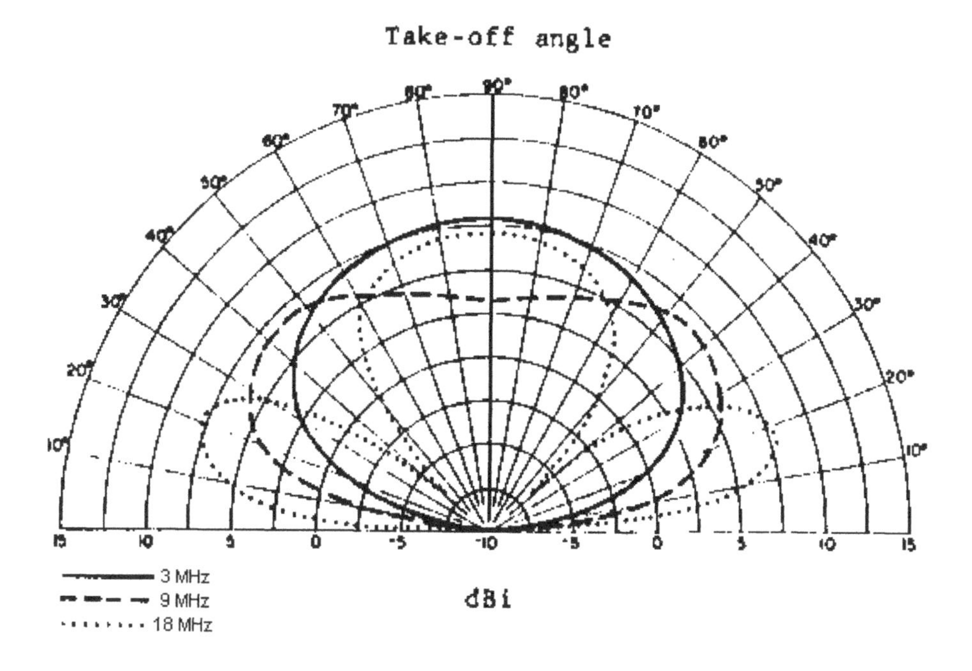

Half-wave dipole antenna vertical pattern, height 12 meters.

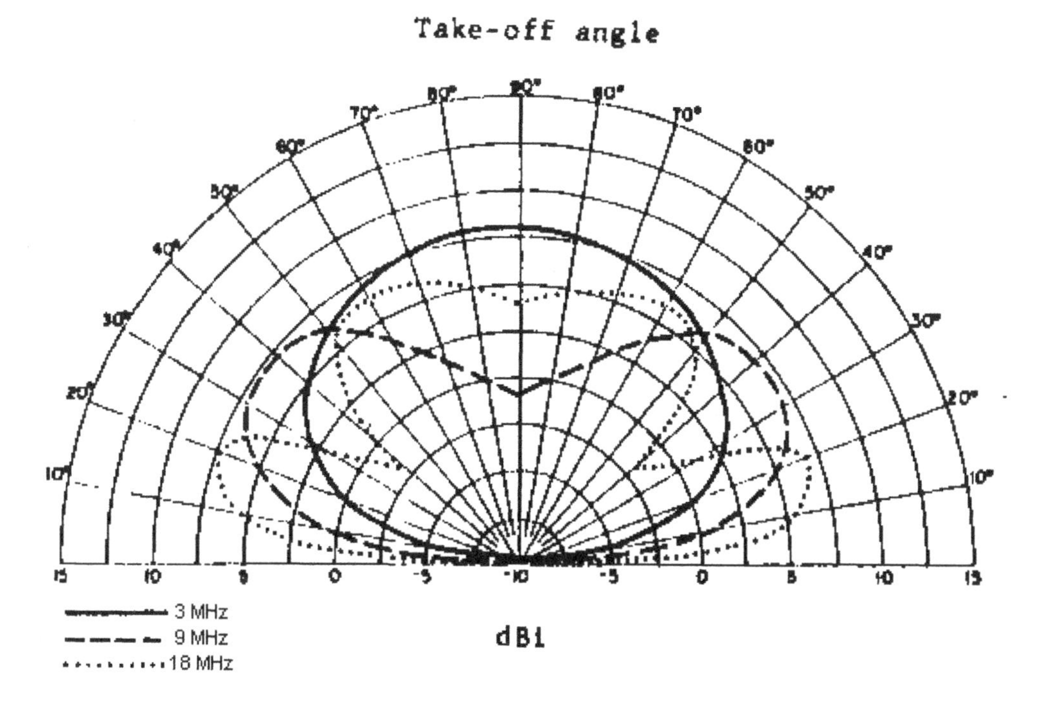

Half-wave dipole antenna vertical pattern, height 14 meters

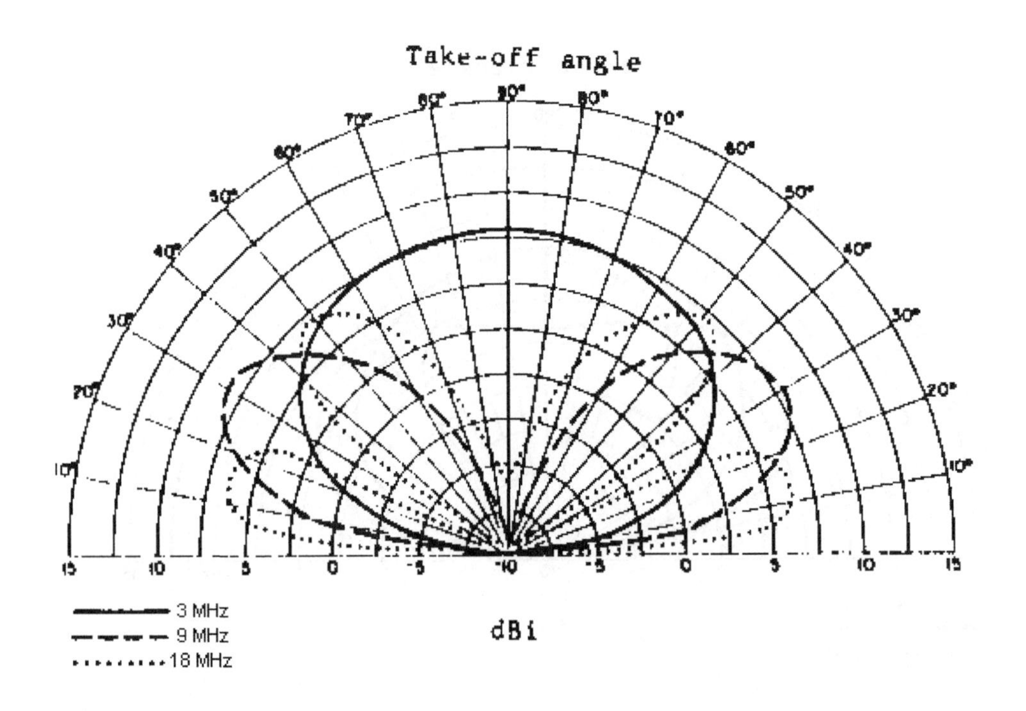

Half-wave dipole antenna vertical pattern, height 16 meters.

(3) The Inverted L is a combination, short range, resonant antenna with a vertical section 1/8-1/4 wavelength high and a horizontal section 1/4-1/2 wavelength long. Because the horizontal section radiates vertically, the inverted L provides omni-directional NVIS. The vertical element radiates a directional ground-wave signal better than the 10' whip. Although somewhat less efficient than a dipole, it is easier to tactically erect and is good for NVIS when stationary and where natural supports such as trees are available.

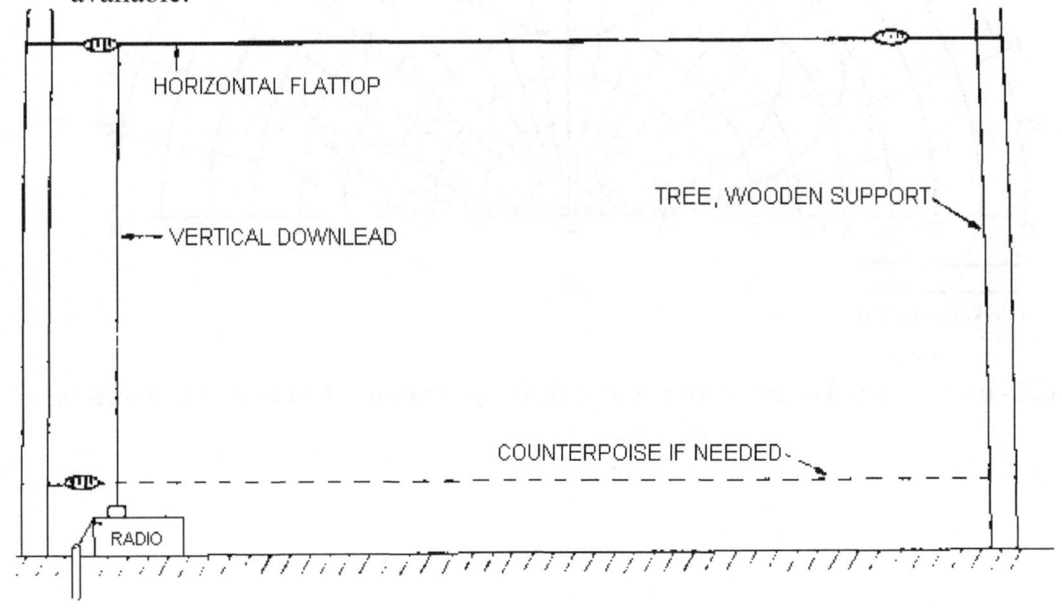

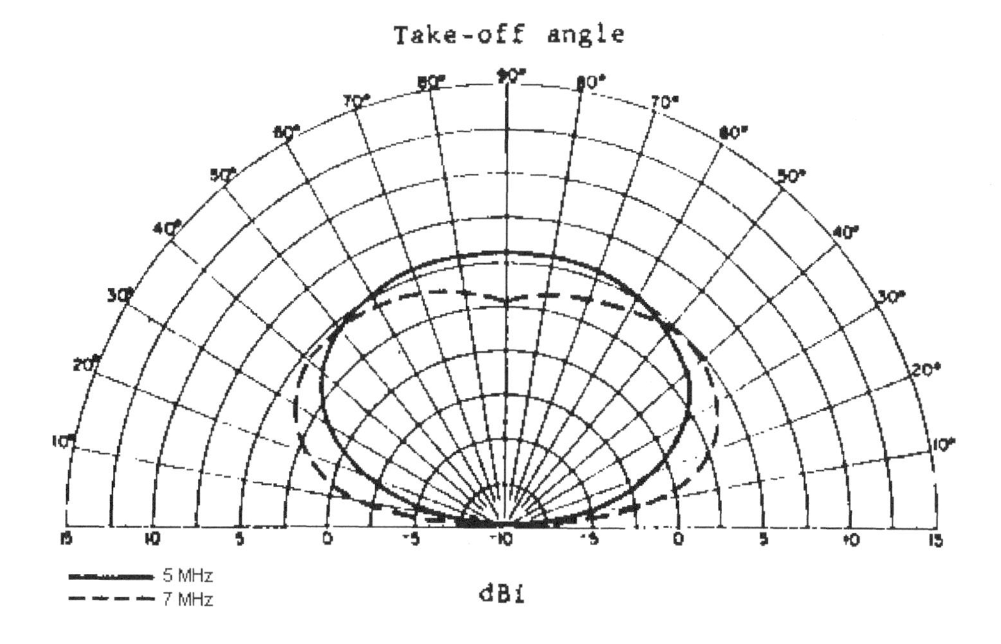

Inverted L antenna vertical pattern, height 40 feet, length 80 feet.

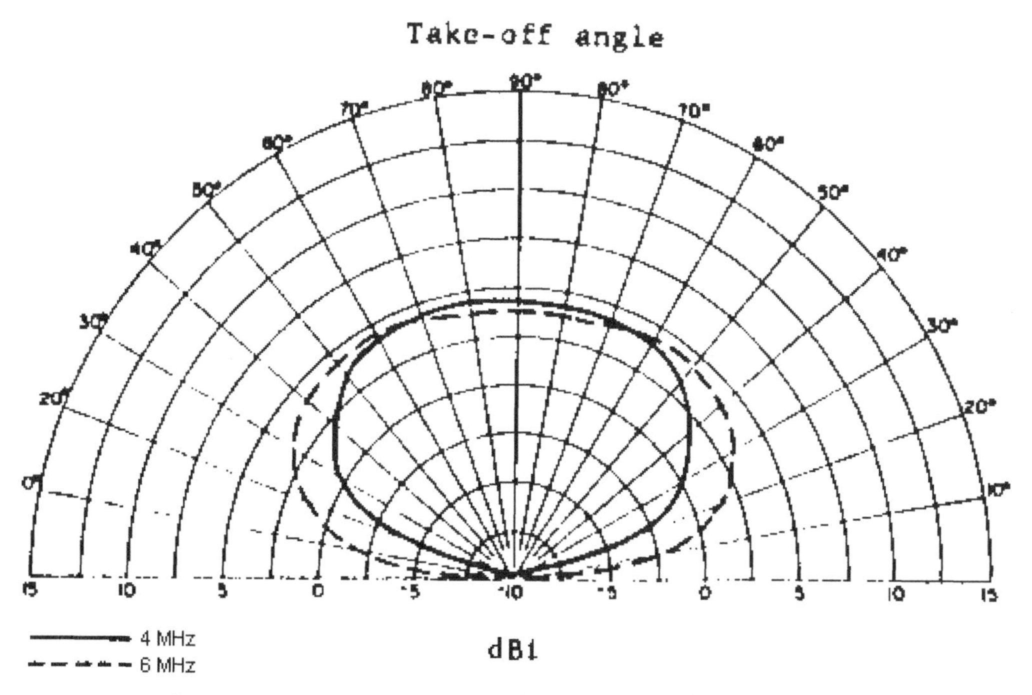

Inverted L antenna vertical pattern, height 40 feet, length 100 feet.

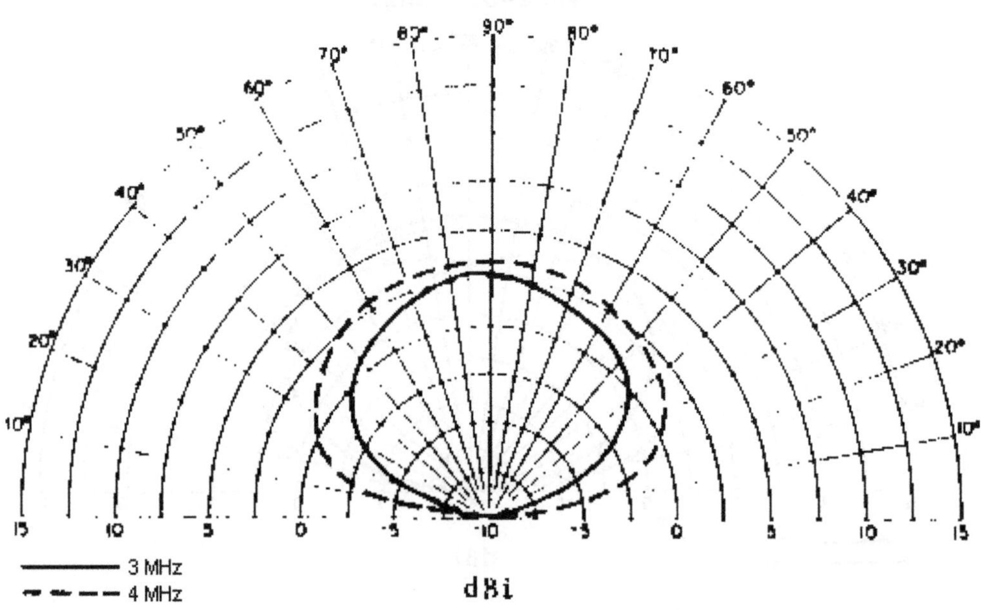

Inverted L antenna, vertical pattern, height 40 feet, length 150 feet.

Frequency Range (MHz)	Horizontal Length (Feet)
2.5 to 4.0	150
3.5 to 6.0	100
5.0 to 7.0	80

(4) The Inverted V is basically a drooping dipole which requires only a single center support. A NVIS, omni-directional, horizontally polarized antenna, it is easier to construct and more compact than either the dipole or inverted L. Furthermore, because it is a resonant antenna, its NVIS performance is superior to the AS-2259, particularly at night in the 3 MHz range. To build, simply attach one end of a feed-line (coax cable) to the radio and attach the other end to a center-fed connector (cobra head). After attaching 1/4 wavelength of copper wire to each terminal of the cobra-head, raise it at least 1/8 wavelength in the air. Next extend the two radiating elements on a 45 degree angle towards the intended station and 90-120 degrees away from each other. Ensure these elements are insulated from the ground. Each leg is half of what the total wavelength is. All construction factors of a dipole are the same here.

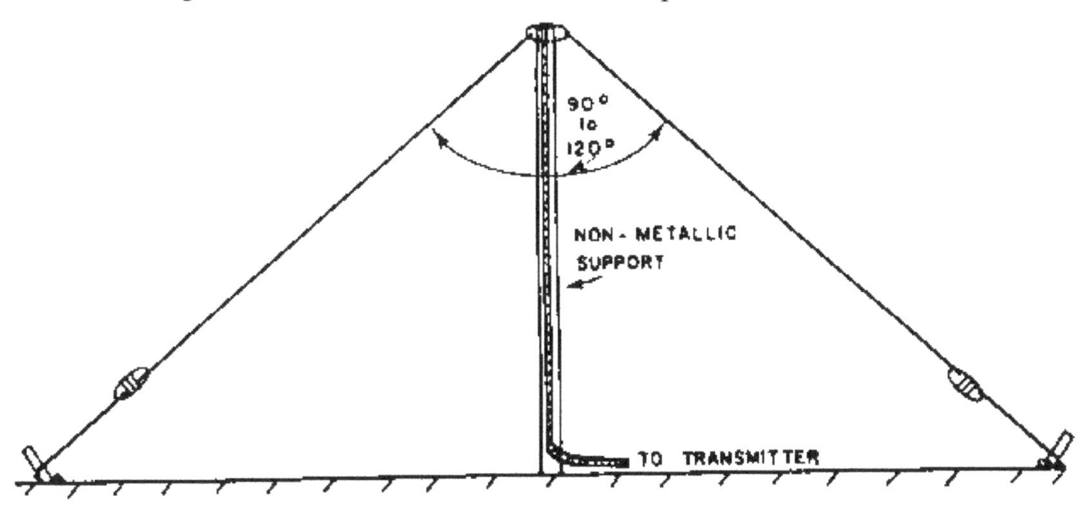

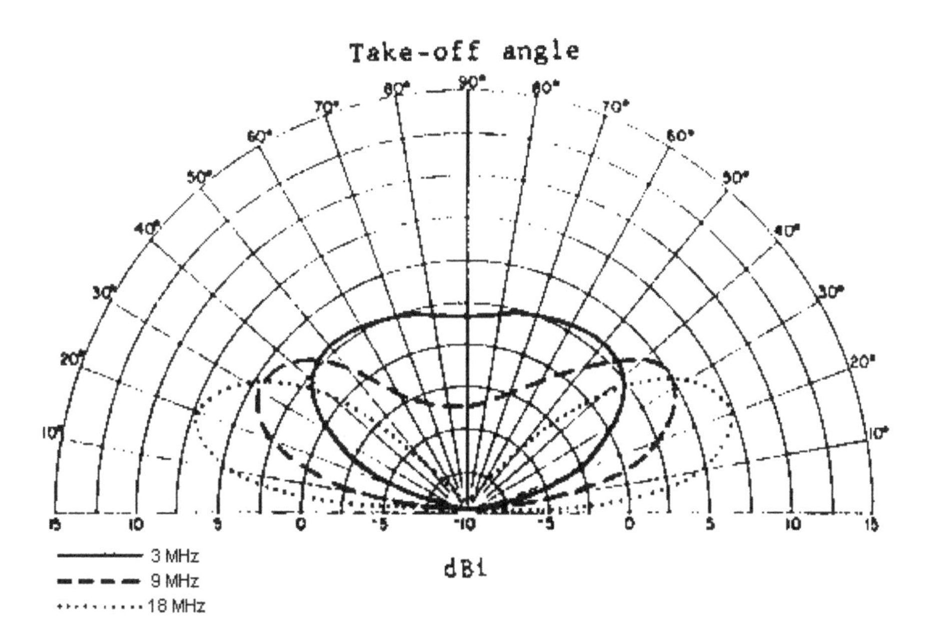

Inverted vee antenna vertical pattern.

(5) The Long Wire is resonant, vertically polarized antenna. Insulated and held at least 10' off the ground, the sloping wire can provide a rapid means to transmit NVIS with some sacrifice in signal strength. If a 600 ohm resistor is added to the end of the transmission line, the antenna should be pointed in the direction of the receiving station (with offset). Grounding the radio should also improve the signal. The antenna must be strung in as straight a line as the situation permits. The more wavelength antenna it is, the more it radiates off the ends, the more gain it has, and the lower the take-off angle.

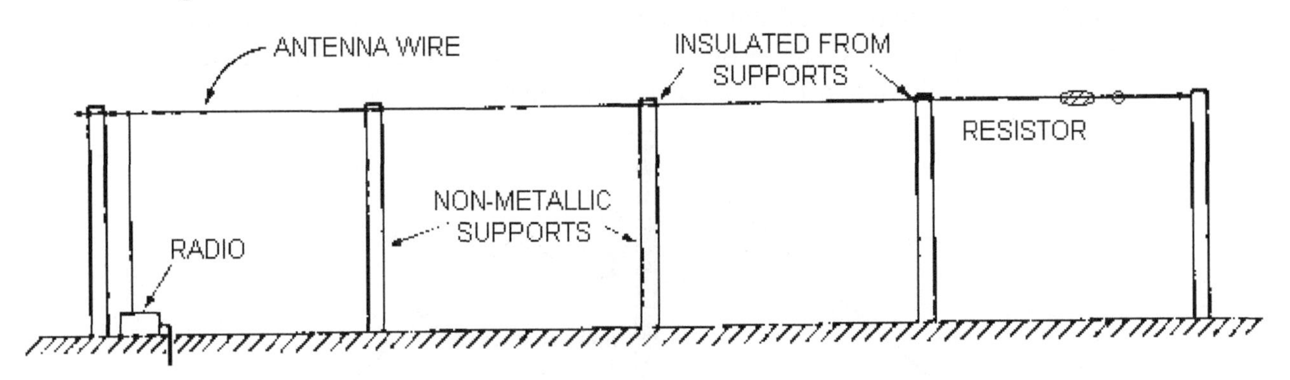

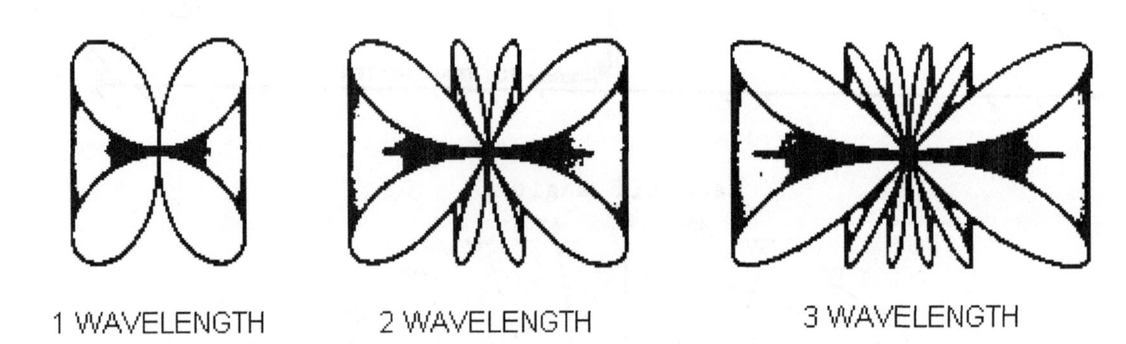

The above diagrams show how the energy radiates more off the ends with the increase of the antenna length.

	Wire Length in Wavelength					
	2	3	4	5	6	7
Radiation Off Axis Angle	30	20	13	10	10	10
Take-off Angle (Degrees)	25	20	15	10	5	5

(6) The Sloping V antenna consists of two long wires arranged to form a V which slopes

down towards the ground. It is bi-directional to directional and produces primarily sky-waves. It is used for medium to long-range communications and is horizontally polarized. The apex angle is the angle between the two legs. For improved performance, use resistors on the ends of the legs.

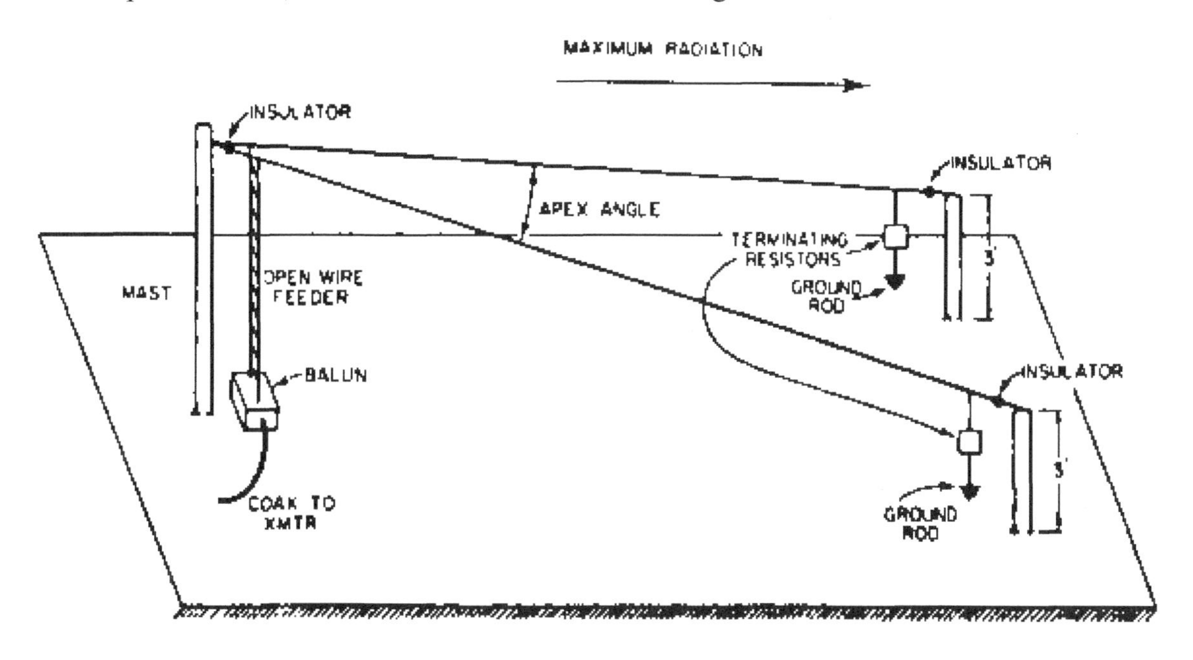

Path Length	Apex Angle
700 to 1000 miles	60 degrees
1000 to 1500 miles	45 degrees
Over 1500 miles	30 degrees

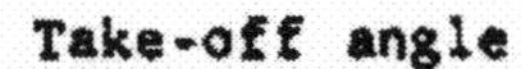

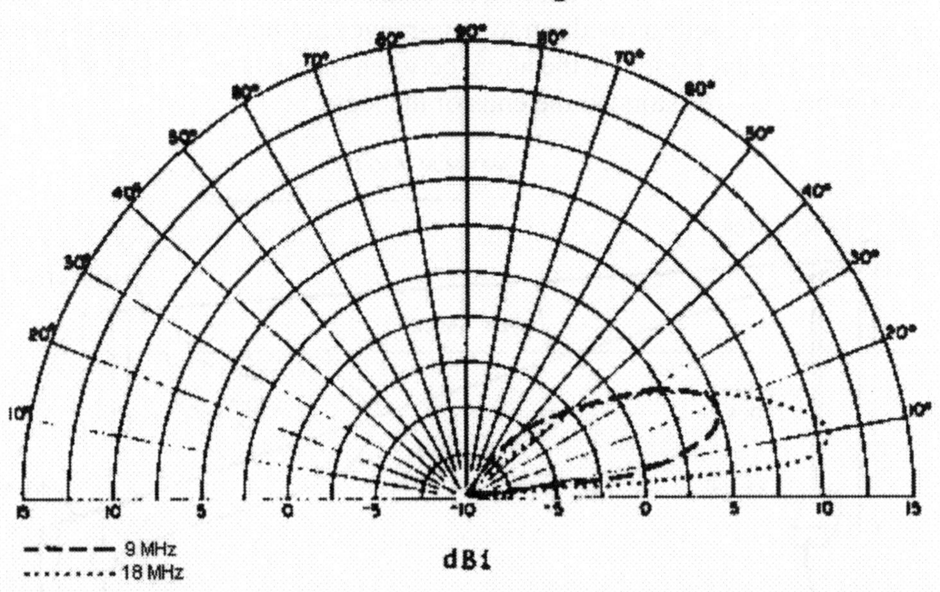

Terminated sloping vee antenna vertical pattern, height 40 feet, length 500 feet, apex angle 30°.

Take-off angle

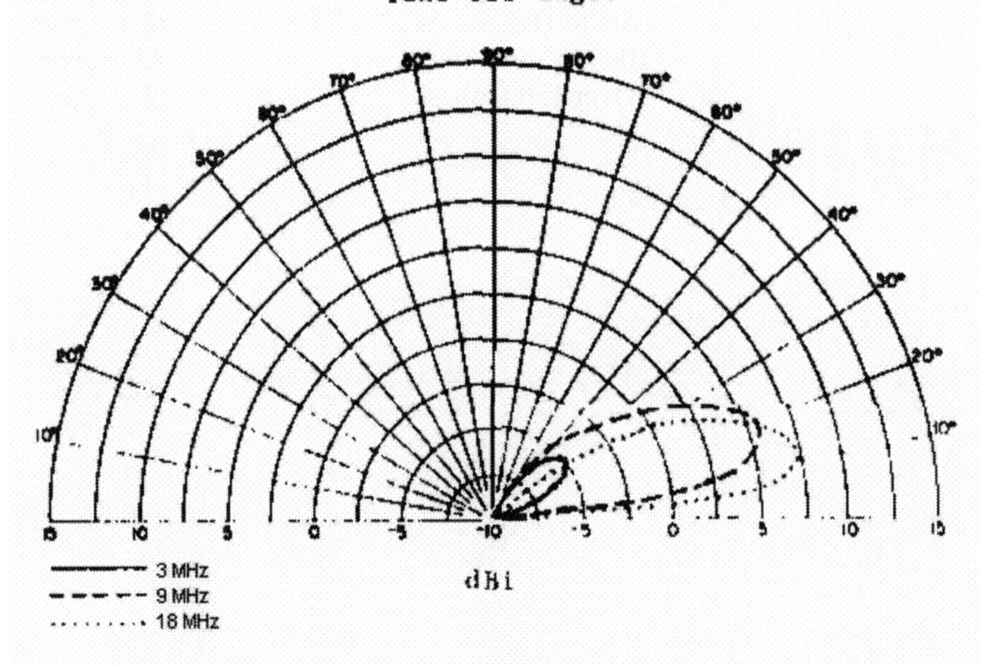

Terminated sloping vee antenna vertical pattern, height 40 feet, length 500 feet, apex angle 45°.

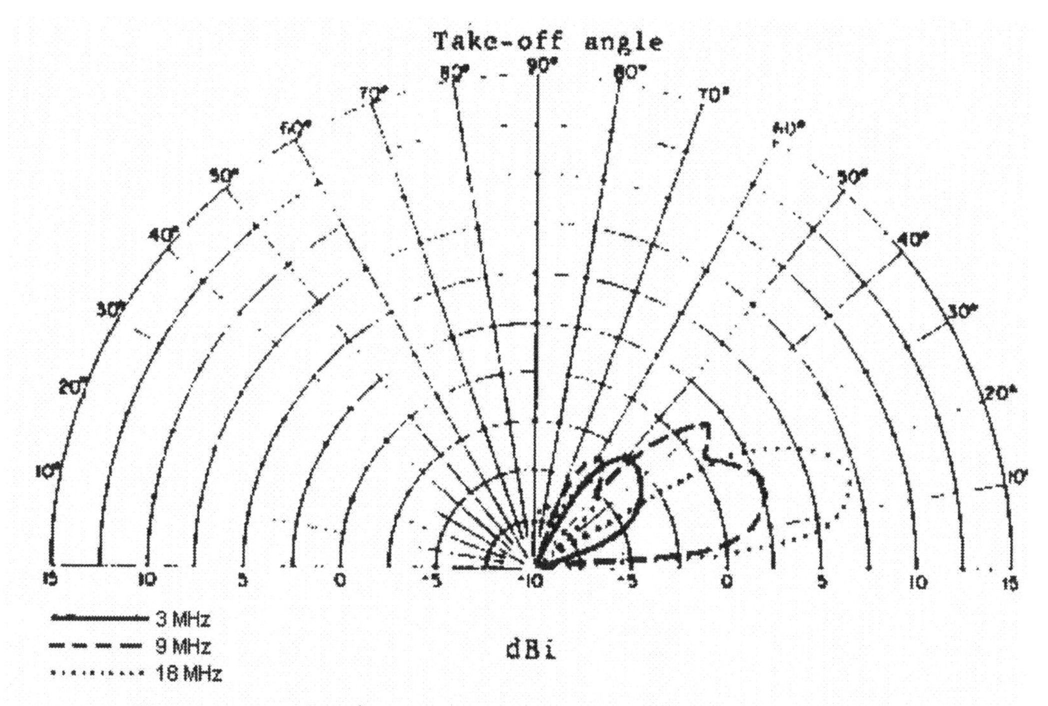

Terminated sloping vee antenna vertical pattern, height 40 feet, length 500 feet, apex angle 60°.

(7) The Sloping Wire is a resonant, vertically polarized, and the simplest and most expedient antenna capable of sky-wave propagation. When HF comm is required on the move outside ground-wave range, or when no means is available or tactically prudent to elevate the antenna, a 1/4 to 1/2 wavelength wire can be connected under the 10' whip antenna base, or to the red terminal of a cobra head connected directly to the radio, and extended away from the direction of the distant station (with a 10-15 degree offset). Insulated and at least 6' off the ground, the sloping wire provides a rapid means to transmit NVIS with some sacrifice in signal strength. If a resistor is added to the end of the transmission line, the antenna should be slanted in the opposite direction of the receiving station (with offset). The low end of the transmission line should always be pointed towards the receiving station. It is bi-directional without a resistor and directional with one.

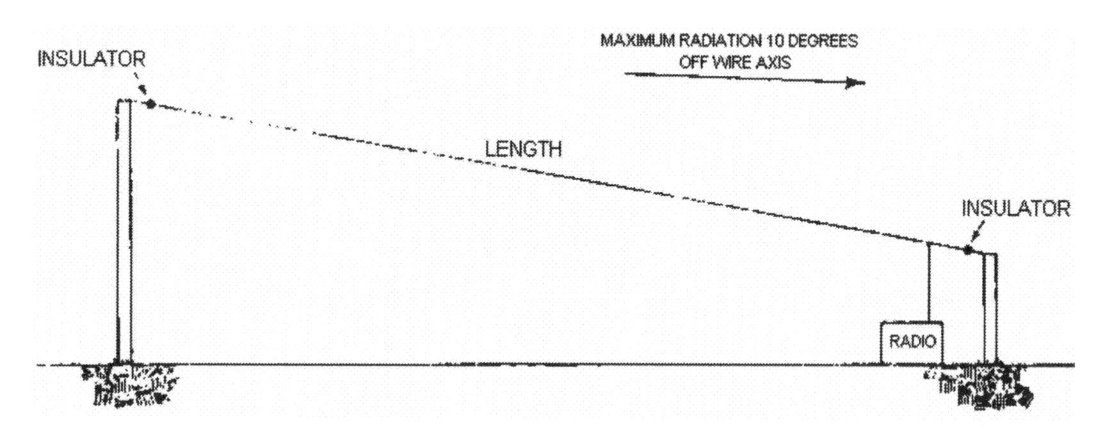

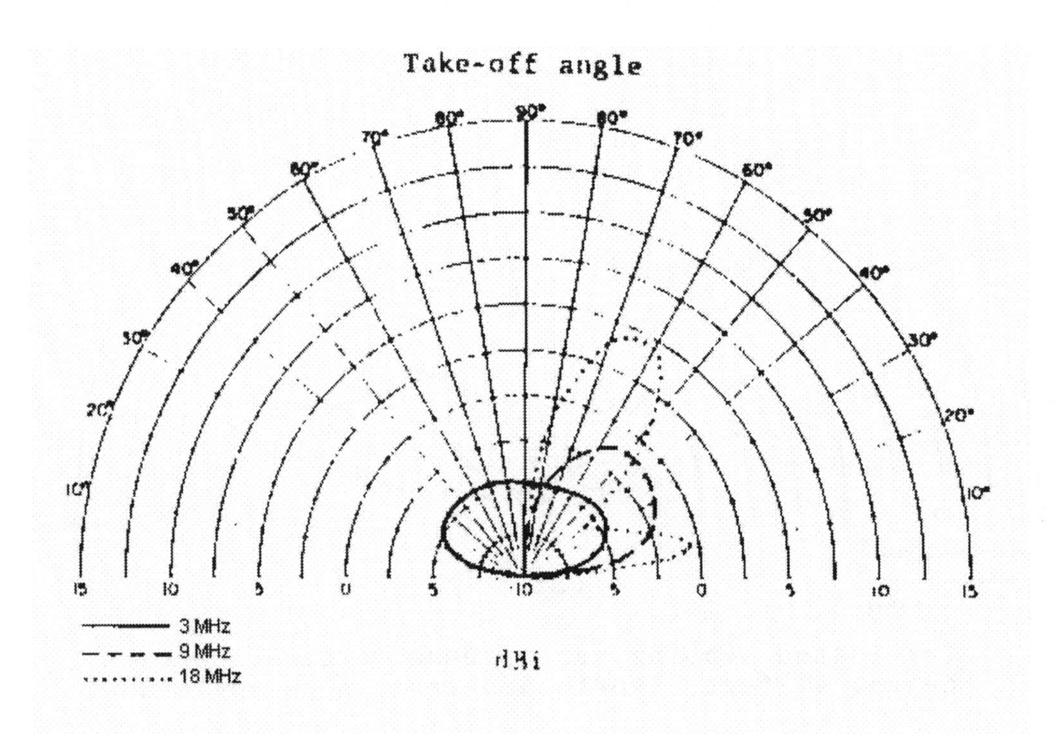

Sloping wire antenna vertical pattern, length 100 feet.

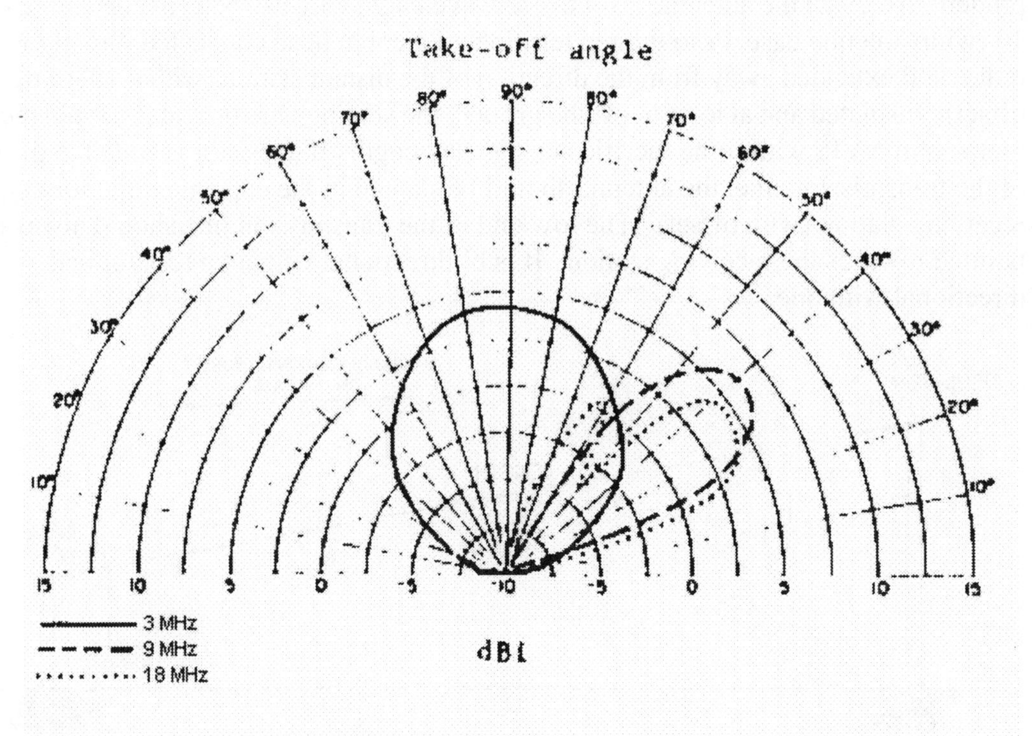

Sloping wire antenna vertical pattern, length 250 feet.

(8) The Vertical Half Rhombic antenna is a version of the long wire except it uses a single center support. It is a vertically polarized, directional antenna. It uses a resistor on one end of the antenna and that is the side that you point at the distant station (with a 10 degree offset).

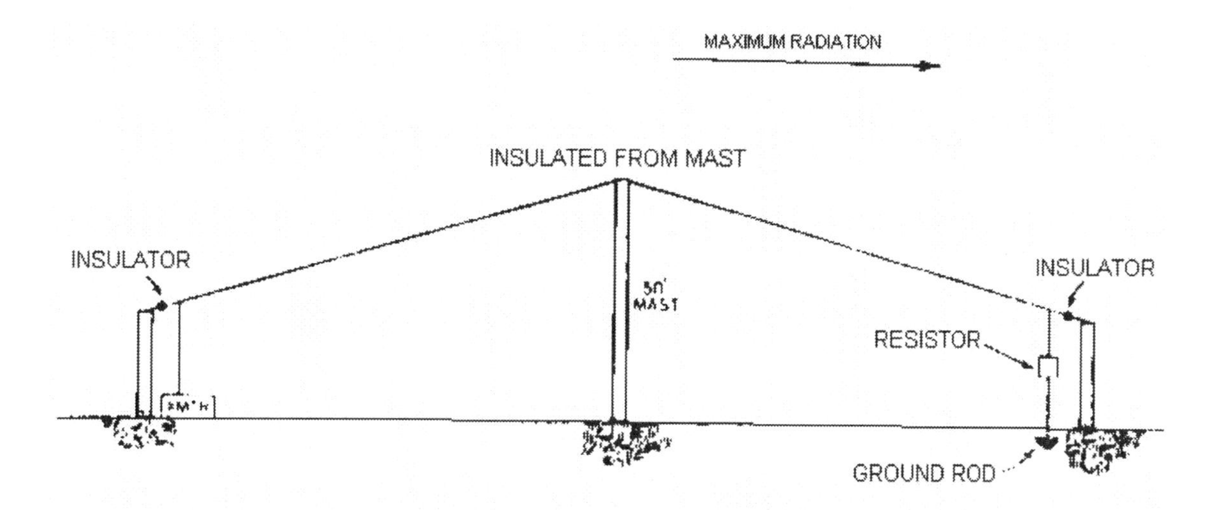

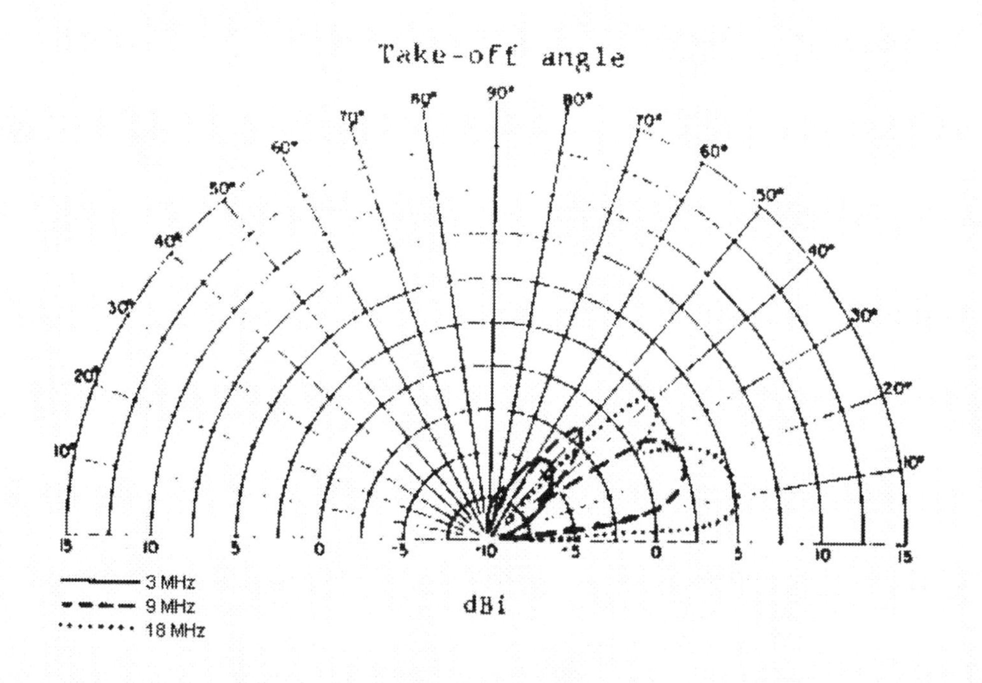

Vertical half-rhombic antenna, vertical pattern, antenna height 50 feet, length 500 feet.

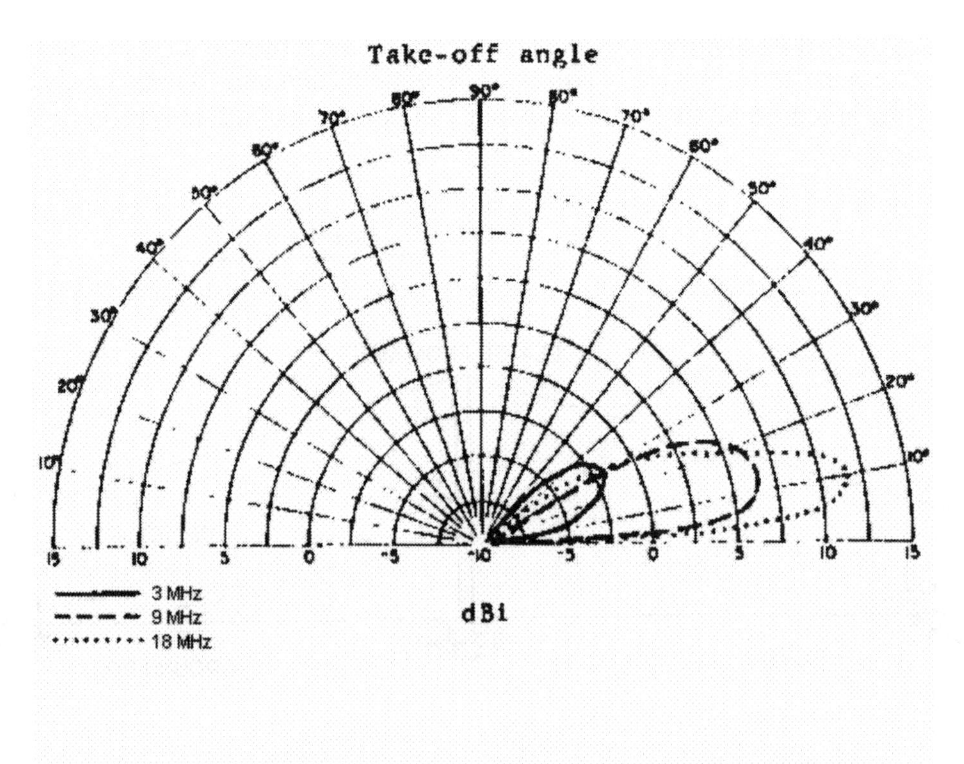

Vertical half-rhombic antenna vertical pattern, antenna height 50 feet, length 1000 feet.

c. <u>VHF Antennas</u>

(1) The <u>Vertical Whip</u> antenna is part of all radio sets, and is the most commonly used VHF antenna because of its simplicity, and because it is the only antenna that can be used on the move. When stationary, however, the vertical whip is a poor choice, because it radiates an inefficient omni-directional signal and does not stand high enough for good VHF-LOS propagation. To improve its performance, ensure the antenna remains vertical when in use, and use a ground radial, reflector, or director as discussed earlier with HF antennas. You can also make a resonant insulated wire and put it up vertically for improved signal. This is known as the <u>Vertical Wire</u>.

(2) A <u>Field Expedient 292</u> can be created by cutting four copper wires to 1/4 wavelength of the operating frequency and then connecting the vertical element to the red terminal of a cobra head (connected to a feed-line) and the ground radial elements to the black terminal. To position the ground radials, tie three 6' long sticks together to form a triangle, the tie the end of a ground radial to each corner. Insulators are placed between the vertical element and the hoist line and the vertical element and the ground elements. The hoist line is a long piece of 550 cord that hoists the antenna up in the air. It is an omni-directional, vertically polarized antenna.

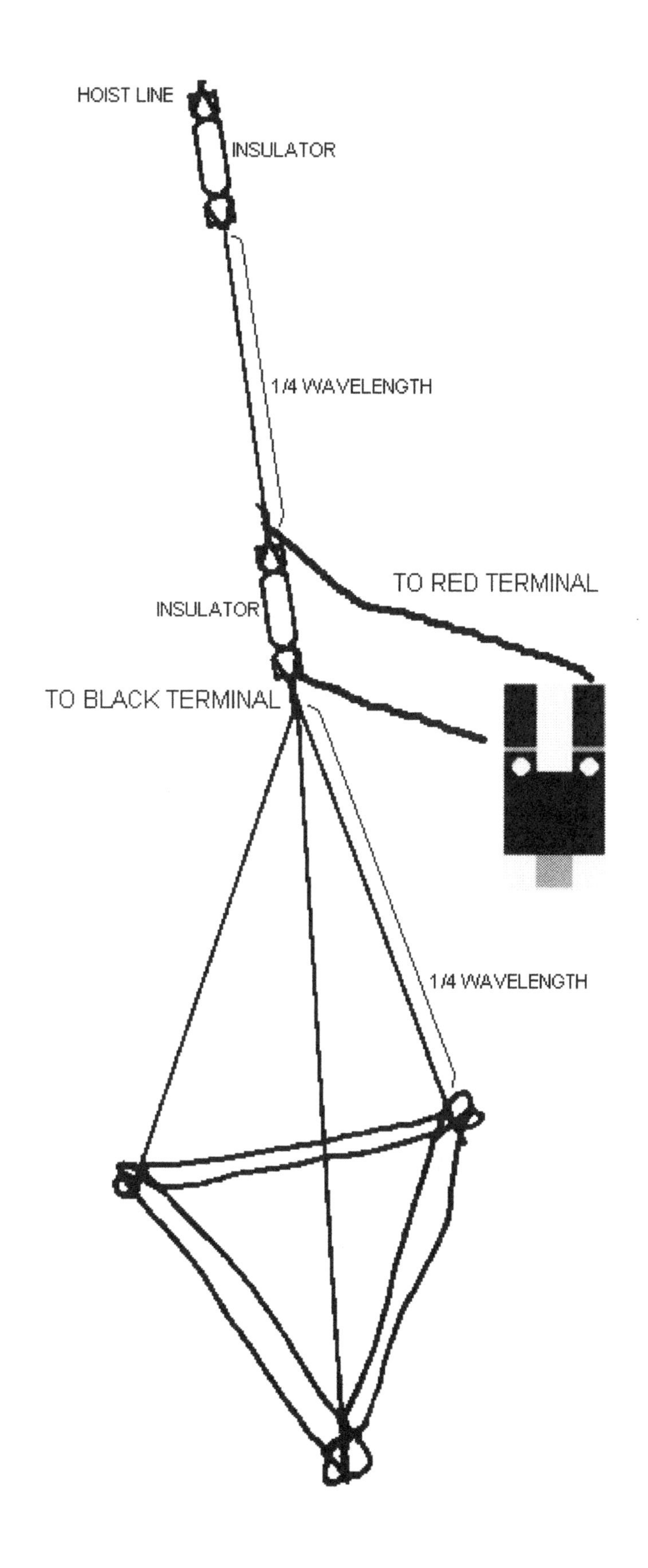
HOIST LINE
INSULATOR
1/4 WAVELENGTH
TO RED TERMINAL
INSULATOR
TO BLACK TERMINAL
1/4 WAVELENGTH

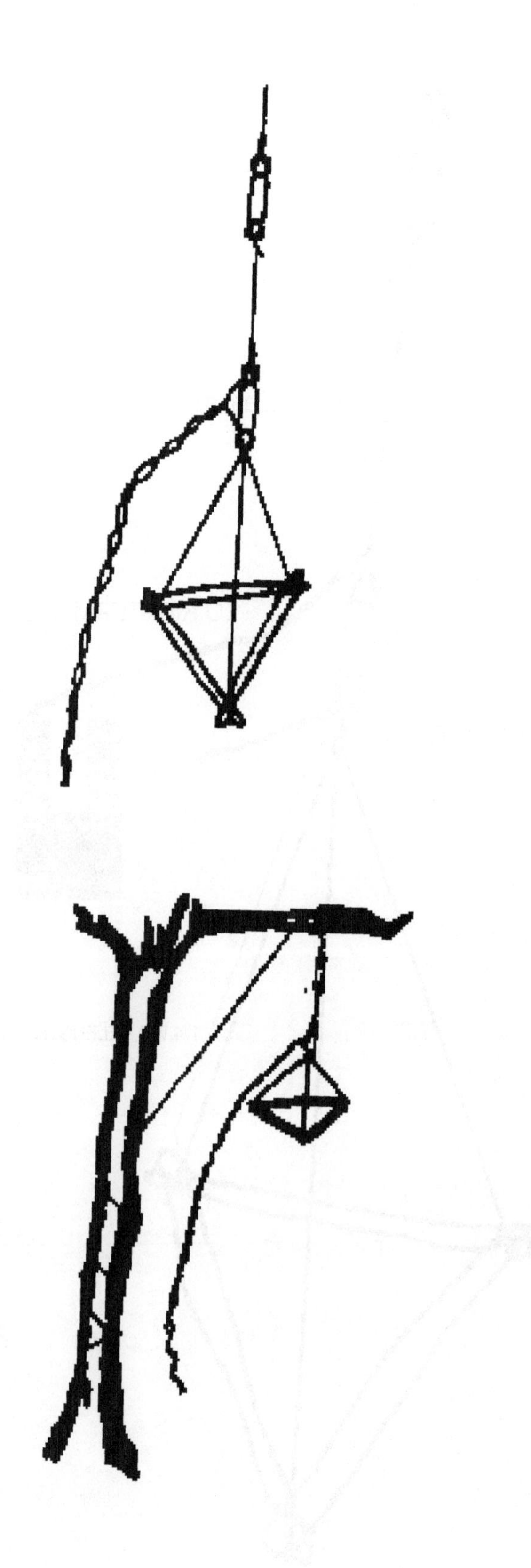

(3) The Vertical Half Rhombic is a vertically polarized, bi-directional antenna, which can be made to work satisfactorily across the entire VHF spectrum, but requires a large installation area. This antenna consists of a of copper wire, connected to the radio under the whip antenna base, supported in the middle by a support, and augmented by a counterpoise. Efficiency is maximized, however, by cutting the antenna length to have a low SWR at the operating frequency, and by making it directional with a resistor. The antenna should be oriented perpendicular to receiving station. When made directional, the resistor should be aimed at the distant station. The OE-303 is a factory produced version of the vertical half rhombic antenna.

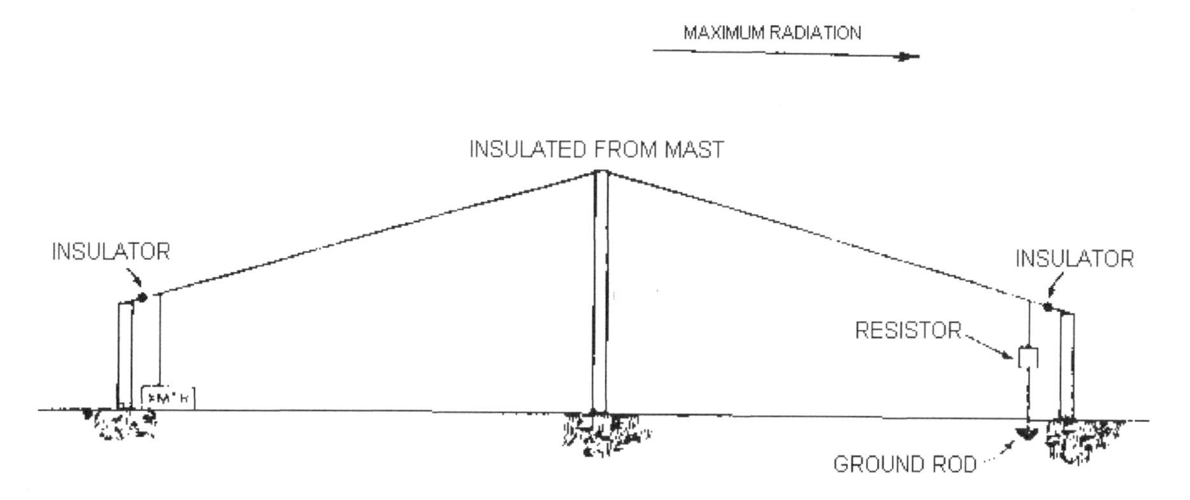

(4) The Dipole or half wave center fed antenna, is constructed in the same fashion for VHF as it is for HF propagation. The VHF version has the tactical advantage of being considerably more compact, (1/2 wavelength at 40 MHz = 23'4") but still requires the use of two supports, roughly perpendicular to the desired signal path.

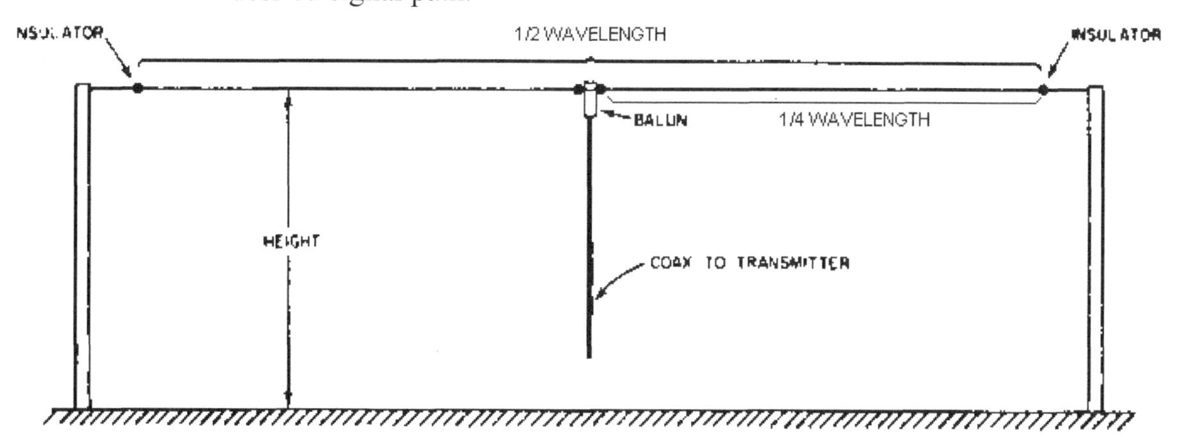

(5) The long wire is used when you need more distance and directivity than your whip antenna. The overall length of your antenna must be between three to seven wave lengths of the operating frequency. This antenna is bi-directional and you point the antenna in the direction you want to talk.

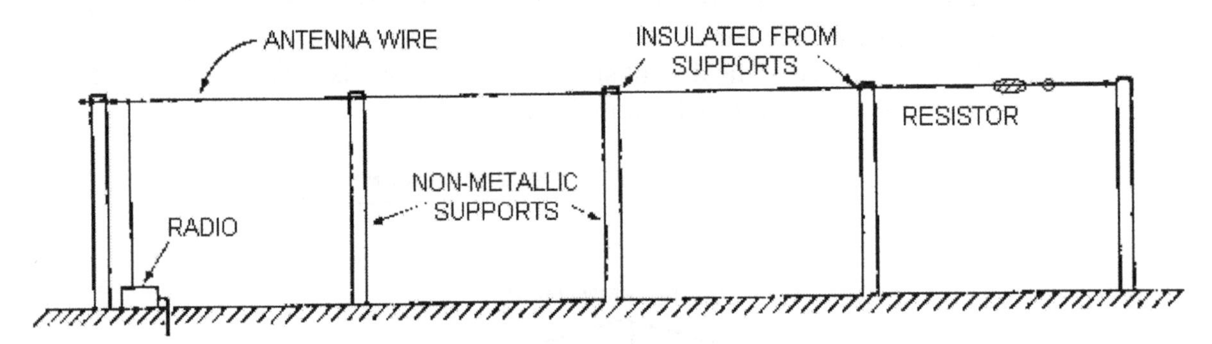

The radiating patterns are the same as the HF Long Wire. The more wave-lengths the antenna is, the more it radiates off the ends of the antenna. When aiming at a distant station, use the same offset angle as before in the HF Long Wire.

6. (5 Min) **FIELD EXPEDIENT ANTENNA KITS (FEAK)**

a. Field Expedient Antenna Kits (FEAK) are crucial for effective recon communications, enabling teams to customize antennas for maximum performance under difficult conditions. At a minimum the following materials should be included in each team's FEAK:

(1) 10-12 Gauge copper wire or firing wire. Any wire is ok but copper wire conducts electricity better than other wire.

(2) 40' of coaxial cable (feed-line)

(3) Resistors. Should be one-half of the power output of the radio it is being used on (watts) and 600-800 ohm.

(4) Center fed connectors (cobra-heads)

(5) Electrical tape

(6) 1' long copper grounding stakes

(7) Insulators

(8) Plastic bags or waterproof covers

(9) Multi-meter (optional). It tests batteries for power, feed-lines, and gives the ohm factor of resistors.

7. (5 Min) **ESTABLISHING COMMUNICATION**

a. Select the best comm site the tactical situation allows. Pay special attention to the elevation of the sight and proximity to obstructions, particularly in the intended direction of your signal, location of power lines and other metal objects, antenna height, ground conductivity, time of day, and weather.

b. Attempt to communicate using the 10' whip on both HF and VHF radios, first with the antennas vertical, then at 45 and 90 degrees. If no comm but radio tunes and functions properly, erect field expedient (FX) antenna(s).

c. Attempt comm with FX antenna. If radio tunes and functions properly but no comm, attempt CW radio check. If still no comm execute the following as applicable, attempting comm after each change:

(1) Adjust take-off angle

(2) Adjust orientation of antenna, to the left and then to the right.

(3) Without switching antennas, switch to the alternate frequency and attempt to transmit. If no comm, attempt to relay message via another friendly unit.

(4) Return to primary frequency and add on a resistor (directional), ground radial, counterpoise, reflector, or director to the FX antenna(s) in use and repeat the above steps.

(5) If still no comm, switch to the secondary FX antenna and repeat the above steps.

(6) If still no comm, move to another sight, which is either closer, farther away, or higher than your present location and repeat all the above steps. Unless specified by Higher HQ, comm has priority over the mission after a comm window is missed. Continue to repeat this cycle at different locations until comm is reestablished or you are forced to execute the no-comm plan.

8. (5 Min) **REMARKS**

a. A counterpoise is much like a reflector except that a counterpoise is directly above the ground. It is totally insulated from all parts of the radio and antenna and is placed directly underneath the antenna.

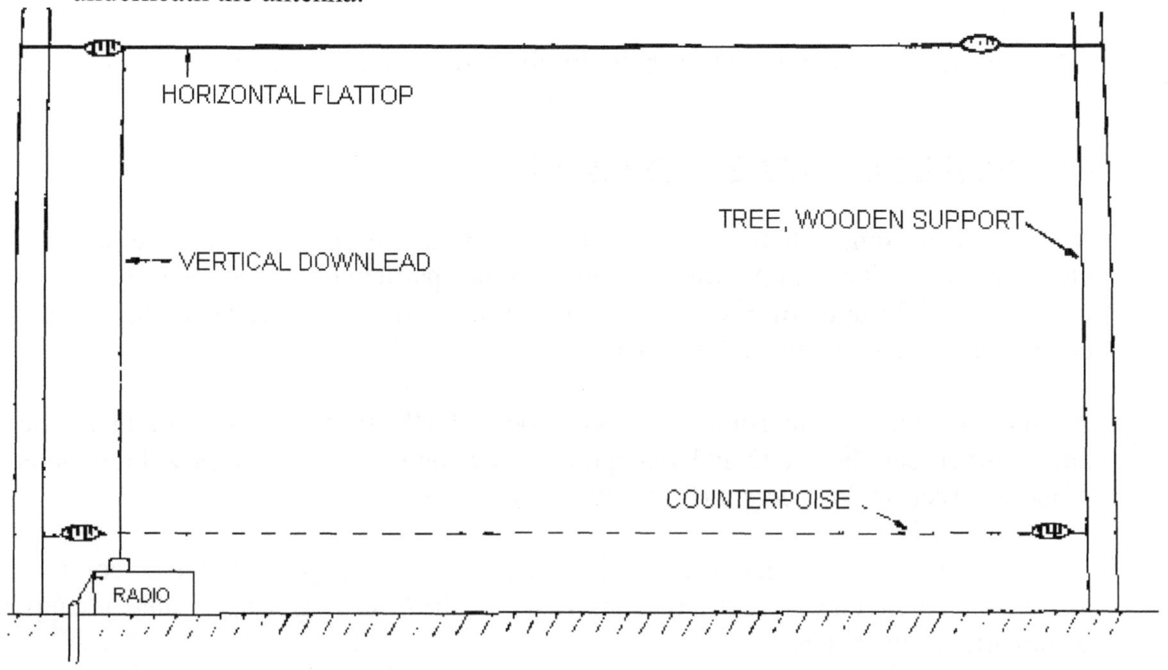

b. Resistors can be attached to antenna in three ways. First resistors should have alligator clips placed on its ends for easy attachment to things. When placing the resistor on the wire, it must be placed near the insulator and in between the insulator and the center fed connector (cobra-head).

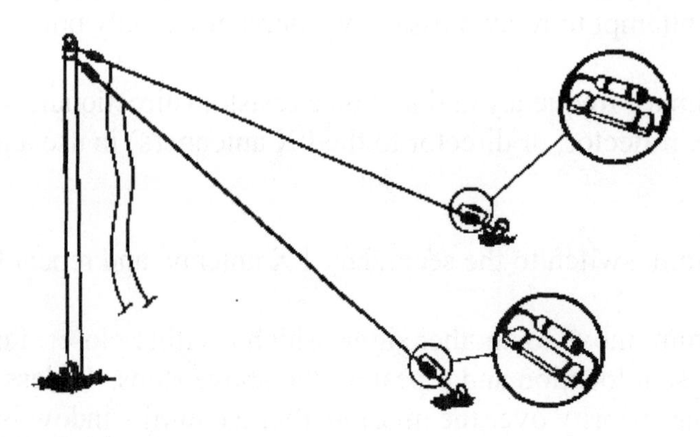

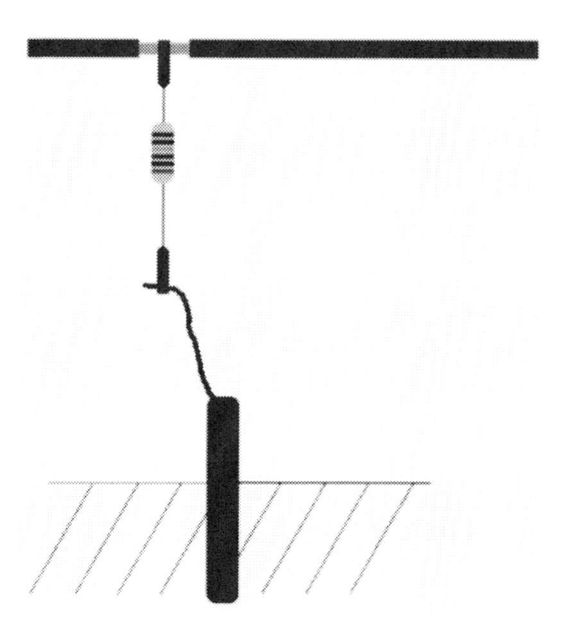

This is the most preferred method. Expose the wire from it's coating and attach the resistor. The other end of the resistor is attached to a grounding rod.

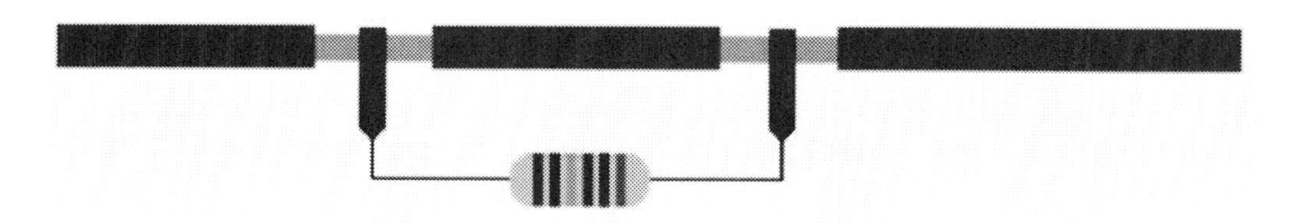

Expose two parts of the wire from the protective coating and attach both ends of the resistor to it.

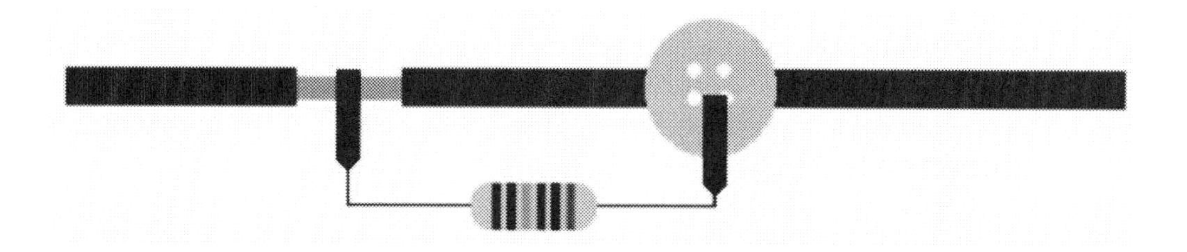

This is the least preferred method. Attach one end of the resistor to an exposed portion of the wire and the other end of the resistor is attached either to the insulator or to the protective coating of the wire that's beyond the insulator.

9. (5 Min) **SPECIAL CONSIDERATIONS**

a. Cold Weather and Mountainous terrain

(1) It's more tactically feasible to use the terrain in a mountainous environment to your advantage by using it to gain height in your antennas. For example, use a cliff face or an angled hill for adjusting your take off angles instead of having to climb trees and adjusting the angle from there.

(2) Vertical antennas are preferred for ground wave propagation in the HF band, but the use of fractional wave whip antennas is not recommended except for short distances.

(3) Because of extended distances and difficulty in radio transmissions in arctic/sub-arctic areas, radio relays, retransmission sites or message pick up and drop techniques may have to be employed.

(4) Teams must utilize terrain masking techniques in order to not become compromised by the enemy using direction finding equipment.

(5) The combined effects of terrain, cold, ice, dampness, and dust on communications equipment increase maintenance.

(6) Emphasis must be placed on the selection and use of proper frequencies, radio propagation graphs, charts, and prediction data presently available and they must be utilized.

(7) If snow or ice covers the pressure release cover on the battery box, ice may form over it which may restrict it from air exchange.

b. Ionospheric disturbances

(1) HF transmissions and reception, while capable of spanning the extended distances dictated by tactical requirements, are subject to interference by magnetic storms, aurora borealis, and ionospheric disturbances, which may completely black out reception for hours or even days in northern areas. These "blackouts" occur throughout the HF spectrum down to 25 KHz.

(2) The greater the ionization of the upper atmosphere, which takes place during aurorals, will on occasion increase the range of the tactical FM radio. The use of VHF equipment will provide the greatest degree of reliability for multi-channel means of communications in northern areas.

(3) Precipitation static is interference caused by snow, ice, and rain particles driven against metal objects and antennas. The result is static noise that will interfere with HF, VHF, and UHF signals. Relocating radio equipment away from vehicles, buildings, trees can reduce precipitation static. It is associated with metal antennae, high power radios, and

sensitive receivers exposed to rain or wet snow. Covering the antenna with polystyrene masking tape will reduce this effect, but it will only be effective if there are no other adjacent metal surfaces against which the discharge can take place.

(4) The greatest auroral activity occurs between 60 degrees and 70 degrees north latitude and occurs at intervals of 27 to 28 days.

(5) Above a certain frequency (40 to 60 MHz) the greater amount of energy in radio waves passes through the ionosphere and is not reflected back down to the receiving station

c. Frozen ground

(1) Copper grounding rods may be difficult to drive into the earth due to the strength of ice and the depth of the snow/ice. The best ground is obtained by driving a spike through ice and into water. The ground spike should be treated with salt, Epsom salt, or salt peter. In areas like Southern Norway or Northern Germany where trees are plentiful, a nail may be driven into a tree and the equipment will be grounded by using the tree's root system.

(2) When grounding of comm equipment is necessary, grounding stakes should penetrate 1 meter (3') below the surface. When a grounding rod cannot be driven that length, it should be driven at least 30 cm (12") and water poured around the base to increase conductivity. You can use water from streams or areas next to streams or lakes that abound the area in the northern latitudes. Epsom salt or other additives should be added to help ground. If you have to ground using this method, you must also ground your radio onto it's battery box for added grounding performance. In many instances it will be impracticable to secure a ground and it will be necessary to install a counterpoise as a substitute.

(3) Pitons are considered excellent anchors for antenna guys in frozen earth, ice, or rocky soil.

d. Field Expedient Antennas

(1) All large horizontal antennas should be equipped with counter weights arranged so as to give before the wire or poles break from the pressure of the ice or wind.

TRANSITION: Are there any questions over what we have just covered?

PRACTICE (CONC)

a. Students will practice what was taught during upcoming field exercises.

PROVIDE HELP (CONC)

a. The instructors will assist the students when necessary.

OPPORTUNITY FOR QUESTIONS (3 Min)

1. QUESTIONS FROM THE CLASS

2. QUESTIONS TO THE CLASS

Q. What are some items that we can set up with our antenna to help improve performance?

A. Insulators, resistors, counterpoise, ground radials, reflectors, directors, feed-line, balun, ground, vertical supports, or a center-fed connector (cobra-head).

Q. What are the steps for establishing communication?

A. (1) Select the best comm site the tactical situation allows. Pay special attention to the elevation of the sight and proximity to obstructions, particularly in the intended direction of your signal, location of power lines and other metal objects, antenna height, ground conductivity, time of day, and weather.
(2) Attempt to communicate using the 10' whip on both HF and VHF radios, first with the antennas vertical, then at 45 and 90 degrees. If no comm but radio tunes and functions properly, erect field expedient (FX) antenna(s).
(3) Attempt comm with FX antenna. If radio tunes and functions properly but no comm, attempt CW radio check. If still no comm execute the following as applicable, attempting comm after each change:
(a) Adjust take-off angle
(b) Adjust orientation of antenna, to the left and then to the right.
(c) Without switching antennas, switch to the alternate frequency and attempt to transmit. If no comm, attempt to relay message via another friendly unit.
(d) Return to primary frequency and add on a resistor (directional), ground radial, counterpoise, reflector, or director to the FX antenna(s) in use and repeat the above steps.
(e) If still no comm, switch to the secondary FX antenna and repeat the above steps.
(f) If still no comm, move to another sight, which is either closer, farther away, or higher than your present location and repeat all the above steps. Unless specified by Higher HQ, comm has priority over the mission after a comm window is missed. Continue to repeat this cycle at different locations until comm is reestablished or you are forced to execute the no-comm plan.

SUMMARY (2 Min)

a. During this period of instruction we have covered HF and VHF propagation fundamentals, optimizing communications, antenna fundamentals, directional antennas, antenna selection, field expedient antenna kits, establishing communication, some remarks and special considerations for wave propagation, antenna theory, and field expedient antennas

b. Those of you with IRF's please fill them out at this time and turn them into me at this time We will now take a 10 minute break.

UNITED STATES MARINE CORPS
Mountain Warfare Training Center
Bridgeport, California 93517-5001

WML
WMO
02/21/02

LESSON PLAN

MOUNTAIN LOGISTICAL CONSIDERATIONS

INTRODUCTION (5 Min)

1. **GAIN ATTENTION**. Operating supply trains and other logistical operations in the mountains is considerably different than in temperate climates.

2. **OVERVIEW**. The purpose of this period of instruction is to introduce the student to those factors peculiar to mountainous terrain that has an effect on logistical planning. This lesson relates to Cold Weather Operational Planning.

3. **METHOD/MEDIA**. The material in this lesson will be presented by lecture. You will practice what you learned during upcoming field training exercises. Those of you with IRF's please fill them out at the conclusion of this period of instruction.

4. **EVALUATION**. There is no formal evaluation over this period of instruction, however you will be evaluated by the instructors throughout the training package.

TRANSITION: Are there any questions over the purpose, learning objectives, how the class will be taught, or how you will be evaluated? Let's first look at the factors effecting logistics.

BODY (60 Min)

1. (5 Min) **FACTORS EFFECTING LOGISTICS WITHIN MOUNTAINOUS REGIONS**

    a. Long and difficult terrain over which support has to be provided.

    b. Lack of ground road systems, even in populated areas.

    c. Overall lack of civil and industrial facilities that can be adapted to military purposes.

d. Environmental Factors, including:

(1) High altitudes.

(2) Weather.

(3) Vegetation cover.

(4) Terrain barriers (mountains, rivers, rocky slopes, glaciers, snowfields, quagmires from snowmelt, bog, etc.).

TRANSITION: Now that we have discussed the factors effecting logistic support, are there any questions? Lets now go into the planning portion of the class.

2. (5 Min) **PLANNING**

a. Supply requirements include the procurement and distribution of many additional specialized sub-classes of supply, which are needed for cold weather operations. This includes items found in the Contingency Training Equipment Pool (C/TEP) and/or in war stock. These requirements must be determined for each exercise/operation according to terrain, weather, time of year, and nature of the operation. All classes of supply are effected by the cold, some to much greater extent than others. Through anticipation, redundancy, imagination and innovation solutions, the logistics officer can meet the MAGTF's needs.

b. The capability of combat service support elements to provide adequate support may be the determining factor in deciding if an operation is feasible.

c. In addition, commanders must prepare alternate plans in case environmental conditions make it impossible for the combat service support element to provide the required support by the desired means i.e., the need to use roads instead of air.

d. During pre-deployment planning, a detailed evaluation of the area of operations must be made in order to identify suitable lines of communications.

(1) Roads.

(2) Airfields.

(3) Ports.

(4) Railways.

e. Combat service support planning begins with the initial stage of operational planning and is a continuing staff process. Early identification of mountain-unique requirements is necessary to ensure timely ordering and delivery.

f. In addition to common items of issue, there are going to be special equipment requirements and these should be determined for each operation according to terrain, weather, and nature of the operation and its planned duration. Such equipment may include:

(1) Ropes.

(2) MAC Kits.

(3) Skis and/or snowshoes.

(4) MCCWI Kits.

(5) Specialized clothing.

g. Host-nation procurement can often be used for such things as snow removal or clearance equipment along MSR's or at airfields. Ports or railway systems may also be available. Every item procured through HNS is one item less that will require shipping, handling, and storing.

TRANSITION: Now that we have covered the planning, are there any questions? Lets also discuss some considerations with mobility.

3. (10 Min) **MOBILITY** (Primary/Alternate Route)

a. Main Supply Route (MSR)

(1) Generally used for large units.

(2) Problems

(a) Uneconomical for small units (may not have enough engineer assets due to mud slides, snow, high water washing out bridges, etc.).
(b) Easily ambushed or cut by a small enemy unit.

b. Waterways

(1) Advantages

(a) Good avenue of transport, if you have access to boats and the water is ice-free (spring breakup).

(2) Disadvantages

(a) Must have boats.

(b) Requires knowledge of local conditions (through thorough reconnaissance), such as ice, waterfalls, flooding, etc.

(c) Waterways may be unpredictable.

(d) It may require additional loading/unloading from the waterway to the unit.

c. Air

(1) Advantages

(a) Much faster than other transportation methods.

(b) Does not need the forward engineer assets required for MSR's and waterways.

(c) Does not burden road networks.

(2) Disadvantages

(a) Require waterproof warming/maintenance areas (if you expect snow or freezing rain).

1. Can use a Herman Nelson as an external source of heat.

2. Should use covers on helos (if you plan to use them in a wet-cold environment) De-icing fluid should be used on the blades prior to covering the helos to prevent covers from freezing to the blades.

(b) Must have a higher visibility level than the other methods.

(c) Air support will have to carry a reduced load in mountainous areas due to the altitude.

(d) Reduced flying distances due to the altitude (75% fuel efficiency factor).

(e) May have smaller drop zones due to ravines, steep slopes, etc.

(f) May be hard to recover air drops due to the terrain.

(g) Helo Support Teams (HST) may need longer straps (40 feet extension) due to smaller drop zones and recovery points.

(h) Unreliable because of weather.

d. Other Considerations

(1) Mobility is greatly enhanced by pre-positioning (air, waterways or roads).

(2) Effort should be made to deceive the enemy as to the routes and the supply off-loading point.

(a) Off-load points should be staggered along roads and different routes should be used for the return.

(b) Fixed routes will increase the vulnerability to an ambush.

(3) Fill runs both ways (i.e. empty gas/water cans and trash should be extracted at the same time).

(4) Skilled drivers should be used both day and night (rough terrain conditions can be disastrous for an unskilled driver).

(5) Drivers should carry chains at all times due to mud or snow (it can snow at any time over 5,000 feet).

(6) MP's should patrol MSR roads to ensure proper utilization, i.e., for convoy control, one way traffic on designated routes and authorized usage of the road).

TRANSITION: Now that we have covered the mobility, are there any questions? Let's talk about how is our supply of gear and equipment effected in the mountains.

4. (10 Min) **SUPPLY**

a. Class I (Subsistence). There will be an increased caloric demand in mountainous environments due to exertion, increase in body heat production requirements, etc. More information on required caloric intake is found in MOUNTAIN HEALTH AWARENESS. Close coordination is necessary with the receiving units so that water can be distributed before it freezes.

b. Class II (Clothing, Individual Equipment, and Tents)

(1) Units should plan on sufficient extra individual cold weather clothing items to replace lost or damaged clothing. Units should be able to react to both dry cold and wet cold conditions.

(2) The need for tent heaters and stoves in storage areas must be emphasized. Heated tents will be necessary for the storage of some classes of supply.

(3) Unique requirements must be anticipated at the unit level; ski wax, tire chains, candles, etc. The primary source for this equipment is the C/TEP or war stocks. Marines will

have to carry their own fuel for squad/personal stoves. This will require special V2-liter bottles. These bottles must be filled using either small funnels or special fuel cans with pour spouts. Sunglasses, special fuel containers, cold weather clothing items, and tentage may be available from C/TEP.

c. Class III (Petroleum, Oil, Lubricants)

(1) If water is provided by melting snow with fuel-fired stoves, more fuel for these stoves must be provided.

(2) Travel over snow or tundra can increase fuel consumption by over 25%. Diesel fuel will need additives to prevent freezing and gelling of fuel. This is a bulk fuel responsibility. A substitute is JP-5. Correctly mixed fuel will normally be available in the cold weather operating area but not at the embarkation point. Standard diesel fuels will not work in sub-zero temperatures. Marine Corps vehicles have multi-fuel capability. If fuel with additives is not available at point of embarkation, use JP-5.

(3) Multi-viscosity oil (15W-40) is recommended for most rolling stock in the cold. Use of 15W-40 will preclude a need to frequent oil changes in an environment with wide, temperature changes. Vehicles should be changed to multi-viscosity oil before embarkation. Only in sustained extreme cold conditions will 10W oil be required. Units should embark a block of 10W in sustained cold.

(4) Special fuels may be needed if using host nation equipment. For example, the BV-202 requires gasoline of 95% octane. The standard DOD 80 octane gasoline will foul its carburetor jets. Generally this fuel is available through host nation support.

(5) Higher levels of Class II (containers) to carry or store the fuel should into be overlooked. Up to double the normal number of fuel cans may be needed if carrying fuel to the vehicle instead of bringing the vehicle to the refueling points. Even at extremely low temperatures, fuel can be delivered to 500 gallon fuel bladders (drum, fabric, collapsible liquid fuel, 500 gallon).

(6) References for fuel problems include:

(a) Military Handbook, Mobility Fuel Users Handbook 114

(b) FM 10-69, Petroleum and Fuel Operations

(c) TI 10340-15/1A, Fuel Compatibility for Engines, Motor Transportation and Ordnance Tracked Vehicles

(d) TM 3835-15/1, Fuel Testing Standards

d. Class IV (Construction Materials)

(1) The lack of ground systems such as roads, airfields, ports, and railways places greater demands on providing and moving building supplies.

(2) To develop and maintain MSRs over both land and water, there will be an initial demand to construct both storage and living spaces.

(3) Supplies needed to maintain MSRs, roads, and bridges must be determined for each operation according to terrain, weather, and the nature of operation.

(a) Great quantities of gravel will be needed for roads and airfields to cover vegetation of permafrost areas. Bulk gravel or sand must be covered with plastic sheeting or canvas so they can be worked below freezing temperatures. Salt also serves to prevent freezing, however it is highly corrosive.

(4) The cold weather environment requires more engineering supplies than normal because of the uniqueness of the area.

(5) Staging areas in prospective destinations are nonexistent/inadequate.

(6) Offloading must be selective and highly organized.

(7) Supplies must be combat loaded.

(8) Dunnage needs to be embarked for it is not available in remote arctic areas. Indigenous shipping and receiving sites may be a source of dunnage.

(9) Construction supplies placed directly on the ground may freeze in place. Expendable material such as canvas, cardboard, or dunnage should be placed between items or pallets to be stored on the ground.

(10) The use of trees, when available, is a good way of supplementing Class IV.

e. Class V (Ammunition)

(1) The CSSE can anticipate handling, storing, and moving greater volumes of ammunition and demolitions. The need for more ammunition is generally due to:

(a) Difficulty adjusting fire (visibility constraints and variable burning rates, etc).

(b) Faulty ammunition from cold temperatures.

(2) Special storage for ammunition will not be required but it should be stored in its original containers.

(3) Preparing ammunition dumps will be more difficult because of freezing and mud conditions.

f. Class VI (Personal Demand Items)

(1) Morale in the cold can be significantly enhanced by paying attention to the health, comfort, person demand, and PX-type items found in the Class IV block.

(2) Demand for this class of supply remains constant in cold weather operations as in any other operation.

(3) Care must be taken to avoid stocking items that will freeze.

g. Class VII (Major End Items)

(1) During cold weather operations, an emphasis must be placed on preventive and corrective maintenance, vice replacement. First and second echelon maintenance are unit responsibilities and will become more important. It may not be feasible to evacuate equipment to the rear because of the limited and/or narrow road systems. It may be advisable to assign a CSS liaison to advance with the forward elements and to send a forward mobile CSSD in trace of the GCE. Marines in these billets need to be highly skilled in supply and cold weather operations. They must constantly monitor events to anticipate a unit's re-supply requirements.

(2) Replacement of major end items is preferred when possible. However, movement of deadline equipment over snow and ice will further stretch the maintenance capabilities of a unit. This may clog the limited MSR.

(3) An increase demand for power generators, Herman-nelson heaters, and rough terrain loaders will a snow removal capability should be anticipated.

h. Class VIII (Medical Supplies)

(1) Must be given high priority of movement and should be air delivered when possible.

(2) High consumption rates of medical supplies must be anticipated to include an increased need for such items as chapstick, cough syrup, and decongestants.

(3) Heated storage areas (warming tents, vehicles, and containers) are required for storing liquid medications and whole blood. Solid medications and freeze-dried material instead of liquids can be used when building the AMAL to minimize freezing, storing, and handling problems.

(4) Perishable materials must be packaged and marked for special handling.

(5) Procedures must be established and followed for special handling of the AMAL from embarkation through its final destination. (Supplemental AMAL pending).

(6) Oxygen should not be used or stored in a tent with an open flame heating device, i.e., forced-air heaters that emit 250 K to 400 K BTUS.

i. Class IX (Repair Parts)

(1) The cold weather, unimproved roads and rugged terrain will significantly effect the increased need for spare parts. Equipment will be under unusual stress. Tires wear from chains and battery cranking power is reduced and eventually fail. Metals and synthetics become brittle. Tools break and get lost in the snow, especially when operating in long periods of sustained, extreme cold (below –25°F). The need of spare fuel can gaskets must also be anticipated. Units should have repair parts prepositioned as far forward as possible.

(2) Complete equipment LTIs must be conducted before embarkation.

(3) Greater attention must be given to requesting Class IX before deploying. Especially important are high demand items like starters and alternators.

(4) Batteries should be new and replacement factors closely examined. Wet cell batteries must be closely monitored.

(5) Only combat essential repair parts should be carried by the unit.

(6) Cannibalization may be required but should not be encouraged.

j. Class X (Material to Support Nonmilitary Programs)

(1) The cold weather effects all personnel, including enemy prisoners of war, indigenous inhabitants, and refugees. The same protective measures and commodities necessary to protect friendly forces are needed.

(2) Considerable effort and expenditure will be necessary to provide for native inhabitants and refugees taken into custody.

(3) The numbers of nonmilitary personnel should be kept to a minimum whenever possible.

(4) All classes of supplies to care for noncombatants should be anticipated to meet this requirement in cold weather operations.

k. Storage

(1) Location

(a) Must be close to all-weather roads.

(b) Transportation to supported units must be considered since time and distance are key factors.

(c) The location should be as far forward as possible.

1. Con-x boxes are ideal for storage of items needing protection from the weather (but not the heat).

2. Site selection must consider the effects of periodic or seasonal thaws on the accessibility and condition of stored supplies. Storage boxes that provide dry storage capabilities for items needing protection from the weather must be procured.

3. Items such as medical supplies, water, batteries, special fuses, Class VII and class VIII, etc. that are subject to damage by freezing, should be stored in heated shelters.

4. When possible, Class I, III, V and IX items should be mobile-loaded (pallets, con-x boxes).

5. Critical items should be dispersed to prevent total loss, in the chance of something damaging them (such as fires or shelling).

6. When selecting a site, consider the effects of spring runoff, flash flooding, etc., on the accessibility and condition of stored supplies.

7. When setting up storage areas, have circular traffic flow with the high use areas on the end.

(2) Staging and marshalling areas are extremely limited in maritime cold weather operating areas. Mountains begin below sea level and leave little flat/rolling terrain at the waters edge. Beaches are extremely limited and may be clogged with ice or windblown snow. Development of staging areas will be difficult because of frozen ground, frost, permafrost, poor drainage patterns, and previous development of almost all of existing terrain for civil use. Previously developed civilian facilities may be considered for development as staging and marshalling areas.

(3) Supplies stored in open areas that may be subjected to drifting snow should be marked with poles and small flags. Particular attention must be given to marking medical supplies.

TRANSITION: Are there any questions over the supply of gear and weapons? Lets now talk about general maintenance.

8. (10 Min) **MAINTENANCE**

   a. General. If corrective maintenance is above the unit's echelon or capabilities, you should attempt to use contact teams vice evacuation (because evacuation increases the load on the transportation requirements).

   b. Specific Mountainous Area Preventive Maintenance

      (1) Keep tire stems capped.

      (2) Check air filters and cleaners for dirt and dust daily (special care should be noted to the oil bath filters).

      (3) Keep dry batteries warm and wet batteries fully charged.

         (a) Dry batteries life-span gets shorter as the temperature gets colder (below 32°F). You can wrap them in an insulating material to prolong the use life.

         (b) Wet batteries freeze easier as the charge gets weaker, i.e. fully charged batteries freeze at -95°F; a battery that is 1/3 charged will freeze at 18°F.

TRANSITION: Last of all we will discuss some problems with medical in the mountains.

9. (10 Min) **MEDICAL**

   a. Major problems will involve evacuation and care of a casualty, as well as the storage of medical supplies.

      (1) Evacuation will tie-up large amounts of personnel and equipment.

         (a) Helos should be the primary source of evacuation.

            (1) Speed

            (2) Eliminates further traffic on an already burdened MSR.

            (3) Unreliable due to weather.

         (b) Must have trained litter bearers (and relief teams). There may be a need for warming tents, augmentation points which can also be used for re-supply.

         (c) There must be a detailed evacuation plan.

b. Other Considerations

(1) Inflatable splints function poorly in cold weather and at higher altitudes.

(2) Low temperature thermometers (subnormal), and EKG monitors will help in diagnosing hypothermia cases.

(3) BAS's should have forced-heat, electricity and adequate space for supplies and patients.

(4) There will probably be a significantly higher number of fractures, sprains and shock due to the terrain, mobility requirements (on skis, snowshoes), weather, temperature and exertion.

TRANSITION: All real military minds think about logistics long and hard before they engage in an operation, because if you cannot support an operation you may as well not do it. Are there any questions? If there are none for me, then I have some for you.

PRACTICE (CONC)

a. Students will plan mountain logistics.

PROVIDE HELP (CONC)

a. The instructors will assist the students when necessary.

OPPORTUNITY FOR QUESTIONS (3 Min)

1. QUESTIONS FROM THE CLASS

2. QUESTIONS TO THE CLASS

Q. What are two of the three advantages of air travel?

A. (1) Much faster than other transportation methods.
(2) Does not burden road networks.

SUMMARY (2 Min)

a. During this period of instruction we have covered some of the considerations for mountain logistics. As you can see, planning and carrying out a mission in the mountains may not be an easy task. But if you follow these guide lines, you my encounter fewer problems.

b. Those of you with IRF's please fill them out at this time and turn them in to the instructor. Take a short break until the next class.

UNITED STATES MARINE CORPS
Mountain Warfare Training Center
Bridgeport, California 93517-5001

SML
SMO
02/21/02

LESSON PLAN

COLD WEATHER AND MOUNTAIN HELICOPTER OPERATIONS

INTRODUCTION (5 Min)

1. **GAIN ATTENTION**. Survive, move, and fight; the three basic requirements to success on any battlefield. In cold weather or in the mountains the ability to move rapidly and with little expenditure of energy may greatly reduce the demands on the individual rifleman and significantly enhance his ability to fight.

2. **OVERVIEW**. The purpose of this period of instruction is to familiarize the student with those aspects of cold weather and mountainous terrain that effect usage of helicopters. This lesson relates to operational planning.

3. **METHOD/MEDIA**. This class will be taught by the lecture method.

4. **EVALUATION**. There will be no formal evaluation for this period of instruction.

TRANSITION: Are there any questions over the purpose how the class will be taught? Let's take a look at a general view of helicopter operations.

BODY (65 Min)

1. (5 Min) **GENERAL**. The helicopter is the single best tactical mobility asset available to Marines during cold weather operations. It can move you farther and faster than any other means of transportation. The helicopter has limitations. The greatest of which are the lack of dependability due to unpredictable weather and the extreme difficulty of performing maintenance in the cold. Additional maintenance personnel and maintenance shelters may be required. This means that unit leaders must always have an alternate movement plan to get to the destination in time to accomplish the mission.

2. (5 Min) **CONSIDERATIONS EFFECTING HELIBORNE OPERATIONS**. The following items should be considered when deploying helicopters in a cold weather and mountainous environment.

a. Reduction in Operational Tempo. Remember that everything takes longer in a cold weather environment. It takes the mechanics longer to fix, fuel, and do routine maintenance on the aircraft. The aircraft may also have more maintenance problems due to the cold weather.

b. Vulnerability in the LZ. Delays in the LZ will make helicopters particularly vulnerable targets to both direct and indirect fire. Helicopters often create large snow signatures when conducting landings and takeoffs in snow covered terrain.

c. Temperature and Altitude. As temperature and altitude increase, helicopter performance decreases. This effects not only payload capability but also time on station, airspeed, and maneuverability. Decreased temperature will not offset the effect of increased altitude when operating in high mountainous terrain.

d. Weight/Bulk Load. Fewer men can be carried because he has more equipment and takes more space in the aircraft. It will take 1 1/2 normal seating space for a Marine with a full cold weather combat load.

e. Weather. Mountainous or arctic terrain is compartmentalized and is characterized by rapid change. Weather may be good in the pickup LZ and bad in the destination LZ. Consequently, commanders must have alternate plans for insertion and extraction if possible.

f. Rotor Wash Identification and Visibility. On landing and takeoffs, helicopters re-circulate large snow clouds that can be observed from considerable distances.

TRANSITION: Now that we've covered the environmental considerations effecting helicopter operations, are there any questions? Let's review assembly areas.

3. (5 Min) **ASSEMBLY AREAS**

a. Assembly areas should provide security, concealment, dispersion, and a windbreak for troops.

b. Anytime Marines wait longer then 40 minutes; they should erect warming shelters.

c. This period may be substantially shorter in extreme cold temperatures or under severe wind-chill conditions.

4. (5 Min) **SAFETY CONSIDERATIONS**. Marines must understand safety considerations to reduce the risk of injury during cold weather and mountainous helicopter operations

a. Frostbite. Frostbite is a constant danger due to the combination of wind-chill and cold temperatures. Use the buddy method to check for possible signs of frostbite.

b. Rotor Blade Hazards. In deep snow-covered LZs, the helicopter may sink into the snow. This reduces the rotor-blades-to-surface clearance. Using the ahkio huddle for loading and unloading reduces the danger of being struck by the rotor blades. In sloping LZs, do not approach the helicopter from the upslope side as rotor-blades to surface clearance is further reduced. Extreme caution must be exercised when operating in close proximity to the tail rotor.

c. Cargo Ramp Hazards. In deep snow the crew chief may not be able to lower the cargo ramp completely. Marines must be aware that this significantly reduces head clearance. Caution needs to be observed when operating near the cargo ramp; it is hydraulically operated and can easily crush personnel. Hydraulic fluid or ice can also cause the ramp to be slick causing a hazard for debarking Marines.

d. Ice Shedding. Under various conditions ice may accumulate on the rotor blades of a helicopter. When it sheds it produces a myriad of flying projectiles. The safest place to be is on the ground with your face covered.

e. Unprepared LZs. When landing in an unprepared LZ, the fuselage will float on the snow's surface. Landing points should be probed and tramped down to determine possible obstacles.

f. Dynamic Rollover Damage. A helicopter will normally settle through the snow surface. If the ground is uneven or there are obstacles beneath the surface of the snow, it may cause the helicopter to contact the ground at an angle. In this condition the helicopter may be in danger of dynamic rollover. Dynamic rollover is a condition where the helicopter could rollover onto itself due the landing gear/skids coming in contact with the surface while power is being applied to the aircraft. Time permitting, the landing spot should be probed for any obstacles.

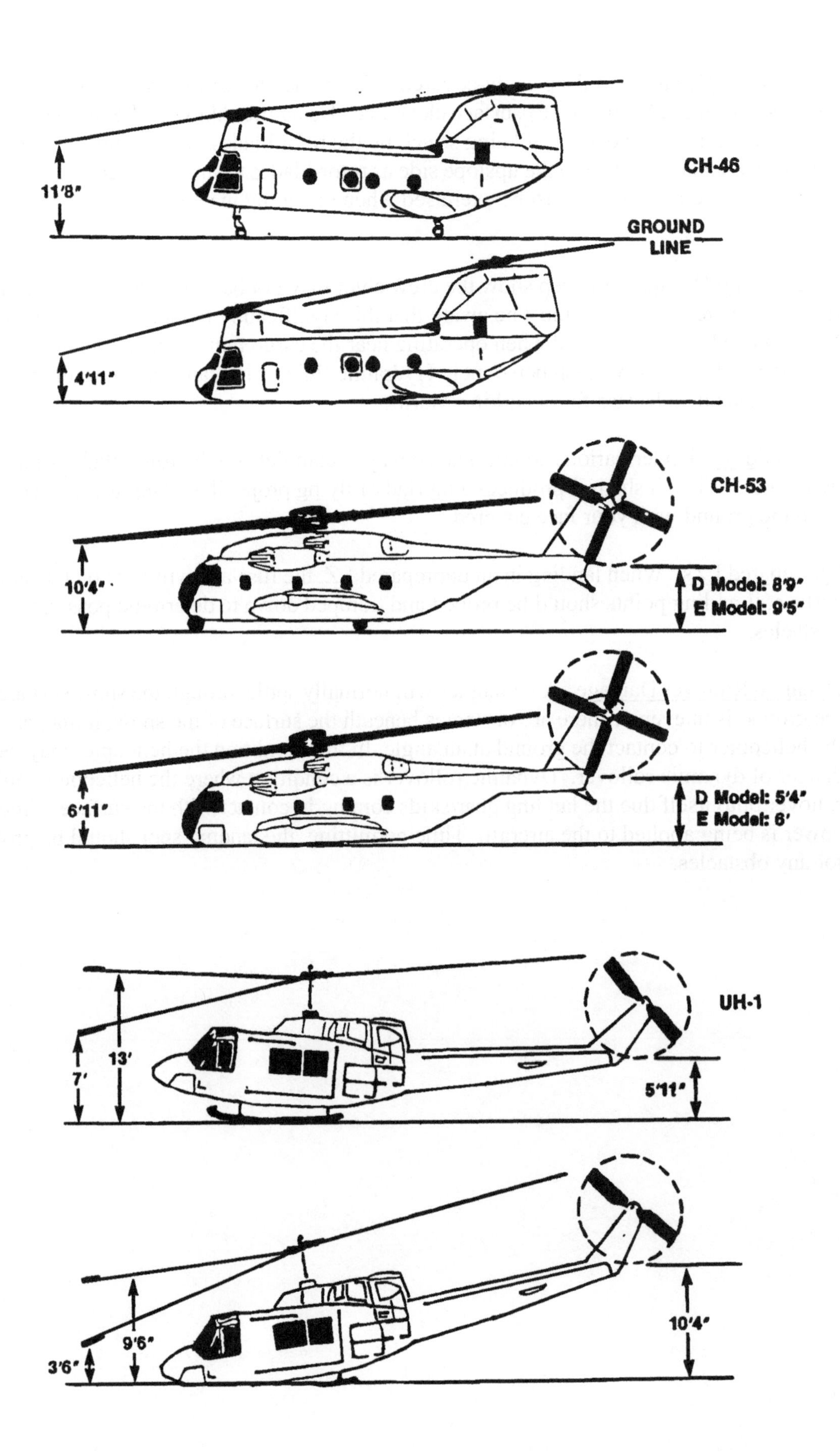

CH-46
11'8"
GROUND
LINE
4'11"
CH-53
10'4"
D Model: 8'9"
E Model: 9'5"
6'11"
D Model: 5'4"
E Model: 6'
UH-1
13'
7'
5'11"
9'6"
3'6"
10'4"

TRANSITION: Now that we talked about safety considerations, let us discuss LZ selection and preparations.

5. (5 Min) **HELICOPTER LANDING ZONE SELECTION AND PREPARATION**.

a. LZ Selection. LZ size is determined by the number and type of helicopters to be employed.

(1) Size. A LZ that is 50m by 50m is generally sufficient to land any helicopter. Consideration should be given though, to the altitude that operations are being conducted. As operating altitudes increase toward 10,000 ft. MSL, the size of the LZ should also increase. This is based on the performance loss that a helicopter will experience while operating at higher altitudes. At higher altitudes (close to 10,000 ft. MSL) a LZ that measures approximately 100m by 100m will provide the necessary clearance for a safe approach and departure.

(2) Approaches and Exits. If the LZ in question has significant obstacles, surrounding the zone or on the approach /departure path, size considerations should be increased.

NOTE: In all situations, a face to face brief with the pilots in the operating area will give you an idea of the capability/limitations of the aircraft and what size zones they will be looking for. The Air Officer/FAC should be the leading authority on this matter.

(3) Wind Direction. The wind direction determines the approach and departure directions. Helicopters normally take off and land into the wind.

(4) Ground Surface. Debris/snow/ice will be kicked up when the helicopter comes in to land in the zone. The majority of the danger from this flying debris is to the Marines on the ground. If rocks/ice or other debris exists, ensure Marines use caution when the aircraft is coming into land. Time permitting, packing down the area where the Ahkio huddle will be located will minimize the effects that blowing snow and ice will have on the Marines.

(5) Ground Slope. Terrain which slopes more than 8° is usually considered too steep for helicopter landing due to dynamic rollover characteristics of all helicopters. Individual helicopters have different limitations for sloped landings. A briefing with aircrew in the operating area will serve to clarify what the capabilities/limitations of the aircraft.

(6) Concealment. LZs should be selected which will conceal both the helicopter and the snow signature from direct enemy observation. The signature that develops from the rotor wash can be observed up to 30 kilometers away.

(7) Obstacles. The unit to be loaded should look for obstacles that may be hidden under snow. Obstacles that are hidden are potentially dangerous to the helicopter. Probing the LZ should be conducted to locate tree stumps, large rocks, etc., which could place the helicopter in a dynamic rollover situation or rupture the skin on the belly of the aircraft.

(8) Snow. Depth and consistency of snow will have a major impact on LZ operations. Loose snow may cause the pilots to white-out, lose reference to the horizon, and have to conduct a wave off. Hard or crusted snow may break up and become a hazard to Marines on the ground.

(9) Lakes and Rivers as LZs. Commanders should consider using frozen lakes and rivers as alternate LZs. Frozen lakes and rivers do make excellent LZs since they are level and have little loose snow due to the scouring winds. CH-53 and CH-46s need 15 inches of ice to conduct operations. UH-1N and AH-1W need 8 inches of ice thickness.

b. LZ Preparation. Marines should make every effort to walk through the LZ to determine snow depth and appropriate locations for helicopter landing points.

(1) Packing the LZ. Packing the LZ makes it easier for a pilot to find the landing point and for Marines to move about. This consideration is particularly important when conducting external operations. Packing does take more time and the possibility of detection by the enemy may be increased.

(a) Time, conditions, and tactical situation permitting, the first area to be packed should be the area around the Ahkio huddle. The next area will be for the landing point, which should be approximately 50 meters square. This will decrease the amount of snow that will be kicked up by the rotor wash.

(b) If an LZ is in a safe area, and will be used frequently, a request for Engineer Support to pack the snow should be made. Over the snow vehicles are most effective for packing landing points quickly. Marines on snowshoes, ski, or the boot packing method can also be used but is more time intensive and exhausting.

(2) Marking the LZ. Marking the LZ and the landing points is critical. The white snow-covered zones can provide a difficult background for the pilots. Blowing snow can cause a white-out condition and may cause the pilots to lose reference to the horizon. A reference point must be visible at all times. Any object that will contrast with the snow and does not move will provide a reference point.

(a) Air Panels. Air panels contrast in color with the snow. They must be secured to ensure that they are not blown away by rotor wash.

(b) Smoke Grenades. These should be used to mark the LZ only and not the landing points. When used in snow covered LZs use a platform to prevent the smoke grenade from sinking in the snow.

(c) Chemical lights. Chemical lights provide good close-in lighting at night but are hard to see beyond ½ mile.

(d) Tree Boughs. Lay or stick these into the snow to provide a contrasting reference for pilots to orient on.

(e) Sled teams. Ahkio huddles are the primary method of marking landing points. The huddle should contrast in color to the background in the LZ. Individuals should remove overwhites, wear a protective face mask, and be sure that no bare skin is exposed to the rotor wash.

(3) The following is the minimum required information for a landing zone brief.

(a) Your call sign

(b) LZ Location

(c) LZ Marking

(d) Wind Direction and Velocity

(e) LZ Size

(f) LZ Elevation

(g) Obstacles/Snow Conditions

(h) Visibility

(i) Approach/Retirement Direction (Recommended)

TRANSITION: Now that we have covered LZ selection and preparation, are there any questions? Next, we'll discuss preparations for embarkation in cold weather snow covered LZs.

6. (5 Min) **PREPARATIONS FOR EMBARKATION**.

a. Planning. Helicopters will often have reduced payloads when operating at higher altitudes. In addition, high temperatures, high humidity, and high Density Altitude will degrade helicopter performance. Consequently, helicopter payloads may change significantly due to both the current and forecasted weather and LZ altitudes.

b. Payload. The following chart is estimates only and should be used as a guideline. Actual lift capacity will vary depending on fuel consumption, ordnance on board, time of flight, weather, etc.

HELICOPTER	SEA LEVEL	5000 FT MSL	10,000 FT MSL
UH-1N	6 Pax and gear	4 Pax and gear	2 Pax and gear

CH-46E	16 Pax and gear	8 Pax and gear	4-6 Pax and gear
CH-53E	37 Pax and gear	24 Pax and gear	18 Pax and gear

c. Personnel. A major hazard to personnel operating around helicopters in the cold weather is the wind chill generated by the rotor wash. Exposed skin should be kept to a minimum. If a long wait is expected, warming tents should be erected.

d. Equipment

(1) The team sled should be staged as near the landing point as possible. To prevent the team sled from being moved by rotor wash, the Marines embarking on the aircraft should lay on top of the sled. This procedure is known as the Ahkio Huddle and will be discussed later.

(2) Weapons should be in Condition 4 when embarking the aircraft. Muzzles should be pointed down on CH-46 and CH-53s. Muzzles should be pointed up or outward on UH-1Ns.

(3) No equipment (skis, poles, radio antennas) should be allowed to protrude above the height of a man. This is to prevent any equipment from going into the rotor blades.

(4) Packs should not be worn aboard helicopters due to the restricted movement and the requirement to fasten seat belts before departure. Packs and team sleds should be staged at the center of the aisle on assault aircraft.

7. (10 Min) **AHKIO HUDDLE PROCEDURES**. The Ahkio huddle is designed to get Marines on and off a helicopter as quickly as possible with minimum exposure to wind chill. This must be rehearsed so that it can be accomplished in extreme weather and reduced visibility.

a. Ahkio Huddle.

(1) The ahkio huddle is established around the sled/tent equipment on the landing point. Packs off and skies are bound together. Marines group together on top of the equipment, face down, to keep the equipment from blowing away.

(2) All of the tent teams equipment necessary for survival against the environment is loaded on the same aircraft as the personnel.

(3) The helicopter will land so as to place the sled team huddle under its rotor arc at the 2 o'clock position.

b. Rehearsals. Before conducting sled team huddle operations, all Marines, including pilots, aircrews and troops to be lifted must be properly trained in ahkio huddle procedures. This will:

(1) Eliminate the dangers of troops walking into the helicopter blades.

(2) Reduces the problems of wind chill.

(3) Reduces the amount of time the helicopter must remain in zone by: providing the pilots a solid reference point, reducing the distance Marines must move through the snow to the aircraft, and reduce the loading/unloading time.

8. (10 Min) **EMBARKING AND DEBARKING THE HELICOPTER**. The ahkio team leader supervises the loading of the sled and any other equipment.

a. The Ahkio Team Leader

(1) Loads first, moves to the front of the helicopter, and secures his gear by the left most forward seat.

(2) Should immediately communicate with the pilot.

(3) When in flight, observe from between the pilots to maintain orientation.

(4) Designates Marines to load / unload equipment.

(5) Straps in for take off and landing.

(6) Maintains accountability of the ahkio team.

b. The Ahkio Team

(1) Loads the helicopter only when directed by the crew chief who will direct the team to load through either the side or rear door.

(2) Enters the aircraft quickly and moves to pre-assigned seats.

(3) Hand carries their snowshoes on board. Once seated, they hold their snowshoes and weapons between their legs.

(4) Bind skis and poles together in pairs. When loading or unloading, keep them parallel to the deck at waist level. Once loaded, place skis on the deck of the aircraft beneath the feet.

c. The Assistant Ahkio Team Leader

(1) The assistant ahkio team leader supervises the loading and unloading.

(2) Ensure that all gear and Marines have boarded.

(3) He boards last and signals thumbs up to the crew chief.

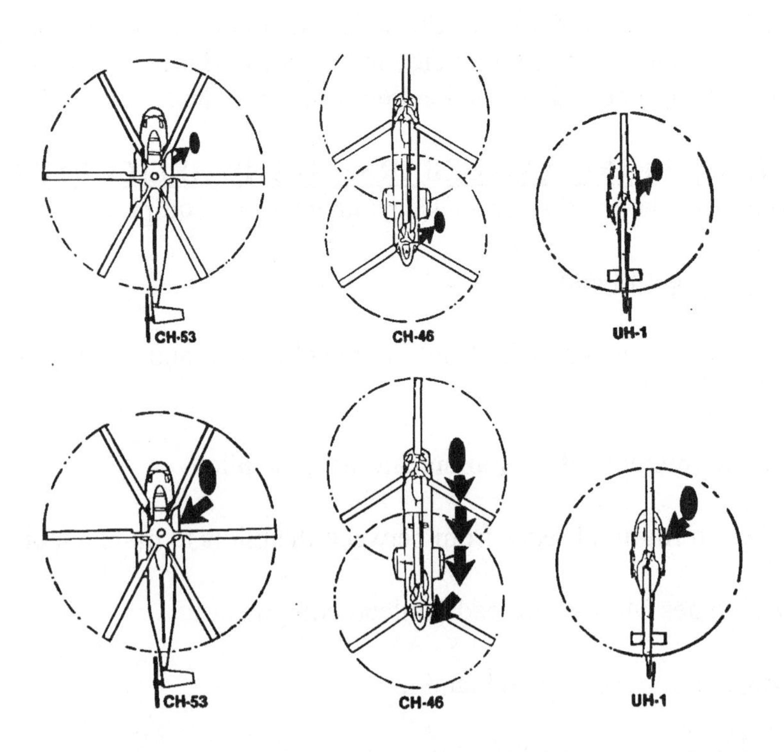

EMBARKING AND DEBARKING BY THE AHKIO HUDDLE METHOD

a. Debarking. As with embarking, the object during debarking is efficiency and safety. The off load generally follows a reverse order of the on load.

(1) Sequence

(a) The assistant ahkio team leader supervises the off load.

(b) Unload the sled and any other equipment first.

(c) Then all remaining Marines exit in reverse order of embarking.

(d) Assume the ahkio huddle just off the ramp or outside the door under the rotor arc.

(2) Tactical Deployment. Only after the helicopter lifts do ahkio teams tactically deploy.

TRANSITION: Now that we've discussed embarking and debarking procedures, are there any questions? Let's talk casevacs.

9. (5 Min) **CASEVAC CONSIDERATIONS IN COLD WEATHER**. Combat casualties are complex enough for a small unit leader to handle; the cold weather will make casevac extremely critical. Some points to remember:

a. If possible, pre-plan your casevac LZs and have a dedicated casevac helicopter on alert.

b. Ensure your Marines are cautious when loading a patient aboard a helicopter in deep snow. Remember the reduced rotor clearance.

c. Protect the patient from the rotor wash. Any exposed skin will be subject to frostbite.

d. Establish a warming tent for your patients and your loading teams. Handling casevac is physically and mentally exhausting.

TRANSITION: Now that we have discussed safety and casevacs, are there any questions? If there are none for me, than I have some for you.

PRACTICE (CONC)

a. Students will plan a helicopter operation.

PROVIDE HELP (CONC)

a. The instructors will assist the student when necessary.

OPPORTUNITY FOR QUESTIONS (3 Min)

1. QUESTIONS FROM THE CLASS

2. QUESTIONS TO THE CLASS

Q. What happens to helicopter lift performance as altitude increases?

A. Helicopter lift is greatly reduced.

Q. Should snow be packed down in the LZ prior to helicopter landing?

A. Yes (time permitting)

SUMMARY (2 Min)

a. We have looked at some of the considerations for using helicopters in cold weather and mountainous environments. Your battalion Air Officer should be able to provide additional information as well as arrange for helicopter support to help train your Marines to the needed level.

b. Those of you with IRF's please fill them out and turn them in. We will now take a short break.

UNITED STATES MARINE CORPS
Mountain Warfare Training Center
Bridgeport, California 93517-5001

SMO
02/21/02

LESSON PLAN

INTRODUCTION TO ANIMAL PACKING

INTRODUCTION (5 Min)

1. **GAIN ATTENTION**. On 15 December 1956 at Fort Carson, Colorado, the U.S. Army deactivated the last two operational mule units - the 35th Quartermaster Pack Company and Battery A of the 4th Field Artillery Pack Battalion. The Army, after having had mules on the payroll for 125 years had finally succumbed to complete mechanization, but there have been pack animals used in almost every conflict by forces around the world to date. The benefit of this class is to introduce the capabilities and limitations of pack animals.

2. **OVERVIEW**. The purpose of this period of instruction is to familiarize the student with animal packing and its use in mountainous terrain.

3. **METHOD/MEDIA**. The material in this lesson will be presented by lecture and demonstration. You will practice what you have learned in upcoming field training exercises. Those of you with IRF's please fill them out at the conclusion of this period of instruction.

4. **EVALUATION**. This is a lesson purpose class.

TRANSITION: Are there any questions over the purpose how this class will be taught? Now let's talk about the mission of the animal packing program at MWTC.

BODY (75 Min)

1. (5 Min) **MISSION**. To aid ground combat units with an alternative method of transporting crew-served weapons, ammunition, supplies to and from areas inaccessible to mechanized or mobile transportation.

TRANSITION: Are there any questions over the mission? Next we will discuss the definitions of personnel and animals.

2. (10 Min) **DEFINITIONS OF PERSONNEL AND ANIMALS**

a. Pack Train Personnel

(1) Pack Master. Officer in charge of the pack train. Responsible for the training of handlers and pack animals.

(2) Assistant Pack Master/Corrigador. SNCO or NCO responsible for supervision and routine inspections of the handlers and pack animals.

(3) Handler. Marine responsible for the handling, grooming, watering and feeding of his mule and proper care of pack equipment.

(a) Successful handling of animals in the field demands at least a limited knowledge of the basic principles of animal management. All Marines assigned to pack animals must know the symptoms and treatment of common diseases, first aid treatment of injuries, to keep the animals in such training and health as will enable them to do their work to the best advantage.

b. Animals

(1) Bell mare. Mules have a tendency to follow a leader. Often a white or gray mare is used for this purpose.

(2) Pack mule. A hybrid animal. A cross between the male donkey (ass-burro) known as a JACK and the female horse, known as the MARE. Ideal animal for carrying heavy leads over long distances. They need not be beautiful but should be strong, alert, well conformed, with good strong bone, weight carrying ability and a good disposition and are usually 1,000 lbs. and up in weight. There is no height limit, but packers prefer short mules to a tall one as being easier to load. The mule has certain advantages over the horse which fit him for this work, namely"

(a) The mule withstands hot weather better, and is less susceptible to colic and founder than the horse.

(b) A mule takes better care of himself, and will seldom get injured. Mules are extremely intelligent equines and will never engage of his own free-will in anything that causes him pain unless he is convinced the alternative is worse.

(c) The foot of the mule is less subject to disorders.

(d) Age and infirmity count less against a mule than a horse.

TRANSITION: Are there any questions over the mission of definitions of personnel and animals? Let's now take a look at the history of the animal in use as pack transportation.

3. (10 Min) **HISTORY**

   a. From the earliest times, humans have used the ass and its offspring to do much of the heavy work, which exceeds the strength of man. Thus the pack animal appeared frequently in ancient and biblical history and many different animals were employed as beasts of burden.

   b. One of the first recorded military application of horses and mules was by Genghis Khan. In the Arizona Territory, General Crook perfected the transportation system and created an efficient mobile mule pack train. During the winter campaign of 1872-73 in the Tonto Basin of the Arizona Territory, Crooks mules each transported a net weight of 320 pounds and averaged thirty miles a day over a thirty day period.

   c. Mules have served along side the American fighting man from the Second Seminole War, 1835-1842 to the Korean War. In august 1953 at Guantanamo Bay, Cuba, the U.S. Marine Corps said "So long mules" and deactivated its last mule unit.

   d. The reality of a need for mules surfaced with the Secretary of the Army in 1984, and a committee was formed with serious considerations of equipping the famous 10th Mountain Infantry Division with pack and artillery mules. The decision rendered was "not to use them". The excuse was "too expensive".

   e. The American pack mule was once again used in 1987 by the Afghanistan rebels to carry "stinger" missiles to the tops of mountains from which a few unskilled men cleared the air of Russian aircraft.

TRANSITION: Now that we have discussed some of the historical uses of mules in armed conflict, lets discuss the conformation of a sound pack animal.

INSTRUCTOR NOTE: Bring out 1 horse and 1 mule for students to see.

4. (10 Min) **CONFORMATION**. Conformation is defined as the components of an animal that make up a whole. (Official specifications from the 1917 Manual of Pack Transportation.)

   a. Conformation of a good pack animal are as follows:

      (1) Head – Medium size, well formed, broad between the eyes, eyes clear, large and full, ears long and flexible, teeth and tongue free of blemishes, muzzle well rounded and firm.

      (2) Neck – Stocky, broad and full at crest, and inclines to arch.

      (3) Withers – Low and broad, indication strength in shoulders.

      (4) Chest – Low and broad, with division well defined, holding the fore legs apart, showing good lung power.

(5) Knees – Large, wide in front and free of blemishes.

(6) Back – Short, straight indicating strength in back over region of the kidneys.

(7) Barrel – Deep and large, indicating a good feeder- not hard to please in either food or water- an essential requisite in selection of pack mules.

(8) Hips – Broad and well rounded.

(9) Dock – Low and stiff, offering resistance, showing endurance.

(10) Hocks – Standing well apart and strongly made, showing well developed buttocks.

(11) Pasterns – Muscled, short and strongly shaped.

(12) Hoofs – Sound, broad, full, with frog well developed, elastic and healthy.

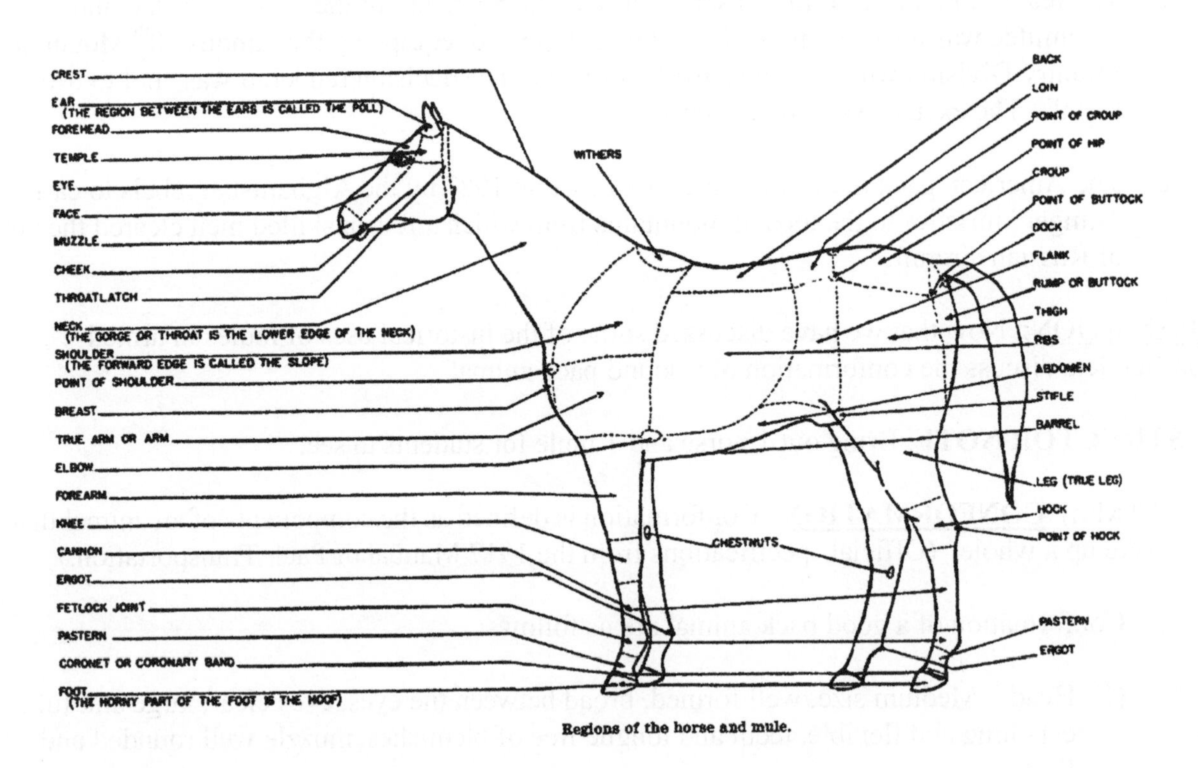

Regions of the horse and mule.

<u>TRANSITION</u>: Now that we have discussed the conformation of a pack animal, lets talk about procurement of animals and pack equipment.

INSTRUCTOR NOTE: Set out pack equipment for demonstration.

5. (10 Min) **<u>PROCUREMENT OF MULES AND TACK</u>**. In the event the Marine Corps is sent into a combat environment, it is highly unlikely that mules and tack will be sent into the theater of operations with you. Thus, you will have to procure animals and pack equipment

locally. The types of pack equipment, in common use by inhabitants of countries where animals form a basic part of transportation, vary in different areas. This native equipment, though crude, can be converted to military purposes and, when no other equipment is available must be used.

a. Advantages of using native equipment:

(1) Generally available in quantities in or near the zone of action.

(2) Relatively cheap, and light in weight.

b. Disadvantages of using native equipment:

(1) Highly skilled specialists are required to use it satisfactorily.

(2) Due to its crude construction it is injurious to the animal.

(3) Many military loads are extremely difficult to pack on this equipment.

c. Types of pack equipment used at MWTC:

(1) Sawbuck pack saddles.

(2) Decker pack saddles.

TRANSITION: Are there any questions over the procurement of mules and tack? Next we will discuss the difficulties in employing animals.

6. (10 Min) **DIFFICULTIES IN EMPLOYING ANIMALS**. The employment of animals by Marine Corps units should be attended by careful planning, intensive training, and close attention to detail. Some of the handicaps which must be faced by units employing animals are:

a. Lack of personnel experienced in handling animals and pack equipment.

b. Lack of necessary specialists, i.e. farriers, veterinarians and pack masters.

c. Absence of animal procurement or remount service.

d. Necessity of subsisting animals off the country due to practical difficulties of providing adequate grain and forage to units.

e. Lack of personnel experienced in tactical handling of mounted units.

TRANSITION: Are there any questions over the difficulties of employing animals? Now that you understand some of the problems with procurement of animals and equipment, lets discuss how to handle our animals.

7. (10 Min) **RULES FOR HANDLING ANIMALS**. All men connected with the care and handling of animals must be taught, and must thoroughly understand, the following rules for care of animals.

   (1) Animals require gentle treatment.

   (2) Never punish an animal except at the time he commits an offense.

   (3) Animals must be cooled down gradually by walking after being heated by use.

   (4) Never feed grain or fresh grass to animals when heated.

   (5) Never water an animal when heated unless the march or exercise is to be immediately resumed.

   (6) Animals must be thoroughly groomed after work.

   (7) Water before feeding.

   (8) Feed in small quantities and often and feed hay before grain.

TRANSITION: It is extremely important to understand the proper treatment of your pack animals, so they will be in the best shape and condition for military operations. Are there any questions at this time? If you have none for me, then I have some for you.

PRACTICE (CONC)

a. The students will utilize the pack animals in an upcoming field training exercise.

PROVIDE HELP (CONC)

a. The instructors will assist the students if necessary.

OPPORTUNITY FOR QUESTIONS (3 Min)

1. QUESTIONS FROM THE CLASS

2. QUESTIONS TO THE CLASS

Q. Name one advantage a mule has over a horse?

A. A mule takes better care of himself, and will seldom get injured. Mules are extremely intelligent equines and will never engage of his own free will in anything that causes him pain unless he is convinced the alternative is worse.

Q. What is one disadvantage of using native equipment?

A. Many military loads are extremely difficult to pack on this equipment.

SUMMARY (2 Min)

a. During this period of instruction, we have talked about the mission of the pack program, definitions of the personnel and animals, history, conformation, procurement of mules and tack, difficulties in employing animals, and rules for handling animals.

b. Those of you with IRF's please fill them out at this time and turn them in to the instructor. We will now take a short break.

UNITED STATES MARINE CORPS
Mountain Warfare Training Center
Bridgeport, California 93517-5001

SMO
02/26/02

LESSON PLAN

MOUNTAIN PATROLLING CONSIDERATIONS

INTRODUCTION (5 min)

1. **GAIN ATTENTION**. Due to the nature of mountainous terrain and unpredictable weather, you must be able to adapt your patrolling SOPs. Standard patrolling techniques apply, but additional considerations need to be used for successful operations in the mountains.

2. **PURPOSE**. The purpose of this period of instruction is to teach the considerations needed to patrol in a mountainous environment. To do this we will discuss the acronym METT-TSL as it pertains to a mountainous environment and we will talk about danger areas, patrol bases and ambushes as they pertain to the mountains. This period of instruction pertains to mountain operations.

3. **INTRODUCE LEARNING OBJECTIVES**

   a. TERMINAL LEARNING OBJECTIVE. In a mountainous environment, conduct patrolling in accordance with the references.

   b. ENABLING LEARNING OBJECTIVES.

      (1) Given a scenario, write a warning order in accordance with the references.

      (2) Given a scenario, write a patrol order in accordance with the references.

      (3) In mountainous terrain, conduct an ambush in accordance with the references.

      (4) In mountainous terrain, conduct a security patrol in accordance with the references.

      (5) In mountainous terrain, conduct a reconnaissance patrol in accordance with the references.

(6) In mountainous terrain, operate from a patrol base in accordance with the references.

4. **METHOD/MEDIA**. This lesson will be presented by the lecture method. Those of you with IRF's please fill them out at the conclusion of this period of instruction.

5. **EVALUATION**. This lesson is not formally evaluated, however, during upcoming field evolutions you will practice using METT-TSL, patrol bases, ambushes and crossing danger areas, using the considerations for a mountainous environment.

TRANSITION: Are there any questions on how this class is going to be taught or what will be covered? As stated before, we will discuss the considerations of a mountainous environment using the acronym METT-TSL. Let us start by discussing the considerations of the Mission.

BODY (50 Min)

1. (15min) PLANNING CONSIDERATIONS USING METT-TSL

a. **M - Mission:**

(1) Be realistic in time and distances for the patrol. Refer to Route Planning and Selection in Mountainous Terrain. Patrolling missions and conduct are the same.

b. **E - Enemy**

(1) What are the enemy's limitations and capabilities in a mountainous environment? Does he have mountain training/experience? What mountaineering equipment does he have? Are the enemy units from the local area or a distant city? What is his resupply capability (can he sustain units on a ridge, or within a limited distance from the MSR)? Is he using vehicles, animals, air or just man-pack for resupply? Does he have comm. Retrans sites? Are anti-air assets at high or low stations? Does he have light mortars (man-packing ammo limitations), where is his arty (out of range fan, masked, arty dead space)? How does he employ patrols, OP/LPs, pickets, security (distance from main body/base, altitude limitations, terrain limitations, comm. limitations, equipment limitations, etc.)?

(2) The mountains themselves can turn into the enemy. Only knowledge, proper gear, proper training, proper planning and good leadership can defeat this enemy.

c. **T - Terrain and Weather**

(1) Terrain - In a mountainous environment, terrain can either impede movement or be beneficial to the mission. Planning must reflect these difficulties of the terrain to best support the mission. Preparation must be made for obstacles such as steep slopes, vertical cliff faces, rivers and fast moving streams, and transportation can be limited to foot movement due only to the terrain. Select your route so that the high ground is to

your advantage in the event of contact. Do not engage the enemy unless you have a rapid egress route to defilade and a covered withdrawal. Stay to the high ground to avoid being ambushed from above (never transit right below a cliff or rock wall outcrop, grenades alone could destroy your unit).

(2) Weather - The weather in a mountainous environment is very unpredictable and changes rapidly. Planning for the different types of weather that might be encountered could help eliminate casualties, save time, and contribute to the success of the mission. Over-planning can impede mission accomplishment (i.e. – burdening the troops with too much clothing and equipment…just in case, or not enough clothing for bad weather so that there are hypothermia cases after a long halt). Using the weather correctly can have great advantages to the mission, such as; the effectiveness of enemy observation devices is greatly reduced, tracks can be eliminated, noise can be covered, and enemy alertness can be diminished by the effects of bad weather.

d. **T-Troops and Fire Support**

(1) Troops - Time of reinforcements will be longer in the mountains. Small units normally operate with more ease in the mountains therefore the commanding officer depends on having strong small unit leaders (decentralized control is needed). Are the troops acclimatized? What is the physical conditioning? What is the morale? Who are the assault climbers and summer (and winter) mountain leaders? How many of these specialists are there (enough for task-organized units, such as mountain pickets, long range patrols, etc.)? What is the mountain experience/training of your unit and also those adjacent?

(2) Fire Support - Fire support elements are sometimes limited to where they can move to support forward units due to being road bound. Due to the high mountain ridges, high angle firing is often used to overcome masking (this slows down the gun line due to breach loading). Infantry units, especially patrols, can quickly move out of the artillery range fan when the route is forward and away from the MSR. This is due to the limited road network and limited suitable firing positions for artillery along that 1 MSR. Aerial observation can be used to adjust fire. Forward observers will have difficulty adjusting fire in compartmented terrain (refer to Weapons and Fire Support Considerations outline). There is more artillery dead space for air to cover in the mountains. Air is also good for covering reverse slopes. However, there are usually not a lot of high value targets (HVTs) where an infantry unit is moving. HVTs are on the MSR and in the rear, low lying areas with infrastructure. Suppression fires in support of an attack will need to be longer due to the rugged terrain slowing down troop movement.

e. **T- Time**

(1) Time-Distance Formula (TDF). Each route should show a completed TDF on the route overlay. Commanders must be patient since the actual execution across that route may include hidden difficulties. This is especially true of routes across streams, roads, or cross- compartment.

Time-Distance Formula:

3 kph + 1 hour for every 300 meters ascent; and/or + 1 hour for every 800 meters descent

(2) Commanders Log is most accurate, but requires time to accumulate. Expect initial time estimates to be wrong until experience is gained in the AO.

(3) Do not set time precedent missions in the mountains.

(4) Speed is relative only to the enemy, not the clock.

NOTE: Refer to the Route Planning and Selection outline for blank Route Cards and Commander's Logs.

e. **S- Space**

(1) Space – Tactical boundaries should be wider to allow sufficient freedom of maneuver. Tactical boundaries should follow valley/ridge complexes rather than grid lines (assign the top of a ridge to one or the other adjacent unit, so both do not have to put patrols on it).

f. **L-Logistics**

(1) Logistics - The logistics train is limited to the MSR and air (weather dependent), therefore limiting the range and speed of advanced forces. Forward push log resupply will require imagination on the part of the S4 and Company Gunnys to make it happen from the MSR to the units on the ridges. Fuel, food, and water requirements will also rise in a mountainous environment. Casevac plans are difficult, air is the best option if available. About a squad is required to move one litter casualty to a road or LZ, if the distance is short (within 2 km). Multiple casualties would mean hold for air extract, as it would be impossible for a platoon to litter carry 8 casualties …even if the ground was abandoned! POW collection points need to be on a road for proper turn over. It will take more time and manpower to escort POWs safely to the rear for turnover.

TRANSITION: Those were extra planning considerations according to the acronym METT-TSL. Are there any questions? Now let us talk about danger areas.

2. (5 Min) **DANGER AREAS**. Danger areas unique to the mountains include vertical cliff faces, glaciers, snow fields, scree/talus slopes, and various water obstacles.

a. Linear Danger Areas. Rivers and streams are numerous in the mountains. These are the steps that a platoon would use to tactically cross a stream with a one-rope bridge. The size of these elements can be adjusted so that this technique can be used for any size patrol.

(1) Patrol halts short of the crossing site and occupies a security perimeter. The Patrol Leader (PL) and bridge team leader, reconnoiter the crossing site. The PL could send a scout swimmer across the river to check the far side. The PL selects the security positions.

(2) When the PL confirms the crossing site, security teams position themselves on each flank in the positions the PL has selected. The downstream security teams will have an additional responsibility as a safety swimmer, selecting a security position that is also a suitable rescue point. Rally points are designated on the near and far side. Both rally points should be out of sight from the crossing site. The security perimeter is designated as a rally point. The Assistant Patrol Leader (APL) positions himself in the rally point so that he can supervise the final check and preparation of equipment.

(3) The bridge team begins to set up the rope bridge. The first squad has the mission of security on the far side and moves to the crossing site when all equipment is prepared for crossing. The PL positions himself at the most critical place, usually at the bridge site.

(4) When the first squad is across and security is established on the far side, near side security teams are relieved by the Marines with gear prepared from the security perimeter. The original security teams prepare to cross. Far side must have communication with he near side.

(5) Squads continue to cross. The PL crosses just before the bulk of his patrol reaches the far shore. Squads occupy the rally point on the far shore. Squads occupy the rally point on the far side as a security perimeter as they cross. The APL clears the near side rally point, pulls in security and insures that all personnel are accounted for. The APL crosses with the last squad.

(6) When the entire patrol is across, the last man crosses and retrieves the bridge. As the last man arrives at the security perimeter, security teams withdraw and rejoin the patrol.

NOTE: These steps will work for any of the stream crossing techniques you have been taught. Using stream crossing methods vice the one rope bridge will expedite movement across the linear danger area.

b. Small Danger Areas. Small open danger areas can be small lakes. The following considerations apply when you must swim across a body of water.

(1) When crossing a body of water in the mountains, the temperature will be colder due to elevation, and therefore the risks of hypothermia will be higher.

(2) Stripping all clothing off except a thin camouflage layer and ensuring all other clothing is water proofed, is the preferred method to cross a body of water in the mountains. This allows for dry clothing to be worn after the crossing is made.

TRANSITION: Now that we have talked about danger areas, are there any questions over the material. Now let us talk about patrol base considerations in a mountainous environment.

3. (10 Min) **PATROL BASES**. Patrol bases are extremely effective in mountainous terrain. They enable patrols to stay in their Area of Operation (AO) longer with more combat power.

   a. Terrain selection is essential to have a secure patrol base.

      (1) Key terrain. Patrol bases are used for re-consolidation and rest for Marines, therefore they should be established off of key terrain and likely avenues of approach.

      (2) Difficult terrain. Consider locating the patrol base on difficult terrain that would make it hard for the enemy to reach. When moving through the brush and encountering difficult terrain, what would the enemy attempt to do? They would most likely move around that area. Placing the patrol base in this type of terrain would discourage visitors.

      (3) Dense vegetation. Dense vegetation will contribute to making it difficult for the enemy to move through your area, but more important it will provide good concealment necessary to hide the site for the patrol base. It also provides early warning due to the noise of movement.

      (4) Defensible for a short period of time. While the intent is to avoid having any contact with the enemy while occupying the patrol base, it is possible for enemy contact. Select terrain that will give the advantage if attacked/discovered and that will allow for a safe and fast egress if contacted by a large unit.

      (5) A safe distance. Establish the patrol base well away from near, known, or suspected enemy positions. Civilian populated areas should also be avoided, this includes grazing areas for sheep, cattle, etc.

      (6) Avenues of Approach. Avenues of approach must be avoided. In a mountainous region, natural lines of drift are considered avenues of approach.

      (7) Water source. Re-supply your patrol's water while you are occupying the patrol base. This can be done by, establishing your patrol base near but not on a water source. Avoid placing the base right at the water source because the enemy may decide to utilize the same source for their water supply.

      (8) Swamp areas and steep slope areas. Swamp areas and steep slope areas may be good to establish a patrol base. Avoid a swamp/bog area if Marines cannot select/rig flat, dry sleeping spots. Slopes that are too steep for adequate rest should be avoided, but often there are shelves that can be utilized (sleep on the high side of a boulder or tree).

      (9) Re-supply point. If re-supply is needed, place the patrol base near but not at the re-supply point.

(10) Communications. Mountainous regions have many dead spaces for communications. Select the patrol base location that will allow continuous communications. Ensure radio checks are done at the tentative patrol base prior to moving the patrol in. Because it is important to maintain contact with higher headquarters you have to ensure that the location you select for your patrol base will allow you to continue to communicate.

(11) Use of claymores. Rocky terrain can cause claymore projectiles to ricochet, select the mine site with caution. Other mines are pretty useless, as they cannot be dug in and camouflaged effectively (except in dirt roads).

TRANSITION: Now that we have talked about establishing a patrol base, are there any questions so far? Now let us talk about offensive actions in the mountains.

4. (10 Min) **MOUNTAIN AMBUSH TECHNIQUES**. Some additional ambush considerations are used to augment those used in a temperate environment.

a. L-shaped Ambush. When setting up the L-shaped ambush the long side (assault element) should always be placed on the high ground. The short end of the L-shaped ambush (support element) should be placed where it can also cover the down slope of the kill zone.

b. V-Shaped Ambush. In a mountainous environment, the advantage is having the higher ground to place the legs of the V-shaped ambush. This reduces the chance of one leg firing on the other leg because the kill zone is lower than the ambush lines. The wider separation of elements makes this formation difficult to control and there are fewer sites that favor its use. Its main advantage is that it is difficult for the target to detect the ambush until well into the kill zone.

c. Use of Waterways as Ambush Sites. In mountainous areas, rivers and streams are often used as avenues of approach, thus making them possible ambush sites.

(1) Demolitions work well as an initiator when placed under the water. Explosive charges can destroy boats effectively. Ensure charges are waterproofed, anchored to the bottom in swift current and dual-primed. As always, have an immediate back-up ambush signal in the event the demolitions fail to detonate. A claymore lined up for grazing fire over the water surface is very effective. The automatic weapon should be set up level with the water line to allow grazing fire.

(2) Place a belay line below the water's surface at a 45 degree ferry angle. This angle may allow enemy personnel to be caught in the line and ferried to one side of the stream. It also acts as a safety line for the members of the enemy prisoner of war (EPW) team to safely recover and search enemy personnel and gear that float downstream after the ambush. The POW search team should be covering the entire line by fire.

(3) Expedient underwater obstacles can be constructed to detain or puncture boats, such as sharpened wooden or metal stakes secured underwater slanted opposite of the direction of enemy movement.

d. Using the natural surroundings to aid in ambushes:

(1) Abatis. The use of an abatis can stop armor and wheeled vehicles in your kill zone effectively. Trees are often large, and roads are often channeled, which make a good combination for the use of an abatis in the mountains. A lot of trees need to be dropped for this obstacle, use engineers to construct it.

(2) Rock, snow and ice avalanches. Avalanches that are initiated by demolitions or artillery can be used to exploit the terrain to enhance the ambushes. Dislodging rock, snow, or ice can destroy or isolate the enemy. There are many choke points on a mountain road that are suitable for cutting this way.

TRANSITION: Now that we have discussed considerations of a mountainous environment according to the acronym METT-TSL, danger areas, and ambushes, are there any questions?

PRACTICE (5 Hrs)

a. The students will practice mountain patrolling in an upcoming field training exercise.

PROVIDE HELP (CONC)

a. The instructors will assist the students as necessary.

OPPORTUNITY FOR QUESTIONS (3 Min)

1. QUESTIONS FROM THE CLASS

2. QUESTIONS TO THE CLASS

Q. What is the Time-Distance Formula?

A. 3 kph + 1 hour for every 300 meters ascent: and/or + 1 hour for every 800 meters of descent

Q. What are two types of ambushes that can be used in the mountains?

A. (1) L-shaped.
(2) V-shaped.

SUMMARY (2 Min)

a. This period of discussion has been on those considerations that are specific to a mountainous patrolling environment according to the acronym METT-TSL, danger areas, patrol bases, and ambushes.

b. Those of you with IRF's please fill them out at this time and turn them in to the instructor We will now take a short break and prepare for the practical application portion of this lesson.

UNITED STATES MARINE CORPS
Mountain Warfare Training Center
Bridgeport, California 93517-5001

SML
ACC
02/21/02

LESSON PLAN

WEAPONS CONSIDERATIONS IN A MOUNTAINOUS ENVIRONMENT

INTRODUCTION (5 Min)

1. **GAIN ATTENTION**. To win on the battlefield, Marines must employ their weapons effectively. A mountainous environment will have a serious effect on weapons capabilities and employment techniques. Failure to understand these considerations will reduce effectiveness on target and may cause mission failure.

2. **PURPOSE**. The purpose of this period of instruction is to familiarize the student with the techniques and considerations used to employ personal and crew-served weapons in a mountainous environment. This lesson relates to mountain patrolling.

3. **INTRODUCE LEARNING OBJECTIVES**.

   a. TERMINAL LEARNING OBJECTIVE. In a summer mountainous environment and given a weapon, employ the weapon in accordance with the references.

   b. ENABLING LEARNING OBJECTIVES.

      (1) Without the aid of references, state in writing how to compensate for slope angle when judging target range sight data, in accordance with the references.

      (2) Without the aid of references, state in writing the techniques for moving heavy crew-served weapons and ammunition through the mountains when away from organic support vehicles, in accordance with the references.

      (3) Without the aid of references, describe in writing how a Forward Observer is affected when adjusting indirect fire in compartment terrain, in accordance with the references.

53

4. **METHOD/MEDIA**. The material in this lesson will be presented by lecture and

demonstration. You will practice what you have learned in upcoming field training exercises. Those of you with IRF's please fill them out at the end of this period of instruction.

5. **EVALUATION**. You will be tested by written and performance evaluation.

TRANSITION: Are there any questions over the purpose, learning objectives, how the class will be taught or how you will be evaluated? Let's discuss effects common to all weapons first.

BODY (25 Min)

1. (10 Min) **CONSIDERATIONS COMMON TO ALL WEAPONS**

   a. Slope Angle: When shooting uphill or downhill, the trajectory, or drop, of the round is the same and therefore the sight setting is also the same. The trajectory of the round is the HORIZONTAL DISTANCE between shooter and target NOT LINE-OF-SIGHT DISTANCE. This is because gravity effects the flight (trajectory) of the round only in the horizontal plane. The greater the angle between shooter and target the higher the round will impact. There are several ways to compensate for this.

      (1) The USMC sniper manual and pocket guides have High Angle of Fire tables and the cosines for slope angles from 0 to 90 degrees. Measure the slope angle using an inclinometer. Find the cosine for that angle and multiply the estimated (line-of-sight) range. This will give you the flat range, which is used for sight settings. This is the trigonometry method.

      (2) Estimate the horizontal distance to target rather than line-of sight (Caution; laser range finders determine line-of-sight distance at this time. Future models will compensate for slope angle.). This is the most inaccurate method … guesstimation by the spatially agile.

      (3) Your map provides horizontal distance, use it. The map method is the easiest and most accurate field expedient method. (GPS will help greatly!).

      (4) Aim at 6 o'clock...whether shooting uphill or downhill, and adjust from impact. When in doubt, aim low.

      (5) When making range cards, use reference points that are on the map so you have the horizontal distance (sight setting) or shoot and adjust on to T.R.P.s and log the sight settings on the range card. This is the range card method.

INSTRUCTOR NOTE: Use right-triangle diagram to show distance difference between slope and horizontal.

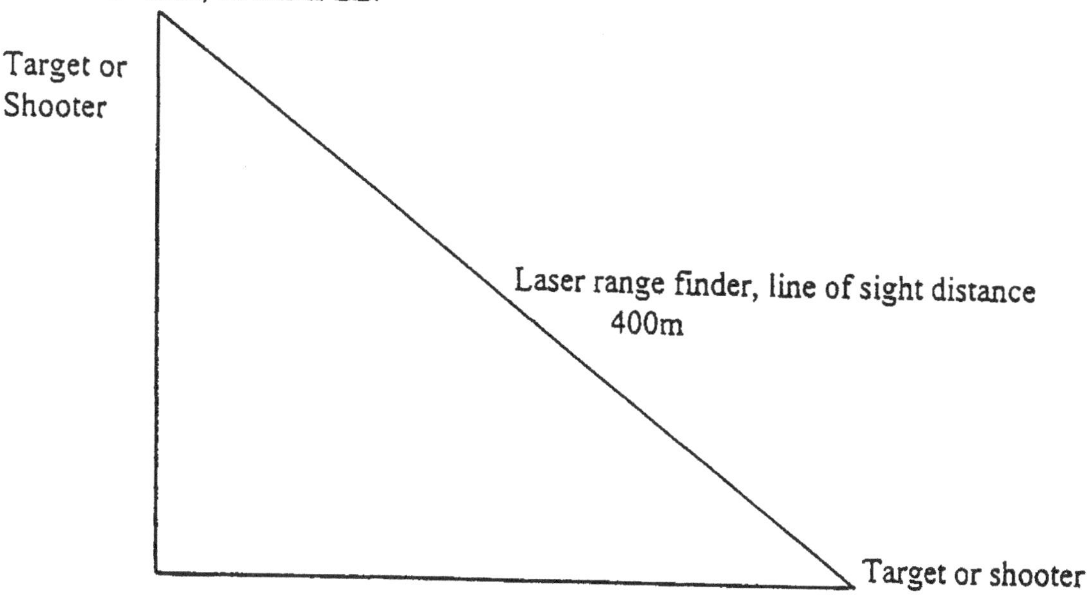

b. Altitude: The higher the altitude, the lower the air pressure. The lower the air pressure, the less dense the air is. The less dense the air is, the less drag is applied on a projectile moving through the air. Reduced drag on a projectile increases the range of it. So the higher in altitude one goes, the more range one's weapon will have... all other factors being equal. However, there are many conflicting factors in weather.

c. Weather: Weather can change rapidly, from sunny to snow or rain and back too sunny in a day. This occurs more often on windward sides than leeward due to the effects of orographic uplift. Erratic variations in solar heating create and dissipate clouds suddenly, causing short, violent thunderstorms. Fronts may be trapped by terrain, causing rain or snow of long duration. Snow and ice must be kept out of the weapon; protect all entry points to the weapon and clean more often than normal. Ammunition should be kept in boxes/containers until needed to protect it from moisture and precipitation.

   (1) Temperature effects range. The colder air is, the denser it is, and therefore, the higher the amount of drag on a projectile which reduces range (all other things being equal). Temperature changes by the adiabatic lapse rate (2.0 - 5.5 degrees) per 1,000' of elevation change. At high altitudes, solar heating is responsible for the greatest temperature contrasts (i.e. very hot in the sun and very cold in the shade, up to a 40-degree difference). The colder the temperature, the slower the burn rate of

propellants. This reduces chamber pressure and muzzle velocity, and therefore range. A slower burn rate also means an increased amount of carbon deposits in the weapon, which will require more weapons maintenance.

(2) Humidity effects range. The higher the humidity, the denser the air which means more drag and less range. Of course, dry air is less dense and so range is increased.

(3) Barometric pressure effects range. Pressure is density and so drag (high pressure, denser air, less range...or low pressure, less dense air, more range).

(4) Winds are stronger and gust more in the mountains. Winds are highly erratic at different altitudes because of the channelization of the terrain that the air mass is being forced over. Adjusting for wind will be difficult due to the varying directions and speeds of the winds.

d. Communications: Comm will be restricted by high ridges, low valleys and high iron ore deposits in many mountainous areas. Terrain and weather will require an increased amount of relay sites and use of field expedient antennas to maintain comm. However, it can also aid with directional comm.

e. Visibility: Visibility is increased at high altitude due to less air density. However, it is often reduced by low clouds, fog, precipitation, etc. Depth perception problems (over and under-estimation) and compartmented terrain will make target spotting difficult. Ridgelines may restrict observation in the next valley even though observation of the nearest valley is outstanding. It may be necessary to use multiple OPs or mountain pickets or an increased number of patrols in order to observe all of one's tactical area of responsibility.

f. Mobility: Moving in the mountains is very difficult. This is true for all, but especially for crew-served weapons. Helicopters are great when you can get them. But the limited number of LZs, reduced lift at high altitude, effects of weather, etc. all hamper the availability of helos. They cannot be depended upon.

(1) Using MSRs. An improved MSR can provide fast and easy mobility. An unimproved MSR can provide the same access, but is highly susceptible to weather (i.e. erosion, mud, etc.), so vehicle types may become limited. Interdiction of MSRs by direct and indirect fire is very high since there are a limited number of roads. Ambush, natural or manmade rock slides or avalanches are a threat. Depending on the importance of an MSR, a team of engineers should be on call to clear any obstacles that may occur.

(2) Vehicles. The type of vehicle used to move Marines/crew-served weapons from point A to B will be in direct relation to the type of terrain to be crossed (i.e. snow covered, rocky, steep, etc.). Common sense and the type of vehicles available are the keys here. The HMMWV w/winch is a tough and capable vehicle with a low center of gravity, wide wheelbase and good ground clearance. Do not automatically rule out the use of HMMWVs in the mountains off road. Careful reconnaissance and driving can provide

surprising results. Drivers will need experience at hill climbing and descent techniques, driving over rock fields/slopes and brush running. This terrain will cause an increase in flat tires, plan accordingly. The BV-206 is an outstanding all-terrain vehicle, not just over-the-snow. Due to base driving restrictions, drivers trained on the BV-206 never see the potential of it (unless they have seen it slogging through brush and bog in Europe).

(3) Pack Animals. Controlling valley floors means occupying ridges. Keeping this in mind, most or all of our crew-served weapons will be on high ridges and peaks where vehicle access is extremely limited. Therefore, the best mode of transportation (all terrain, all weather) will be the pack animal. The Soviet Union was beaten in Afghanistan by rebels being supplied almost exclusively by strings of pack animals from Iran and Pakistan. There are many types of pack animals (horse, mule, camel, llama, etc.). Utilize the local pack animal. The mule is a popular example. A mule can carry up to 200 lbs. of cargo and traverse slopes up to 60 degrees. Panniers can be rigged to carry just about everything in the battalion inventory of weapons and ammo. Carts can also be used. A team of mules requires very little maintenance. If natural grazing material is available, mules need only graze 4 hours a day. If unavailable or inaccessible, mules can pack in their own grain and water.

(4) The Marine. Remember that the Marine is the most dependable and available "pack animal".

(a) Gun carts (like in WWII).
(b) Bicycles (like in Russo-Finnish War, WWII, Vietnam).
(c) Carry less weapon systems and thus more ammo (1 gun team with gun other with only ammo).
(d) Attach Marines as ammo humpers to Weapons from line platoons/H&S.
(e) Spread load weapons' ammo to all Marines (1 mortar round per man).

Planning is critical! Rehearsal should include moving with the load to ensure it is realistic (not overburdening). You must weigh a reduced number of guns and a limited amount of ammo in support of your mission, even if you are the point of main effort. Here are 2 quick examples. A .50 cal weighs 128 lbs. mounted on a tripod plus each 100 round belt of ammo weighs 45 lbs. The sustained rate of fire is under 40 rpm, rapid over 40 rpm. So 10 minutes of suppression is 400 rounds, which is 180 lbs. A 60mm mortar fires 8 rpm sustained and 16 rapid. 3 tubes firing a 5-minute suppression, 2 rapid plus 3 sustained, would equal 168 rounds. Do the math in the planning phase!

g. Resupply: This is probably the most important single factor to winning the battle in the mountains. This gives a unit staying power high up on the dominant terrain features. Helicopters are choice 1, but are undependable. Always have a no air plan. Vehicle use is limited. Even if they get the load part way up a ridge, part way is better than none. Pack animals and manpower are going to be the primary means of resupply. Planning for logistical resupply, with its use of time and manpower, is going to be critical to operational tempo and mission accomplishment.

h. Fighting positions: In a fixed position, it is often impossible to dig in. So you must build up with rocks, gravel, dirt and timber. A gun position of only rocks is called a sangar. The dimensions are the same as for an earth and log position. Be sure to fill in pockets between large rocks with small rocks, gravel, etc. Try to place the rocks so that they overlap and interlock forming a solid wall. Use rounded rocks on the inside and save flat rocks for the outer layer. Place the flat rocks at angles to help ricochet rounds up and away from the position. The rocks used to form the embrasure should angle inward, not outward in order to prevent ricochets into the position.

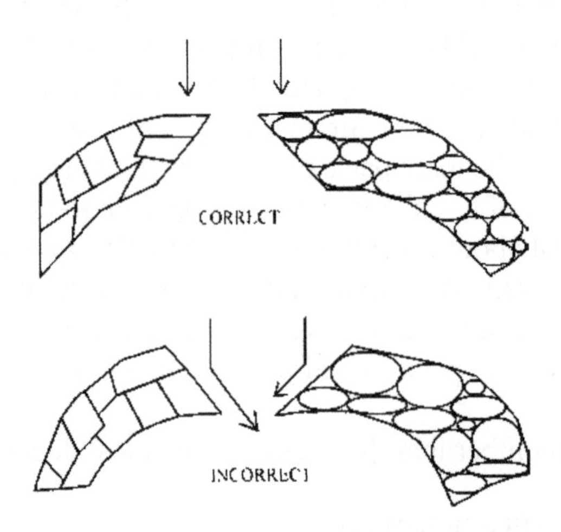

TRANSITION: Now that we've discussed considerations common to all weapons, let's talk about some specific considerations for crew-served weapons.

2. (10 Min) **SPECIFIC CONSIDERATIONS FOR CREW-SERVED WEAPONS**. The following methods are guidelines only. Common sense and imagination are the only limiting factors to a gun team trying to solve an employment problem in the mountains.

   a. Mortars: Mortars are at times the only indirect fire asset available in a mountainous environment, due to arty being tied to MSRs.

      (1) In rocky terrain, it is hard to find a good base for the baseplate. The baseplate must be buffed to prevent cracking or shattering it. Sand bags half filled, pine boughs, etc. can be used (too little buffing can result in breakage, too much will cause the baseplate to shift or collapse after each round). Always carry empty sandbags.

      (2) In a snowfield, it is best to dig down to the ground. If the ground is frozen, it must be buffed or thawed (burning increments on the deck works well, some ammunition is pre-set to charge 1,2,3 and left at 4). The cold and frozen ground will break a baseplate just like rocky ground. If set on snow, a platform must be constructed. Layers of pine boughs and snow, sand bags filled with snow, etc. can be used. Buffing considerations are the same as rocky ground (too much-it shifts, too little it breaks).

(3) Positions are likely to be on the valley floors, so it will be likely that a retrans site will be needed on top of a piece of surrounding high ground to establish comm between the FDC and the FOs. This increases displacement time for mortar sections. Also, training in field expedient antennas should be conducted and gear needed to make them should be kept with each mortar section.

(4) In the attack, it can be difficult to provide sufficient ammo for all tubes. In this situation, it is better to advance fewer tubes and more ammo. The 60mm mortar is usually in this boat. The 81mm only runs into it when they must separate from their HMMWV due to rugged terrain demands.

(5) FOs will take more time and ammo to adjust onto target due to the compartmentalization of terrain. This will cause an increase in the number of lost rounds, more rounds in adjustment due to spatial problems (2d Vs 3d), reduced effect on target because a sheaf will spread out more as the slope angle increases, destruction missions will rarely be worth the expenditure of ammo (if even possible at all). The opportunity to call and adjust fire in mountainous or compartmented terrain in peacetime should never be passed up. Expect an inexperienced FO to call some ineffective fire missions at first. Try to use suppression missions only, if possible.

(6) Rocky terrain will increase the effect of fragmentation. A mix of airburst and quick will produce a better fragment pattern. VT will have an increased rate of malfunction due to weather patterns. Mechanical time is better than variable time for airbursts.

b. Machine-Guns: When employing machine-guns, grazing fire becomes difficult to achieve due to terrain and plunging fire becomes more common.

(1) Leveling the tripod in expedient firing positions on moderate to steep slopes is difficult. Zero the T&E and align to the FPL/PDF if the slope is too steep to level with a canteen cup of water. This means indirect fire/use of firing tables can not be used. Often the tripod can not sit level in the 2 dimensional plane, which means the gunner will need rapid and deft manipulation of the T&E in both traverse and elevation (i.e. on a steep slope firing to the front oblique valley floor).

(2) When possible, dig into the hillside or build up, to level the gun in relation to the target or anticipated direction of the target with the T&E zeroed instead of in relation to gravity. It takes a lot of practice to become proficient at setting in a gun on a moderate to steep slope, ready to fire accurately.

(3) When firing from a rocky position, use sand bags or pine bough to build a stable firing platform. For the M240G (or any medium machine-gun), the A-gunner's pack makes a great field expedient platform in rocks or snow.

(4) In a fixed position, it is often impossible to dig in. So you must build up with rocks, gravel, dirt and timber. A gun position of only rocks is called a sangar. The

dimensions are the same as for an earth and log position. Be sure to fill in pockets between large rocks with small rocks, gravel, etc. Also, try to place the rocks so that they overlap and interlock forming a solid wall.

(5) The Heavy Machine-gun Platoon will be in the same situation as the mortars. When going foot mobile, less guns and more ammo will normally be the call. Establish SOPs that anticipate this kind of employment.

(6) Defilade and partial defilade positions can be used when firing across a compartment to the opposite ridge, but cannot be used to hit the valley floor because the round does not drop that fast. Positioning for reverse slope defenses is no problem.

(7) Hmg Plt displacement will take longer than in moderate terrain. HMMWVs will move faster than foot and carry the load, but it will still be slower than on gentle to moderate terrain. If moving steadily by pack animal or foot, expect to be unable to maintain continuous support throughout the attack (such as during exploitation of success).

c. Anti-Tank Guided Missiles: The range of the TOW and Dragon will not be effected by altitude.

(1) Dragons, being manportable, work very well in the mountains. Getting a stable platform in the rocks or on a slope is not hard. However, tracking a moving target on a steep slope at an oblique angle is difficult. These awkward positions can be practiced with the LET system in suitable terrain. Select a position that has minimum dust in both front and rear to reduce target obscuration and signature during firing. Be cautious of backblast reflection from steep or vertical terrain. When shooting low over lakes, bogs, snowfields, etc., there is an increased risk of erratic missile flight.

(2) TOWs are bound to MSRs. That's not so bad since their primary target (heavy armor) is also road bound. The TOW2 system will operate properly up to 10,000' in elevation. Above 10,000', the air is not dense enough for the control features to steer properly (increased risk of erratic missile flight).

d. Mines and Demolitions: MSRs are an easy target to mine, and so are trails. There are many cuts, fills, bridges and choke points to target. However, large surveyed minefields are difficult to put in. FASCAM is the best choice for area denial mining in the mountains. Claymores are still great for patrol base perimeter security.

(1) Demolitions are very effective in making obstacles, such as an abatis, on MSRs. They can be used to initiate a rockslide, however, an artillery mission is much easier for rockslide and avalanche initiation. Non-electric blasting caps are preferred at high altitude due to the static electricity in the air.

TRANSITION: We've discussed considerations for all weapons and specific considerations for crew-served weapons. Are there any questions?

PRACTICE

a. Students will practice what they have learned in upcoming field training exercises.

PROVIDE HELP

a. Instructors will assist the students when necessary.

OPPORTUNITY FOR QUESTIONS (3 Min)

1. QUESTIONS FROM THE CLASS

2. QUESTIONS TO THE CLASS

Q. What type of estimated range do you put on your rifle sights when shooting at an angle?

A. Horizontal or map

Q. What is your aiming point when firing uphill or downhill?

A. Six O'clock

Q. What is an expedient platform for the M240G?

A. The A-gunner's pack

Q. True or false. When man packing heavy machine-guns, take all the SL-3 complete guns you can plus as much ammo that you can carry.

A. False. Carry fewer guns and more ammo.

SUMMARY (2 Min)

a. During this period of instruction, we have discussed the effects of slope angle, altitude and weather on weapons, as well as, mobility and resupply considerations. We have also discussed crew-served weapons considerations in the mountains. These considerations must be taken into account when operating in a mountainous environment in order to be successful.

b. Please fill out the IRFs and pass them to the front.

UNITED STATES MARINE CORPS
Mountain Warfare Training Center
Bridgeport, California 93517-5001

SML
SMO
02/26/02

LESSON PLAN

ROUTE PLANNING AND SELECTION IN MOUNTAIN OPERATIONS

INTRODUCTION (3 Min)

1. **GAIN ATTENTION**. Movement in a mountainous environment is in itself difficult; however, when combining the inexperience of Marines in mountains, changes in weather, altitude, snow/ice fields, vertical obstacles, water obstacles, mountainous terrain itself, and the tactical situation, movement becomes a monumental task. To move a large unit across this terrain requires more than just words of encouragement or demands by higher authority. It requires positive leadership on the small unit level and a keen sense of technical skill in route selection.

2. **PURPOSE**. The purpose of this period of instruction is to familiarize the student with the requisite knowledge to plan and execute proper route selection. This includes planning for additional time required for movement due to additional equipment and terrain considerations.

3. **INTRODUCE LEARNING OBJECTIVES**

   a. TERMINAL LEARNING OBJECTIVE. In a cold weather/mountainous environment, plan a route, in accordance with the references.

   b. ENABLING LEARNING OBJECTIVES.

      (1) In a cold weather/mountainous environment, using the route planning considerations, plan a route, in accordance with the references.

      (2) In a cold weather/mountainous environment, execute tactical movements, in accordance with the references.

(3) Given a blank route card, map and a scenario, prepare a route card in accordance with the references.

(4) Given overlay paper, map and a scenario, prepare a route overlay in accordance with the references.

(5) Given necessary equipment in a mountainous environment, navigate in accordance with the references.

4. **METHOD / MEDIA**. The material in this lesson will be presented by lecture and demonstration. You will practice what you have learned during upcoming field training exercises. Those of you with IRF's please fill them out at the conclusion of this period of instruction.

5. **EVALUATION**. You will be tested later in the course by performance evaluation.

TRANSITION: To be able to plan a route, we must first understand exactly what a route is. Let's also look at the definition of selection.

BODY (30 Min)

1. (2 Min) **DEFINITION**. By Webster's Dictionary, a route is "a course, way, or road for travel or shipping." Similarly, Webster has defined selection as "the act or an instance of choosing in preference to another of others." As you can see then, route selection is nothing more than the logical or systematic approach of determining one course over another. As a small unit leader or planner, we take on an awesome responsibility to ensure that the route we choose is the best. There are unfortunately an infinite number of routes across one piece of terrain. However, there is only one logical or "common sense" route, if all the necessary precautions and considerations are taken.

TRANSITION: Let's look at some route considerations that must be seriously analyzed during the planning of a route.

2. (8 Min) **ROUTE CONSIDERATIONS**. In the summer there may be a few or many variables in route selection. Perhaps we can get by with considering only the terrain and tactical situation... but usually NOT! There are several factors, which must not be neglected!

   a. The Route Considerations are:

      (1) Terrain

      (2) Weather

      (3) Avalanche/rock slide potential

      (4) Snow/melt-off conditions

(5) Group's ability

(6) Tactical situation/mission

(7) Equipment

(8) Time

b. Each Consideration Contains Its Own Respective Meaning

(1) Terrain

(a) Contour. The lay of the land in the mountains will have a great deal to do with the way in which we negotiate that piece of real estate. To move directly up or down a slope without regard to gradient could be disastrous. Unnecessary cold/heat casualties could result, not to mention the fact that your Marines will not be effective to fight once on or at the march objective. Routes curve, following a contour line, rather than going straight on an azimuth. Conserve energy and time by following contour lines, not azimuths (cross-compartment movement can become so bad as to cause mission failure). Contouring also reduces slope angle of the actual movement, saving energy.

(b) Natural lines of drift. As their name indicates, these are terrain features that tend to draw a unit into them, due to the ease of movement they provide. All things, water, animals, and even humans, tend to take the path of least resistance whenever given the opportunity. But, to allow yourself to be drawn into this situation, could be leading your unit straight into an enemy ambush. A good leader in this environment will set up an ambush near one of these natural lines of drift, because he knows how easily a less experienced leader can allow the terrain to dictate his route.

(c) Altitude. What is the high and low altitude mark for the route? Will this be a time factor? Will it be a clothing factor due to the lapse rate? Are the Marines acclimatized to the route/target altitude? What's the helo lift capacity at the high point of the route (emergency extract, casevac, etc.)?

(d) Tree line. Above or below the tree line during the route? Deciduous or evergreen trees? This will impact camouflage pattern, deception plan, formation, etc.

(e) Vegetation/rock type. Predominant micro-terrain type(s). This will affect; Assault Climbers' equipment and climbing go/no go selection, fire support plan, camouflage pattern, and route selection (macro and micro).

(2) Weather. The weather is perhaps the greatest hazard in a mountainous environment for the untrained and unprepared unit. Weather factors are temperature, visibility, precipitation and the wind velocity and direction; all of which will hamper your movement. Bad weather causes even the simplest tasks to become a burden,

particularly security awareness. Winter-like conditions can occur suddenly in the middle of summer at elevations above 5,000'.

(3) Avalanche/rock slide. The avalanche/rock slide hazard in the area of operations must not be taken lightly. Moving across such a critical slope without proper prior reconnaissance could be adverse to the attainment of a unit's goal. Avalanche/rock slide initiation is a feasible means by which our adversary can destroy our forces, movement on or near critical slopes must be carefully planned. Advance initiation by supporting arms is an option. Engineer assets will need to be forward to clear the MSR. In the IPB process, an avalanche overlay will be needed for the MCOO in determining go/no go mobility corridors. Collect essential elements of information in order to make a hazard analysis, like; local avalanche maps, local knowledge, current weather patterns, last 2 weeks weather, historical trends and forecast, areas of loose, near-vertical rock (from aerial photos, recon, etc.).

(4) Snow/melt-off conditions. Snow conditions can greatly enhance or deteriorate a unit's ability to negotiate a particular route or segment of that route. The constant and sometimes unpredictable changes in the snow conditions make it very difficult for combat laden Marines to utilize a particular route; especially in areas which are windswept or crusted because of temperature changes. What is the depth? Is it a crust that you break through or frozen hard enough to walk on? What is the melt-freeze cycle (times of day you can walk on it, break through, slip and slide out of control)? What is the slope aspect and angle? Are snow-fields only on north-facing slopes, or in shaded draws, or are they still widespread? These answers will determine if you can avoid using over-the-snow mobility specific equipment. Remember, it can snow anytime above 5,000'. This is primarily an early summer and fall consideration, but is year-round in some mountains. Snow melt in spring and early summer can cause quagmires, wash out roads, turn a field into a bog, raise rivers/streams above fording levels, flood valley floors/fields, etc. Areas that were good for artillery positions, log dumps, bivouac sites, etc. can be useless. Areas that vehicles could easily drive over can be impassable. Do not underestimate the implications of the spring thaw on military operations, routes, resupply, etc.

(5) Group's ability. As we all remember, being aware of the group's ability plays a major factor in cold weather and mountainous operations. To move a unit across diversified terrain without regard to degree of slope, snow/vegetation cover, temperature, weight of combat load and mission-essential gear, etc. (all of which influence ability) could be detrimental to both morale and mission accomplishment. Remember as a leader to not base the difficulty of the route on your ability, but to your unit's ability. What is the unit's morale? Are they acclimatized? What is the physical conditioning of the unit? What is they're training status (mountain experience, knot/rope work, cold weather experience, navigation, night ops, etc.)?

(6) Tactical situation/mission. Although independent of each other, the tactical situation and/or the mission effect route selection in a similar manner. Under normal training circumstances, unit leaders and commanders tend to neglect the tactical scenario while

overburdened with the difficulties of moving the unit from one march objective to another. Route selection is very difficult in these circumstances, since moving on the valley floors and out of the cover and concealment of trees, is easier and faster for inexperienced troops in mountainous terrain. As leaders, we must ensure that Marines are not silhouetted on crests and are not caught out in the open valley floors. Our particular concern then becomes route selection for multiple columns, mountain pickets, and cross-compartment movement. Speed vs. security will be balanced with the threat condition in route selection.

USING TERRAIN AND SURROUNDINGS TO CAMOUFLAGE YOUR TRAIL

(7) Equipment. The equipment and the degree of training which Marines have received in the use of that equipment, will directly influence route selection. What equipment is available? ...Marine Assault Climber Kit (MACK), skis or snowshoes, sleds, gun carts, ropes, pack animals, crampons, ice axes, etc. Is the unit trained on any of it? As a general rule, the more equipment that you carry - the easier the route must be. When Marines are humping crew-served weapons, the slope gradient cannot be much more than 25 degrees if traversing that slope. To negotiate excessively steep terrain, which should have been avoided, may require belaying Marines to the bottom or by top roping or short roping Marines to the top. The over-all weight of equipment will determine how aggressive a route one can choose.

(8) Time. When route selecting, we tend to look at a piece of terrain in relation to the amount of time which will be required to negotiate it. Considering all the factors mentioned thus far, picking a route which may appear to temporarily take us away from our march objective and ultimately converge with our original axis of advance, may in the long run gain time. The tendency, however, is to sacrifice proper route selection to gain time, when in actuality time is lost due to subsequent problems, perhaps manifested as a casevac. Appreciate time requirements. Tempo of ops is relative to the enemy, not the clock. Time precedence missions are usually unrealistic in the mountains. They can quickly lead to disaster as a unit loses cohesiveness as the straggler column grows. If mission dictates a time precedent mission, assign Summer Mountain Leaders to key leadership billets to ensure the presence of key leaders at the objective in time.

TRANSITION: We've looked at route considerations, now let's discuss the actual route planning phase.

3. (7 Min) **ROUTE PLANNING**. All the considerations already discussed must be considered for proper planning of any size unit. Failure to consider the above could mean disastrous results in unit movements. When planning for unit movement in a temperate zone, axis of advances, boundaries, and other control measures, are routinely established without regards for weather conditions, and certainly not track discipline. In mountainous terrain, operations officers and other planning communities must consider any and all factors which influence mobility. These considerations must then be analyzed through map and photo studies and ultimately depicted by an overlay or other description. Let us look at the varied planning assets and documents which are influenced by the route selection considerations.

a. The route planning tools are:

(1) Map/photo reconnaissance

(2) Overlays/orders

(3) Time-Distance Formula

(4) Track discipline, if on snow

b. Each route planning tool has its own meaning:

(1) Map/Photo Reconnaissance. Before any route description or overlays can be established, a thorough map/photo study must be conducted. Realizing that this type of study is normally done during temperate climatic operations, the need to only discuss those items peculiar to a cold weather/mountainous environment exists. The prevailing weather patterns that influence the snow pack can be correlated with terrain aspects in a map study. For instance, knowing that our weather generally comes from the W/SW, we can conclude that most north and east facing slopes will have the greatest deposition

of snow, and subsequent formations of cornices. For planning purposes, we must be careful when establishing routes which cross sharp ridgelines oriented with northeasterly facing slopes. Convex and concave slopes can be determined. Another common area of neglect through improper map reconnaissance is cross-compartment movement. Realizing that the tactical situation and the mission at hand may dictate establishing cross-compartment movement to contact, attention should be directed toward skirting/avoiding avalanche prone slopes, scree, talus, rock slide areas, staying in the tree line, etc. From standard 1:50,000 or 1:25,000 maps, critical slopes can easily be determined by looking for a few key indicators such as type of slope (either convex or concave), gradient of the slope, and vegetation on that slope, i.e., obvious cuts in the tree line. Finally, when establishing routes utilizing a map or photo, wider than normal boundaries will have to be established, or a greater degree of flexibility, allowing the unit leader more freedom to select his route. The large contour interval in mountains can hide many obstacles. Aerial photos show actual vegetation type and coverage, as well as the hidden obstacles in between contour lines (such as cliffs). However, snow can obscure linear terrain features. An actual flyover will give you a good feel for the difficulty of the terrain.

(2) Orders/Overlays. When writing the operational order, particular attention should be directed towards allowing the commander of the subordinate units plenty of room for freedom of action; not only in the freedom of maneuver, but also for route selection. On the overlay, critical slopes, cliffs, avalanche paths, and any other natural hazard need to be identified, since they may be part of the adversary's barrier plan as well as a natural obstacle/hazard. Include the TDF (time-distance formula) in the margin of the overlay. Considerations to be plugged into an op order will be covered in Mountain Patrolling.

(3) Time-Distance Formula (TDF). Each route should show a completed TDF; however, commanders must be patient since the actual execution across that route may include hidden difficulties. This is especially true of routes across streams, roads, or cross-compartments.

(a) Time-Distance Formula

1. 3 kph + 1 hour for every 300 meters ascent, and/or + 1 hour for every 800 meters descent.

2. The TDF is made for acclimatized troops on foot in the summertime or experienced troops on skis in the wintertime. If on snowshoes, multiply the total time by 1.33. If on foot in deep (1 to 3 feet) snow, multiply the total time by 2.0. Over 3 feet of snow on foot is "no go" criteria.

3. It is a good idea to leave space on your route cards to write in your actual and planned time for that leg. Continuously doing this will enhance your ability to estimate the actual time it takes your unit to cover a certain distance. The TDF is totaled on the Route Card and on the overlay. The Commander's Log is the most accurate method for estimating time for a unit's movement, but requires time and unit experience to develop into a planning tool.

INSTRUCTOR NOTE: A blank copy of the Commander's Log and Route Card are attached at the back of the outline. These are to be used as master copies for reproduction back at their units.

TRANSITION: The final stage in route planning is to execute the movement. Here are some other general considerations that should be taken into account.

4. (5 Min) **GENERAL CONSIDERATIONS**. All of the planning in the world cannot make up for poor performance. It is the actual Marine on the ground which selects the route to either success or annihilation. Here are some other general considerations when selecting a route:

   a. Terrain Negotiation. Once we have determined where we are going, the physical route must be selected by the lead element.

      (1) By traversing slopes early on, we eliminate the necessity for sudden gains or losses in elevation, which will ultimately add several minutes or hours to a previously short move.

      (2) It may be more efficient to contour around an object (i.e., slope, draw or finger) than to travel in a straight line. This will prevent unnecessary loss or gain in elevation.

      (3) When descending slopes of scree or snow, utilize the "plunge-step".

   b. Narrow Depressions. Contour in and out of them. Establish a short simple fixed rope or hand line or use buddy aid for the heavily laden.

   c. Waterways/Lakes. If frozen, a body of water makes an ideal avenue of approach. Otherwise, it is a large obstacle. They can be by-passed, bridged, or swum. Scout swimmers need to take warming layers (waterproofed) with them for recovery. The banks around these areas are frequently very steep and may be difficult to ascend. See Stream Crossing for fording guidelines.

5. (10 Min) **MARCH DISCIPLINE**

   a. Before March. Route overlays and route cards are completed and briefed to all Marines in movement. The unit leader dictates the outer layer for purposes of identification and camouflage, not the number or type of insulating layers. Inspection of packs, mountaineering equipment, weapons/equipment/radio weatherproofing is completed.

   b. During March. Security is paramount! Continuous supervision is required due to the tendency of Marines to focus on the ground or heels of the Marine in front of them. An initial vent/adjustment break is recommended after 10-15 minutes. Frequent short halts are better in the cold or when ascending steep slopes, such as 25 & 5 rather than 50 & 10. This helps with preventing overcooling on breaks and overheating on the move. Security discipline is maintained during all halts. Clothing drills, frost bite checks, food/water drills are performed by buddy team. Security is maintained throughout these drills by a port/starboard security SOP. Do not allow Marines to sit or lie directly on the snow or cold ground during halts. March formation is usually a column in snow, to avoid multiple trail breaking and visible tracks. Use of existing tracks can be incorporated into the deception plan or used for speed when balanced with the danger of ambush/booby trapping. Perform avalanche hazard assessment along the route, if needed. Overwatch techniques and IA drills are the same in the mountains. Camouflage pattern changes must be made when the terrain dictates.

c. After March. Security is maintained. The mission starts. A successful route can be determined by the energy level of the troops...are the Marines alert and ready to attack? Or are they ready to collapse from fatigue?

TRANSITION: Remember, a good plan brilliantly executed is better than an excellent plan poorly executed. Are there any questions?

PRACTICE (CONC)

a. Students will practice what was taught in upcoming field evolutions.

PROVIDE HELP (CONC)

a. The instructors will assist the students when necessary.

OPPORTUNITY FOR QUESTIONS (3 Min)

1. QUESTIONS FROM THE CLASS

2. QUESTIONS TO THE CLASS

Q. What are the route selection considerations?

A. (1) Terrain
(2) Weather
(3) Avalanche/rock slide potential
(4) Snow/melt-off conditions
(5) Group's ability
(6) Tactical situation/mission
(7) Equipment
(8) Time

Q. What are the route planning tools?

A. (1) Map/photo reconnaissance
(2) Overlays/orders
(3) Time-distance formula

Q. What are the general considerations when executing the route?

A. (1) Terrain navigation
(2) Narrow depressions
(3) Waterways/lakes

SUMMARY (2 Min)

a. This period of instruction has discussed the selection and planning of routes in a cold weather environment.

b. Those of you with IRF's please fill them out at this time and turn them in to the instructor. We will now take a short break.

COMMANDER'S LOG

FOR TIME-DISTANCE ESTIMATION IN SNOW-COVERED TERRAIN
(attach route overlay and/or route card, if possible)

Fill out a copy for every route done by the whole unit or any subordinate unit. The more sheets accumulated, the more accurate the time-distance estimation will become.

UNIT:

DISTANCE (map):

ELEVATION GAIN, TOTAL:

ELEVATION LOSS, TOTAL:

WEATHER CONDITIONS:
(winds, precipitation, humidity, day/night)

TEMPERATURE (high and low):

ALTITUDE (high and low):

SNOW CONDITIONS:
(depth, hardness/flotation, dry/wet, etc)

OVER-THE-SNOW MOBILITY EQUIPMENT:
(skis, snowshoes, sleds, combat boots, VB boots, ski/march boots, skins, wax, etc)

REMARKS:

ROUTE CARD

UNIT ID		UNIT CMDR		# OF PERS		DTG		MAP REF
CHECK POINT	AZIMUTH	DIST.	GRID LOCATION	ELEV GAIN	ELEV LOSS	ETA	ETD	DESCRIPTION

TOTALS: DISTANCE__________ ELEV. GAIN__________ ELEV. LOSS__________ EST. TIME__________

UNITED STATES MARINE CORPS
Mountain Warfare Training Center
Bridgeport, California 93517-5001

BIBLIOGRAPHY OF REFERENCES

Mountain Safety

a. TC 90-6-1, Military Mountaineering
b. USMC Battle Skills Training Book
c. Mountaineering, The Freedom of the Hills, 6th Edition, The Mountaineers, Seattle WA 1996.
d. MCRP 3-35.1, Commander's Guide to Cold Weather Operations
e. MCRP 3-35.1A, Small Unit Leader's Guide to Cold Weather Operations
f. MCRP 3-35.2B Military Mountaineering
g. MCWP 3-35.2, Mountain Operations
h. The Royal Marine's Mountain and Arctic Warfare Handbook, 1972

Mountain Health Awareness

a. Wilderness and Travel Medicine, Eric A. Weiss M.D., Adventure Medical Kit, Oakland, CA 1997.
b. Wilderness Medicine, 4th Edition, Wm. Forgey M.D., ICS Books Inc., Merrillville, IN 1994.
c. Mountain Sickness, Peter H. Hackett M.D., The American Alpine Club, Golden, CO 1980.
d. J.A. Wilkerson, Medicine for Mountaineering, Third Edition.
e. MCRP 3-35.1, Commander's Guide to Cold Weather Operations.
f. MCRP 3-35.1A, Small Unit Leader's Guide to Cold Weather Operations.
g. M. J. Lentz, Mountaineering First Aid, Third Edition.

Cold Weather Mountain Leadership Challenges

a. Cold Weather Operations Manual, U.S. Army Alaska, NWTC, December 1999

Mountain Weather

a. FMFRP 3-29, U.S. Navy Oceanographic and Meteorological Support Syetem Manual
b. Jeppesen Sanderson, Private Pilots Manual, 1983, Jeppesen Sanderson, Inc. Englewood, CO
c. Lehr, Paul E; Burnett, R. Will; Zim, Herbert S. Weather-1975, Western Publishing Company, Inc., Racine, WI.

Summer Warfighting Load Requirements

a. The Soldier's Load and the Mobility of a Nation, Col. S.L.A. Marshall, MCA Quantico, VA 1980.
b. MCRP 3-35.1, Commander's Guide to Cold Weather Operations
c. MCRP 3-35.2A, Small Unit Leader's Guide to Mountain Operations
d. FM 31-70, Basic Cold Weather Manual.

Marine Assault Climber's Kit

a. Marine Corps Stocklist, Marine Assault Climbers Kit MACK, SL-3-10161A
b. Marine Assault Climber's Kit Care and Maintenance Manual

Nomenclature and Care of Mountaineering Equipment

a. Marine Assault Climber's Kit Care and Maintenance Manual
b. MCRP 3-35.2B Military Mountaineering
c. TC 90-6-1, Military Mountaineering
d. Mountaineering, The Freedom of the Hills, 6th Edition, The Mountaineers, Seattle WA 1996.

Rope Management

a. MCRP 3-35.2B Military Mountaineering
b. TC 90-6-1, Military Mountaineering
c. Mountaineering, The Freedom of the Hills, 6th Edition, The Mountaineers, Seattle WA 1996.
d. Luebben, Craig, Knots for Climbers-1995, Chockstone Press Inc, Evergreen, CO
e. Wheelock, Walt, Ropes, Knots and Slings for Climbers-1992, La Siesta Press, Glendale, CA
f. Padgett, Allen; Smith, Bruce, On Rope-4th Printing, National Speleological Society, 1992

Movement in a Mountainous Terrain

a. MCRP 3-35.2B Military Mountaineering
b. TC 90-6-1, Military Mountaineering
c. Mountaineering, The Freedom of the Hills, 6th Edition, The Mountaineers, Seattle WA 1996.

Mountain Casualty Evacuation

a. MCRP 3-35.2B Military Mountaineering
b. TC 90-6-1, Military Mountaineering
c. M. J. Lentz, Mountaineering First Aid, Third Edition.
d. Manual of US Cave Rescue Techniques-2nd Edition, Speleological Society, 1988

Natural and Artifial Anchors

a. MCRP 3-35.2B Military Mountaineering
b. TC 90-6-1, Military Mountaineering
c. Mountaineering, The Freedom of the Hills, 6th Edition, The Mountaineers, Seattle WA 1996.
d. Long, John, Climbing Anchors-1993, Chockstone Press Inc, Evergreen, CO
e. Robbins, Roy, Basic Rock Craft-1971, La Siesta Press, Glendale, CA

Stream Crossing

a. MCRP 3-35.2B Military Mountaineering
b. TC 90-6-1, Military Mountaineering
c. MCRP 3-35.2A, Small Unit Leader's Guide to Mountain Operations
d. Mountaineering, The Freedom of the Hills, 6th Edition, The Mountaineers, Seattle WA 1996.
e. Bechdel, Les; Ray, Slim, River Rescue 2nd Edition, Appalachian Mountain Club Books, Boston, MA 1989.

Mechanical Advantage

a. MCRP 3-35.2B Military Mountaineering
b. TC 90-6-1, Military Mountaineering
c. Bechdel, Les; Ray, Slim, River Rescue 2nd Edition, Appalachian Mountain Club Books, Boston, MA 1989.

One Rope Bridge

a. MCRP 3-35.2B Military Mountaineering
b. TC 90-6-1, Military Mountaineering
c. MCRP 3-35.2A, Small Unit Leader's Guide to Mountain Operations
d. Bechdel, Les; Ray, Slim, River Rescue 2nd Edition, Appalachian Mountain Club Books, Boston, MA 1989.
e. Padgett, Allen; Smith, Bruce, On Rope-4th Printing, National Speleological Society, 1992

A-Frames

a. MCRP 3-35.2B Military Mountaineering
b. TC 90-6-1, Military Mountaineering
c. Mountain Operations Manual 2000, Northern Warfare Training Center, US Army
d. MCRP 3-35.2A, Small Unit Leader's Guide to Mountain Operations

Vertical Hauling Lines

a. MCRP 3-35.2B Military Mountaineering

b. TC 90-6-1, Military Mountaineering
c. MCRP 3-35.2A, Small Unit Leader's Guide to Mountain Operations
d. Manual of US Cave Rescue Techniques-2nd Edition, Speleological Society, 1988
e. Padgett, Allen; Smith, Bruce, On Rope-4th Printing, National Speleological Society, 1992

Suspension Traverse

a. MCRP 3-35.2B Military Mountaineering
b. TC 90-6-1, Military Mountaineering
c. MCRP 3-35.2A, Small Unit Leader's Guide to Mountain Operations
d. Manual of US Cave Rescue Techniques-2nd Edition, Speleological Society, 1988
e. Padgett, Allen; Smith, Bruce, On Rope-4th Printing, National Speleological Society, 1992

Rappelling

a. MCRP 3-35.2B Military Mountaineering
b. TC 90-6-1, Military Mountaineering
c. Mountain Operations Manual 2000, Northern Warfare Training Center, US Army
d. MCRP 3-35.2A, Small Unit Leader's Guide to Mountain Operations
e. Mountaineering, The Freedom of the Hills, 6th Edition, The Mountaineers, Seattle WA 1996.
f. Frank, James A.; Patterson, Donald E.; Rappel Manual, CMC Rescue, Inc. Santa Barbara CA.

Balance Climbing

a. MCRP 3-35.2B Military Mountaineering
b. TC 90-6-1, Military Mountaineering
c. Mountain Operations Manual 2000, Northern Warfare Training Center, US Army
d. Mountaineering, The Freedom of the Hills, 6th Edition, The Mountaineers, Seattle WA 1996.
e. Long, John, Rock Climb 2nd Edition 1993, Chockstone Press Inc, Evergreen, CO.

Top Roping

a. MCRP 3-35.2B Military Mountaineering
b. TC 90-6-1, Military Mountaineering
c. Mountain Operations Manual 2000, Northern Warfare Training Center, US Army
d. Mountaineering, The Freedom of the Hills, 6th Edition, The Mountaineers, Seattle WA 1996.
e. Long, John, Climbing Anchors-1993, Chockstone Press Inc, Evergreen, CO

SIT Harness

a. MCRP 3-35.2B Military Mountaineering
b. TC 90-6-1, Military Mountaineering

c. Mountaineering, The Freedom of the Hills, 6th Edition, The Mountaineers, Seattle WA1996.
d. Long, John, Climbing Anchors-1993, Chockstone Press Inc, Evergreen, CO

Placing Protection

a. MCRP 3-35.2B Military Mountaineering
b. TC 90-6-1, Military Mountaineering
c. Mountain Operations Manual 2000, Northern Warfare Training Center, US Army
d. Mountaineering, The Freedom of the Hills, 6th Edition, The Mountaineers, Seattle WA1996.
e. Long, John, Climbing Anchors-1993, Chockstone Press Inc, Evergreen, CO
f. Long, John, Rock Climb 2nd Edition 1993, Chockstone Press Inc, Evergreen, CO.
g. Robbins, Roy, Basic Rock Craft-1971, La Siesta Press, Glendale, CA

Belaying for Party Climbing

a. MCRP 3-35.2B Military Mountaineering
b. TC 90-6-1, Military Mountaineering
c. Mountain Operations Manual 2000, Northern Warfare Training Center, US Army
d. Mountaineering, The Freedom of the Hills, 6th Edition, The Mountaineers, Seattle WA1996.
e. Long, John, Climbing Anchors-1993, Chockstone Press Inc, Evergreen, CO
f. Long, John, Rock Climb 2nd Edition 1993, Chockstone Press Inc, Evergreen, CO.
g. Robbins, Roy, Basic Rock Craft-1971, La Siesta Press, Glendale, CA

Military Aid Climbing

a. MCRP 3-35.2B Military Mountaineering
b. TC 90-6-1, Military Mountaineering
c. Mountain Operations Manual 2000, Northern Warfare Training Center, US Army
d. Mountaineering, The Freedom of the Hills, 6th Edition, The Mountaineers, Seattle WA1996.

Alternative Belays and Anchors

a. MCRP 3-35.2B Military Mountaineering
b. TC 90-6-1, Military Mountaineering
c. Manual of Military Mountaineering 1992, Mountain and Arctic Warfare Cadre Royal Marines
d. Mountain Operations Manual 2000, Northern Warfare Training Center, US Army
e. Mountaineering, The Freedom of the Hills, 6th Edition, The Mountaineers, Seattle WA1996.
f. Long, John, Climbing Anchors-1993, Chockstone Press Inc, Evergreen, CO
g. Long, John, Rock Climb 2nd Edition 1993, Chockstone Press Inc, Evergreen, CO.
h. Robbins, Roy, Basic Rock Craft-1971, La Siesta Press, Glendale, CA

Rescue for Party Climbing

a. MCRP 3-35.2B Military Mountaineering
b. TC 90-6-1, Military Mountaineering
c. Manual of Military Mountaineering 1992, Mountain and Arctic Warfare Cadre Royal Marines
d. Mountaineering, The Freedom of the Hills, 6th Edition, The Mountaineers, Seattle WA1996.

Steep Earth Climbing

a. MCRP 3-35.2B Military Mountaineering
b. TC 90-6-1, Military Mountaineering
c. Manual of Military Mountaineering 1992, Mountain and Arctic Warfare Cadre Royal Marines
d. Mountain Operations Manual 2000, Northern Warfare Training Center, US Army

Tree Climbing Techniques

Fixed Rope Installation

a. MCRP 3-35.2B Military Mountaineering
b. TC 90-6-1, Military Mountaineering
c. Manual of Military Mountaineering 1992, Mountain and Arctic Warfare Cadre Royal Marines
d. Mountain Operations Manual 2000, Northern Warfare Training Center, US Army

Cliff Reconnaissance

a. MCRP 3-35.2B Military Mountaineering
b. TC 90-6-1, Military Mountaineering
c. Mountain Operations Manual 2000, Northern Warfare Training Center, US Army

Cliff Assault

a. MCRP 3-35.2B Military Mountaineering
b. TC 90-6-1, Military Mountaineering
c. Mountain Operations Manual 2000, Northern Warfare Training Center, US Army

Geology and Glaciology

a. MCRP 3-35.2B Military Mountaineering
b. TC 90-6-1, Military Mountaineering
c. Manual of Military Mountaineering 1992, Mountain and Arctic Warfare Cadre Royal Marines

d. Mountain Operations Manual 2000, Northern Warfare Training Center, US Army

Ice Axe Technique

a. MCRP 3-35.2B Military Mountaineering
b. TC 90-6-1, Military Mountaineering
c. Manual of Military Mountaineering 1992, Mountain and Arctic Warfare Cadre Royal Marines
d. Mountain Operations Manual 2000, Northern Warfare Training Center, US Army
e. Mountaineering, The Freedom of the Hills, 6th Edition, The Mountaineers, Seattle WA1996.
f. March, Bill, Modern Snow and Ice Techniques-1988, Cicerone Press Hunter Publishing Inc, Edison, NJ

Step Kicking and Cutting

a. MCRP 3-35.2B Military Mountaineering
b. TC 90-6-1, Military Mountaineering
c. Manual of Military Mountaineering 1992, Mountain and Arctic Warfare Cadre Royal Marines
d. Mountain Operations Manual 2000, Northern Warfare Training Center, US Army
e. Mountaineering, The Freedom of the Hills, 6th Edition, The Mountaineers, Seattle WA1996.
f. March, Bill, Modern Snow and Ice Techniques-1988, Cicerone Press Hunter Publishing Inc, Edison, NJ

Crampon Technique

a. MCRP 3-35.2B Military Mountaineering
b. TC 90-6-1, Military Mountaineering
c. Manual of Military Mountaineering 1992, Mountain and Arctic Warfare Cadre Royal Marines
d. Mountain Operations Manual 2000, Northern Warfare Training Center, US Army
e. Mountaineering, The Freedom of the Hills, 6th Edition, The Mountaineers, Seattle WA1996.
f. March, Bill, Modern Snow and Ice Techniques-1988, Cicerone Press Hunter Publishing Inc, Edison, NJ

Glassading and Arrest Positions

a. MCRP 3-35.2B Military Mountaineering
b. TC 90-6-1, Military Mountaineering
c. Manual of Military Mountaineering 1992, Mountain and Arctic Warfare Cadre Royal Marines
d. Mountain Operations Manual 2000, Northern Warfare Training Center, US Army

e. Mountaineering, The Freedom of the Hills, 6th Edition, The Mountaineers, Seattle WA1996.
f. March, Bill, Modern Snow and Ice Techniques-1988, Cicerone Press Hunter Publishing Inc, Edison, NJ

Glacier Travel

a. MCRP 3-35.2B Military Mountaineering
b. TC 90-6-1, Military Mountaineering
c. Manual of Military Mountaineering 1992, Mountain and Arctic Warfare Cadre Royal Marines
d. Mountain Operations Manual 2000, Northern Warfare Training Center, US Army
e. Mountaineering, The Freedom of the Hills, 6th Edition, The Mountaineers, Seattle WA1996.
f. March, Bill, Modern Snow and Ice Techniques-1988, Cicerone Press Hunter Publishing Inc, Edison, NJ
g. Selters, Andy; Glacier Travel and Crevasse Rescue-1990, The Mountaineers, Seattle WA

Specific Snow and Ice Equipment

a. MCRP 3-35.2B Military Mountaineering
b. TC 90-6-1, Military Mountaineering
c. Manual of Military Mountaineering 1992, Mountain and Arctic Warfare Cadre Royal Marines
d. Mountain Operations Manual 2000, Northern Warfare Training Center, US Army
e. Mountaineering, The Freedom of the Hills, 6th Edition, The Mountaineers, Seattle WA1996.
f. March, Bill, Modern Snow and Ice Techniques-1988, Cicerone Press Hunter Publishing Inc, Edison, NJ

Establishing Anchors on Snow and Ice

a. MCRP 3-35.2B Military Mountaineering
b. TC 90-6-1, Military Mountaineering
c. Manual of Military Mountaineering 1992, Mountain and Arctic Warfare Cadre Royal Marines
d. Mountain Operations Manual 2000, Northern Warfare Training Center, US Army
e. Mountaineering, The Freedom of the Hills, 6th Edition, The Mountaineers, Seattle WA1996.
f. March, Bill, Modern Snow and Ice Techniques-1988, Cicerone Press Hunter Publishing Inc, Edison, NJ
g. Selters, Andy; Glacier Travel and Crevasse Rescue-1990, The Mountaineers, Seattle WA

Requirements for Survival

a. MCRP 3-02F, Survival

Water Procurement

a. MCRP 3-02F, Survival

Survival Navigation

a. Cold Weather Operations Manual, U.S. Army Alaska, NWTC, December 1999
b. MCRP 3-35.1A, Small Unit Leader's Guide to Cold Weather Operations
c. MCRP 3-02F, Survival

Expedient Shelters and Fires

a. MCRP 3-02F, Survival
b. MCRP 3-02H, Survival, Evasion and Recovery

Survival Diet

a. MCRP 3-02F, Survival

Signaling and Recovery

a. MCRP 3-02F, Survival
b. MCRP 3-02H, Survival, Evasion and Recovery

Communications Considerations in a Cold Weather Environment

a. Cold Weather Operations Manual, U.S. Army Alaska, NWTC, December 1999
b. MCRP 3-35.1, Commander's Guide to Cold Weather Equipment
c. MCRP 3-35.1A, Small Unit Leader's Guide to Cold Weather Operations
d. FMFRP 3-34, Field Antenna Handbook

Wave Propagation, Antenna Theory, and Field Expedient Antennas

a. Cold Weather Operations Manual, U.S. Army Alaska, NWTC, December 1999
b. MCRP 3-35.1, Commander's Guide to Cold Weather Equipment
c. MCRP 3-35.1A, Small Unit Leader's Guide to Cold Weather Operations
d. FMFRP 3-34, Field Antenna Handbook

Mountain Logistical Considerations

a. FM 31-72, Mountain Operations
b. MCRP 3-35.1, Commander's Guide to Cold Weather Operations
c. MCRP 3-35.1A, Small Unit Leader's Guide to Cold Weather Operations

Cold Weather and Mountain Helicopter Operations

a. MCRP 3-35.1, Commander's Guide to Cold Weather Operations
b. Cold Weather Operations Manual, U.S. Army Alaska, NWTC, December 1999

Introduction to Animal Packing

Mountain Patrolling Considerations

a. Cold Weather Operations Manual, U.S. Army Alaska, NWTC, December 1999
b. MCRP 3-35.1, Commander's Guide to Cold Weather Equipment
c. MCRP 3-35.1A, Small Unit Leader's Guide to Cold Weather Operations
d. MCRP Small Unit Leaders Guide to Mountain Operations

Weapons Considerations in Mountainous Terrain

a. Cold Weather Operations Manual, U.S. Army Alaska, NWTC, December 1999
b. MCRP 3-35.1, Commander's Guide to Cold Weather Equipment
c. FM 31-72, Mountain Operations
d. TC 90-6-1, Military Mountaineering
e. MCRP Small Unit Leaders Guide to Mountain Operations

APRIL 1996

SL-3-10161A

MARINE CORPS STOCKLIST

COMPONENTS LIST

FOR

MARINE ASSAULT CLIMBERS KIT-MACK

DEPARTMENT OF THE NAVY
Headquarters, U.S. Marine Corps
Washington, DC 20380-0001

15 April 1996

Marine Corps Stocklist, SL-3-10161A, TAMCN K45232E, is effective upon receipt.

BY DIRECTION OF THE COMMANDANT OF THE MARINE CORPS

OFFICIAL

GARLAND W. ROWLAND
Director, Logistics Data Management Division
Marine Corps Logistics Bases
Albany, Georgia

DISTRIBUTION: PCN 123 101610 00

PREFACE

SCOPE

1. This publication lists all components and accessories for collection-type supply items, such as major combinations, systems, groups, outfits, kits, sets or assortments. The components to be issued with the end item are identified under the heading of "SUPPLY SYSTEM RESPONSIBILITY" and when required, under the heading "COLLATERAL MATERIEL." End items requiring collateral materiel are governed by whether the end item is initial or replacement issue. The Commander, Marine Corps Logistics Bases, 814 Radford Boulevard, Albany, Georgia 31704-1128, will direct whether the issue of the end item is with collateral materiel or without collateral materiel. Those items listed under "USING UNIT RESPONSIBILITY" heading are to be requisitioned separately through the supply system when applicable. Using Units are also responsible for requisitioning the required publications to support the end item identified by the ID number shown on the cover of this SL-3. The end item will be complete when the total quantity of items, as applicable, shown in the SL-3 is on hand.

LIST OF COMPONENTS

2. This listing comprises the major portion of this publication. The data, arranged in columnar form, presents the information needed to identify the item and determine its type of issue.

3. Item Number (Column 1). This column specifies a number assigned to each item as it appears in the list. Numbers are assigned in sequence and are for reference purposes only.

4. Stock Number (Column 2). This column furnishes the National Stock Number (NSN) assigned to the item.

5. Reference Designator/Figure-Key (Column 3). This column contains alphabetical and/or numerical designators for referencing individual component or item to an illustration. The absence of a reference designator indicates there is no illustration for the item.

6. Model (Column 4). This column indicates in alphabetical code the specific application of components, or assemblies when more than one model of the end item is contained in this publication.

7. Item Identification (Column 5). This column provides the item name and description listed under the heading of either "SUPPLY SYSTEM RESPONSIBILITY," ' "COLLATERAL MATERIEL" or "USING UNIT RESPONSIBILITY." (See paragraphs 10, 11, and 12.)

8. Unit of Measure (Column 6). This column gives the measurement standard of each item. It may or may not be the same as the unit of issue. For example, the unit of issue of a certain wire is coil but only 4 feet are required for the end item. Therefore, the unit of measure shown will not be used for requisitioning purposes. For the proper unit of issue and other required management data, refer to the applicable Management Data List (ML) when requisitioning.

9. Quantity Used in Unit (Column 7). This column lists the total quantity of an item, according to the unit of measure, required for full functional operation of the end item.

SUPPLY SUPPORT CATEGORIES

10. Supply System Responsibility. A list, in alphabetical sequence, of items that are furnished with and must be turned in with the end item. Any item requiring replacement is the responsibility of the holding organization or using unit.

11. Collateral Materiel. A list, in alphabetical sequence, of items that are supplied with the initial issue of the end item and are retained by the unit. The "9999-00-000-0000" NSN shown under the heading of "COLLATERAL MATERIEL" is for control within the distribution system only, and is not authorized for requisitioning purposes.

12. Using Unit Responsibility. A list, in alphabetical sequence, of items that will not be issued with the end item. They must be requisitioned, as required, through the supply system by the using unit or holding organization.

5TH ECHELON REHABILITATION PROGRAM

13. Major items returned under this program will be evacuated under the provision(s) of the applicable Marine Corps Order(s) with the items listed under Supply System Responsibility. Rebuild and replacement under a 5th Echelon rehabilitation program will be

limited to these items only. Those items under the heading "COLLATERAL MATERIEL" and using unit items shall be held by using organizations for application to the replacement end items.

CHANGES

14. Changes to this publication will be kept current by and issued as Page Inserts and/or Replacement Pages.

REQUISITIONING OF PUBLICATIONS

15. Publications stocked by the Marine Corps shall be requisitioned as set forth in the current editions of MCO P5600.31, Marine Corps Publications and Printing Regulations and MCO P4400.150, Marine Corps Consumer-Level Supply Policy Manual. Failure to comply with these instructions may result in return of the requisition or delay in processing.

PUBLICATIONS FEEDBACK

16. Technical publications play a critical role in achieving system and equipment readiness. Because of this factor, the currency and accuracy of the data published in these documents are essential. Form NAVMC 10772, Recommended Changes to Publications/ Logistics-Maintenance Data Coding provides a medium for accelerating information feedback to effect the necessary corrections, changes, and/or revisions, as appropriate. Typographical errors need not be reported. Using units shall requisition Form NAVMC 10772 through the normal Marine Corps supply channel. If this form is not adequate to cover a particular situation or recommendation a letter should be directed to the Commander (Code 853), Marine Corps Logistics Bases, 814 Radford Boulevard, Albany, Georgia 31704-1128.

MISCELLANEOUS

17. For full information concerning the Marine Corps Stocklist publications, see the current edition of MCO P5215.17, The Marine Corps Technical Publications System.

INVENTORY SHEET

18. Inventory Sheet is to be used for monthly inspection and may be reproduced locally by the user as required. The user may indicate in the first column the month of the first inventory.

SPECIAL NOTE

19. 19. Items without NSN's assigned should be requestioned by part number and/or drawing number. When NSN's are assigned, changes to this SL-3 will be published.

8465-01-394-8325	MARINE ASSAULT CLIMBERS KIT-MACK	EA

TECHNICAL DATA

FUNCTIONAL DESCRIPTION:

The Marine Assault Climbers Kit is a comprehensive collection of climbing equipment that will enable a Marine rifle company to negotiate an average 300 foot vertical danger area.

LIST OF COMPONENTS

1 ITEM NO	2 STOCK NUMBER	3 REF DESIG FIG-KEY	4 MODEL	5 ITEM IDENTIFICATION	6 UNIT OF MEAS	7 QTY USED IN UNIT
				SUPPLY SYSTEM RESPONSIBILITY		
1	8465-01-415-5135			ASCENDER: sst cams; blk; str 5000 lb; 17 oz. wt; P/N ULTAS01(BLACK) CAGE 66165	EA	8
2	8465-01-412-0148			BAG, CLIMBING GEAR: lg cordura nyl tac rope bag, 20 oz.; P/N 30-AT81M CAGE 0TC88	EA	20
3	8465-01-408-8193			BIGGER BRAKE: two slot breakplate with conical spring; P/N BBWS CAGE 66165	EA	8
4	8465-01-412-3106			CAMMING DEVICE: red; sm; quadcam-4 cam, spring loaded, 10.5-16.5mm; P/N HBQ00 CAGE 66165	EA	4
5	8465-01-412-0151			CAMMING DEVICE,: orn; sm; quadcam-4 cam, spring loaded, 13.0-19mm; P/N HBQ0 CAGE 66165	EA	4
6	8465-01-412-0149			CAMMING DEVICE: yel; sm; quadcam-4 cam, spring loaded, 15.0-23mm; P/N HBQ.5 CAGE 66165	EA	4
7	8465-01-412-0150			CAMMING DEVICE: grn; sm; quadcam-4 cam, spring loaded, 20.0-30mm; P/N HBQ1.0 CAGE 66165	EA	4
8	8465-01-412-0152			CAMMING DEVICE: blu; sm; quadcam-4 cam; spring loaded; 24.0-36mm; P/N HBQ1.5 CAGE 66165	EA	4
9	8465-01-412-0144			CAMMING DEVICE: red; lge; camalots; no. 1; size-range 1.10-2.00 in.; 2.37 in. w; P/N 260101 CAGE 0SZC3	EA	4
10	8465-01-412-0146			CAMMING DEVICE: gld; lge camalots; no. 2; size-range 1.45-2.65 in.; 2.37 in. w; P/N 260102 CAGE 0SZC3	EA	4
11	8465-01-412-0147			CAMMING DEVICE: blu; lge camalots; no.3; size-range 2.00-3.65 in.; 2.75 in. w; P/N 260103 CAGE 0SZC3	EA	4
12	8465-01-411-0574			CAMMING DEVICE: blk; lge camalots; no.4; size-range 2.75-5.00 in.; 3.00 in. w; P/N 260104 CAGE 0SZC3	EA	4
13	8465-01-415-5134			CARABINERS, LOCKING, MODIFIED: stl; blk; 7.5 oz. wt, 4-1/2 in. w, 2-3/4 in. h; P/N MD716S3-5 CAGE 78848	EA	240
14	8465-01-415-5137			CARABINER, OMEGA: blk; non-locking; oval shape; str 4400 lb; P/N OMEGA OVAL CAGE 78848	EA	336
15	5130-01-414-5896			CUTTER, ROPE, ELECTRICAL: P/N 55-AT10A CAGE 66165	EA	1
16	8465-01-415-5136			HARNESS, VARIO SIT: blk nyl webbing; P/N 650067 CAGE 0SZC3	EA	8
17	8465-01-415-7020			HEXENTRICS: nut no. 1w; 62g wt; 0.75 in. lg, 0.44 in. w; P/N 220001 CAGE 0SZC3	EA	4
18	8465-01-415-7446			HEXENTRICS: nut no. 2w; 62g wt;, 0.88 in. lg, 0.56 in. w; P/N 220002 CAGE 0SZC3	EA	4
19	8465-01-415-7447			HEXENTRICS: nut no. 3w; 67g wt; 1.00 in.lg, 0.63 in. w; P/N 220003 CAGE 0SZC3	EA	4
20	8465-01-415-7448			HEXENTRICS: nut no. 10; 140g wt; 3.00 in. lg, 2.25 in w; P/N 220010 CAGE 0SZC3	EA	4
21	8465-01-415-7449			HEXENTRICS: nut no. 7; 56g wt; 1.88 in. lg, 1.38 in. w; P/N 220007 CAGE 0SZC3	EA	4

LIST OF COMPONENTS

1 ITEM NO	2 STOCK NUMBER	3 REF DESIG FIG-KEY	4 MODEL	5 ITEM IDENTIFICATION	6 UNIT OF MEAS	7 QTY USED IN UNIT
22	8465-01-415-7445			HEXENTRICS: nut no. 5; 28g wt; 0.38 in. lg, 0.88 in. w; P/N 220005 CAGE 0SZC3	EA	4
23	8465-01-415-7444			HEXENTRICS: nut no. 6; 39g wt; 1.63 in. lg, 1.13 in. w; P/N 220006 CAGE 0SZC3	EA	4
24	8465-01-415-7443			HEXENTRICS: nut no. 4; 20g wt; 1.13 in. lg, 0.75 in. w; P/N 220004 CAGE 0SZC3	EA	4
25	8465-01-415-7452			HEXENTRICS: nut no. 8; 70g wt; 2.25 in. lg, 1.63 in. w; P/N 220008 CAGE 0SZC3	EA	4
26	8465-01-415-7453			HEXENTRICS: nut no. 9; 84g wt; 2.63 in. lg, 1.88 in. w; P/N 220009 CAGE 0SZC3	EA	4
27	4240-01-409-2006			LADDER, ESCAPE: stl cable with al rungs; collapsible; cross-piece 6.000 in. lg; 33 ft oa lg; P/N 65-AT33 CAGE 66165	EA	10
28	8465-01-414-1807			NUT, STOPPER: blk; str side; no. 2; str 770 lb; 14g wt; 0.40 in. lg, 0.20 in. w; P/N 225002 CAGE 0SZC3	EA	4
29	8465-01-414-1808			NUT, STOPPER: grn; str side; no. 1; 8g wt; 0.34 in. lg, 0.17 in. w; P/N 225001 CAGE 0SZC3	EA	4
30	8465-01-414-1809			NUT, STOPPER: prp: crv; no. 4; 26g wt; str 1430 lb; 0.47 in. lg, 0.28 in. w; P/N 225004 CAGE 0SZC3	EA	4
31	8465-01-414-1810			NUT, STOPPER: dk orn; str side; no.3; str 1430 lb; 22g wt; 0.48 in. lg, 0.24 in. w; P/N 225003 CAGE 0SZC3	EA	4
32	8465-01-414-2747			NUT, STOPPER: grn; crv; str 2420 lb; no. 10; 67g wt; 0.97 in. lg, 0.69 in. w; P/N 225010 CAGE 0SZC3	EA	4
33	8465-01-414-2748			NUT, STOPPER: cherry; crv; no. 6; str 2420 lb; 36g wt; 0.57 in. lg, 0.40 in. w; P/N 225006 CAGE 0SZC3	EA	4
34	8465-01-414-2749			NUT, STOPPER: lime; crv; no. 5; str 1430 lb; 32g wt; 0.52 in. lg, 0.33 in. w; P/N 225005 CAGE 0SZC3	EA	4
35	8465-01-414-2750			NUT, STOPPER: lt orn; crv; no. 9; str 2420 lb; 62g wt; 0.86 in. lg, 0.60 in. w; P/N 225009 CAGE 0SZC3	EA	4
36	8465-01-414-2751			NUT, STOPPER: royal; crv; no. 7; str 2420 lb; 40g wt; 0.65 in. lg, 0.46 in. w; P/N 225007 CAGE 0SZC3	EA	4
37	8465-01-414-2752			NUT, STOPPER: yel; crv; no.8; str 2420 lb; 51g wt; 0.75 in. lg, 0.53 in. w; P/N 225008 CAGE 0SZC3	EA	4
38	8465-01-414-2753			NUT, STOPPER: red; crv; no. 11; str 2420 lb wt; 81g wt; 1.10 in. lg, 0.79 in. w; P/N 225011 CAGE 0SZC3	EA	4
39	8465-01-414-2755			NUT, STOPPER: navy; crv; no. 12; str 2420 lb; 95g wt; 1.25 in. lg, 0.90 in. w; P/N 225012 CAGE 0SZC3	EA	4
40	5120-01-410-7213			NUT TOOL, CLIMBER'S: stl; heat treated; tapered working end; P/N 620055 CAGE 0SZC3	EA	8
41	3020-01-411-1132			PULLEY ASSEMBLY, CLIMBERS EQUIPMENT: sst axle and ndl brg; 5000 lb wt cap.; 2.500 in. dia; P/N RP103 CAGE 66165	EA	8
42	8465-01-412-0145			RACK PACK: cordura nyl bag; P/N 30-AT81RACK PACK CAGE 0TC88	EA	8
43	4020-01-411-9059			ROPE, FIBROUS (Dynamic): olive drab; 11mm; 165 ft dry; P/N 30-AT366 CAGE 66165	EA	32

LIST OF COMPONENTS

1 ITEM NO	2 STOCK NUMBER	3 REF DESIG FIG-KEY	4 MODEL	5 ITEM IDENTIFICATION	6 UNIT OF MEAS	7 QTY USED IN UNIT
44	4020-01-400-2853			ROPE, FIBROUS (Static): 7/16 in.; 0.04375 in. crcmf; 600.00 ft lg; P/N E931 CAGE 6W002	EA	5
45	4020-01-400-5862			ROPE, FIBROUS: 5.5mm; P/N RBAC07 CAGE 66165	EA	1
46	4020-01-400-5864			ROPE, FIBROUS: 7mm; P/N RBACK55 TYPE 2 CAGE 66165	EA	1
47	4020-01-411-9053			ROPE, FIBROUS (Static): nyl; 7/16 in.; 0.438 in. crcmf; P/N 30-AT367 CAGE 66165	EA	8
48	8145-01-406-1427			SHIPPING AND STORAGE CONTAINER, MISCELLANEOUS EQUIPMENT: plstc; rect; 37.320 in. lg, 27.370 in. w, 19.600 in. h; P/N AL3424-1205 CAGE 11214	EA	4
49	8305-01-409-0889			WEBBING, TEXTILE (Tubular): nyl; blk or olive drab; str; 4400.00 lb; 1.000 in. w; P/N 7901XX CAGE 62645	EA	2

USING UNIT RESPONSIBILITY

NOTE: Using units should refer to the SL-1-2 and requisition the required publications to support the item identified by the ID number shown on the cover of this SL-3.

INVENTORY SHEET

NAME OF EQUIPMENT MARINE ASSAULT CLIMBERS KIT-MACK

ITEM NO	STOCK NUMBER	ITEM IDENTIFICATION	UNIT OF MEAS	QTY USED IN UNIT	MONTH												REMARKS
1	8465-01-415-5135	ASCENDER	EA	8													
2	8465-01-412-0148	BAG	EA	20													
3	8465-01-408-8193	BIGGER BRAKE	EA	8													
4	8465-01-412-3106	CAMMING DEVICE	EA	4													
5	8465-01-412-0151	CAMMING DEVICE	EA	4													
6	8465-01-412-0149	CAMMING DEVICE	EA	4													
7	8465-01-412-0150	CAMMING DEVICE	EA	4													
8	8465-01-412-0152	CAMMING DEVICE	EA	4													
9	8465-01-412-0144	CAMMING DEVICE	EA	4													
10	8465-01-412-0146	CAMMING DEVICE	EA	4													
11	8465-01-412-0147	CAMMING DEVICE	EA	4													
12	8465-01-411-0574	CAMMING DEVICE	EA	4													
13	8465-01-415-5134	CARABINERS	EA	240													
14	8465-01-415-5137	CARABINER	EA	336													
15	5130-01-414-5896	CUTTER	EA	1													
16	8465-01-415-5136	HARNESS	EA	8													
17	8465-01-415-7020	HEXENTRICS	EA	4													
18	8465-01-415-7446	HEXENTRICS	EA	4													
19	8465-01-415-7447	HEXENTRICS	EA	4													
20	8465-01-415-7448	HEXENTRICS	EA	4													
21	8465-01-415-7449	HEXENTRICS	EA	4													
22	8465-01-415-7445	HEXENTRICS	EA	4													
23	8465-01-415-7444	HEXENTRICS	EA	4													
24	8465-01-415-7443	HEXENTRICS	EA	4													
25	8465-01-415-7452	HEXENTRICS	EA	4													
26	8465-01-415-7453	HEXENTRICS	EA	4													
27	4240-01-409-2006	LADDER	EA	10													
28	8465-01-414-1807	NUT	EA	4													
29	8465-01-414-1808	NUT	EA	4													
30	8465-01-414-1809	NUT	EA	4													
31	8465-01-414-1810	NUT	EA	4													
32	8465-01-414-2747	NUT	EA	4													
33	8465-01-414-2748	NUT	EA	4													
34	8465-01-414-2749	NUT	EA	4													
35	8465-01-414-2750	NUT	EA	4													
36	8465-01-414-2751	NUT	EA	4													

INVENTORY SHEET

NAME OF EQUIPMENT MARINE ASSAULT CLIMBERS KIT-MACK

ITEM NO	STOCK NUMBER	ITEM IDENTIFICATION	UNIT OF MEAS	QTY USED IN UNIT	MONTH												REMARKS
37	8465-01-414-2752	NUT	EA	4													
38	8465-01-414-2753	NUT	EA	4													
39	8465-01-414-2755	NUT	EA	4													
40	5120-01-410-7213	NUT TOOL	EA	8													
41	3020-01-411-1132	PULLEY ASSEMBLY	EA	8													
42	8465-01-412-0145	RACK PACK	EA	8													
43	4020-01-411-9059	ROPE	EA	32													
44	4020-01-400-2853	ROPE	EA	5													
45	4020-01-400-5862	ROPE	EA	1													
46	4020-01-400-5864	ROPE	EA	1													
47	4020-01-411-9053	ROPE	EA	8													
48	8145-01-406-1427	SHIPPIG AND STORAGE	EA	4													
49	8305-01-409-0889	WEBBING	EA	2													

MARINE ASSAULT CLIMBER'S KIT (MACK)

CARE AND MAINTENANCE MANUAL

CONTRACT #M67854-93-C-3009

MAY 1994

TABLE OF CONTENTS

PURPOSE

This manual is provided to guide Marines through basic preventative maintenance and serviceability checks for components of the Marine Assault Climber's Kit (MACK). These maintenance steps and serviceability inspections are required to ensure that each MACK component functions properly and maintains its safety features. With the exception of the rope bag and climbing rack-bag, no attempt should be made to repair the components of the MACK. ***Any item that becomes unserviceable or shows excessive wear must be replaced immediately.*** A damaged or broken component should be disposed of using standard supply procedures.

IMPORTANT SAFETY INFORMATION

Any Marine using components of the Marine Assault Climber's Kit (MACK) must be supervised by a Marine NCO, SNCO, or officer who has received formal military mountaineering training. This formal instruction must be provided by the Marine Corps Mountain Warfare Training Center, a Marine Expeditionary Force Special Operations Training Group, or another training agency authorized by the Marine Expeditionary Force Commander.

SUPPLY

Although no initial issue provisioning was conducted for the Marine Assault Climber's Kit (MACK), a spare parts package was procured under the original contract and will be held at the SASSY Management Units (SMUs) of the Force Service Support Groups (FSSGs). These spares are available through the SMUs until components have National Stock Numbers (NSNs). Once NSNs are assigned, kit components may be ordered via the Military Standard Requisitioning and Issue Procedures (MILSTRIP) process.

The SL-3 for the MACK may be used to identify component stock numbers for ordering purposes. SL-3s will be forced fed to using units; additional copies of the SL-3 will be placed in stock at Marine Corps Logistics Base, Albany and may be ordered by Publication Control Number (PCN).

The MACK comes packaged in four separate containers. On the inside of each container lid is a listing of the contents and quantities contained therein. The outside of each container is marked with the kit stock number, manufacturer's name and address, contract number, and container number.

Replenishment procurement to the Source of Supply should be accomplished utilizing the MILSTRIP process. Kit components not available through the MILSTRIP process may be procured through local purchase. Detailed item specifications for each MACK component are contained in this manual should an open purchase become necessary. When open purchasing replacements for any of the items listed below (items used for protection or direct aid), they must be Union of International Alpine Association (UIAA) approved, meet National Fire Protection Association (NFPA) 1983, or be manufacturer certified or qualified third party certified that it meets UIAA, NFPA 1983, and/or the requirements in accordance with the specifications provided in this manual:

Carabiner, Aluminum Non-Locking	Ascender
Carabiner, Steel Locking "D"	Harness, Climbing
Nut, Wedge, Wire	Rope, Dynamic
Nut, Hexcentric	Rope, Static
Camming Device, 4 Cam (Small)	Cord, Accessory Type I
Camming Device, 4 Cam (Large)	Cord, Accessory Type II
Pulley, Rescue	Webbing, Nylon
Plate, Belay	Ladder, Cable

MARINE ASSAULT CLIMBER'S KIT (MACK) OVERVIEW

Description:

The Marine Assault Climber's Kit (MACK) is a comprehensive collection of climbing equipment that enables a Marine rifle company (reinforced), approximately 200 men with organic equipment, to negotiate an average 300-foot vertical danger area. The kit contains sufficient climbing equipment to equip four two-man climbing teams plus the additional items necessary to supply the remainder of the rifle company. The climbing teams use their equipment to conduct two-party climbs over vertical obstacles and establish fixed rope installations and vertical hauling lines to facilitate movement of the remainder of the company.

Care and Maintenance:

All components of the MACK must be inspected for serviceability before, after, and frequently during each use. All items must also be cleaned after each use per the following instructions. Because any excess moisture could lead to corrosion of the metal parts and mildewing of the ropes and bags, every item must be dried thoroughly before being returned to the MACK container for storage.

CARABINER, ALUMINUM NON-LOCKING

Description:

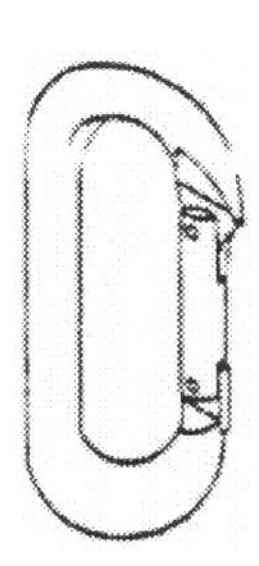

The non-locking carabiner is oval in shape and constructed of aircraft quality aluminum alloy. It has a matte black finish and a maximum weight of 2.3 ounces. Its gate is spring loaded and made of a material that is the same or better than the carabiner body material. The carabiner has a slight protrusion at the gate latch to easily identify its working end. The gate rivets are made from stainless steel and the gate opening clearance is greater than 0.65 inches. Gate-open strength is greater than 1,300 pounds. Minimum carabiner breaking strength is at least 4,400 pounds.

Care and Maintenance:

Inspect the non-locking carabiner before, after, and frequently during use for signs of wear. These signs include nicks and grooves. If a carabiner appears worn or bent, replace the carabiner.

Hairline fractures normally occur after being dropped onto hard surfaces, particularly after long falls. These fractures are not always readily visible. If a non-locking carabiner is dropped it may lose it structural integrity, so examine it carefully for hairline fractures. If any fractures are discovered, replace the carabiner immediately.

Gate action on the carabiner should always be smooth. If it becomes difficult to move the gate or the gate becomes "sticky," clean the gate hinge with soapy water or a non-corrosive solvent and once the carabiner is dry lightly apply a standard lubricant to the hinge. Ensure that all excess lubrication is removed from the carabiner so as not to contaminate the rope. If the carabiner gate continues to stick, replace the carabiner.

CARABINER, STEEL LOCKING "D"

Description:

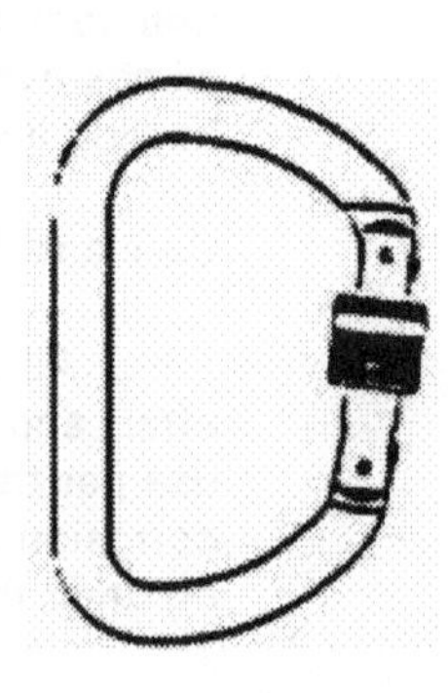

The steel locking "D" carabiner is "off-set D" in shape and constructed of cadmium plated, zinc plated, or stainless steel. It has a matte black finish and a maximum weight of 7.5 ounces. The carabiner has a locking screw gate to prevent inadvertent opening. The gate is spring loaded and made of a material that is the same or better than the carabiner body material. Gate opening clearance is greater than 0.93 inches. Gate-open strength is greater than 1,300 pounds. Minimum carabiner breaking strength is at least 7,000 pounds.

Care and Maintenance:

Inspect the locking carabiner before, after, and frequently during use for signs of wear. These signs include nicks and grooves. If a carabiner appears worn or bent, replace the carabiner.

Gate action on the carabiner should always be smooth. If it becomes difficult to move the gate or the gate becomes "sticky," clean the gate hinge with soapy water or a non-corrosive solvent and once the carabiner is dry lightly apply a standard lubricant to the hinge. Ensure that all excess lubrication is removed from the carabiner so as not to contaminate the rope. If the carabiner gate continues to stick, replace the carabiner.

The locking sleeve action of the carabiner gate should always be smooth. If it becomes difficult to turn the locking sleeve or the sleeve action becomes "sticky," clean the sleeve and threaded portion of the carabiner with soapy water or a non-corrosive solvent and once the carabiner is dry lightly apply a standard lubricant to the sleeve and threads. Ensure that all excess lubrication is removed from the carabiner so as not to contaminate the rope. The threads must be free of burrs, deformations, and debris. If the carabiner sleeve continues to stick, replace the carabiner.

NUT, WEDGE, WIRE

Description:

The wire wedge nuts come in twelve sizes, ranging from widths of 0.16 inches ± 0.02 inches to 0.90 inches ± 0.04 inches. The wedged portion of the nut is constructed of 6061-T6 aluminum or equal material. The wire portion is made of stainless or galvanized steel and, depending on the nut size, is 1/32 to 1/8 inch in diameter. The wire protrudes between 6.5 to 7.5 inches from the wedged nut depending on the nut size. Breaking strengths range from a minimum of 550 pounds for the smallest nut to a minimum of 2,800 pounds for the largest.

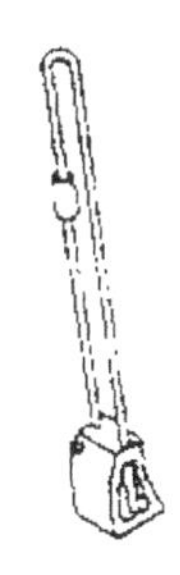

Care and Maintenance:

'Inspect the wedge wire nuts before, after, and frequently during use for signs of wear. These signs include nicks or grooves in the nuts, fraying or parting of the wire, and cracks or chips in the swages and soldering. If any part of the wedge nut appears worn, replace it.

NUT,-HEXCENTRIC

Description:

The MACK contains both wired and non-wired hexcentric nuts. The hexcentric portion of each nut is constructed of 6061-T6 aluminum or equal material. They come in ten sizes, ranging from widths of 0.44 inches ± 0.01 inches to 2.25 inches ± 0.01 inches. At a minimum, the three smallest sized hexcentric nuts are wired. The wire portion is made of stainless or galvanized steel, approximately 1/8 inch in diameter for all the nut sizes, and it protrudes 7.5 to 8 inches from the nut. Each non-wired hexcentric nut has two holes that allow it to be threaded with 5.5 millimeter Type I accessory cord. Minimum breaking strength for both types and all sizes of nuts is 2,800 pounds.

Care and Maintenance:

Inspect the hexcentric nuts before, after, and frequently during use for signs of wear. These signs include nicks or grooves in the nuts, fraying or parting of the wire, and cracks or chips in the swages and soldering of the wired nuts. If any part of a wired hexcentric nut appears worn, replace it. For non-wired hexcentric nuts, inspect the accessory cord for knot slippage as well as any fraying or parting of the outer sheath, especially where the cord is threaded into the nut. If the nut portion of a non-wired hexcentric nut appears worn, replace it. If the accessory cord portion appears worn, retain the nut but replace the accessory cord. This nut must be rethreaded by personnel with formal military mountaineering training using the Type I accessory cord from the MACK. It must be inspected by at least one other properly trained military mountaineer before use.

CAMMING DEVICE, 4 CAM (SMALL)

Description:

The small 4 cam camming device is a spring loaded mechanism made with two parallel flexible steel handles that are welded to the cam pivot pin frame. A finger pull-ring expands and contracts the cams and the finger pull-ring assembly permits one-finger operation. The camming device is designed for one-hand use. The surface of the 4 cams are serrated to provide better gripping and reduce "walking." The entire device has a clean profile with no side projections that could dislodge the device. The camming devices come in five sizes for use in cracks ranging in size from 0.36 inches ± 0.06 inches to 1.4 inches ± 0.05 inches. Any crack within this range must be safely filled by at least one device. The devices are color coded by size for easy identification. Minimum breaking strength for all the small camming devices is 2,600 pounds.

Care and Maintenance:

Inspect each small camming device before, after, and frequently during use for signs of wear. These signs include nicks or grooves in the face of the cam, cracks or fractures in the cams or finger pull-ring, fraying or parting of the wire, and cracks or chips in the soldering. Additionally the serrations on the cam face can be worn down over time and the cam face may become smooth. If any part of a small camming device appears worn, replace it.

The action of the finger pull-ring and cams on the camming device should always be smooth. If it becomes difficult to expand or contract a cam or the action becomes "sticky," clean the pull-ring, cam hinge, and cam springs with soapy water or a non-corrosive solvent and once the device is dry lightly apply a standard lubricant to these areas. Ensure that all excess lubrication is removed from the camming device so as not to contaminate the rope. These same areas must be free of burrs, deformations, and debris. If the pull-ring or cams continue to stick, replace the camming device.

Inspect the sewn-on sling of the small camming device for thread failures at the sewn seams as well as for any fraying or parting of the sling itself. If any part of the sling appears worn, remove the sling from the camming device. Thereafter attach carabiners to the small camming device in the same manner that you would attach them to a large 4 cam camming device.

CAMMING DEVICE, 4 CAM (LARGE)

Description:

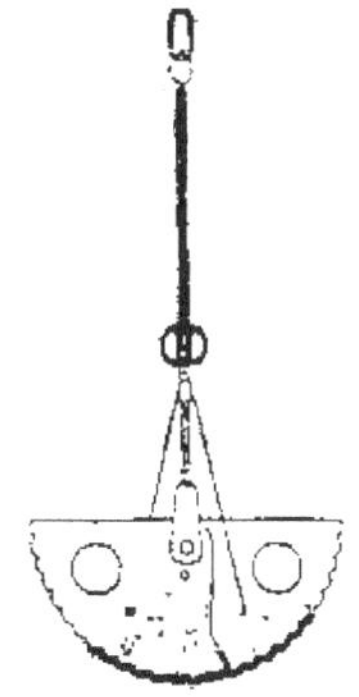

The large 4 cam camming device is a spring-loaded mechanism made with two parallel flexible steel handles that are welded to the cam pivot pin frame. The trigger bar extends around flexible handles to prevent failure and the trigger assembly permits one-finger operation. The camming device is designed for one-hand use. The surface of the 4 cams are serrated to provide better gripping and reduce "walking." The entire device has a clean profile with no side projections that could dislodge the device. The camming devices come in four sizes for use in cracks ranging in size from 1.0 inch ± 0.2 inches to 5.0 inches ± 0.5 inches. Any crack within this range must be safely filled by at least one device. The devices are color coded by size for easy identification. Minimum breaking strength is 2,400 pounds for all the large camming devices.

Care and Maintenance:

Inspect the large camming device before, after, and frequently during use for signs of wear. These signs include nicks or grooves in the face of the cam, cracks or fractures in the cams or trigger bar, fraying or parting of the wire, and cracks or chips in the soldering. Additionally the serrations on the cam face can be worn down over time and the cam face may become smooth. If any part of the large camming device appears worn, replace it.

The action of the trigger bar and cams on the camming device should always be smooth. If it becomes difficult to expand or contract a cam or the action becomes "sticky," clean the trigger bar, cam hinge, and cam springs with soapy water or a non-corrosive solvent and once the device is dry lightly apply a standard lubricant to these areas. Ensure that all excess lubrication is removed from the camming device so as not to contaminate the rope. These same areas must be free of burrs, deformations, or debris. If the trigger bar or cams continue to stick, replace the camming device.

PULLEY, RESCUE

Description:

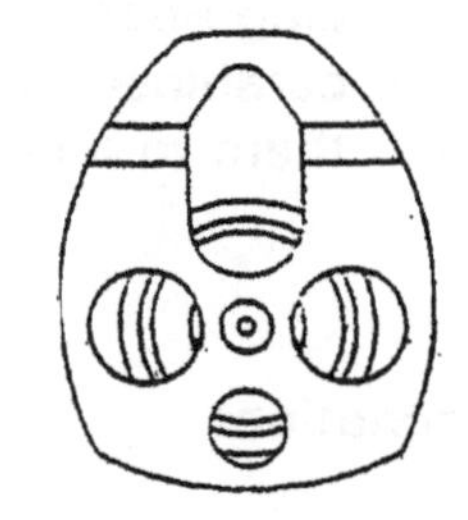

The rescue pulley has two independent side plates that enable a user to insert the rope onto the wheel without having to thread the rope through the pulley. The side plates are fabricated from aluminum and are black in color. The wheel is made from aluminum and has a permanently lubricated sealed bearing. The axle is made of stainless steel. The pulley is large enough to accommodate a 1/2-inch diameter rope and has an eye-hole large enough to accommodate two of the steel locking "D" carabiners included in the MACK. Minimum breaking strength is 5,000 pounds. Maximum weight is 0.5 pounds.

Care and Maintenance:

Inspect the rescue pulley before, after, and frequently during use for signs of wear. These signs include nicks, grooves, cracks, or burrs in the side plates or wheel face. Cracks or nicks in the side plates could cause a pulley failure. Nicks, grooves, or burrs on the wheel face could damage and weaken ropes. If the rescue pulley appears worn, replace it.

Because the rescue pulley is constructed with a sealed bearing, the wheel does not require lubrication. If the wheel begins to stick or fails to roll smoothly, replace the pulley.

PLATE, BELAY

Description:

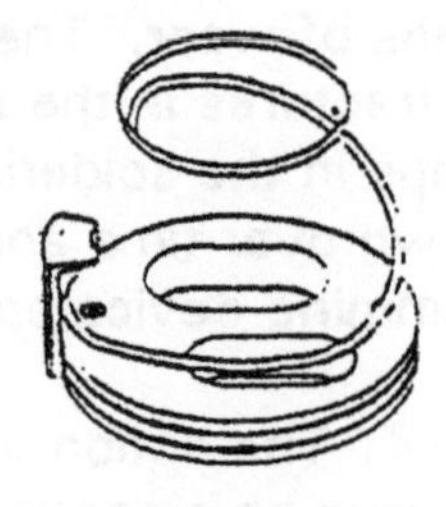

The belay plate is circular in shape with a maximum diameter of 2.5 inches. It is fabricated from aluminum. It contains two slot-shaped holes that can each accommodate an 11 millimeter rope. A spring is attached to the plate to prevent a rope from locking up. Maximum weight of the belay plate is 3.0 ounces.

Care and Maintenance

Inspect the belay plate before, after, and frequently during use for signs of wear. These signs include nicks, grooves, cracks, or burrs. Cracks or nicks in the belay plate could cause the device to fail. Nicks, grooves, or burrs on the rope-hole surfaces could damage and weaken ropes. If a belay plate appears worn, replace it.

ASCENDER

Description:

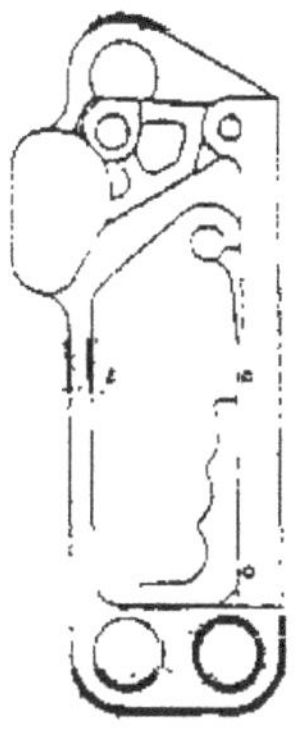

The ascender frame is made from sheet or extruded aluminum. It is black in color and is finished in a way that permits easy inspection for cracks or fractures in the frame. The handle has a large, molded plastic handgrip that accommodates a gloved hand. The top of the ascender has an attachment hole large enough to accommodate the steel locking "D" carabiner. The ascender permits easy placement and removal from a rope with one hand. A stainless steel camming device allows the rope to run through the device in one direction, but grips it in the other. A safety device is incorporated to insure that the cam only releases a rope when the trigger is pressed and the cam are sprung out of position. At a minimum, the camming device securely holds ropes sized from 8 through 11 millimeters. Minimum ascender breaking strength is 1,500 pounds. Maximum weight of the ascender is 18 ounces.

Care and Maintenance:

Inspect the ascender before, after, and frequently during use for signs of wear. These signs include nicks, grooves, cracks, or hairline fractures in the body of the ascender as well as any cracks in the cam or trigger. If any part of the ascender appears worn, replace it.

The trigger and cam action should always be smooth. If it becomes difficult to move the trigger, to engage or release the cam, or if these actions become "sticky," clean those components and their hinges with soapy water or a non-corrosive solvent and once the ascender is dry lightly apply a standard lubricant to these areas. Ensure that all excess lubrication is removed from the ascender so as not to contaminate the rope. If the trigger or cam continues to stick, replace ascender.

TOOL, NUT

Description:

The nut tool assists in the removal of protection devices (primarily wired wedge nuts and hexcentric nuts) from cracks. It is constructed of steel with a length between 7 to 9 inches and is a dark grey or black in color. The handle end has at least one eye-hole that is large enough to accommodate the aluminum non-locking carabiner and the working end has a hook to aid in protection removal. The nut tool has a maximum weight of 2.0 ounces.

Care and Maintenance:

Inspect the nut tool before and after use for signs of wear. If a nut tool develops a significant bend or crack, it is unserviceable and must be replaced.

HARNESS, CLIMBING

Description:

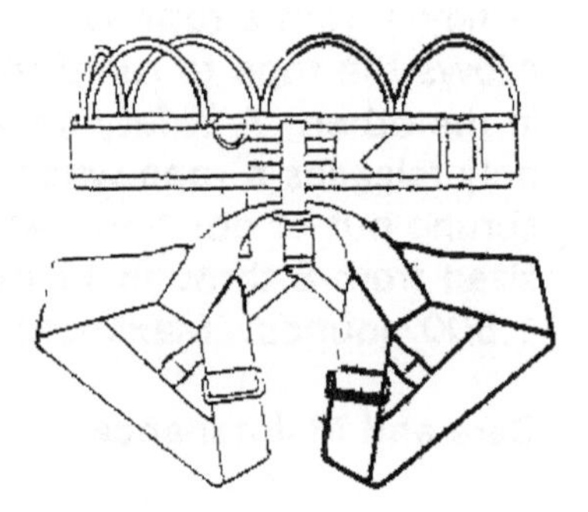

The climbing harness is used for climbing and belaying during two-party climbs. It is made of black nylon webbing. Adjustable waist and leg loops to accommodate a Marine with a waist size of 28 to 38 inches. The harness has a reinforced belay loop or "D" ring in the front and at least two gear loops. The padding or webbing is wide enough to keep the harness comfortable during extended multi-pitch climbs. The harness is designed to keep the climber upright after a fall has occurred. Minimum breaking strength at all points in the harness (except gear loops) is 3,500 pounds.

Care and Maintenance

Inspect the climbing harness before, after, and frequently during use for signs of wear. These signs include thread failures at any sewn seam and any fraying or parting of the harness webbing itself. If any part of the climbing harness appears worn or exhibits loose, fraying, or missing thread, replace the harness.

Clean the harness as needed. Wash it in mild soapy water, rinse it thoroughly, then dry it completely. If possible, hang the harness on a wooden peg to dry. Do not dry it in direct sunlight or hang it on a nail or other metal object.

BAG, ROPE.

Description:

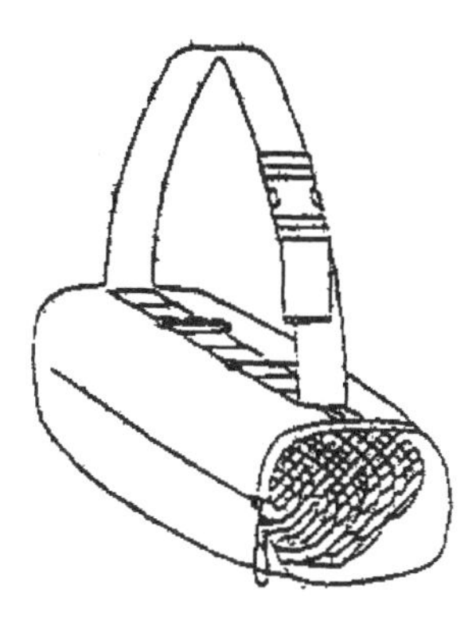

The rope bag is made from water resistant cordura nylon or equal material that is black, olive drab, or woodland camouflage in color. The bag accommodates 165 feet of dynamic or static rope. An attachment loop inside the bag secures one end of a rope. The bag has a drawcord-type closure with a cord lock or other equal device. Adjustable exterior straps enable the user to strap the bag to his leg or waist, or to wear the bag over his shoulder. The rope bag has All-Purpose Lightweight Carrying Equipment (ALICE) clips or a similar devices to attach it to the exterior of the Field Pack, Medium and the Field Pack, Large with Internal Frame.

Care and Maintenance:

Inspect the rope bag before, after, and frequently during use for signs of wear. These signs include thread failures at any sewn seam as well as any fraying or parting of the bag material. If any part of the rope bag appears worn, attempt to repair the failing area by sewing it with strong nylon thread. If worn area of the rope bag cannot be fixed by sewing, replace it.

Clean the rope bag as needed. Wash it in mild soapy water, rinse it thoroughly, then dry it completely. If possible, hang the bag on a wooden peg to dry. Do not dry it in direct sunlight or hang it on a nail or other metal object.

ROPE, DYNAMIC

Description:

The dynamic rope is kernmantle in construction with a water resistant treatment to reduce friction and increase rope life. It is certified by the UIAA as a single rope. The rope is able to survive a minimum of ten falls when subjected to the standard UIAA dynamic fall test with an 80-kilogram weight. Maximum rope elongation is 6% with an 80-kilogram weight per UIAA standards or U.S. Federal Test Standard 191A. The dynamic rope is 11 millimeters ± 0.7 millimeters in diameter per UIAA standards and 165 feet ± 5 feet long. Maximum weight is 79 grams per meter. The dynamic rope is olive drab in color.

Care and Maintenance:

Inspect the dynamic rope before, after, and frequently during use for signs of wear. These signs include cuts in the rope surface and any fraying or parting of the outer sheath. If the rope appears worn, replace it.

The dynamic rope must be replaced periodically during frequent use based on visual signs of wear. A rope is weakened after stopping an extremely hard fall by a Marine or piece of equipment. Replace it after two to four of these falls, depending on their severity.

After every period of use, wash the rope in mild soapy water, rinse it thoroughly, then dry it completely. If possible, coil the rope and hang it on a wooden peg to dry. Do not dry the rope in direct sunlight or hang it on a nail or other metal object.

To store the rope, loosely coil it with all knots removed and return it to the MACK container. The rope must be dry when placed in the container.

Only cut the rope with the rope cutter. The cutter's heated blade should be used to fuse the rope ends which will prevent the ends from fraying.

ROPE, STATIC

Description:

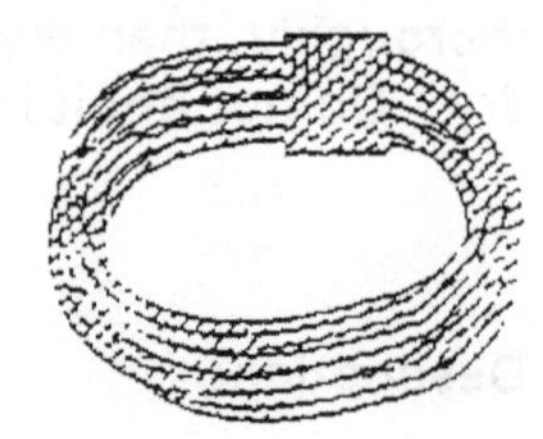

The static rope is kernmantle in construction. It has a minimum tensile strength of 6,500 pounds per U.S. Federal Test Standards. Maximum working elongation is 1.5% with an 80-kilogram weight per UIAA standards or U.S. Federal Test Standard 191A. Maximum allowable shrinkage when wet is 8% per Federal Test Standard 191A. The static rope is 7/16 inches in diameter and the two lengths used are 165 feet ± 5 feet and 300 feet ± 10 feet. Maximum weight is 0.92 ounces per foot. The static rope is black in color.

Care and Maintenance:

Inspect the static rope before, after, and frequently during use for signs of wear. These signs include cuts in the rope surface and any fraying or parting of the outer sheath. If a rope appears worn, replace it.

Replace the static rope periodically during frequent use based on visual signs of wear. A rope is weakened after stopping an extremely hard fall by a Marine or a piece of equipment. Replace a rope after two to four of these falls, depending on their severity.

After every period of use, wash the rope in mild soapy water, rinse it thoroughly, then dry it completely. If possible, coil the rope and hang it on a wooden peg to dry. Do not dry the rope in direct sunlight or hang it on a nail or other metal object.

To store the rope, loosely coil it with all knots removed and return it to the MACK container. The rope must be dry when placed in the container.

Only cut the rope with the rope cutter. The cutter's heated blade should be used to fuse the rope ends which will prevent the ends from fraying.

CORD, ACCESSORY TYPE I

Description:

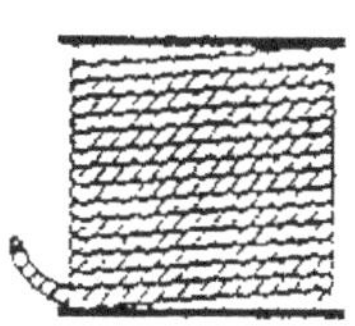

The Type I accessory cord is used to string non-wired hexcentric nuts. It has a sheath made of nylon and a core composed of Spectra, Kevlar, or equal fibers. It has a minimum tensile strength of 4,400 pounds per Federal Test Standard 191A. The Type I accessory cord is 5.5 millimeters wide and 150 feet ± 10 feet long. It is subdued in color.

Care and Maintenance:

Inspect the accessory cord before, after, and frequently during use for signs of wear. These signs include cuts in the cord's surface and any fraying or parting of the outer sheath. If the cord appears worn, remove it from the hexcentric nut and replace it with new cord. A nut must be rethreaded by personnel with formal military mountaineering training. It must be inspected by at least one other properly trained military mountaineer before use.

After every period of use, wash the cord (do not untie) in mild soapy water, rinse it thoroughly, then dry it completely. If possible, coil the cord and hang it on a wooden peg to dry. Do not dry the cord in direct sunlight or hang it on a nail or other metal object.

To store the cord, leave all unused cord on its original spool and return it to the MACK container. The cord must be dry when placed in the container.

Only cut the cord with the rope cutter. The cutter's heated blade should be used to fuse the cord ends which will prevent the ends from fraying.

CORD, ACCESSORY TYPE II

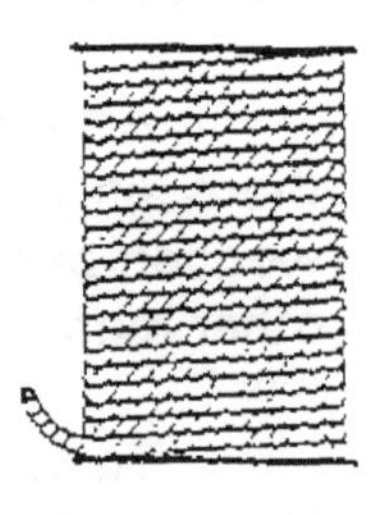

Description:

The Type II accessory cord serves as a prussik cord. The accessory cord is kernmantle in construction. It has a minimum tensile strength of 2,200 pounds per Federal Test Standard 191A. The Type II accessory cord is 7 millimeters wide and 300 feet ± 10 feet long. It is subdued in color.

Care and Maintenance:

Inspect the accessory cord before, after, and frequently during use for signs of wear. These signs include cuts in the cord's surface and any fraying or parting of the outer sheath. If the cord appears worn, replace it.

After every period of use, wash the cord used in mild soapy water, rinse it thoroughly, then dry it completely. If possible, coil the cord and hang it on a wooden peg to dry. Do not dry the cord in direct sunlight or hang it on a nail or other metal object.

To store the cord, loosely coil it with all knots removed and return it to the MACK container. The cord must be dry when placed in the container.

Only cut the cord with the rope cutter. The cutter's heated blade should be used to fuse the cord ends which will prevent the ends from fraying.

WEBBING, NYLON

Description:

The nylon webbing is tubular in construction. It has a minimum breaking strength of 4,400 pounds per Federal Test Standard 191A. The webbing is 1 inch wide and 300 feet ± 10 feet long. It is black or olive drab in color.

Care and Maintenance:

Inspect the nylon webbing before, after, and frequently during use for signs of wear. These signs include cuts in the webbing's surface and any fraying or parting of the outer sheath. If the webbing appears worn, replace it.

After every period of use, wash the webbing in mild soapy water, rinse it thoroughly, then dry it completely. If possible, coil the webbing and hang it on a wooden peg to dry. Do not dry the webbing in direct sunlight or hang it on a nail or other metal object.

To store the webbing, loosely coil it with all knots removed and return it to the MACK container. The webbing must be dry when placed in the container.

Only cut the webbing with the rope cutter. The cutter's heated blade should be used to fuse the webbing ends which will prevent the ends from fraying.

LADDER, CABLE

Description:

The cable ladder is constructed with stainless steel cables and aluminum crossbars. Each cable end has a steel ring large enough to accommodate the largest carabiner in the MACK to allow the user to connect cable ladders to one another. Minimum breaking strength of each individual ladder section is 1,322 pounds (600 kilograms). Minimum breaking strength of entire ladder when two or more sections are connected together is also 1,322 pounds (600 kilograms), assuming the device used to connect ladders is not the weakest point in assembly. The minimum width of the ladder crossbars is 6 inches and the length of an individual ladder section is 30 feet ± 5 feet. A minimum of 300 feet ± 5 feet of cable ladder is provided in each MACK.

Care and Maintenance:

Inspect the cable ladder before, after, and frequently during use for signs of wear. These signs include fraying or parting of the wire cables and nicks, grooves cracks, or fractures in the steps or tie-in points. If any part of the cable ladder appears worn, replace it.

CUTTER, ROPE

Description:

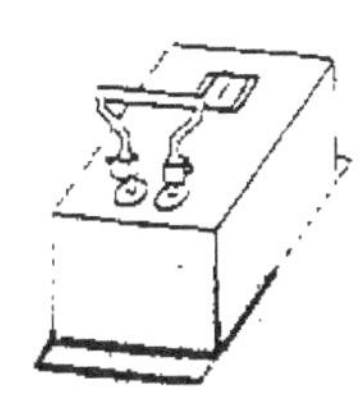

The rope cutter uses an electrically heated blade to cut rope lengths and seal the rope ends. The cutter is issued with two replacement blades, for a total of three blades. The cutter operates on 110 volts AC. It has an on/off switch and a light to indicate when the unit is on. The rope cutter is Underwriter's Laboratory (UL) approved.

Care and Maintenance:

Replace the rope cutter blade after frequent use if cutting or fusing becomes difficult. If the blade is damaged or fails to heat, cut, or seal properly, remove it from the rope cutter and replace it. If the rope cutter unit fails to perform properly, replace it.

BAG, CLIMBING RACK

Description:

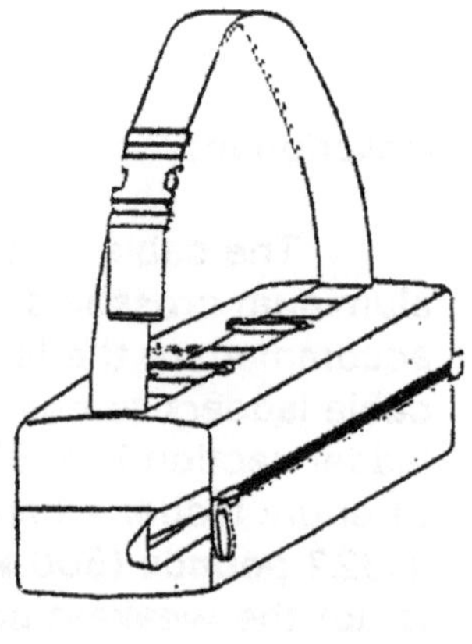

Eight climbing rack bags are provided in the MACK to allow each two-man climbing team the capability to store and transport all their individual rack items, not including the 165-foot dynamic line. Each rack bag is large enough to accommodate half of a team's equipment, but no larger than a size sufficient to carry all a team's gear. The bag is made of water resistant cordura nylon or equal material and is sufficiently durable to be used as a haul bag. It has a quick and easy closure system that prevents items of the kit from inadvertently falling from the bag when closed. The bag has ALICE clips or similar devices to attach it to the exterior of the Field Pack, Medium and the Field Pack, Large with Internal Frame. The climbing rack bag is black, olive drab, or woodland camouflage in color.

Care and Maintenance:

Inspect the climbing rack bag before, after, and frequently during use for signs of wear. These signs include thread failures at any of the sewn seams as well as any fraying or parting of the bag material. If any part of the rack bag appears worn, attempt to repair the failing area by sewing it with strong nylon thread. If the damaged area of the climbing rack bag cannot be fixed by sewing, replace the bag.

Clean the climbing rack bag as needed. Wash it in mild soapy water, rinse thoroughly, then dry it completely. If possible, hang the bag on a wooden peg to dry. Do not dry it in direct sunlight or hang it on a nail or other metal object.

CONTAINER, MACK

Description:

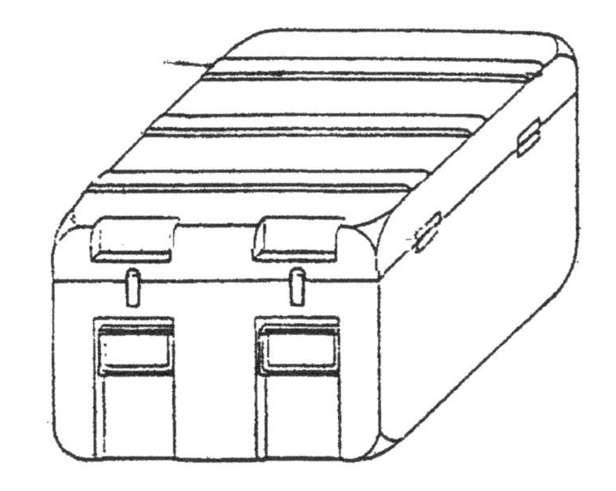

The MACK container is used to store elements of the MACK when the kit is not in use. Four containers hold all the items contained in the MACK and have features that facilitate the organization and accountability of MACK items. Each container protects the contents from degradation due to sunlight and moisture during storage periods of up to five years. The lid's interior has a permanently affixed list of the components and quantities stored within that container. Container #1 contains the climbing team equipment. Containers #2, #3, and #4 contain the company climbing equipment. Each MACK container is marked with the following information in accordance with MIL-STD-129: item nomenclature, contract number (or National Stock Number (NSN) when assigned), manufacturer name and address, and container number. The containers are olive drab in color.

Cate and Maintenance:

Inspect each MACK container occasionally for signs of wear. These signs include cracks in the container material and failure of the hinges, handles and clasps. If discrepancies cannot be repaired, replace the container.

If a container becomes dirty, wash it with soapy water and dry it thoroughly

CPSIA information can be obtained at www.ICGtesting.com
Printed in the USA
LVOW011224270612
287873LV00001B/4/A